AUG - - 2018

R

REFERENCE
DO NOT CIRCULATE

Principles of Physics

Principles of Physics

Editor

Donald R. Franceschetti, PhD

University of Memphis

SALEM PRESS

A Division of EBSCO Information Services, Inc.
Ipswich, Massachusetts

GREY HOUSE PUBLISHING

Copyright © 2016, by Salem Press, A Division of EBSCO Information Services, Inc., and Grey House Publishing, Inc.

All rights in this book are reserved. No part of this work may be used or reproduced in any manner whatsoever or transmitted in any form or by any means, electronic or mechanical, including photocopy, recording, or any information storage and retrieval system, without written permission from the copyright owner. For permissions requests, contact proprietarypublishing@ebsco.com.

∞ The paper used in these volumes conforms to the American National Standard for Permanence of Paper for Printed Library Materials, Z39.48–1992 (R2009).

**Publisher's Cataloging-In-Publication Data
(Prepared by The Donohue Group, Inc.)**

Names: Franceschetti, Donald R., 1947- editor.
Title: Principles of physics / editor, Donald R. Franceschetti, PhD, the University of Memphis.
Description: [First edition]. | Ipswich, Massachusetts : Salem Press, a division of EBSCO Information Services, Inc. ; [Amenia, New York] : Grey House Publishing, [2016] | Series: Principles of | Includes bibliographical references and index.
Identifiers: ISBN 978-1-61925-946-1 (hardcover)
Subjects: LCSH: Physics.
Classification: LCC QC7 .P75 2016 | DDC 530--dc23

PRINTED IN THE UNITED STATES OF AMERICA

CONTENTS

Publisher's Note vii
Editor's Introduction ix
Contributors xv

Aberrations 1
Absorption 4
Accuracy and precision 7
Alpha radiation 10
Amplitude 12
Angular forces 15
Angular momentum 19
Antenna 22
Arago dot 24
Aperture 26
Archimedes's principle 27

Band theory of solids 30
Bernoulli's principle 32
Beta radiation 34
Blackbody radiation 36
Bohr atom 39
Bose condensation 42
Bra-ket notation 44
British thermal unit (BTU) 46

Calculating system efficiency 48
Circular motion 51
Closed systems and isolated systems 54
Concave and convex 55
Conservation of charge 57
Conservation of energy 60
Convection and conduction 64
Cosmic radiation 67
Crest and trough 69

Density and specific gravity 71
Diode 73
Distance and area 75
Distortion 77
Doppler effect 80
Dynamic systems theory 83

Edison effect 86
Eigenvectors 88
Elastic and inelastic collisions 90

Electric circuits: Parallel vs series,
 diagrams and components 94
Electric potential 97
Energy and power 100
Enthalpy 104
Entropy 107
Euler's laws of motion 110

Falling bodies and physics 114
Faraday's law 117
Feynman diagrams 120
Flywheel 122
Focal point 125
Free body diagram 127
Frequency 128

Gamma radiation 132
Gauss's law 134
Gluon 137
Gravitational potential energy 139

Harmonic oscillator 143
Harmonics 146
Heisenberg uncertainty principle 150
Higgs boson 152
Horsepower 155

Ideal gas law 157

Joule 162

Kepler's laws 164
Kilowatt-hour 167
Kinetic and potential energy 168

Lenz's law 173
Light waves 176
Linear motion 178
Load 181
Loudness and sound intensity 184

Magnification 189
Mass and weight 191
Mass spectrometry 195
Mechanical or electrical load and work .. 197
Mössbauer effect 200

Net force 203
Newton-meter 206
Normal force 207
Nuclear fission, fusion, and mass defect 211

Open systems 214

Particle accelerators 216
Particle detectors 220
Phase 222
Phonon 224
Photon 225
Pitch 227
Planck's constant 229
Potential energy 231
Power 236
Prisms 240
Projectiles 241
Psychophysics 245

Quantum chromodynamics 248
Quantum field theory 251
Quantum mechanics 253
Quantum statistics 256
Quarks 259

Radians and degrees 262
Radiation 264
Resonance: Nuclear magnetic resonance
   and electron spin resonance 267
Revolutions per minute (rpm) 270

Simple, linear, and complex systems 272
Simple machines and mechanical advantage ... 274
Solenoid 279
Sound amplitude 282
Speed 285
Springs 286
String theory 289

Strong nuclear force 292
Superconductors 294
Switches 296
Symmetry 298
Système International (SI) Units 299

Temperature and internal energy 302
Term symbol 306
Torque 307
Turbines 311

Ultrasound 315
Ultraviolet radiation 317

Velocity of sound 320
Velocity vs. speed 322
Volume and capacity 325

Watt 328
Wavelength 329
Wave properties 332
Weak force 333
Work and force 335
Work-energy theorem 339

X-ray radiation 342

Appendices
The Standard Model 347
Nobel Notes: Sub-atomic and
   Sub-nuclear Physics 351
Nobel Laureates 354
Pre-Nobel Notables 364
Physics Constants 367
Physics Laws 369
Glossary 374
Bibliography 394
Subject Index 405

Publisher's Note

Grey House Publishing is pleased to add *Principles of Physics* to its Salem Press collection, the second of four titles in a new *Principles of* series: Chemistry, Physics, Astronomy, and Computer Science. This new resource introduces students and researchers to the fundamentals of physics using easy-to-understand language, giving readers a solid start and deeper understanding and appreciation of this complex subject.

The 142 entries range from Aberrations to X-rays and are arranged in an A to Z order, making it easy to find the topic of interest. Entries include the following:
- Related fields of study to illustrate the connections between the various branches of physics, including acoustics, high energy physics, psychophysics, quantum electrodynamics, and nanotechnonlogy;
- A brief, concrete summary of the topic and how the entry is organized;
- Principal terms that are fundamental to the discussion and to understanding the concepts presented;
- Illustrations that clarify difficult concepts via models, diagrams, and charts of such key topics as blackbody radiation, Bernouilli's principle, and Higgs boson;
- Equations that demonstrate how to determine mechanical advantage, understand the ideal gas law, the fundamentals of quantum mechanics, and Einstein's famous mass-energy equation—$E = mc^2$;
- Photographs of significant contributors to the study of physics;
- Sample problems that further demonstrate the concept, law, or constant presented;
- Bibliography lists that relate to the entry.

This reference work begins with a comprehensive introduction to the field, written by editor Donald R. Franceschetti, PhD. It starts with the ancient Greeks' quest to understand motion, both on earth and in the heavens; includes Einstein and the *annus mirabilis*; J. Robert Oppenheimer, the father of the atom bomb; a discussion of how vacuum tubes and digital processesors have fundamentally changed the way we live; and, finally, ends with a discussion of the challenge and the promise of physics education in today's high schools.

The book's backmatter is another valuable resource and includes:
- The Standard Model, a discussion of the key discoveries and concepts that led to a growing understanding of subatomic particles and the electromagnetic, strong, and weak interaction;
- Nobel Notes that explain the significance of the prizes in physics to the study of the science and their interdisciplinary nature;
- Nobel Prize Laureates in the area of physics from the first awards in 1901, given to William Conrad Röntgen to the prize awarded in 2015 to Takaaki Kajita and Arthur B. McDonald "for the discovery of neutrino oscillations, which shows that neutrinos have mass";
- Pre-Nobel Notables, recognizing some of the important figures in physics who did not receive a Nobel Prize;
- Physics Constants, showing their symbols, names, and values;
- Physics Laws;
- Glossary;
- General bibliography; and
- Subject index.

Grey House Publishing extends its appreciation to all involved in the development and production of this work. The entries have been written by experts in the field. Their names and affiliations follow the Editor's Introduction.

Principles of Physics, as well as all Salem Press titles, is available in print and as an e-book. Please visit www.salempress.com for more information.

Editor's Introduction

Natural philosophy, an antecedent to the field of study we now know as physics, was the term applied by the Ancient Greeks to the study of nature. Whenever they were asking questions or making direct observations about the nature of the world in which they lived–whether they were trying to understand why an arrow falls or why water rises when it is displaced by an object–they were engaged in a form of study we can readily identify as the direct precursor of all of the natural sciences. Natural philosophy itself gave direct rise to the study of physics; when one considers how long we have been studying and questioning our physical world, it is easy to see that the study of physics is actually far older than one might suppose. Physics is often referred to as the "foundation science," not merely because it is one of the oldest of the sciences but also because it has become essential to the study of so many of the natural sciences, from chemistry to geology to biology.

But what exactly is physics? Clearly physics is a science, so we might first attempt to answer the question, "What is a science?" Over 50 years ago, Nobel Laureate Richard P. Feynman, lecturing to Caltech students, attempted to answer that question by suggesting that science is like watching a complex game that is being played at high speed, chess perhaps, and trying to deduce the rules of the game from what you see. While you are permitted to rearrange the game board a little (i.e., do an experiment), there are stringent limits on the observations that can be made.

Physics — The Foundation Science

We can think of physics as a search for nature's rule book and an attempt to learn the principles that underly observable events and phenomona. While people have wondered about and tried to understand falling bodies and the paths taken by arrows on the way to their targets since earliest times, the ancient Greeks were among the first to think about "motion" in the abstract. Some historians of science see the origins of physical theory in the debate between two pre-Socratic Greeks philosophers: Heraclitus, who held that all was in flux, and Parmenides, who claimed that all change is illusory.

Many people would argue that the beginnings of physics as a science may be traced back to Plato's *Timaeus*, one of his famous dialogs, written in 360 C.E. That work speculated on the ultimate constituents of matter. But it was Aristotle, Plato's student, who introduced the term *physics* and stressed the importance of observation. While Aristotle deserves credit for careful analytical thought about motion, he also deserves some level of blame for misleading conclusions. Perhaps most far-reaching and long-lasting of these incorrect conclusions was the distinction between motion in the heavens and motion on earth. As Aristotle watched the stars parade across the night sky, he made a natural assumption that the stars were somehow attached (fixed) to a celestial sphere. As he watched the planets move against the background of these "fixed" stars, it seemed only natural to conceive a cosmology based on spheres within spheres. Even Copernicus, writing *On the Revolution of the Heavenly Orbs* two thousand years later, was still influenced, at least to some degree, by these same mental images.

We now know that the Greeks made several natural, but serious, errors in classifying motion. The Greeks, masters of mathematics, made a distinction between motions in the celestial and terrestrial realms. They assumed that motions in the celestial realm were perfect and therefore different from motions in the terrestrial realm, which they believed were necessarily imperfect. They also believed that mathematics, being perfect, could not be applied to terrestrial motions. While their mathematical prowess allowed Greek astronomy to flourish, the practical result of drawing a distinction between a "perfect" celestial realm and an "imperfect" terrestrial realm meant that Greek physics languished.

Yet another problem arose from the fact that the Greeks interpreted motions teleologically, i.e. as resulting from an object's desire to achieve a certain end. Thus, they assumed that the path taken by falling bodies represented the desires of the bodies. Heavy objects "wanted" to be as near as possible to the center of the earth. The weight of an object represented the strength of its desire to be on the earth's surface. This may seems reasonable enough but, as Galileo would point out centuries later, it fails to explain the constant acceleration of a falling body when air resistance can be removed from consideration.

The Scientific Revolution

It would be an error to assume no progress occurred in physics during the Middle Ages, but the birth of physics in the modern sense can rightly be assigned to the period of time referred to as the late Renaissance and is most especially associated with the work of a Florentine Italian, Galileo Galilei, and an Englishman, Sir Isaac Newton.

Galileo was born in 1564, twenty-one years after the death of Copernicus, whose deathbed publication of *De Revolutionibus Orbium Coelestim* proposed the heliocentric model of the solar system as a computational aid for the calendar. Galileo was professor of mathematics at Pisa, and later Padua, and one of the first to embrace the use of experiments to expand our knowledge of the physical world. Galileo differed from Copernicus in that he wrote in the dialog form made famous by Plato and in the vernacular Italian, making it easier to appeal to the broader (but still educated) public, rather than the more limited world of academics capable of reading Corpernicus in Latin. Like Copernicus, however, Galileo eventually incurred the displeasure of Church leaders of the time, who put him on trial in 1633. By the time Galileo died, he was under house arrest and his theories had been officially condemned by the Church.

Isaac Newton was born in 1642, less than a year after Galileo's death. As Protestant, Newton did not have to worry about his reception in Catholic countries. Newton was a unique and most complex personality. He never married, but wrote voluminously on religion, the Bible, history, alchemy and physics. He did not write in vernacular English but rather, for the most part, in scholarly Latin. There is a story that Newton, who had been appointed a Professor of Mathematics at Cambridge University, was not a very good teacher. He would often lecture (in Latin) to an empty room, since the condition of his employment was that he lecture three times a week, whether or not anyone attended his classes. His *magnum opus*, the *Principia Mathematica Philosophiae Naturalis*, was not fully published in English until after his death. Newton began with Galileo's principle of inertia, and then he added a second and third law of motion and a law of universal gravitation. These additional laws allowed him to deal with both celestial and terrestrial motions.

Over the succeeding years, mathematicians developed methods to apply Newton's laws in numerous cases: The anomalies they saw in the orbit of Uranus allowed them to calculate the position of a new planet, Neptune. In 1930, careful measurements of Neptune's orbit made it possible for them to locate Pluto. In more recent days, some 2000 extra-solar planets have been found using those same mathematical techniques, coupled with the more highly accurate data provided by the Hubble telescope.

It is difficult to fully appreciate the impact of Newton's laws on the evolution of human culture. Consider the Church's arguments for the existence of God: Writing in the thirteenth century, St. Thomas Aquinas included a proof from motion among his five proofs for the existence of God. Aquinas claimed that every object that moved was either moved by itself or moved by another agent. In Aquinas's thinking, God was—among a great many other things—the unmoved mover who was ultimately responsible for motion in the universe. But according to Galileo's principle of inertia and Newton's second law, an explanation was required only for *changes* in motion. Further, the law of universal gravitation said that there was nothing special about the center of the earth and that all masses should be attracted to it since every mass in the universe attracts every other mass with a force that is proportional to the product of the masses divided by the square of the distance between them.

Newton was one of the first individuals elected to membership in the Royal Society, so named because it functioned under the patronage of the King. Its charter eschewed "meddling in metaphysics, rhetoric, etc." and instead adhered to testable propositions. Thus, the physics of Newton became one of the pillars of the Industrial Revolution.

Impact of the Steam Engine & Thermodynamics

In addition to the theory of testable propositions, the steam engine was also responsible for driving the Industrial Revolution. This new power source, first devised by Thomas Savery in 1689 and then much improved by a number of individuals, most notably James Watt, was the first source of motive force that was independent from human or animal muscle power or a force of nature like the wind or falling water. Steam engines were, therefore, relatively location independent. This advance came with some significant environmental costs, including widespread deforestation of the countryside as well as coal-mining that removed coal from the ground at an alarming rate. Eventually, mineshafts that had been stripped of their coal began to fill with rainwater and a major technological challenge became finding an answer to the pressing question of how to pump the water out. It rapidly became apparent that a column of water more than thirty feet in length cannot be drawn up a vertical pipe. This challenge spurred research into the behavior of gases, ultimately leading to the serious scientific study of the gas laws and thermodynamics.

At the time, scientists were still arguing about whether there was any value in a theory dealing with entities too small to be seen, so the atomic theory was not an initial part of thermodynamics. At first, then, heat was seen as an imponderable fluid that flowed from hot to cold, filling objects according to their "heat capacity." Today, we understand heat to be random atomic motion, but it was not until the twentieth century that this understanding would become fully established.

Voltaic Cells & the Study of Optics

The voltaic cell, developed just before the year 1800 C.E., offered scientists yet another source of power. Although electric and magnetic phenomena had been known since 600 B.C.E., and a magnetic compass had been used by Chinese sailors since around the year 1100 C.E., electric and magnetic forces continued to be a mystery until the early nineteenth century. Once the voltaic cell was invented, however, discovery in the area of electromagnetism proceeded at a breathtaking pace. New elements could be isolated due to electrolysis. Discoveries were made about the behavior of electromagnets and the induction of an electric current when the magnetic flux through a coil was changed.

James Clerk Maxwell, a professor of physics at a Cambridge University in England, summarized what was known about electromagnetism in four differential equations, known today as Maxwell's equations. Maxwell then found that if he restricted the equations to empty space, electromagnetic disturbances could travel nonetheless, provided that they moved at the speed of light traveling in a vacuum. Thus the study of optics became a matter of electromagnetic theory. Equally important, there was no limitation to the frequency that light exhibited. Vast regions of the electromagnetic spectrum still had yet to be explored.

Nuclear Physics

The study of nuclear physics dates back little more than a century. The first three decades of the twentieth century, however, have been termed "the thirty years that shook physics." The year 1905 was particularly significant and has been termed the *annus mirabilis* (the year of miracles). It was in that year that Albert Einstein noted that it was not necessary that observers in relative motion with respect to each other agree on their length and time interval measurements. Indeed it would be impossible for them to do so, if the speed of light is to be independent of the velocity of the light source, as Maxwell's equations required. Further there were some incompatibilities involved in the idea that atoms, located at some point in space, could emit electromagnetic waves.

Analysis of the process used to locate a particle using light waves revealed that one could not precisely specify both the position and velocity of the particle at the same time. This led to the Heisenberg uncertainty principle and the emergence of quantum mechanics as the proper description of the interaction of atomic systems with the electromagnetic field.

By 1930, the three elementary particles—protons, neutrons, and electrons—had been discovered, as well as the first particle of antimatter, the positron. By that time, Ernest Rutherford had already scattered alpha particles from a sheet of gold leaf and found that, contrary to expectations, the positive charge in the atom was concentrated in a very tiny region known as the nucleus.

The community of physics researchers was quite small in 1930, but by 1939, Albert Einstein's discoveries had brought physics and physicists to the attention of President Franklin D. Roosevelt, which dramatically changed the situation. The natural radioactivity of the element uranium had been discovered by Henri Becquerel in 1896. The discovery first made by Lisa Meitner and Otto Hahn in 1934, that nuclei of 235-uranium could spontaneously break apart into smaller nuclei, and that particles emitted by a fissioning uranium nucleus could collide with other nuclei and stimulate their decay, led physicists to contemplate a chain reaction that could release immense power. In fact, it was that precise discovery that eventually triggered a massive program of military spending by the American government and ultimately resulted in the destruction of two Japanese cities and the loss of hundreds of thousands of lives in 1945.

The development of the atomic bomb had clearly changed the nature of warfare and state of international politics dramatically. Following World War II, many of the physics and chemistry professors who had been working for the government returned to their campuses to find things quite changed. The ability of the community of physicists to respond to a national crisis by producing a weapon unique in its destructive potential was a lesson not lost on the physics community, the military, or the politicians. The bombs dropped on Japanese cities with their incredible destructive power did not escape the notice of other nations around the world. Although the United States had a monopoly on atomic bombs for about four years following the bombing of Hiroshima and Nagasaki, the Soviet Union exploded its first atomic bomb in 1949, thus beginning its arms race against western countries.

The power of the first atomic bombs could be described in terms of equivalent tons of TNT. The first atomic explosion in New Mexico was described as having an explosive yield of more than 20 thousand tons of TNT. Scientists already knew that more energy was available from the *fusion* of smaller nuclei than the *fission* of large ones and eventually, the United States exploded its first thermonuclear, or hydrogen, bomb in 1952. The Soviet Union followed with their first hydrogen bomb the following year, in 1953. Now, instead of bombs with an explosive yield measured in the tens of thousands of tons of TNT, the explosive yields would be measured in tens of millions of tons of TNT.

Many Americans were suspicious of the speed with which the Soviets moved forward in the arms race. Those suspicions had significant ramifications that exposed the uneasy relationship between science, politics, and the government and which came to a head during the so-called McCarthy era, a time when people thought to have Communist sympathies were often persecuted without much evidence of wrongdoing. J. Robert Oppenheimer was one such person.

J. ROBERT OPPENHEIMER & THE MCCARTHY ERA

Today, we know Oppenheimer as the "father of the atomic bomb" because of the work he did as the director of the Los Alamos Laboratory during World War II. Born to a wealthy New York family in 1904, Opje (as he was called by all who knew him) was an extremely intelligent person. He completed his education in Europe and then took faculty positions at Caltech and Berkley upon his return to the United States. When he was not working, he busied himself with liberal political causes.

When the decision was made to pursue the development of an atomic bomb during World War II, the project was given the code name Manhattan Project. It was assigned to the U. S. Army under the supervision of Brigadier General Leslie Groves, who sought out Oppenheimer to be part of the effort. Oppenheimer's standing in the community of physicists was such that General Groves considered him indispensible. His propensity to get involved with liberal political causes made some of the lower-ranking officers involved in the project doubt his loyalty to the United States. Groves overruled them, however, and Oppenheimer was granted a security clearance.

The Manhattan Project demanded the utmost secrecy. It was so closely guarded that most people working on it were unaware of the project's ultimate goal. It was only at the Los Alamos Laboratory, a facility under the direction of Oppenheimer, that staff were aware that they were building a new weapon based on nuclear fission.

After the war, most of the project structure was dismantled and many of the leading physicists either returned to academia or took up lucrative appointments as government consultants for the Defense Department or Atomic Energy Commission (AEC), an agency established after the World War II by presidential order. Among the AEC's responsibilities was oversight of the system of national laboratories, seventeen in all. Staff members at those laboratories were charged with conducting scientific research, particularly in areas where research was too expensive and complex and had too many national security implications to be conducted on college campuses. The AEC instituted its own security clearance program and began to systematically screen almost all applicants for federal employment in the atomic energy area.

Oppenheimer himself assumed the directorship of the Institute for Advanced Study at Princeton, New Jersey, in 1947. At the same time, he took up the position of Chairman of General Advisory Committee of the AEC. Then, in December of 1953, at the height of the McCarthy Era, his security clearance was revoked by order of President Eisenhower. Oppenheimer requested an administrative hearing but, even though he was exonerated of any wrongdoing, the suspension of his clearance was upheld. Ultimately, President Lyndon Johnson presented him with the Enrico Fermi Award of the Atomic Energy Commission in 1963, a gesture that was widely interpreted as symbolically reinstating him to the good graces of the government.

By 1974, the AEC itself had come under attack, and it was dissolved by Congress. The Nuclear Regulatory Commission (NRC) was established by the Energy Reorganization Act of 1974 and began its operations on January 19, 1975. Today, the focus of the NRC is nuclear reactor safety as well as oversight and reviewing applications for new licenses.

Many comparisons have been made between the Oppenheimer case and that of Galileo four centuries earlier. While Oppenheimer was not tortured, he was branded a traitor in the minds of many. Other physicists saw the potential risks to their own careers and turned to nongovernmental employment. There were more than a few scientists who felt that the loss of life at Hiroshima and Nagasaki could not be justified, despite the fact that it won the war for the Allies. As Oppenheimer put it, physicists had "known sin" and the world would never be quite the same.

VACUUM TUBE ELECTRONICS, THE COMPUTER, SOLID STATE DEVICES, & PHOTOLITHOGRAPHY

Vacuum tube electronics and the digital computer have had a major effect on the development of physics as a science. Thomas Edison noticed that, if one heats a metal filament in vacuum, the filament tended to be surrounded by electrons that have boiled off the filament's surface. He recorded this in his laboratory notebook in 1883 and but did not pursue it further.

In 1904 Sir John Ambrose Fleming, who had visited Edison, was awarded a patent for a vacuum tube diode, which could rectify current. When an American scientist, Lee De Forrest, added a third element—a grid between the heated cathode and the anode—he was able to obtain a device in which a small voltage on the grid could control the current.

Meanwhile at Cambridge University, a young mathematical logician, Alan M. Turing, had developed the notion of a programmable digital computer while trying to solve a problem in pure mathematics. A few computers had been built using vacuum tubes in the years preceding World War II. These machines showed some promise but there were practical problems in using them for large-scale computation; the tubes burned out with some regularity. If each tube had a probability of burning out at the rate of only 0.01 per hour and if a machine had 1000 tubes, then the probability that the entire machine would have at least one tube fail in the first hour was greater than 50 percent.

For the computer to come into its own, it had to be more reliable and efficient than vacuum tube electronics allowed. An alternative to vacuum tube devices was provided by solid-state physics. Physicists had discovered that the electrical conductivity of a material like silicon was extremely sensitive to the presence of impurities. Adding an electron donor greatly enhanced the number of electrons that participated in current flow, while the presence of electron acceptors introduced mobile electron holes in the conduction band. In 1947 the transistor (short for *transfer resistor*) was invented. It was capable of acting like a vacuum tube triode because a small voltage applied to the middle section had a large effect on the current flowing between the outer regions.

Transistors can be reduced in size (miniaturized) in a way that vacuum tubes cannot. It is possible to grow nearly perfect silicon crystals with impurity atom concentrations of just one part per trillion. By means of masking and ion implantation, it is possible to create thousands of copies of the same circuit. The monolithic integrated circuit, first developed in 1958, made possible such marvels as the hand-held calculator and the desktop computer, as well as enormous computer memories that make it possible to simulate numerous processes.

The computational resources made possible by the integrated circuit along with establishment of the national laboratories in the United States and other countries, made further progress inevitable and physicists were able to explore the atomic nucleus at last. They learned that the short-range nature of the forces within the nucleus prevented the formulation of a simple force law like Coulomb's for the strong force. It became apparent, however, that the hadron family of particles were in fact composite. Each hadron consisted of three particles, which Murray Gell-Mann dubbed quarks (a nonsense word from James Joyce's *Finnegan's Wake*).

To explain why no free quarks had ever been detected, Gell-Mann assumed that the force between quarks got stronger as the distance between them increased, and that collisions with enough energy to break interquark bonds had enough energy, in fact, to create an additional quark-antiquark pair. That meant that the quarks were in effect confined to being either a tri-color combination that added up to white or a meson, a medium-mass particle that combined a quark and a complementary colored antiquark.

THE CHALLENGE OF PHYSICS EDUCATION
Because physics is so essential to so many fields of human endeavor, and because there are a diversity of physics curricula to draw from, there are conflicting opinions concerning which curricula is best. In the United States, the recommendations of the Committee of Ten, written over a hundred years ago, still holds sway. That curriculum assumed single-year courses in biology, chemistry, and physics, in that sequence, so that students could reach a certain level of mathematical skill before attempting to solve physics problems. That approach assumes that it is possible to have an understanding of biology without a working knowledge of physics and that the problem-solving skills demanded by chemistry can be developed without knowledge of physics. An alternative to this sequential ordering of science study is found in the "Physics First" movement, which encourages students to get the fundamentals of physics in the first year, before moving on to biology and chemistry, and then offers a more in-depth course in the junior or senior year to students who intend to pursue careers in science or engineering.

Yet another question deserving thought concerns how much modern physics education has to offer the non-specialist student. Although many students find the traditional physics course boring, they might find some of the recent developments in the field intriguing. It takes only a trip to the local bookshop or perusal of quality newspapers to come in contact with happenings at the forefront of research that are being presented to the general, non-specialized public as exciting, relevant information. This volume combines the latest topics with more traditional concepts to present a balanced view of this exciting and continuously evolving science.

REFERENCES:
Calder, Nigel. *The Key to the Universe : A Report on the New Physics*. New York: Viking, 1977. Print.

Feynman, Richard P., Matthew L. Sands, and Robert B. Leighton. *The Feynman Lectures on Physics : New Millenium Edition*. New York, NY: Basic , A Member of the Perseus Group, 2010. Print.

Griffiths, David J. *Introduction to Elementary Particles*. 2nd ed. Weinheim [Germany]: Wiley-VCH, 2008. Print.

Pais, Abraham. *Inward Bound : Of Matter and Forces in the Physical World*. Oxford [Oxfordshire]; New York: Clarendon ; Oxford UP, 1986. Print.

Schlager, Neil, and Josh Lauer. *Science and Its times : Understanding the Social Significance of Scientific Discovery*. Vol. 1-7. Detroit: Gale Group, 2000. Print.

Taylor, Edwin F., and John Archibald, Wheeler. *Spacetime Physics: Introduction to Special Relativity*. San Francisco: W.H. Freeman, 1966. Print.

Contributors

Angel G. Fuentes, MS
Chandler-Gilbert Community College

Casey M. Schwarz, PhD
University of Central Florida

Donald R. Franceschetti, PhD
University of Memphis

Gina Hagler, MBA
Writer/Author

J. D. Ho, MFA
PhD candidate, Carnegie-Mellon University

Kenrick Vezina, MS
Science Writer/Editor

Lindsay Brownell, MS
Science Writer

Marianne M. Madsen, MS
University of Utah

Nathan Olsson, MEd
Science Editor/Educator

Richard M. Renneboog, MSc
University of Western Ontario

Randa Tantawi, PhD
Senior Content Editor, EBSCO Information Services

Principles of Physics

ABERRATIONS

FIELDS OF STUDY

Optics; Relativity; Quantum Physics

SUMMARY

Aberrations in the motion of light and other wave phenomena produce distortions in the perception of those phenomena. Waves moving at low velocities are described by Newton's laws of motion. Waves moving at relativistic velocities are described using relativity theory and the Lorentz transformation.

PRINCIPAL TERMS

- **chromatic aberration:** a distortion created by the separation of white light into its component wavelengths when passing from one medium into another, such as through a lens.
- **coma:** the extended geometric image formed along the optical axis of a lens by light entering the lens obliquely.
- **inertial reference frame:** a means of describing relative motions through space according to Newtonian mechanics.
- **Lorentz transformation:** a means of describing relative motions through space according to the mathematics of relativity.
- **refraction:** a change in the direction of light due to the different speeds of light when passing through various media.
- **relativistic beaming:** the effect in which a luminous beam appears brightest when pointing directly at an observer.
- **special relativity:** the theory that physical laws are constant for matter with a uniform motion.
- **spherical aberration:** the blurring or distortion of an image produced by a spherical lens or mirror due to differences in the refraction of light that enters the optical system along the optical axis and light that enters closer to the edge of the lens.

Aberrations

The image from a lens is produced by refraction of light rays passing from one medium, such as air, into another medium, such as glass. A convex lens refracts light rays in such a way that they converge at the focal length of the lens. The focal length is determined by the radius of curvature of the lens and the lens material's index of refraction. The focal length describes how strongly a lens converges or diverges light. Errors in images produced by lenses and other optical systems are called aberrations. Some common lens aberrations include coma, chromatic aberration, and spherical aberration.

Coma results when light rays enter a lens at an oblique angle. Coma creates an elongated image rather than a clear image at the focal length of the lens. A simple example of coma can be seen using an ordinary magnifying glass and holding it at an angle to a point light source such as the sun or a light bulb. A coma produces several images of various sizes. Rather than a clear image of the light source, the image will appear to resemble a comet, with a distinct head and a blurred tail spreading out behind it. This aberration occurs because the magnification of a lens is different for light that enters the lens near its center and light that enters near the edge.

Chromatic aberration results when light rays pass through two different media, such as air and the lens. Each medium has a specific index of refraction. The index of refraction is the ratio of the speed of light in a vacuum to its speed in a material. Light is diffracted by the medium in relation to its wavelength. As light passes through a lens, each wavelength of light changes direction by a slightly different amount. For example, sunlight that passes through a prism and becomes separated into bands of colored light. Chromatic aberration occurs when light entering a lens is diffracted, producing colored outlines, halos, and rings around the image. This is often seen in low-cost optical devices of poor quality, but it is also a major concern for even advanced optical telescopes and microscopes.

Spherical aberration is a problem that occurs in optical systems with spherical lenses or mirrors. For a curved surface, the reflected or refracted rays from each radial location on the surface converge along a central axis. However, the spherical geometry requires them to converge at different foci along that axis. The clearest image from a spherical lens is found at a point that is effectively the average value along the central axis. This point is termed the circle of least confusion.

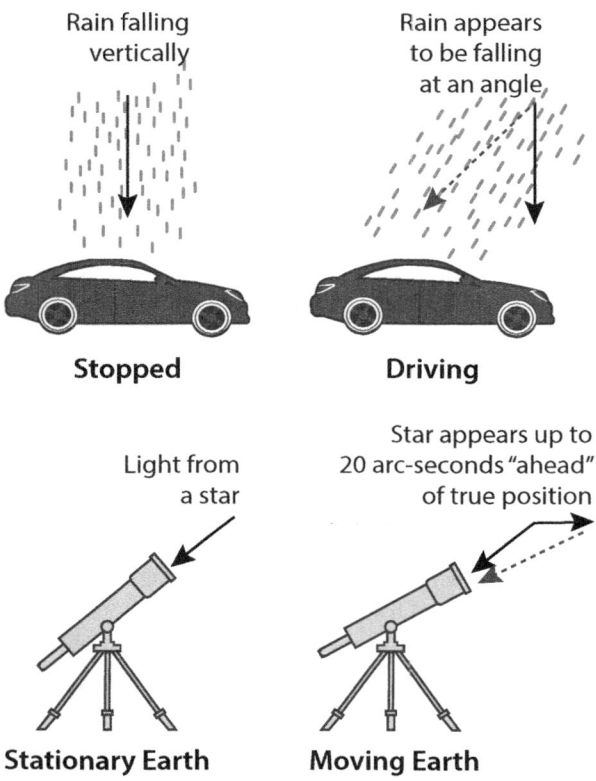

The earth's rotation causes a change in the apparent position of the stars, called stellar aberration. Because our position is constantly moving, we interpret the light from stars to be coming at a different angle. This same phenomenon occurs when a person drives in the rain. Rain will fall vertically when a car is stopped, but when a car is in motion the rain appears to be coming toward us at a different angle due to an aberration.

DISTORTIONS

Light and sound depend on the movement of waves. Sound waves require a physical medium such as air to propagate. On the other hand, light propagates as electromagnetic waves that do not require a physical medium. They can thus travel easily through a vacuum. Each light wave has a specific wavelength and corresponding frequency that defines its essential properties. Any aberration, or change in the motion of the light waves, causes a distortion of the light as perceived by an observer. Aberrations and distortions are most commonly associated with lenses and mirrors, such as those used in powerful telescopes and similar optical devices.

Light coming through space from distant stars undergoes aberration due to the relative motion of Earth as it moves through its orbit around the sun. The sun itself is also moving through space relative to the other stars and galaxies in the universe. This relative motion can cause observers to see the stars as though they are ahead of their actual positions. The effect is similar to how a person sees raindrops through the window of a car. When the car is stationary, the raindrops are seen to fall straight down, but when the car is moving, the raindrops appear to be falling on an angle. The faster the car moves in relation to the raindrops, the steeper is the angle at which the raindrops appear to be falling.

FRAMES OF REFERENCE

The stationary car in the example above provides the view from an inertial reference frame. An inertial reference frame must be moving at a constant speed relative to the observed phenomena. The direction of the raindrops seen when the car is stationary can be described by the laws of motion English physicist Isaac Newton (1642–1727) formulated. However, when the car is moving, any mathematical description of the raindrops must account for the effect of the relative motion of the car and the raindrops. Similarly, the motion of light through space relative to Earth must be accounted for mathematically. This is something that Newton's laws of motion are not capable of describing. Instead, this is a job for the mathematics of relativity.

RELATIVITY

Newtonian mechanics, which describes motion and energy relationships, can account for the motion of objects with nonrelativistic velocities. The simple equations describe force as the product of mass and acceleration, and momentum as the product of mass and velocity. These equations are quite sufficient for objects that are moving very slowly relative to the speed of light. However, as velocity increases, the ability of Newtonian mechanics to accurately describe motion decreases. Therefore, one must use the Lorentz transformation to accurately describe its properties. The Lorentz transformation accounts for the differences in observations of the same phenomenon from two inertial reference frames that are moving at constant velocities with respect to one another. Any phenomenon that two observers want to describe can be converted from one reference frame to the other using the Lorentz transformation. In the theory of special relativity, formulated by German-born scientist Albert Einstein (1879–1955) in 1905, two assumptions are made. For one, it is held that the laws of physics are constant in all inertial reference frames. The second and most important assumption is that the velocity of light in a vacuum is constant regardless of the velocity of the light source. Because of this, an observer always sees light traveling at the same speed regardless of their reference frame.

This is a difficult concept to understand, because it is counterintuitive to the perception of human-scale motion governed by Newtonian mechanics. In everyday life, the motions of wave phenomena such as sound and water are seen to be affected by the motion of the source of the waves. Sound increases in pitch when the sound source is moving toward the observer and decreases when it is moving away. This is called the Doppler effect. What is observed for light sources in motion, such as stars, is the redshift and blueshift of the wavelength that they emit. A light source that is red-shifted appears to have a longer wavelength because it is moving away from the observer. A blue-shifted light source appears to have a shorter wavelength because it is moving toward the observer.

This perception masks the fact that the velocity of sound through some conductive medium is not affected by the motion of the medium. Similarly, the velocity of light through space or some other transparent medium is not affected by the relative motion of the medium and is thus constant. If the velocity of light is not affected by the motion of the medium that it travels through, then the velocity of light from a source moving through space is not affected by the motion of the source. For luminous beams traveling at relativistic velocities, such as those emitted by certain stars, an effect called relativistic beaming can be observed. The beam and its apparent source will appear brightest when the beam is oriented directly toward the observer. This is sometimes referred to as the headlight effect because the headlights of a vehicle moving toward an observer appear brightest when pointed directly at the observer but are barely visible from other angles.

—*Richard M. Renneboog, MSc*

BIBLIOGRAPHY

Bodanis, David. *E=mc²: A Biography of the World's Most Famous Equation.* New York: Walker, 2000. Print.

Einstein, Albert, Robert W. Lawson, Robert Geroch, and David C , Cassidy. *Relativity: The Special and General Theory.* New York: Pi, 2005. Print.

Jha, A. K., and A. K. Jha. *A Textbook of Applied Physics.* New Delhi: I.K. International Pub. House Pvt., 2009. Print.

Koks, Don. *Explorations in Mathematical Physics: The Concepts Behind an Elegant* Koks, Don. *Explorations in Mathematical Physics: The Concepts behind an Elegant Language.* New York: Springer, 2006. Print.

Nave, R. "Aberration of Starlight." *HyperPhysics.* Georgia State University, n.d. Web. 21 Aug. 2015.

Sasian, Jose M. *Introduction to Aberrations in Optical Imaging Systems.* Cambridge: Cambridge UP, 2013. Print.

Woodhouse, N. M. J. *Special Relativity.* London: Springer, 2003. Print.

ABSORPTION

FIELDS OF STUDY
Acoustics; Electromagnetism

SUMMARY

The mechanisms by which acoustic and electromagnetic waves carry energy are briefly described. Absorption is the transfer of energy from a wave to the medium through which it propagates. Absorption is one main component of wave attenuation, the other being scattering. Each material has a unique absorption coefficient that characterizes how much energy it can absorb from a sound or light wave.

PRINCIPAL TERMS

- **absorption coefficient:** a value characteristic of a particular medium that represents the amount of light or sound it absorbs from a wave passing through it.
- **acoustics:** the study of sound; also, the qualities of a space that affect how sound is heard within that space.
- **albedo:** the portion of electromagnetic energy that is reflected when its waves encounter a surface or boundary; often used to describe solar radiation reflecting off Earth or another body in space.
- **attenuation:** the loss of energy from a wave passing through a medium due to absorption or scattering.
- **Beer-Lambert law:** a formula that relates the attenuation of an electromagnetic wave in a given medium to the thickness of that medium and the concentration of attenuating materials within it.
- **light wave:** an oscillation in an electromagnetic field.
- **reflection:** the rebounding of a wave from a surface or boundary between two mediums, causing it to travel back through the original medium.

Wave Energies

Energy comes in many different forms. Some types of energy, such as sound energy and radiant energy, travel in the form of waves. Sound energy is kinetic energy that is carried by sound waves, also called acoustic waves. Sound waves are a type of mechanical wave, meaning that they travel, or propagate, by oscillating the molecules of the surrounding medium. A mechanical wave must have a medium in order to propagate.

Radiant energy is the energy carried by electromagnetic radiation, which travels in the form of electromagnetic waves. These waves are sometimes called light waves because visible light is a kind of electromagnetic radiation. However, the electromagnetic spectrum extends far beyond the range of visible light. Electromagnetic waves do not require a medium, although the presence of one can influence how they propagate. Instead, they travel as oscillations in the electromagnetic field. This is why light can travel through the vacuum of space.

Different types of waves share a number of common physical properties. For example, all waves have a wavelength, frequency, and amplitude. Wavelength is the distance between identical points on two successive wave cycles. Frequency is the number of wave cycles per unit time. Amplitude is the distance between a wave's highest point (crest) and its lowest point (trough).

In addition, all waves experience reflection, refraction, diffraction, and interference. Reflection occurs when a wave bounces off a boundary between two mediums and changes direction. In the case of radiant energy, albedo refers to the amount of waves that are reflected. Refraction is a change in wave direction caused by the wave passing through the boundary rather than rebounding from it. Diffraction occurs when a wave bends around an obstacle or spreads out after passing through a small opening. Interference is the superposition of two or more waves to form a single wave with an amplitude equal to the sum of those of the contributing waves at the points where they meet.

Acoustic and electromagnetic waves are also subject to scattering and absorption. Scattering occurs when a portion of a wave's energy is reflected by irregularities in the medium. Absorption is the transfer of energy from a wave into the medium through which it is traveling. The medium takes up energy from the wave and transforms it into another kind of energy, such as heat.

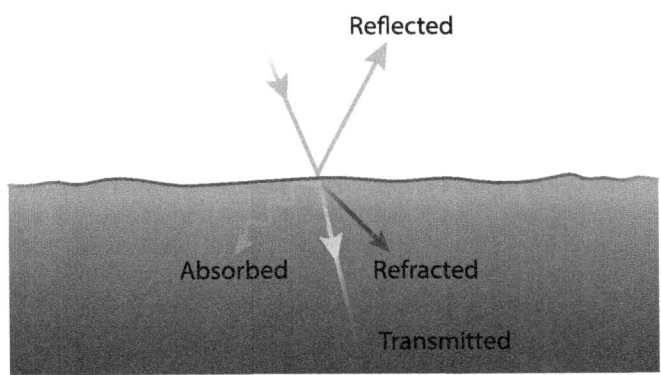

Sound waves and light waves can move through a homogenous medium in a consistent manner, but when a wave interacts with a change in medium, it can either reflect, refract, transmit, or be absorbed. When a light wave is absorbed, the material absorbing it heats up. When a sound wave is absorbed, the sound seems to disappear.

ENERGY AND MATTER

Energy is observed and measured in terms of its effects on matter. When an acoustic wave passes through matter, the sound energy pushes its particles together, forming an area of compression, or increased particle density. Once the wave is no longer exerting pressure, this creates an area of rarefaction, or decreased particle density, just behind the area of compression. The rarefaction allows the compressed particles to return to their original position. These back-and-forth movements of particles are the oscillations that make up an acoustic wave. As they oscillate, the particles bump into each other, producing friction. The heat generated by this friction is sound energy that has been absorbed by the medium and transformed.

If light is directed in a beam through a medium, the energy of the beam can be measured on both sides of the medium and compared. The amount of energy transmitted per unit time is called "radiant power" or "radiant flux." It is measured in watts (W), the International System of Units (SI) derived unit of power. One watt is equal to one joule of energy transmitted per second (J/s). Transmitted radiant flux (φ_t) is the amount of radiant flux that exits a medium. The initial radiant flux (φ_i) is the amount that entered the medium. The ratio between the transmitted and initial radiant flux produces a value known as the transmittance (T):

$$T = \frac{\varphi_t}{\varphi_i}$$

Transmittance can be used to calculate both the attenuance (D) of a material and its optical depth (τ):

$$D = -\log_{10} T$$

$$\tau = -\ln T$$

Attenuance measures the radiant flux lost to attenuation. Optical depth is the opacity of a material to electromagnetic radiation. The function ln is the natural logarithm, or $\log_e$, where e is the mathematical constant known as Euler's number, roughly equal to 2.71828. The relationship between transmittance, absorbance, and optical depth can also be expressed using the inverse functions of the logarithms:

$$T = e^{-\tau} = 10^{-D}$$

There is some debate over the use of the term *attenuance* versus *absorbance*. The quantity defined here as attenuance is also commonly called absorbance (A), even though it measures energy lost by scattering as well as absorption. The International Union of Pure and Applied Chemistry (IUPAC) has recommended using absorbance only when attenuation due to scattering is negligible or otherwise not taken into account. Transmittance calculated using absorbance alone is known as internal transmittance, as opposed to total transmittance:

$$A = -\log_{10} T_{int}$$

$$T_{int} = 10^{-A}.$$

ATTENUATION AND ABSORPTION COEFFICIENTS

A wave may pass through a medium with little to no interaction. Or, it may lose some or all of its energy to that medium. This loss of energy is called attenuation. It results in a decrease in the wave's intensity, or its power per unit area.

The two main components of attenuation are scattering and absorption, both of which depend on the characteristics of the medium. A given material is characterized by an attenuation coefficient (μ). This unique value represents how easily the material can be penetrated by a wave. Just as attenuation is the sum total of energy lost to scattering and to absorption, a material's attenuation coefficient is the sum of its scattering coefficient and its absorption coefficient.

For electromagnetic waves, one usually specifies either a molar absorption coefficient or a linear absorption coefficient. The molar absorption coefficient (ε) is typically used in chemical analysis of solutions. It relates absorbance (A) to path length (l)—that is, the distance the wave travels through the medium—and the concentration (c) of absorbing materials in the medium, as described by the Beer-Lambert law:

$$A = \varepsilon c l.$$

The linear absorption coefficient (a) is defined as the absorbance (A) per unit path length (l):

$$a = \frac{A}{l}$$

For acoustic waves, the absorption coefficient (α) of a material is the ratio of absorbed sound intensity (I_a), in watts per meter squared (W/m^2), to initial sound intensity (I_i):

$$\alpha = \frac{I_a}{I_i}$$

Its value ranges from 0 (no sound absorbed) to 1 (all sound absorbed). This value can be used to calculate the total sound absorption (A) of either a single surface or an enclosure of multiple surfaces, such as a room, according to the following equation:

$$A = \alpha_1 S_1 + \alpha_2 S_2 + \ldots + \alpha_n S_n.$$

Here, S is the surface area of a given material, and A is measured in sabins. The sabin, named after American physicist Wallace Clement Sabine (1868–1919), is a unit of sound absorption equal to the absorbing ability of a quantity of material with an absorption coefficient of 1. It can be either metric (one square meter of completely absorbing material) or imperial (one square foot), depending on whether S is measured in meters or feet squared.

APPLICATIONS OF ABSORPTION

Electromagnetic absorption is an important factor in numerous fields. In medicine, x-ray imaging works because different tissues absorb different amounts of x-rays; in meteorology, temperature is affected by the absorption of solar radiation by Earth's atmosphere and surface. In chemistry and materials science, the Beer-Lambert law can be used to identify unknown solutions.

Acoustic absorption is an everyday concern for architects, engineers, and anyone else who designs buildings or other structures where sound propagation is an issue. An engineer may want to design a recording studio with soundproofed booths or an auditorium that can project sound to distant audience members. In either case, it is necessary to use materials with appropriate absorption coefficients. The studio material should have a higher coefficient, to prevent sound from entering from outside. The auditorium material should have a lower coefficient, to increase the reverberation time of sound from the stage. Engineers who deal with acoustics frequently consult tables showing the absorption coefficients of common building materials.

—*Nathan Olsson, MEd, and Randa Tantawi, PhD*

BIBLIOGRAPHY

"Attenuation of Sound Waves." *Attenuation of Sound Waves.* NDT Resource Center, n.d. Web. 9 July 2015.

Band, Yehuda Benzion. *Light and Matter: Electromagnetism, Optics, Spectroscopy and Lasers.* Chichester: John Wiley, 2006. Print.

Band, Yehuda Benzion. *Light and Matter: Electromagnetism, Optics, Spectroscopy and Lasers.* Chichester: John Wiley, 2006. Print.

Everest, F. Alton, and Ken C. Pohlmann. *Master Handbook of Acoustics.* N.p.: n.p., 2015. Print.

"Fundamentals of Waves." *Science Learning Hub RSS.* University of Waikato, n.d. Web. 31 July 2015.

Henderson, Tom. "Waves." *Waves.* Physics Classroom, n.d. Web. 4 Aug. 2015.

"Propagation of Sound Indoors." *Propagation of Sound Indoors.* Engineering ToolBox, n.d. Web. 25 Aug. 2015.

"Wave Behaviors." *Mission: Science.* NASA, n.d. Web. 24 Aug. 2015.

ACCURACY AND PRECISION

FIELDS OF STUDY

Classical Mechanics

SUMMARY

This article defines the relationship between accuracy and precision in physics. Accuracy is how well a method of measurement can determine the real value of a property. Precision is how consistent the measurements of that property are. A method that is accurate is not always precise, and measurements that are precise are not always accurate.

PRINCIPAL TERMS

- **bias:** the extent to which a measurement differs from the real value of the property being measured, or an intrinsic factor in a method that consistently causes such deviation.
- **discrepancy:** the difference between the measured value of a property and its real value, or between nonidentical measurements of the same property.
- **measurement:** the numerical value of a physical property according to a standard relative scale, or the act of determining that value.
- **random error:** a measurement error that is due to unpredictable and inconsistent factors that do not affect all measurements equally.
- **systematic error:** a measurement error that is due to intrinsic, mechanical, or environmental factors that affect all measurements equally.

Accuracy versus Precision

The terms "accuracy" and "precision" are often incorrectly used to mean the same thing. While it may seem logical to think that to be accurate and to be precise are the same thing, they are not. In physics, every property has one real value. That value can usually be determined, or at least estimated, by some form of measurement using a standard scale. Temperature, for example, is usually measured using an accepted standard temperature scales, such as kelvin, Celsius, or Fahrenheit.

Accuracy refers to how close the measured value of a property such as temperature comes to the real value of that property. The closer a measured value is to the real value, the more accurate the measurement is, and thus the method used to take that measurement. Precision, on the other hand, is a measure of the discrepancy between measurements of the same property. The closer those values are to one another, the more precise they are. Ideally, measurements should both precise and accurate, so that different measurements of the same property will have very similar values that are all very close to the property's real value.

The difference between accuracy and precision can be explained using the example of an archer shooting arrows at the bull's-eye of a target. The "real value" being measured is represented by the exact center of the bull's-eye. The closer the arrows land to the bull's-eye, the more accurate they are. A tight cluster of arrows within a small circle around the center would be a very accurate and precise result. A larger circle around the center, with the arrows farther apart, would be less accurate and less precise. Imagine the archer fires several arrows directly at the bull's-eye when a strong wind blows across the field, causing the arrows to land slightly off-center but just as clustered as before. In that case, the precision of the shots is just as high, because they are still very close to one another. However, the accuracy of the shots is low, because they are not near the center. Thus, arrows fired with high precision may not be accurate, and arrows that are accurate may not be precise.

In all physical systems and methods of measurement, both systematic errors and random errors occur. In the above example, the wind is a systematic error: it affects all of the shots equally, shifting each arrow roughly the same distance in the same direction away from the bull's-eye. This represents the bias of the system. The wind drives the shots toward their new impact point instead of their intended target. A random error, on the other hand, might be if the archer sneezes just as an arrow is released. Another would be if the archer fails to draw back the bowstring the same distance each time, making each arrow travel at a slightly different velocity. A systematic error affects all arrows in the same way, while a

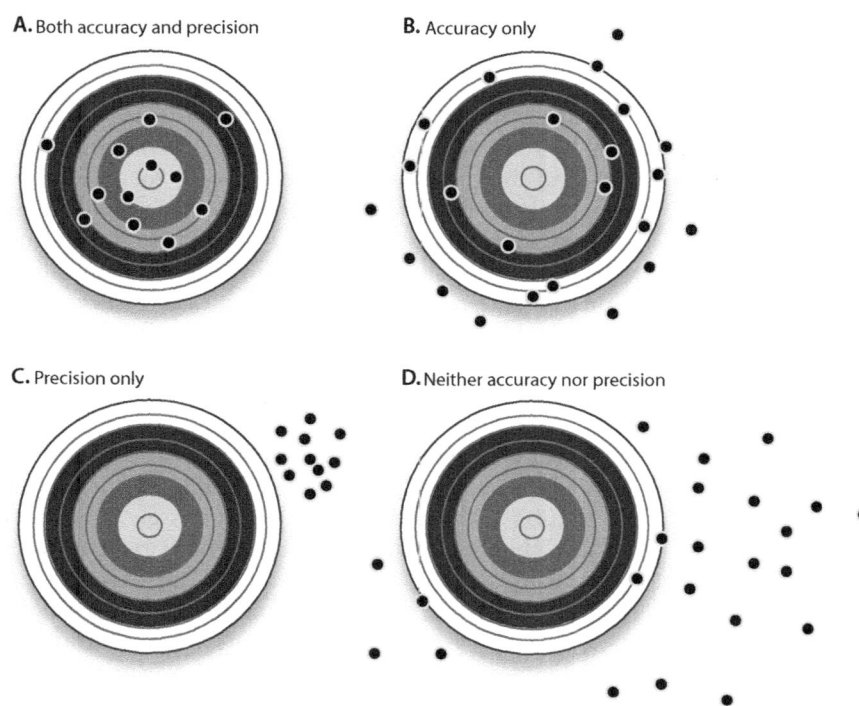

Points on a bull's-eye may be clumped or scattered and centered around the bull's-eye or some other location off to the side. Clumped points are more precise, scattered points are less precise. Points centered around the bull's-eye are more accurate and points centered off to the side are less accurate.

random error might affect only one arrow or several arrows in different ways.

MEASUREMENT AND ERROR LIMITS

The goal of measurement is to find the real value of the property being measured. In science, every effort is made to prevent both systematic and random error. Researchers design experiments so that the value of one, and only one, factor can be measured. They calibrate, or adjust, the tools and methods to be used in the experiment against known standard values and outcomes. This ensures that the measured results will be both as accurate and as precise as possible.

There are physical limits on how well the value of any particular measurement in any situation can be known. This normally depends on the researcher's visual perception— specifically, on the eye's ability to estimate small distances. When reading the graduated scale of a measuring device, the standard rule of thumb is that the measurement can be read accurately only to one-tenth of the smallest scale division. For example, the length of an object is measured with a ruler whose smallest units are centimeters, the length can only be accurately determined to the nearest tenth of a centimeter, or one millimeter (1 mm). If the ruler is marked in millimeters, the length can only be determined to the nearest tenth of a millimeter (0.1 mm).

Digital electronic devices have greatly improved both precision and accuracy. However, because electronic circuits and sensors control the process by which devices take measurements, they have their own unique requirements for ensuring that the device is functioning properly. Electronic measuring devices must be calibrated often to specific standards, mainly because their parts are dynamic systems. Those parts are subject to drift as they are affected by environmental factors. Thus, the physical limits of the parts themselves, rather than the visual abilities of their users, determine the error limits in digital measurements.

CALIBRATION AND CALCULATING ACCURACY

To maintain both the accuracy and the precision of various measuring devices, those devices must be calibrated by measuring properties whose values are both known and stable. Such standards serve as the reference base for the values taken from different devices. Comparing measurements against standard reference values may enable the measuring device to be calibrated and made more accurate. Or, it may simply allow the accuracy and error limits of the measuring device to be defined so that future measurements may be corrected. For example, if the time given by a wall clock is five minutes ahead of that shown on an atomic clock (the world's most accurate timepiece), the owner of the wall clock may set it back five minutes so that it is more accurate, or simply remember to subtract five minutes from the displayed time in future.

SAMPLE PROBLEM

A very accurate device for measuring the pH of water-based solutions is used to measure the pH of ten different samples. Because the device is known to be accurate, these measurements are used as reference standards. The standard pH measurements from the first device are then used to test the accuracy of a second such device used outside of the laboratory. Calculate the accuracy of the second device (as a percentage), given the following data:

	Standard pH	Measured pH
Sample 1:	2.34	2.28
Sample 2:	3.17	3.09
Sample 3:	4.28	4.17
Sample 4:	5.55	5.41
Sample 5:	6.82	6.65
Sample 6:	7.12	6.94
Sample 7:	8.61	8.39
Sample 8:	9.22	8.98
Sample 9:	9.87	9.62
Sample 10:	11.2	10.91

Answer:

To calculate the accuracy of a measurement, first subtract the standard or true value from the measured value. Then divide the absolute value of the difference by the standard value and multiply the result by 100. This calculation gives a value known as the percent error. The percent error can then be subtracted from 100 to determine the percent accuracy. For example, the percent accuracy of sample 1 is calculated as follows:

$$\frac{|2.28 - 2.34|}{2.34} \times 100 = 2.564 \, (\text{rounded})$$

$$100 - 2.564 = 97.436$$

The measurement of the pH of sample 1 is 97.436 percent accurate. Performing the same calculation for the other samples gives the results below:

- 97.436
- 97.476
- 97.430
- 97.477
- 97.507
- 97.472
- 97.445
- 97.397
- 97.467
- 97.411

Thus, the accuracy of the second device is between 97.4 and 97.5 percent. However, because all measurements made with the same device can only be considered as accurate as the least accurate measurement, the lowest percent-accuracy value represents the functional accuracy of the device. Thus, the device can be said to be 97.397 percent accurate.

ACCURACY AND PRECISION

It is important to remember that accuracy and precision are never the same thing. Each is a different property of a single set of measurement events. A single measurement is precise by default because there are no other measurements whose values may vary. However, that same single measurement can be anything from perfectly accurate to wildly inaccurate, depending on the discrepancy between the measured and real values. For example, a device used to measure the temperature of an ice-water mixture may give a precise reading of 10.633 degrees Celsius. That would be a very inaccurate reading if the mixture has a real temperature of 0 degrees Celsius.

The terms "accuracy" and "precision" are most useful when applied to a set of measurements of the same property taken under the same conditions. In such cases, accuracy and precision are given number values and descriptions calculated through statistical analysis. Statistical terms, such as mean, median, and skew, describe the relationships between the different measurements in the set. This in turn makes it possible to determine the accuracy and precision of both the measurements and the methods used to take them.

—*Richard M. Renneboog, MSc*

BIBLIOGRAPHY

Bewoor, Anand K., and Vinay A. Kulkarni. *Metrology & Measurement.* New Delhi: Tata McGrawHill, 2009. Print.

Concise Dictionary of Physics. Hyderabad: V & S, 2012. Print.

Kenkel, John. Analytical Chemistry for Technicians. 4th ed. Boca Raton: CRC, 2014. Print.

Loyd, David H. *Physics Laboratory Manual.* 4th ed. Boston: Brooks, 2014. Print.

Putten, Anton F. P. Van. *Electronic Measurement Systems: Theory and Practice.* Bristol: Institute of Physics Pub., 1996. Print.

Rabinovich, Semyon G. *Evaluating Measurement Accuracy: A Practical Approach.* New York: Springer, 2010. Print.

Wilson, Jerry D., and Cecilia A. Hernández-Hall. *Physics Laboratory Experiments.* 7th ed. Boston: Brooks, 2010. Print.

ALPHA RADIATION

FIELDS OF STUDY

Electromagnetism; Atomic Physics; Nuclear Physics

SUMMARY

Alpha radiation is generated by the release of alpha particles, which are two protons and two neutrons bound together. Alpha radiation is typically emitted by radioactive materials undergoing alpha decay, especially nuclear fission. Alpha radiation was discovered by Ernest Rutherford in 1899.

PRINCIPAL TERMS

- **alpha decay:** a form of radioactive decay in which a radioactive atom's nucleus splits and discharges an alpha particle, made up of two protons and two neutrons.
- **Geiger-Müller probe:** a device that can be used to detect alpha radiation as well as beta radiation, gamma radiation, and x-rays.
- **half-life:** the average time it takes for half of the unstable nuclei in a radioactive element to undergo radioactive decay, transforming into a lighter element and giving off radiation.
- **isotopes:** variants of a chemical element with differing numbers of neutrons; they are often unstable and radioactive.
- **neutrons:** subatomic particles that, with protons, make up the mass of an atom's nucleus; they have functionally the same weight as protons but no electric charge.
- **protons:** subatomic particles that, with neutrons, make up the mass of an atom's nucleus; they have functionally the same weight as neutrons but hold a positive electric charge.

- **radiation:** energy being transmitted via electromagnetic waves (e.g., light, heat, x-rays) or subatomic particles (e.g. alpha particles, beta particles).

Alpha and Beta: Radiation without Waves

"Radiation" often refers to electromagnetic radiation (EMR), which includes visible light, x-rays, and ultraviolet rays. Alpha radiation, however, is different. It is transmitted by subatomic particles ejected at high speed when a radioactive atom's nucleus undergoes alpha decay. The alpha particle and the beta particle, which transmits beta radiation, were detected and described by British physicist Ernest Rutherford in 1899. Alpha radiation can be used in a variety of applications, including cancer treatment and smoke detection.

An alpha particle is made up of four subatomic particles: two protons and two neutrons. Because protons have a positive charge and neutrons are neutral, alpha particles are positively charged. They have an atomic weight of four, and are equivalent to a helium nucleus stripped of its electrons.

Radioactivity and Isotopes

An atom is said to be radioactive if it gives off ionizing radiation. This radiation is the result of an unstable nucleus ejecting energy in the form of subatomic particles in order to reach a more stable configuration. The half-life of a radioactive material is the average time it takes for half of a given sample of the material to undergo decay.

The three main forms of radiation emitted during radioactive decay are alpha, beta, and gamma radiation. All three are considered ionizing radiation. They possess enough energy to change the electrical charge of atoms they pass through. This can in turn lead to chemical changes. Through interaction with the atomic structure of living molecules, ionizing radiation can cause cancer, sickness, and increased mutation rates.

The number of protons in an atom's nucleus determines which element it is. This number is known as the atomic number and is used to sort elements in the periodic table. The nucleus of any one atom of a given element will always have the same number of protons as every other atom of that element. However, the number of neutrons may vary. This variation in number of neutrons creates isotopes. Isotopes have more or less total atomic weight because of the addition or subtraction of neutrons. Normal carbon has 6 protons and 6 neutrons, adding up to a total atomic weight of 12. Carbon occurs in isotopes known as carbon-13 (one extra neutron) and carbon-14 (two extra neutrons). Isotopes can occur naturally or can be created artificially. Some isotopes are stable. Some, called radioisotopes, are inherently unstable and subject to spontaneous radioactive decay. All elements with atomic numbers of 83 or above are inherently unstable and radioactive. Most below that number have at least one radioisotope.

Rutherford, Villard, Geiger, and Müller

In 1899 Ernest Rutherford (1871–1937) performed radiation experiments using samples of uranium, thin sheets of metal foil, and a detection screen coated in zinc sulfide. The screen gave off light when exposed to the radiation given off by uranium. Rutherford found that one type of radiation, alpha radiation, could be stopped by an ultrathin sheet of metal. Another type, beta radiation, could penetrate 100 times as much metal before being absorbed. The following year, French physicist Paul Villard (1860–1934) discovered an even more penetrating form of radiation. Rutherford would later dub these gamma rays.

Almost a decade later, Rutherford and his assistant, German physicist Hans Geiger (1882–1945), used beams of alpha particles as a method of probing the structure of atoms. In 1908, during his tenure with Rutherford, Geiger came up with the idea for a tube filled with gas that would produce an electric pulse in response to the presence of

DECAY TYPE	SYMBOL	PARTICLE EMITTED	EXAMPLE	STOPPED BY	ENERGY RELEASED	SOURCE
Alpha	α	Alpha Particle (helium nucleus)	Am-241 → Np-237	Paper	5 MeV	Uranium-238 Radium-226 Americium-241

Alpha particles are two protons and two neutrons emitted from a radioactive element such as uranium. They can be stopped by paper.

ionizing radiation. Twenty years later, Geiger and Walther Müller (1905–79) worked together to produce a functional Geiger-Müller tube. This tube is the basis of the Geiger-Müller probe (often called a "Geiger counter"), a device that is used to detect ionizing radiation.

USES OF ALPHA PARTICLES

The positive charge and relatively low penetration of alpha particles make them suitable for a variety of applications. Tiny amounts of radium-226 may be used as a targeted form of radiation therapy for cancer, because the alpha particles will not penetrate far beyond the tumorous tissue they are intended to kill. Some smoke detectors use the positive charge of alpha particles emitted by americium-241 to generate an electric current in a sensing chamber. When smoke enters this chamber, it disrupts the current and sets off the alarm. Far and away the biggest impact alpha particles have had on everyday life, however, is through their discovery and use in establishing the basis of modern nuclear physics and with it nuclear energy and weapons technology.

—*Kenrick Vezina, MS*

BIBLIOGRAPHY

"Alpha Particles." *Radiation Protection*. US Environmental Protection Agency, 26 July 2012. Web. 12 May 2015.

Davidson, Michael W. "The Rutherford Experiment." *Molecular Expressions: Electricity & Magnetism*. Florida State U, 28 Feb. 2014. Web. 7 May 2015.

"Electromagnetic Radiation." *Encyclopædia Britannica*. Encyclopædia Britannica, 2015. Web. 7 May. 2015

Errede, Steven. "A Brief History of the Development of Classical Electrodynamics." UIUC Physics 245 Lecture Notes. U of Illinois at Urbana-Champaign, 2007. Web. 7 May 2015.

Lucas, Jim. "What Is the Weak Force?" *LiveScience*. Purch, 24 Dec. 2014. Web. 14 May 2015.

AMPLITUDE

FIELDS OF STUDY

Electromagnetism; Acoustics; Harmonics

SUMMARY

This article describes the basic properties of different types of waves and relates them to a wave's amplitude. Waves are characterized by the basic properties of amplitude and frequency. Amplitude describes the distance between the neutral value of a wave to its maximum displacement.

PRINCIPAL TERMS

- **crest:** the highest point of a wave from its neutral value; the distance between the crest or trough of a wave and the wave's neutral value is called the amplitude.
- **displacement:** the upward or downward extent to which the amplitude differs from the neutral value of a wave.
- **frequency:** the number of complete wavelengths that occur within one unit of time, typically expressed as hertz (Hz; cycles per second).
- **loudness:** the intensity of sound waves, which depends on the wave's amplitude; measurements of loudness or volume are expressed in decibels.
- **oscillation:** a variation between maximum and minimum values of displacement from a neutral value.
- **peak amplitude:** the value of the amplitude at its maximum displacement from the neutral value of the wave.
- **peak-to-peak amplitude:** the absolute value of the sum of the peak positive and peak negative amplitudes; the distance between the crest and the trough of a wave.
- **root-mean-square amplitude:** for sinusoidal wave systems, the square root of the sum of squared amplitude values divided by the number of amplitude values.
- **trough:** the lowest point of a wave from its neutral value.
- **wavelength:** in any wave system, the distance from one point in a wave to the equivalent point in the

next wave, typically measured between successive peak values.

PROPERTIES OF WAVES

A wave is any physical phenomenon that can be described as an oscillation, or an upward and downward displacement that travels through a medium, such as water or air, or through space. Waves in water and the vibrations of a guitar string are just two examples of waves. Visible light and all other wavelengths of light are electromagnetic wave phenomena. Sound is another physical phenomenon that can be described in terms of wave properties. An essential feature of wave systems is their specific wavelength. Wavelength describes the distance between two identical points in successive waves. Wavelengths are typically measured between successive peaks, or crests, of a wave system. Another property of waves is their frequency, or the number of times waves repeat in a single unit of time.

Wave properties can be clearly visualized in water. For example, if the crests of a series of waves approaching a shore are separated by a distance of 3 meters (10 feet) then the wavelength of those waves is 3 meters. If six waves pass the same point in a span of three seconds, then their frequency is two waves per second. Frequency is normally described in units of cycles per second called hertz (Hz). The neutral value of water waves is at the level of perfectly smooth, undisturbed water. The difference between this level and any part of a wave is the displacement. The maximum displacement either above or below the neutral level is the amplitude of the wave. As each wave approaches, half of the wave is above the neutral level and half is below the neutral level. The maximum upward displacement of the wave is the crest, and the maximum downward displacement of the wave is the trough. The terms *crest* and *peak amplitude* are often considered synonyms.

SIMPLE HARMONIC OSCILLATIONS

The pendulum in a grandfather clock is one of the most common examples of a simple harmonic oscillator. The motion of a pendulum can be described in terms of wave functions. At rest, a pendulum hangs straight down, which is its neutral value. The distance that the pendulum is swung away from the neutral value is its displacement. The maximum displacement in either direction is the amplitude. The motion of the pendulum as it swings between its maximum displacement on either side demonstrates a sinusoidal relationship. Sinusoidal motion is a pattern of behavior that can be described by a sine wave function. Once the pendulum reaches its maximum displacement, its motion reverses and it falls toward the neutral position once again. The distance from the peak amplitude in one direction to the peak amplitude in the other direction is called the peak-to-peak amplitude.

SOUND WAVES

Sound waves propagate or move through a medium such as air or water when a force is exerted against the matter that makes up the medium, compressing the atoms or molecules together. This compression travels through the medium as the compressed molecules push against

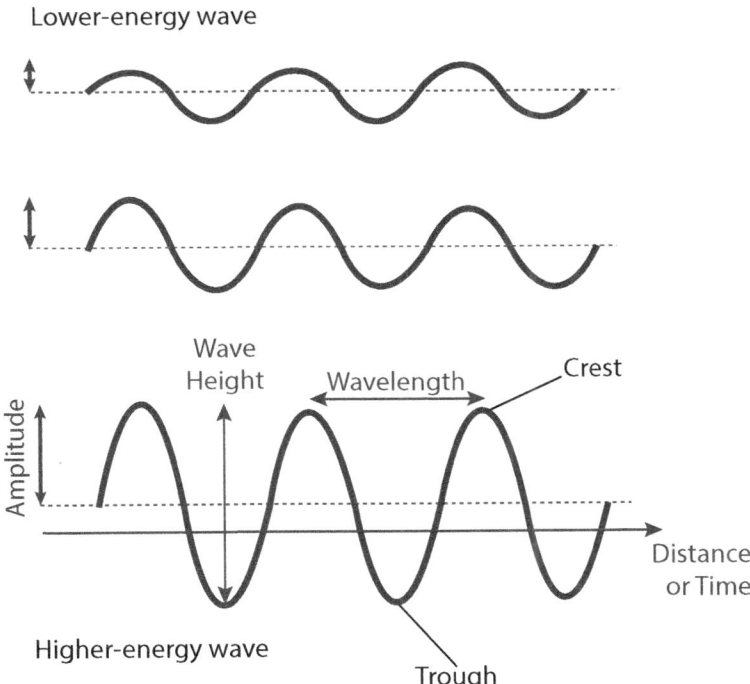

A wave's amplitude is measured from the origin to the wave's crest, or half the wave height. Waves with higher amplitude are higher energy than waves with a lower amplitude. In terms of electricity, amplitude is the maximum or peak voltage of a current.

the molecules next to them. At the same time, the molecules of matter behind the compression are pulled farther apart. The compression is the positive part of the sound wave. The rarefied portion that follows is the negative part of the sound wave. The compression and rarefaction of the particles in the medium displace those particles from their neutral position. The loudness of the sound is determined by the extent that the particles are displaced. The greater the amplitude of a sound wave, the greater the loudness of the sound.

When the compression and the rarefaction stages have passed, the molecules return to their neutral position. The motion of the matter involved in a sound wave can therefore be described by the same sinusoidal wave functions as other waves. The displacement of the medium from its neutral position describes the amplitude of the sound wave. The number of sound waves per unit time defines the frequency of the sound (its pitch). The distance between equivalent points in successive sound waves defines the wavelength of the sound.

Sinusoidal Motion and Amplitude

The amplitude of any wave is the maximum displacement of the wave from its neutral, or equilibrium, value. Smooth and cyclic wave functions can be described by a sine wave function, or sinusoidal motion. In simple terms, sinusoidal motion follows the value of the sine of an angle about a fixed center, just like a point on a rotating wheel. Sine values cycle between a range of 1 and –1, beginning at an angle of 0 degrees, or 0 radians. A radian is the arc length along a circle that is equal to the length of the radius of the circle. An angle in radians is equal to the arc length divided by the radius, which is equal to approximately 57.3 degrees. As the point travels around the circle, it reaches an angle of 90 degrees (or $\pi/2$ radians) relative to its start point. As it travels further and again reaches the neutral level, the angle it makes with its starting point is 180 degrees (or π radians). At the lowest point the angle is 270 degrees (or $3\pi/2$ radians). The full circle is completed at the starting point, after moving through an angle of 360 degrees (or 2π radians) about the center of the circle.

A graph of the sine function exhibits the characteristic sideways S shape of a sine curve. In a sine wave, the amplitude is equal to the sine value of 90 degrees, which is 1, or 270 degrees, which is –1. Note that the trigonometric values of sine, cosine, and tangent have no units, as they are simply ratios and not measurements.

Sines and Cosines

Both the sine and the cosine function are used to describe wave motion. They have the same behavior, but the value of the cosine is shifted 90 degrees ($\pi/2$ radians) from the value of the sine for the same angle. Whereas sine begins at 0, cosine begins at the peak amplitude of 1.

—*Richard M. Renneboog, MSc*

Bibliography

De Pree, Christopher G. *Physics Made Simple*. New York: Random, 2004. Print.

Freegarde, Tim. *Introduction to the Physics of Waves*. Cambridge: Cambridge UP, 2013. Print.

Gilbert, P. U. P. A., and W. Haeberli. *Physics in the Arts*. Burlington: Academic, 2008. Print.

SAMPLE PROBLEM

On a windy day at the beach, waves roll in regularly along a pier that rises 3 meters (10 feet) from the level seabed. The crest of each wave reaches the top of the pier, and the trough reaches only halfway to the top of the pier. Determine the peak amplitude of the waves.

Answer:

The piles of the pier mark a distance of 3 meters (10 feet) from the seabed to the top of the pier. The trough of each wave reaches one-half of that distance, or 1.5 meters (5 feet). The maximum difference between the crest and the trough of a wave, or the peak-to-peak amplitude of the waves, is therefore 1.5 meters (5 feet), corresponding to the distance from the top of the pier to the trough of the wave. The peak amplitude of the waves is equal to the difference in wave height from the neutral point to the crest, or one-half of the peak-to-peak amplitude, which in this case is 0.75 meters (2.5 feet).

$$A = (AP-P) / 2 = 1.5 \text{ m} / 2 = 0.75 \text{ m}$$

Mansfield, Michael, and Colm O'Sullivan. *Understanding Physics*. 2nd ed. Hoboken: Wiley, 2012. Print.

Serway, Raymond A., and John W. Jewett, Jr. *Physics for Scientists and Engineers*. 9th ed. Boston: Cengage, 2013. Print.

Young, David, and Shane Stadler. *Cutnell and Johnson Physics*. 10th ed. Hoboken: Wiley, 2015. Print.

ANGULAR FORCES

FIELDS OF STUDY

Classical Mechanics

SUMMARY

When an object moves in a circular path, as in a planetary orbit, it is subject to an angular force pulling it toward the center of the circle. This is called the centripetal force, and in the case of planetary orbits, it is gravity. Newton's laws for linear motion also describe circular motion and the forces at play.

PRINCIPAL TERMS

- **centrifugal force:** a fictitious force that seems to push a body in circular motion away from the axis of rotation; in reality, objects in circular motion are subject to centripetal force.
- **centripetal force:** a force "toward the center" that, in combination with inertia, generates the curved path of an object in circular motion.
- **cosine:** a trigonometric function describing the relationship between sides of a right triangle; the cosine of an angle is equal to the length of the side adjacent to the angle divided by the length of the hypotenuse.
- **inertia:** the principle that an object at rest tends to stay at rest and an object in motion tends to stay in motion unless acted on by an outside force.
- **perpendicular:** being at a right angle relative to a given line or plane, as in the lines of the letter T.
- **radian:** a nondegree unit of angle measurement, based on the radius of a circle; there are 2π radians (equal to 360 degrees) in one complete circle or revolution.
- **sine:** a trigonometric function describing the relationship between sides of a right triangle; the sine of an angle is equal to the length of the side opposite the angle divided by the length of hypotenuse.
- **vector:** a quantity with both direction and magnitude.

FORCES OF CIRCULAR MOTION

Whenever an object follows a curved path of motion, whether rotating about an internal axis or revolving around an external axis, an angular force is at play. Angular forces are forces that tend to produce circular motion. These forces act in a straight line, as all forces do. However, the result of their influence is a curved path of motion. This is in contrast to linear forces and linear motion, which follow straight lines.

The force that causes objects to follow a curved path is known as centripetal force. *Centripetal* comes from the Latin words for "center-seeking." Any curved motion can be thought of as tracing the circumference of an imaginary circle. The centripetal force always acts toward the center of this circle.

An object moving in a circular path is constantly caught in a tug-of-war between its inertia—that is, its tendency to move in a straight line at a constant speed—and the centripetal force. Centripetal describes a category of force and therefore may refer to a variety of forces, such as gravity (planetary orbits), tension (a ball on a string), or even friction (a car turning around a race track). At any given moment, an object in circular motion will tend to continue traveling in a straight line. The centripetal force acts to change this.

CENTRIPETAL VERSUS CENTRIFUGAL FORCE

The centripetal force (F_c) needed for an object to travel a circular path is described by the following equation, where m is the mass of the object in motion, v is its linear velocity, and r is the radius of the circular path it is following:

$$F_c = mv^2/r$$

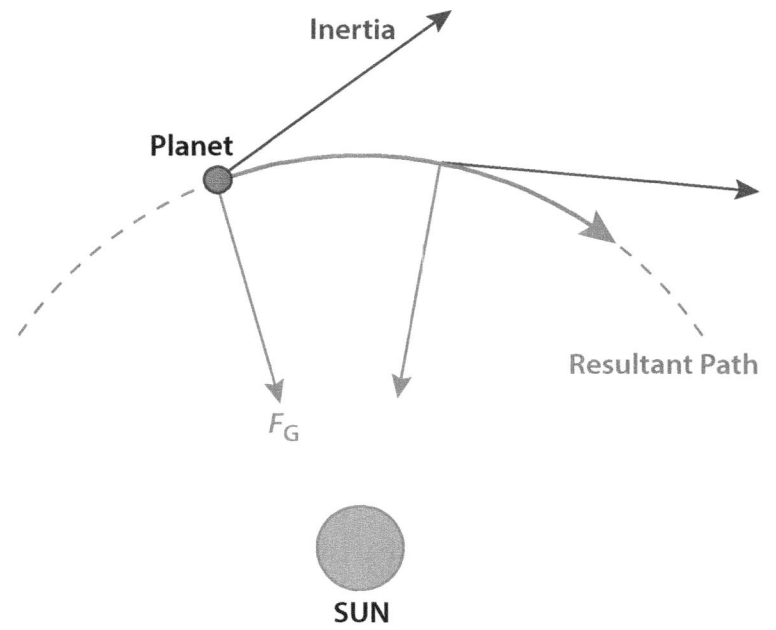

The resultant path of a planet in orbit is dependent on the inertia of the planet and the gravitational force (FG) exerted on the planet by the sun. The angle of the resulting path falls between these two vectors.

Force is measured in newtons (N), an International System of Units (SI) derived unit. In SI base units, one newton is equal to one kilogram-meter per second squared (kg·m/s^2). Be sure to use kilograms (not grams) for the mass when making calculations with this equation. Sometimes centrifugal force is incorrectly used in place of centripetal force. Centrifugal force is a fictitious or illusory force that seems to push outward from an object's axis of rotation. Imagine being the passenger in a car as it makes a sharp left turn and feeling pressed against the right-hand door. This "force" is not a real force, however. In the example of the turning car, the "force" being felt is that of the car pressing against the passengers to keep them from following their individual inertia and continuing forward in a straight line. When an object is set in circular motion and then released, such as a rock on a string that is spun around and then let go, it is not a "centrifugal force" that propels it outward. Rather, it is the lack of a centripetal force (the tension of the string) that suddenly allows the rock to travel unimpeded in a straight line.

Force Vectors

Forces are vector quantities, meaning that they have both a magnitude and a direction. Vectors are typically indicated using arrows, with the length of the arrow corresponding to the magnitude of the force.

When two vectors are combined, their relative directions determine how they interact. Two forces of equal magnitude forces acting in opposite directions will cancel each other out. Two forces of equal magnitude acting in the same direction result in a doubled total, or net, force. When adding two forces, the angle formed between them determines their net magnitude and direction. Imagine hitting a baseball up and away from home plate. Ignoring drag, two forces are at play. First, the collision imparts the force of the bat's swing to the ball, pushing it in a straight line up and away toward some point above the horizon. Second, the force of Earth's gravity pulls the ball downward. The interaction of these forces gives the ball its curved trajectory.

Uniform Circular Motion

In uniform circular motion, an object follows a perfectly circular path at a constant speed. At any given instant, centripetal force acts perpendicular to the linear momentum of the object in motion, forming a right angle between the two competing vectors. These two vectors can be thought of as two sides of a right triangle within the object's circular path of motion, with the apex of the triangle corresponding to a point along the circle's circumference. Therefore, trigonometric functions can be used to describe the relationship a given angle of rotation and the vectors at play in circular motion.

The two most commonly used trigonometric functions are sine (sin) and cosine (cos). The sine of a given angle (θ) in a right triangle is equal to the length of the side opposite that angle divided by the length of the triangle's hypotenuse (the longest side, opposite the right angle):

$$\sin\theta = \text{opposite} / \text{hypotenuse}.$$

The cosine of the angle is equal to the length of the adjacent side divided by the length of the hypotenuse:

$$\cos\theta = \text{adjacent} / \text{hypotenuse}.$$

The preferred unit of angular measure when dealing with circular motion is the radian (rad). Radians are based on the relationship between the radius of a circle and its circumference; there are 2π radians (corresponding to 360 degrees) in one complete revolution. Because circular motion traces the circumference of a circle, radians are also used to describe the angular distance an object travels—that is, how much of its circular path it completes.

Torque is a measure of rotational force acting on an object. Imagine the spoke of a wheel. A hand grips the wheel at some point along the spoke and applies a force perpendicular to the length of the spoke, causing the wheel to rotate around its central pivot point. The radius (r) is the length between that pivot point and the point where the force is applied. Torque (T) is the product of the radius, the force applied (F), and the sine of the angle between their directions (θ):

$$T = rF\sin\theta$$

Another use of these trigonometric functions is in finding the difference between two vectors. This has various applications, such as determining the change in velocity due to a glancing collision. In such a collision, the initial velocity vector and the final (post-collision) velocity vector form an angle at the point of impact, with the initial vector leading toward the impact and the final vector leading away. If the final vector were moved so that its starting point were the same as that of the initial vector, the difference between the two would be equal to a vector leading from

---

SAMPLE PROBLEM

A car is driving around a circular racetrack one kilometer (km) in circumference. It moves at a constant linear velocity of thirty meters per second (m/s) and has a mass of two thousand kilograms (kg). What is the magnitude of the centripetal force acting on the car?

Answer:

Use the equation for calculating centripetal force:

$$F_c = mv^2/r$$

Although the radius (r) of the racetrack is not given, it can be determined from the circumference. The equation for finding the circumference of a circle is as follows:

$$C = 2\pi r$$

Rearrange this equation, substitute in the known value of the circumference (C), and solve for r:

$$C / 2\pi = r$$

$$1 \text{ km} / 2\pi = r$$

$$r \approx 0.16 \text{ km}$$

Convert the radius from kilometers to meters:

$$0.16 \text{ km} \times 1{,}000 \text{ m/km} = 160 \text{ m}$$

Using this value for the radius, as well as the given values for linear velocity (v) and mass (m), calculate the centripetal force (F_c):

$$F_c = mv^2/r$$

$$F_c = (2{,}000 \text{ kg})(30 \text{ m/s})^2 / 160 \text{ m}$$

$$F_c = (2{,}000 \text{ kg})(900 \text{ m}^2/\text{s}^2) / 160 \text{ m}$$

$$F_c = 11{,}250 \text{ kg·m/s}^2 = 11{,}250 \text{ N}$$

The car is subject to a centripetal force of approximately 11,250 newtons.

the end point of the initial vector to the end point of the final vector, forming the third side of a triangle. The law of cosines says that for a triangle with sides a, b, and c, the length c can be found when the angle C between lengths a and b (opposite side c) is known, according to the following equation:

$$c^2 = a^2 + b^2 - 2ab\cos C.$$

This is built into the formula for the cross product, or dot product, of two vectors. (The cross product is the product obtained when two vectors in a three-dimensional space are multiplied, resulting in a third vector perpendicular to both.) That is, the cross product of vectors X and Y is equal to the value $2|X||Y|\cos\theta$. Thus, if the angle (θ) between the initial velocity vector ($v\mathbf{i}$) and the final velocity vector ($v\mathbf{f}$) were known, the length of the difference vector ($v\mathbf{d}$) would be calculated as follows:

$|v\mathbf{d}|^2 = (v\mathbf{f} - v\mathbf{i})^2$
$|v\mathbf{d}|^2 = (v\mathbf{f} - v\mathbf{i})(v\mathbf{f} - v\mathbf{i})$
$|v\mathbf{d}|^2 = (v\mathbf{f} \times v\mathbf{f}) - (v\mathbf{f} \times v\mathbf{i}) - (v\mathbf{f} \times v\mathbf{i}) + (v\mathbf{i} \times v\mathbf{i})$
$|v\mathbf{d}|^2 = |v\mathbf{f}|^2 + |v\mathbf{i}|^2 - 2|v\mathbf{f}||v\mathbf{i}|\cos\theta$

Of course, objects do not always travel with uniform speed. Once an object in circular motion begins to speed up or slow down, the equations above no longer work. As long as the object continues to follow a circular path, the net force acting on the object will always equal the centripetal force, but its magnitude will vary depending on the acceleration of the object (remember, force equals mass times acceleration).

Circular Motion in Everyday Life

It is not difficult to find examples of both rotation and revolution in everyday situations. Understanding torque and rotational motion is a vital part of engineering automobiles so that the wheels are given enough force to roll the car forward. Spin on an object moving through the air dramatically affects its aerodynamics and trajectory. In tennis, topspin is a vital technique that allows a player to make the ball drop much more sharply than it would under the influence of gravity alone. Although these situations may seem more complicated than the more familiar linear motion of, for instance, billiard balls bouncing around a pool table, it is important to remember that the physical principles underpinning linear and circular motion are the same.

—*Kenrick Vezina, MS*

Bibliography

Colwell, Catharine. "Rotational Kinematics." PhysicsLAB. PhysicsLAB, 1997–2015. Web. 13 Sept. 2015.

Cross, Rod, and Crawford Lindsey. "Tennis Ball Trajectories: Aerodynamic Drag and Lift in Tennis Shots." *Tennis Warehouse University*. TWU, 22 Dec. 2013. Web. 13 Sept. 2013.

Graham, T., et al. "Force as a Vector." *Mathcentre*. Mathcentre, 2009. Web. 13 Sept. 2015.

Henderson, Tom. "Motion in Two Dimensions." N.p.: Physics Classroom, 2012. Digital file.

Martinez, Jason. "Angular Kinematics—Solving Circular Motion Problems with Wolfram|Alpha." Wolfram|Alpha Blog. Wolfram Alpha, 20 Mar. 2013. Web. 13 Sept. 2015.

Nave, Carl R. "Law of Cosines." HyperPhysics. Georgia State U, 2012. Web. 21 Sept. 2015.

Weisstein, Eric W. "Law of Cosines." MathWorld. Wolfram Research, 1999–2015. Web. 21 Sept. 2015.

ANGULAR MOMENTUM

FIELDS OF STUDY

Classical Mechanics; Quantum Mechanics; Particle Physics

SUMMARY

In macroscopic systems, angular momentum is the rotational equivalent of momentum due to linear motion. In quantum mechanics, angular momentum is a property ascribed to subatomic particles in relation to the associated quantum property called spin. In both systems the relevant mathematics are essentially the same. The angular momentum vector is related to the direction of rotation by the right hand rule. Angular momentum is a universal property that applies to all aspects of the physical universe, from electromagnetic radiation to entire galaxies.

PRINCIPAL TERMS

- **angular velocity:** the speed and direction of movement of a rotating object.
- **conservation of energy:** a fundamental law of physics that states that the amount of energy in a system remains constant over time. Although the energy can be transformed or transferred, it cannot be created or destroyed.
- **conservation of momentum:** in physics, the principle that the total momentum in a closed system is always constant.
- **kinematics:** a subfield of classical mechanics that studies the motion of objects without reference to the forces that cause this motion.
- **quantum mechanics:** the branch of physics that deals with matter interactions on a subatomic scale, based on the concepts that energy is quantized, not continuous, and that elementary particles exhibit wavelike behavior.
- **quantum state:** the condition of a physical system as defined by its associated quantum attributes.
- **spin:** an intrinsic form of angular momentum carried by elementary particles, composite particles (hadrons), and atomic nuclei.
- **standard model:** a generally accepted unified framework of particle physics that explains electromagnetism, the weak interaction, and the strong interaction as products of interactions between different types of elementary particles.

A Fundamental Property

The property of angular momentum is a fundamental property of the physical universe, regardless of the scale. Momentum is defined as the product of mass and velocity in the branch of Newtonian mechanics called kinematics, which deals with the physics of motion independent of applied forces, and is the basic principle of Newton's second law: objects in motion tend to stay in motion, and objects at rest tend to stay at rest, unless acted upon by an external force. For example, with the minimal resistance to its motion by friction between their two respective surfaces, a rock thrown by a curler slides down the length of a curling rink with an almost perfectly constant velocity until it strikes either another rock or the backboard at the end of the curling rink. The motion of the rock is linear, or one-dimensional.

A thrown baseball also moves through space with the same restrictions, but its motion follows a ballistic trajectory and is therefore two-dimensional. The position of the rock, or any other particle or object undergoing such motion can be described relative to a specified starting point, or origin, in a Cartesian coordinate system as the vector sum of two or three values in as many orthogonal directions. The same motions can be described using polar or spherical polar coordinates that give the particle's location as a radius operating through one or two orthogonal angles relative to the focal point and the principal axis. The property of angular momentum is also the product of mass and velocity. However, the motion is restricted to rotation of the mass about a central axis, and the angular velocity is stated as the angle about the central axis that the rotation passes through in unit time. The standard measurement of angle in this usage is in radians per second, one complete revolution being through an angle of 2π radians (equivalent to $360°$).

As a macroscopic property, the principle of conservation of angular momentum is easily demonstrated by a figure skater who goes into a spin. While rotating about the central axis of the spin, the skater decreases

> **SAMPLE PROBLEM**
>
> A skater enters a spin at a rate of 30 complete revolutions per minute, with her body mass centered in a radius of 1 meter. Calculate her rate of rotation if she then enclosed her body mass within a radius of 0.33 meter by folding her arms in as tightly as she is able.
>
> **Answer:**
> The skater's angular momentum is given by
>
> $$L = I\omega$$
>
> and her moment of inertia is given by
>
> $$I = r^2 m.$$
>
> Conservation of angular momentum requires that the value of L remain constant, and her body mass does not change either. Therefore, the values of r^2 and ω must change in an inverse relationship to maintain the constant value of L:
>
> **initial state:** $r = 1m$ and $r^2 = 1\ m^2$
> $\omega = 30$ rpm or 0.5 revolutions per second = π radians per second
> **final state:** $r = 0.33m$ and $r^2 = (0.33)^2 = 0.1089 m^2$
> $\omega = ?$
>
> $$L = r_i^2 m \omega_i = r_f^2 m \omega_f$$
>
> $$L/m = r_i^2 \omega_i = r_f^2 \omega_f$$
>
> $$r_i^2 \omega_i = r_f^2 \omega_f$$
>
> $$\omega_f = r_i^2 \omega_i / r_f^2$$
>
> $$= (1 m^2)(\pi\ \text{rad.sec}^{-1})/(0.1089\ m^2)$$
>
> $$= 28.85\ \text{rad.sec}^{-1}$$
>
> 1 revolution = 2π radians
>
> Therefore, 28.85 rad.sec^{-1} = 4.59 revolutions per second, or 275 revolutions per minute.<end sample problem>

and increases the rate of rotation by extending his or her arms and legs or drawing them in closer to the central axis, respectively. Since the skater's mass is constant throughout the process, the increases and decreases of the rate of rotation maintain the constant value of the skater's angular momentum. As a vector property, the direction of the angular momentum is along the principal axis of the rotational motion. The angular velocity is the rate of rotation of the mass about the central axis.

ANGULAR MOMENTUM AND QUANTUM MECHANICS

In the standard model based on quantum mechanics atoms consist of an extremely small, dense nucleus composed of protons and neutrons enclosed within a very large, diffuse cloud of electrons. In finer detail, these subatomic particles are composed of even smaller entities, called quarks, in specific combinations. According to the mathematics of the model, one of the properties conferred by a combination of quarks is termed spin, although it does not actually indicate that the particle is actually rotating about an axis. It is instead a term applied as the name of a state. Accordingly, the quantum state is described in terms of angular momentum.

Electrons in an atom are required to occupy specific regions called orbitals that correspond to very specific amounts, or quanta, of energy. The state of the electron within the atom corresponds to the allowed energy levels through the corresponding angular momentum states. This is due to the interaction of the intrinsic angular momentum of a subatomic particle with the angular momentum designation of the orbital. The probability of finding a particle at a specific location in an atom depends on the absolute value of the square of the wave function Ψ. This in turn requires that inverting the sign of one variable in Ψ must convert the wave function to $-\Psi$. This is the basis of the Pauli exclusion principle, which sates that two electrons with the same value of spin cannot occupy the same orbital. The quantum numbers that designate the electron levels in atomic and molecular orbitals are angular momentum quantum numbers.

Mathematical Relationships

When angular momentum is expressed in terms of the angular velocity in radians per second, it is dependent on the sine function. Negation of the sine value of a quantity is equal to the negative sine value of the original quantity, as

$$\sin(-x) = -\sin(x),$$

which is exactly the requirement of the relationship of the wave function Ψ in determining the probability of finding an electron at a specific location within an atom. Accordingly, the mathematics of angular momentum on a macroscopic scale is effectively the same mathematics of angular momentum on the quantum scale. The orbital requirements of angular momentum as it relates to electrons in atoms echo the orbital requirements of much larger objects undergoing motion about a central axis. It is easy to see from this that even though the property of angular momentum on the quantum scale may not actually refer to physical rotation of the particle, the designation of that property as 'angular momentum' is a logical one. Since it is a vector property, the mathematics of angular momentum is necessarily vector calculus, including matrix manipulations for the determination of the cross products and dot products of vectors. The calculation of angular momentum depends on the "moment of inertia" of the rotating entity, as

$$L = I\omega,$$

where L is the angular momentum vector, I is the moment of inertia, and ω is the rate of rotation of the mass. The moment of inertia is determined by the distribution of the mass about the axis of rotation. For a single point mass, the moment of inertia is calculated as

$$I = r^2 m,$$

where m is the mass of the point and r is the distance of the point from the axis of rotation. For a complex mass consisting of multiple point masses, such as the body of a skater, I is the sum of the individual moments of inertia.

A Truly Universal Property

The property of angular momentum is a truly universal property that applies not only to macroscopic systems, but to the most fundamental particles and is even associated with the fundamental nature of electromagnetic radiation. Light has an angular momentum component in its propagation. Electrons in atoms obey the quantum mechanical restrictions of angular momentum. The flow of electrons through a conductor has an associated angular momentum property, as evidenced by the magnetic field produced by current moving through a solenoid. The angular momentum vector obeys the same "right hand rule" as the magnetic field of a solenoid. Turning wheels have an associated angular momentum, as do footballs and baseballs spinning after being thrown. Planets have angular momentum due to their motion as they spin on their respective axes, as well as their orbital motion about the nearest star. Entire planetary systems have angular momentum due to their orbital motion about their central star, as well as due to the motion of that star system about the center of whatever galaxy it is in. Indeed, that entire galaxy has an associated angular momentum. The property, and the need to understand it, therefore extends into all fields and branches of physics. It is an especially central concept in quantum and classic mechanics, particle physics, optics (or the study of electromagnetic radiation) and astronomy, and is finding valid applications in such seemingly unrelated fields as protein biochemistry and nuclear medicine.

—*Richard M. Renneboog M.Sc.*

Bibliography

Andrews, David L. and Mohamed Babiker, eds. *The Angular Momentum of Light.* New York, NY: Cambridge University Press, 2013. Print.

Bartlett, Roger. *Introduction to Sports Biomechanics: Analysing Human Movement Patterns.* 2nd ed., New York, NY: Routledge, 2007. Print.

Brenneman, Laura. *Measuring the Angular Momentum of Supermassive Black Holes.* New York, NY: Springer, 2013. Print.

Chabbay, Ruth W. and Bruce A. Sherwood. *Matter & Interactions.* 3rd ed., Hoboken, NJ: John Wiley & Sons, 2011. Print.

French, A. P. and M.G. Ebison. *Introduction to Classical Mechanics.* Boston, MA: Kluwer Academic Publishers, 2012. Print.

Levi, A.F.J. *Applied Quantum Mechanics.* 2nd ed., New York, NY: Cambridge University Press, 2006. Print.

Rae, Alastair I. M. *Quantum Mechanics.* 5th ed., New York, NY: Taylor & Francis, 2007. Print.

Thompson, William J. *Angular Momentum. An Illustrated Guide to Rotational Symmetries for Physical Systems.* Weinheim, GER: Wiley-VCH Verlag, 2004. Print.

ANTENNA

FIELDS OF STUDY

Electromagnetism; Acoustics

SUMMARY

An antenna receives and interacts with electromagnetic radiation and converts it into an electric current that carries information. An antenna can also convert a varying electrical signal into a corresponding electromagnetic waveform. Ideally, the length of an antenna corresponds to the wavelength that it will receive or transmit. Antennas operate either passively or actively, and are designed in various shapes for different purposes. All types of wireless communication rely on antennas.

PRINCIPAL TERMS

- **attenuation:** the loss of energy from a wave passing through a medium due to absorption or scattering.
- **frequency:** the number of complete wavelengths that occur within one unit of time, typically expressed as hertz (Hz; cycles per second).
- **interference:** the distortion of a wanted signal by an unwanted one, such as an unwanted radio station "bleeding into" another because both are broadcasting on the same wavelength.
- **parabolic:** refers to a shape (a parabola) that can be described by an equation of the form $y = ax^2 + b$.
- **period:** the length of time for one complete cycle of a wave or other cyclic property to occur.
- **radio waves:** electromagnetic radiation with a wavelength between 1×10^{-3} and 1×10^5 meters; able to travel long distances without being broken up by atmospheric interference.
- **signal-to-noise-and-distortion (SINAD) ratio:** the ratio of total signal power received to noise and distortion received, indicating how well a signal or waveform has been reproduced.
- **wavelength:** in any wave system, the distance from one point in a wave to the equivalent point in the next wave, typically measured between successive peak values.

THE ELECTROMAGNETIC SPECTRUM AND ANTENNAS

The electromagnetic spectrum is a continuum of all wavelengths of electromagnetic radiation. Each wavelength is defined by a specific frequency and period of sinusoidal oscillation. The spectrum ranges continuously from the complete absence of electromagnetic waves to a theoretically infinite value. Within the electromagnetic spectrum, wavelengths in certain ranges have been identified, such as infrared light, visible light, ultraviolet light, radio waves, microwaves and X-rays. An antenna is a relatively simple structure that acts to receive electromagnetic radiation and convert it into an electrical signal or to convert an electrical signal into electromagnetic radiation. The most effective size of the antenna is determined by the wavelength(s) of electromagnetic radiation that it is intended to receive or transmit. In this way, an antenna acts as an interface between an electrically conductive material and a medium such as air that does not conduct electrical signals well.

ANTENNA FUNCTION

In operation, an antenna can receive or transmit an electromagnetic signal, depending on the manner in which it is structured. The simplest type of antenna is a plain conductor such as a wire or a metal rod connected to an electronic system. Interaction of the antenna with an incident electromagnetic signal produces an electrical current in the conductor that varies in synchronization with the incident signal.

> **SAMPLE PROBLEM**
>
> Calculate the optimum length of a quarter-wave antenna for a wavelength of 1MHz. Use the value of 2.999×10^8 m.sec^{-1} (9.821×10^8 ft.sec^{-1}) as the speed of light.
>
> **Answer:**
>
> In 1 sec, the signal passes exactly 1 million (10^6) wavelengths and spans a distance of 2.99×10^8 meters (9.821×10^8 feet)
>
> Therefore 1 wavelength spans $(2.99 \times 10^8/10^6) = 2.99 \times 10^2$ meters (9.821×10^2 feet)
>
> One quarter wavelength therefore spans $(299/4) = 74.75$ meters (245.525 feet)

The electronic system can then amplify the current and convert it into an output signal such as sound from the speaker of a radio or cell phone. Conversely, the electronic system can be one that directs an electrical current through the antenna so that it emits the corresponding electromagnetic wave output. If the signal is amplified during transmission as part of the antenna structure itself the structure is referred to as an 'active' antenna. A structure that transmits the signal without amplification is termed a passive antenna. The effectiveness of any antenna is determined by the wavelengths that it is designed to receive or transmit. The most effective length for an antenna is the corresponding wavelength. However, since electromagnetic waves travel at the speed of light, the distance spanned by a single wavelength makes this rule increasingly impractical as wavelength increases and frequency decreases. The optimum antenna length for a frequency of 530KHz, the lower end of the radio wave frequencies, requires a tower that is hundreds of meters (or several hundred feet) tall. In comparison, the corresponding antenna for a frequency of 900MHz, typically the range of cordless telephones and cell phones, requires a length of just about 15 centimeters (6 inches). An alternative structure size that is almost as effective as the full wave antenna is an antenna whose dimension is a specific fraction of the wavelength. Such antennas are termed half wave, quarter wave, and so on. Obviously, any practical antenna must be able to detect its designed wavelength with good sensitivity to compensate for attenuation of the signal and interference from other signal sources. An unavoidable source of interference is the universe itself, as influx of cosmic particles from solar activity and the background radiation produce a constant background noise. An antenna must be effective in facilitating differentiation between the desired signal and the background noise. This is usually stated as the signal-to-noise-and-distortion ratio. For electromagnetic signals that are directional rather than broadcast, parabolic antennas are typically used. Such antennas are commonly seen as satellite dishes and on cell phone relay towers, and are the mainstay structures of radio-telescope arrays. In actuality, the dish structure serves as a reflector to concentrate the signal that is being received or sent at the focal point of the parabolic shape, which is where the actual antenna is located. The large surface area of the dish relative to the antenna acts as a collector to provide a certain amount of amplification for an incoming signal whose power may be just a few nanowatts (1 nanowatt = 1 billionth or 10^{-9} of a watt). In transmission mode, the signal radiating in all directions from the focal point of a parabolic reflector is reflected from the dish in such a way that it travels in just one direction. The deeper is the parabolic shape, the more tightly focused is the transmission coming from it.

WIRELESS COMMUNICATION

The modern world communicates essentially all of its information wirelessly in the present day, allowing people half-way around the world to talk to each other in real time. Vast quantities of electronic data related to economics, politics, current affairs and any number of other activities are transmitted between locations ranging from the personal wireless home network to world capitols, and even into space. While all such transfers depend on antennas of various kinds. astronomy is an area of research that relies exclusively on the antenna for the discoveries made in that field.

—*Richard M. Renneboog M.Sc.*

ARAGO DOT

FIELDS OF STUDY

Optics;

SUMMARY

The Arago dot, also known as the Fresnel or Poisson bright spot, is an artifact of the propagation of light waves about a spherical surface. Diffraction and refraction of light waves passing around a smooth, spherical surface undergo constructive and destructive interference in the shadow of the object, with a bright spot in the center of the shadow. The intensity of the spot can approach the intensity of the light source. The effect occurs in other conditions in which the flow of particles is wave-like.

PRINCIPAL TERMS

- **circumference:** the distance around a circle defined by a radius.
- **collimated rays:** parallel rays of light that propagate with minimal spreading.
- **constructive interference:** when two or more waves of the same phase combine to form a larger amplitude.
- **destructive interference:** when two or more waves of different phases combine to form a smaller amplitude.
- **diffraction:** a change in the direction of a wave as it passes around an obstruction or through an opening.
- **refraction:** the alteration of a wave's path, speed, and wavelength when it passes from one medium to another.

Fresnel, Poisson and the Propagation of Light

In the early 1800s, the general belief of science was that light traveled only in straight lines as collimated rays. There was as yet no concept of light as having a dual nature as both waves and particles. In 1807, experimental evidence was presented that indicated light traveled as a wave, and a competition was proposed by the French Academy of Science to explain the behavior of light. Augustin-Jean Fresnel entered a proposal explaining the propagation of light as waves. Siméon Denis Poisson was a proponent of the particle nature of light, and examined Fresnel's proposal in detail in order to identify a fatal flaw that would support the particle theory. His analysis indicated that if light traveled as waves, then the resulting diffraction and refraction would produce a bright spot in the center of the shadow of a spherical object having an appropriate circumference. The theory was tested experimentally by the head of the competition committee, Dominique-Francois-Jean Arago, using a 2mm glass sphere. His observation of the theoretical bright spot in the shadow of the sphere demonstrated that Fresnel's hypothesis and the wave nature of light were both true.

Theoretical Foundation

The basic principle of light propagating as waves is the Fresnel-Huygens concept of wave source regeneration. In this model every point of an unimpeded wavefront becomes the source of a secondary wave front, effectively describing a continuous field traveling through space. The wave nature of light makes it subject to constructive interference and destructive interference. Propagation of light waves around the surface of an appropriate spherical object has the effect of changing the phase of waves. Images of the shadow of the object clearly show concentric rings of high and low intensity, centered about a small bright Arago dot. Using complex integral calculus, it can be shown that the intensity of the Arago dot approaches the intensity of the light source, in much the same way that the intensity of the image of a light source from a magnifying glass approaches the intensity of the light source itself. The intensity of the Arago dot is given approximately by the relation

$$I_A = b^2 I_O / (a^2 + b^2)$$

where I_A and I_O are the Arago dot intensity and the original intensity respectively, a is the radius of the spherical object, and b is the distance from the center of the spherical object projected along the central axis behind the object.

POTENTIAL APPLICATIONS

The Arago dot itself is little more than a curiosity of optics, given that its existence is highly dependent on the regularity of the spherical surface. However, as technology improves the Arago dot can be used as a highly sensitive alignment device in many applications. Computers that use light to carry data instead of electrons may rely on the Arago dot in this capacity. The Arago dot is also found to occur in the movement of matter that can be described as analogous to wave motion. The effect is observed in electron microscopy and in the flow of electrons from electron beams. The Arago dot effect may therefore become increasingly useful as a fluid flow control mechanism.

—*Richard M. Renneboog M.Sc.*

SAMPLE PROBLEM

Calculate the relative intensity of the Arago dot behind a spherical object having a diameter of 5 mm at distances of 2, 5 and 10 mm from the object's center.

Answer:

The diameter of the object is 5 mm. Therefore, the radius a is 2.5 mm. The values of b are 2 mm, 5 mm and 10 mm. Can the intensity of the Arago dot ever rqual the intensity of the source?

The relative intensity is given by rearranging the equation for I_A, to obtain

$$I_A/I_O = b^2/(a^2 + b^2)$$

For b = 2mm, $I_A/I_O = b^2/(a^2 + b^2) = 2^2/(2.5^2 + 2^2) = 0.39$

For b = 5mm, $I_A/I_O = b^2/(a^2 + b^2) = 5^2/(2.5^2 + 5^2) = 0.8$

For b = 10mm, $I_A/I_O = b^2/(a^2 + b^2) = 10^2/(2.5^2 + 10^2) = 0.94$

For the two intensities to be equal, for I_A to be equal to I_O, then the value of b^2 must be equal to the value of $(a^2 + b^2)$. This can only be true if $a = 0$ (the spherical object does not exist), but the condition can be approached as b becomes much greater than a. This is demonstrated by the increasing relative intensities calculated above as the value of b becomes increasingly larger than the value of a.<end sample problem>

BIBLIOGRAPHY

Erbschloe, Donald Ross. *The Spot of Arago and Its Role in Aberration Analysis.* Defense Technical Information Center, 1983. Print.

Levitt, Theresa. *The Shadow of Enlightenment. Optical and Political Transparency in France, 1789 – 1848.* New York, NY: Oxford University Press, 2009. Print.

Sharma, K.K. *Optics Principles and Applications.* New York, NY: Elsevier, 2006. Print.

Wayne, Randy O. *Light and Video Microscop.* 2nd ed. San Diego, CA: Academic Press, 2014. Print.

Weiner, John and Frederico Nunes. *Light-Matter Interaction. Physics and Engineering at the Nanoscale.* New York, NY: Oxford University Press, 2013. Print.

APERTURE

FIELDS OF STUDY

Optics

SUMMARY

This article defines and describes the function and applications of apertures. An aperture is an opening that controls light transmission. Apertures are used in many optical systems, such as telescopes and microscopes. Photography depends on aperture control to produce desired images and effects.

PRINCIPAL TERMS

- **collimated rays:** parallel rays of light that propagate with minimal spreading.
- **depth of field:** the range of distance over which objects appear sharp or in focus.
- **diaphragm:** a circular structure with an aperture that controls the amount of light entering an instrument.
- **focal length:** the distance from the focal point of a lens or mirror to the center of the lens or mirror.
- **optical system:** a system of mirrors, lenses, or prisms that can be used for imaging in an instrument such as a telescope or microscope.
- **vignetting:** a darkening or shading of an image's edges compared to the center of the image.

Optical Systems

An aperture is an opening or hole that controls the amount of light that enters an optical system. The aperture controls the admission of collimated rays, which in turn determines how focused or sharp an image appears. Small apertures allow mainly collimated light rays to enter, producing sharp focus at the image plane. Larger apertures allow rays that are not collimated to enter, which leads to sharp focusing only for rays that have a particular focal length.

Apertures are an important part of optics, which is the branch of physics that studies the properties of light, how it interacts with matter, and the instruments that use it. Apertures are found in a variety of such instruments, including cameras, microscopes, and telescopes. A diaphragm can be used to control the aperture's size.

Apertures in Imaging

In photography, f-numbers and f-stops are related to aperture size and depth of field. The f-number is the ratio of the focal length to the aperture size, while the f-stop is the corresponding aperture setting. A larger f-number indicates a smaller aperture. The smaller the aperture, the greater the distance over which an image will be in focus at one time. In other words, in images with a large depth of field, both near and far images appear in focus. If the f-number is too small, optical aberrations may result, producing a blurred or distorted image. Optical systems can be made to compensate for the effects of aberrations. Vignetting may also occur when the f-number is small and the intensity of the incoming light decreases toward the edges of the image.

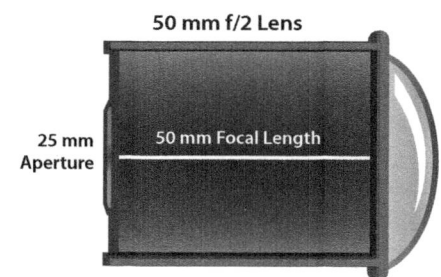

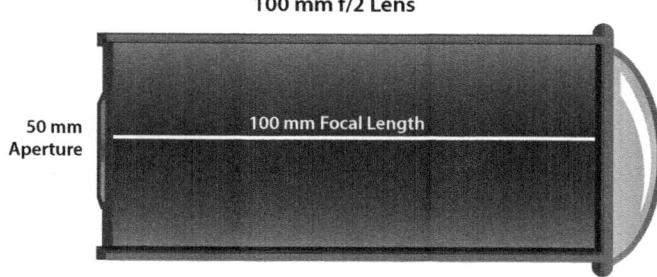

The aperture of a lens is a measure of the opening through which light can pass. The photography term f-stop refers to the ratio of a camera's aperture width to its focal length.

The Human Aperture

Apertures are not found only in optical instruments. The pupil of the human eye is another example of an aperture. Much like a camera aperture, the pupil opening can change size in order to control the amount of light that enters the eye.

—*Casey M. Schwarz, PhD*

Bibliography

Allen, Elizabeth. *The Manuel of Photography*. 10th ed. Oxford: Focal, 2011. Print.

Ang, Tom. *Digital Photography Masterclass*. 2nd Amer. ed. New York: DK, 2013. Print.

Gatcum, Chris. *The Beginner's Photography Guide*. New York: DK, 2013. Print.

Ingledew, John, and Lorentz Gullachsen. *Photography*. 2nd ed. London: King, 2013. Print.

Johnson, Charles S. *Science for the Curious Photographer: An Introduction to the Science of Photography*. Natick: Peters, 2010. Print.

Mansurov, Nasim. "Understanding Aperture—A Beginner's Guide" *Photography Life*. Photography Life, 19 Dec. 2009, Web. 22 Apr. 2015.

ARCHIMEDES'S PRINCIPLE

FIELDS OF STUDY

Fluid Mechanics; Classical Mechanics

SUMMARY

This article discusses the buoyancy principle put forth by the ancient Greek mathematician Archimedes. Archimedes discovered that when an object is placed in a fluid, it will displace a volume of that fluid equal to its own volume. He further determined that the object in the fluid will experience an upward force equal in magnitude to the weight of the displaced fluid. This force is known as buoyancy. Archimedes's principle is a fundamental law of fluid mechanics.

PRINCIPAL TERMS

- **buoyancy:** the upward force exerted by a fluid on a body immersed in that fluid.
- **density:** a measure of the amount of matter in a substance per unit area.
- **displacement:** in fluid mechanics, the process by which a body immersed in a fluid pushes the fluid out of the way and occupies the space in its stead. The volume of the displaced fluid is equal to the volume of the displacing body.
- **mass:** the amount of matter contained in an object.
- **specific gravity:** the ratio of the density of a substance to that of a standard reference substance; also known as relative density.
- **volume:** the amount of three-dimensional space enclosed within a given area.

The Eureka Moment

It is said that while in the public bath one day, the ancient Greek philosopher Archimedes of Syracuse (ca. 287–212 BCE) realized the solution to a difficult problem. A king had given an artisan an amount of pure gold with which to make him a crown. Believing that the artisan had replaced some of the gold in the crown with an inferior metal and kept the extra for himself, the king asked Archimedes to determine if this was the case. Archimedes, however, could not think of a way to determine the purity of the gold without destroying the crown in the process. Simply weighing the crown and comparing it to the original weight of the pure gold would not work. Even if the gold in the crown were of the same weight, it might have a different density. The artisan could simply have added more of a less dense material, or vice versa, to achieve the same weight.

According to popular legend, while pondering this problem, Archimedes visited a public bath to think it over. Once there, he noticed that the water level rose as he immersed himself in the bath and fell again as he stood to leave. Archimedes realized that his body was displacing the water when he was submerged. It is said that upon this realization, Archimedes leapt from the bath and ran through the streets naked, exclaiming, "Eureka!"

Further testing proved that when an object is submerged in water, the volume of the displaced water is equal to the volume of the object causing the displacement. Previously, Greek mathematicians had developed equations to calculate the volume of regular geometric objects, such as spheres and pyramids. However, they had no way to determine the volume of an object that could not be broken down into such shapes. Archimedes's discovery was significant because it allowed him to accurately measure the volume of an irregular object—in this case, the king's crown. Using this method, Archimedes could calculate the crown's density, because an object's density is equal to its weight divided by its volume. (Strictly speaking, density is in fact equal to mass divided by volume, while weight is equal to mass times the acceleration due to gravity. However, on Earth, an object's mass is more or less equivalent to its weight, because the average magnitude of acceleration due to gravity near Earth's surface—a quantity known as standard gravity—is defined as 1.) The density of the crown could then be compared to the density of pure gold. When Archimedes did this, he proved that the artisan had in fact cheated the king.

The discovery proved to be significant in another way as well: it allowed Archimedes to develop the principle that would bear his name. Archimedes's principle states that an object submerged in a fluid experiences an upward force equal to the weight of the fluid displaced by the object. This upward force is called buoyancy. If the weight of the submerged object is less than the weight of the displaced fluid, and thus less than the buoyant force, the object will rise to the top of the fluid; if its weight is greater, the object will sink; and if the weight is the same, the object will neither rise nor sink.

Because the volume of the object and the volume of the displaced fluid are the same, Archimedes's principle can be stated in another way: an object that is less dense than the surrounding fluid will float, while an object that is denser than the fluid will sink. This is related to the concept of specific gravity, also called relative density. An object's specific gravity is equal to the ratio of its density to the density of a standard reference substance, typically water. If the object's specific gravity is less than 1, it is less dense than the reference substance; if its specific gravity is greater than 1, its density is likewise greater.

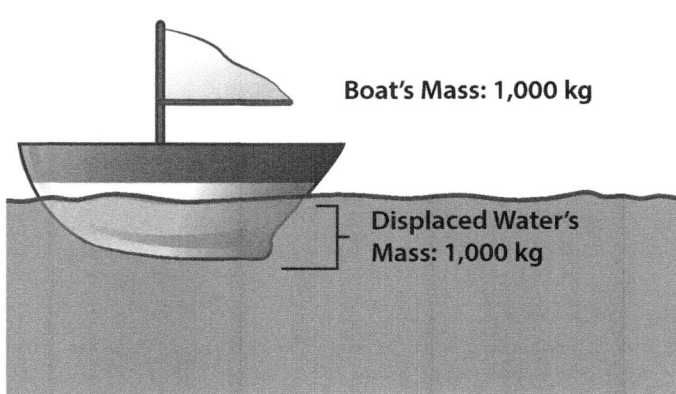

A boat floats in water when the density of the water displaced is greater than the density of the boat. The boat will sink into the water to the point where the volume of water displaced by the boat has an equal mass to the boat.

THE BUOYANCY OF SHIPS

While it is true that Archimedes formalized the concept of buoyancy—a concept that shipbuilders had understood for many years—he did not expand on the idea or undertake additional research in that area. However, the concept later spread throughout the region when the Romans used it to build coin-operated water dispensers. Archimedes's method was used to check if the coins were genuine.

While ancient peoples might not have understood why wood floated, they knew that it did. They also knew that they had to be careful not to put too much cargo on board a ship and cause it to sink. Similarly, the ancient Chinese philosopher Zhuangzi (ca. 369–ca. 286 BCE) was aware that large ships need a great deal of water to float. He also understood that crumbs would float in a small bowl. He used such phenomena as a metaphor for knowledge and understanding. This understanding allowed people to use rocks for ballast in their ships, thus making them more stable by more closely approximating the density of water.

Additional experimentation in a formal setting took place in the model basins of Europe and the United States in the late 1800s and early 1900s. The findings from these early studies inform the computer models that eventually entered into common use. This enhanced understanding of Archimedes's buoyancy principle remains crucial to the design of ships in the twenty-first century.

> **SAMPLE PROBLEM**
>
> A large model ship weighs 5 kilograms (kg) and displaces 0.004 cubic meters (m³) of water. The density of water is approximately 1,000 kilograms per cubic meter (kg/m³). Will the model ship float or sink?
>
> **Answer:**
> Given the density of water, calculate the weight of 0.004 m³ of water. Recall that the density (ρ) of a substance is equal to its mass (m) divided by its volume (V).
>
> $$\rho = \frac{m}{V}$$
>
> $$1{,}000 \text{ kg/m}^3 = \frac{m}{0.004 \text{ m}^3}$$
>
> $$(1{,}000 \text{ kg/m}^3)(0.004 \text{ m}^3) = m$$
>
> $$4 \text{ kg} = m$$
>
> The mass of the displaced water is 4 kg. Mass on Earth is roughly equivalent to weight, so the weight of the water is also 4 kg. Because this is less than the weight of the model ship (5 kg), the model will not float in water, or at least not in freshwater. However, it might float in water that contains some other substance that increases its density, such as the highly saline water of an alkaline lake.

Archimedes's Principle in Action

Archimedes's principle is one of the foundational ideas in fluid mechanics. It allows engineers to design ships that take into account the motion of waves, enabling them to move faster and more efficiently even in rough waters. Archimedes's principle is also at the root of how submarines can ascend and descend in water. A submarine has floodable tanks that allow the operator to increase the vessel's mass by taking in water. This increases the submarine's density and makes it heavier than the water it displaces, causing it to sink to the desired depth.

Fish use a similar mechanism to alter their buoyancy. Many fish have an internal organ known as a swim bladder. They can expand or contract these bladders by taking in or expelling gas. When a fish's swim bladder expands, its body likewise expands to accommodate it, causing its overall volume to increase. As a result, the fish's density decreases, making it more buoyant. Contracting the bladder makes the fish smaller again, increasing its density and decreasing its buoyancy, so that it can swim in deeper water.

—*Gina Hagler, MBA*

Bibliography

Archimedes. *On Floating Bodies: Book I. The Works of Archimedes.* Ed. Thomas Little Heath. 1897. Mineola: Dover, 2002. 253–62. Print.

Archimedes. *On Floating Bodies: Book II. The Works of Archimedes.* Ed. Thomas Little Heath. 1897. Mineola: Dover, 2002. 263–300. Print.

Bemelmans, Josef, Giovanni P. Galdi, and Mads Kyed. "Fluid Flows around Floating Bodies, I: The Hydrostatic Case." *Journal of Mathematical Fluid Mechanics* 14.4 (2012): 751–70. Print.

Burton, Lisa Janelle, and John W. M. Bush. "Can Flexibility Help You Float?" *Physics of Fluids* 24.10 (2012): n.p. Academic Search Complete. Web. 21 July 2015.

Costanti, Felice. "The Golden Crown: A Discussion." *The Genius of Archimedes: 23 Centuries of Influence on Mathematics, Science and Engineering.* Proceedings of an International Conference Held at Syracuse, Italy, June 8–10, 2010. Ed. Stephanos A. Paipetis and Marco Ceccarelli. Dordrecht: Springer, 2010. 215–26. Print.

Eckert, Michael. *The Dawn of Fluid Dynamics: A Discipline between Science and Technology.* Weinheim: Wiley, 2006. Print.

Munson, Bruce Roy, et al. *Fundamentals of Fluid Mechanics.* 7th ed. Hoboken: Wiley, 2013. Print.

Wysession, Michael, David Frank, and Sophia Yancopoulos. *Physical Science: Concepts in Action.* Needham: Prentice, 2004. Print.

B

BAND THEORY OF SOLIDS

FIELDS OF STUDY

Quantum Physics; Materials Physics; Electronics

SUMMARY

The close proximity of atoms in solids allows the combination of atomic orbitals to form delocalized electron states. The atomic energy levels combine to form bands of energy levels closely spaced in energy. The electrical properties of the material are determined by the electron occupancy of the bands. In an insulator the highest energy levels are fully occupied by electrons and there is a substantial band gap separating the occupied levels from the nearest conduction band. In a normal metal there is a partially filled band and electrons are easily promoted to the unoccupied levels. In semiconductors there is only a small band gap, separating the valence and conduction bands and both the excited electrons and the unoccupied (hole) levels both contribute to the electric current.

PRINCIPAL TERMS

- **conductor:** a material that has a low resistance to the flow of electric charges, allowing them to move through it easily.
- **current:** the rate at which an electric charge, usually in the form of electrons, moves through a wire or other conductive material.
- **matter:** anything that can be characterized as having mass and can be measured by some criterion of measurement.
- **quantum field theory:** a theory that explains interactions between subatomic particles as the result of a field extending between them.
- **quantum mechanics:** the branch of physics that deals with matter interactions on a subatomic scale, based on the concepts that energy is quantized, not continuous, and that elementary particles exhibit wavelike behavior.
- **quantum state:** the condition of a physical system as defined by its associated quantum attributes.

ATOMIC STRUCTURE AND BAND STRUCTURE

Solid matter, especially metals, consists of atoms arrayed in a regular lattice structure, rather like so many marbles in a box. If one imagines the atoms to be very far apart, the atoms could be treated as if they were independent . bring them closer together however and the allowed states will form bands of energy with electron density spread over the entire crystal. The collection of states derived from a particular orbital is called a band.

THE CONDUCTION BAND

The atomic orbital overlap that forms bands from the individual orbitals also applies to the vacant orbitals of the atoms. An atomic orbital that is not completely filled with electrons forms a band that is similarly not filled with electrons. Nevertheless, that band has the full number of levels available to contain electrons. The empty levels are termed holes, and electrons from the filled levels can move into the holes and through the band as an electric current. For other materials, bands are also formed from completely vacant atomic orbitals, but there is a "band gap" that must be crossed by electrons in order to enter this conduction band. The smaller is this band gap, the better the material is as a conductor. In semiconductor materials such as carbon and silicon, the atomic orbitals undergo hybridization and do not have the same electron distributions as the orbitals in single atoms. Unlike the electrons in the bands of a conductor, the electrons in the filled bands of semiconductor solids are not able to cross the band gap freely to enter the holes of the conduction band.

higher temperature, which reduces the number of holes available for electron movement in the conduction band. The resistance that a material has to the flow of electrons relative to temperature is called the resistivity (ρ). The reciprocal value is called the conductivity (σ). The resistivity of a material at a particular temperature is calcualated as

$$\rho = \rho_t(1 + a\Delta T)$$

where ρ_t is the resistivity at standard temperature, a is the material-specific resistivity coefficient, and ΔT is the difference in temperature.

SEMICONDUCTORS

Silicon, a semiconductor material, is a very poor conductor of electrical current, yet it is the principal material in the function of all computers and other digital devices. Other materials such as phosphorus and germanium are typically added to introduce more accessible holes into the conduction band. Other materials being studied for use in transistor structures, such as graphene and nanoparticle-impregnated polymers, and materials thought of as superconductors also rely on the understanding and application of band theory.

—*Richard M. Renneboog M.Sc.*

SAMPLE PROBLEM

The resistivity coefficient of copper is 0.00386 per °C, and the resistivity at standard temperature is 1.68 X 10^{-8} Ω.m. Calculate the conductivity of copper at 275°C and at -100°C—conditions that would be encountered in aerospace applications.

Answer:
At 275°C:

$$\rho = \rho_t(1 + a\Delta T)$$

$$\rho = (0.00386)(1 + 1.67 \times 10^{-8} \times (275 - 20))$$

$$= 2.6281 \times 10^{-8} \; \Omega.m$$

$$\sigma = 1/\rho = 1/(2.6281 \times 10^{-8}) \; \Omega^{-1}.m^{-1}$$

$$= 4.409 \times 10^{7} \; \Omega^{-1}.m^{-1}$$

At -100°C:

$$\rho = (0.00386)(1 + 1.67 \times 10^{-8} \times (-100 - 20))$$

$$= -1.2367 \times 10^{-8} \; \Omega.m$$

$$\sigma = 1/\rho = -8.086 \times 10^{-7} \; \Omega^{-1}.m^{-1}$$

The ability of copper to conduct electricity increases considerably as temperature decreases.

TEMPERATURE EFFECT

The conductivity of a solid changes with temperature. The change in conductivity is unique to each element, and typically conductivity decreases as temperature increases. This can be explained as the result of electrons being promoted into the conduction band because of the energy associated with the

BIBLIOGRAPHY

Sirdeshmukh, D B, L Sirdeshmukh, K G. Subhadra, and C S. Sunandana. *Electrical, Electronic and Magnetic Properties of Solids*. New York, NY: Springer, 2014. Print.

Simon, Steven H., *The Oxford Solid State Basics:* New York: Oxford University Press, 2013. Print.

Smart, Lesley, and Elaine Moore. *Solid State Chemistry: An Introduction*. 4th ed. Boca Raton: Taylor & Francis, 2012. Print.

Solymar, Laszlo, Donald Walsh, and Richard R A , Syms. *Electrical Properties of Materials*. 9th ed. Oxford: Oxford University Press, 2014. Print.

Yacobi, B. G. *Semiconductor Materials: An Introduction to Basic Principles*. New York: Springer-Verlag, 2013. Print.

BERNOULLI'S PRINCIPLE

FIELDS OF STUDY

Classical Mechanics; Fluid Mechanics

SUMMARY

This article will discuss Bernoulli's principle, which is derived from the work of Swiss scientist Daniel Bernoulli. Bernoulli observed that the smaller an opening available to a fluid, the greater the rate of flow of that fluid through the opening. He also observed that air flowing up and over a surface moves more quickly than air flowing directly past a flat surface. These observations formed the basis for the modern science of aerodynamics.

PRINCIPAL TERMS

- **Bernoulli's equation:** used to relate and determine velocity, acceleration, and density in fluid mechanics.
- **conservation of energy:** a fundamental law of physics that states that the amount of energy in a system remains constant over time. Although the energy can be transformed or transferred, it cannot be created or destroyed.
- **flow rate:** the amount of fluid that flows in a given period of time; expressed as a numerical quantity.
- **incompressible:** unable to be pressed or squeezed.
- **laminar flow:** the even and stable flow of a fluid; opposite of turbulent flow.
- **lift:** the force that directly opposes the weight of an object and the force of gravity and holds the object aloft.
- **nonviscous fluid:** a fluid that flows without friction.
- **static pressure:** the pressure of a fluid on an object when that object is at rest relative to the fluid.
- **velocity:** a vector quantity that includes both the speed and the direction of motion.

BERNOULLI'S PRINCIPLE

Daniel Bernoulli (1700–1782) was the son of prominent mathematician Johann Bernoulli (1667–1748). While the elder Bernoulli's mathematical interests led him to a thorough investigation of what was then a new field of mathematics called calculus, his son was ultimately more interested in the physics of flowing water. One of his earliest explorations into the properties of fluid flow was in response to his curiosity about the ways in which blood pressure worked. Assisted by former student Leonhard Euler, Bernoulli punched holes in pipes and inserted glass tubes and discovered that the pressure in the pipe was related to the height of the fluid within the tube.

By 1738, Bernoulli had completed his book *Hydrodynamica*, which explained the properties of fluids in motion. The field of hydrodynamics was created based on Bernoulli's work. In *Hydrodynamica*, Bernoulli explored the essential properties of fluid flows. He identified the properties of pressure, density, and velocity. He also described the relationship between these properties in the theory that is today known as Bernoulli's principle, a fundamental observation about the nature of laminar flow and the conservation of energy.

Fluids can be either liquids or gases, and in a laminar flow, the motion of the fluid is continuous and uniform. In other words, all fluid moving through the stream has the same velocity and direction. The law of conservation of energy states that energy is never created or destroyed; it is always either converted from one form to another or it is moved from one object to another. The same theory can be applied to fluids because when a fluid travels through a pipe, for example, its velocity, pressure, and the height of the pipe remain constant. When one of these variables changes, there is a corresponding change in another variable in order to compensate for change in the initial variable.

Bernoulli's principle utilizes the theory of conservation of fluids in stating that a fluid's pressure decreases as its velocity increases. Similarly, a fluid's pressure increases when its velocity decreases. For example, water will exit a small nozzle opening more forcefully than it will a large nozzle opening. The principle also explains why airplanes fly: the air passing up and over the curved surface of an airplane's wings moves more rapidly than does the air passing beneath the flat surface of the wings. The swiftly moving air above the wings has a lower pressure that is relative to the higher pressure of the

Bernoulli's principle

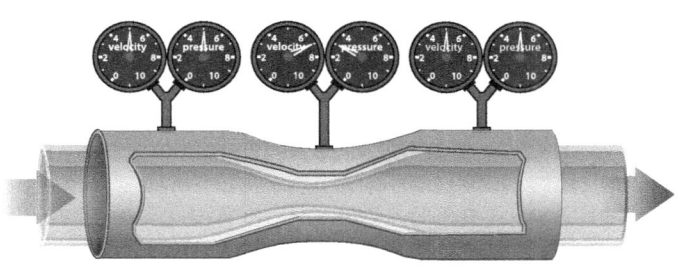

The flow rate of a fluid is influenced by the space through which it passes. Water moving through a pipe with a constriction will have a drop in pressure but an increase in velocity. If the diameter of the pipe expands, the velocity will decrease and the pressure will increase.

slower moving air beneath the wings. This difference in pressure causes the airplane to have lift.

Bernoulli's Equation and Fluid Flows

Fluid dynamics is the study of fluids in motion. Fluid flows are an important consideration in the design of any object that moves through a fluid, such as vessels moving through water, planes moving through air, and spacecraft making their way through frictionless space. When the proper amount of pressure for optimal operation of a booster rocket needs to be known, fluid dynamics must be considered.

Bernoulli's equation, which can be adapted for different types of fluid flow, provides information on pressure, gravity, and velocity of ideal fluids. For a fluid to be ideal, it must meet three criteria. First, it must be incompressible, meaning its density is constant and does not change. Second, the fluid must be nonviscous, meaning there is no resistance or friction to the fluid. Finally, an ideal fluid must have a laminar flow, meaning its flow is constant and steady.

Bernoulli's equation analyzes pressure, speed, and height at two points in the flow of an ideal fluid. In the equation, P is the pressure, p is the fluid's density, v is the fluid's velocity, g is acceleration due to gravity, and h is the height above a specified reference point, with the reference point given a value of zero. The subscript numbers 1 and 2 represent the two points along the flow of the fluid. The equation is expressed as

$$P_1 + \frac{1}{2}pv_1^2 + pgh_1 = P_2 + \frac{1}{2}pv_2^2 + pgh_2$$

The Pitot Tube

Even the most arcane theory can provide the explanation behind an observation made in pursuit of

---

SAMPLE PROBLEM

A distracted gardener drops the end of his hose on the ground with the water running. The normal household pressure is 334 kilopascals. The velocity of the water is 1 meter per second while it is in the hose and 20 meters per second leaving the nozzle. Using Bernoulli's equation, calculate the pressure of the water leaving the hose. Recall that the density of the water is 1,000 kg/m³.

Answer:

To solve this problem using Bernoulli's equation, it is important to first note that because the garden hose is lying on the ground, both point 1 and point 2 are at a height (h) of zero. Thus, the equation can be simplified as follows:

$$P_1 + \frac{1}{2}pv_1^2 + pgh_1 = P_2 + \frac{1}{2}pv_2^2 + pgh_2$$

$$P_1 + \frac{1}{2}pv_1^2 + pg0 = P_2 + \frac{1}{2}pv_2^2 + pg0$$

$$P_1 + \frac{1}{2}pv_1^2 = P_2 + \frac{1}{2}pv_2^2$$

Next, plug in the values for the initial pressure (P_1), the initial and final velocities (v_1 and v_2), and the density of water (p):

$$334 \text{ kPa} + \frac{1}{2}\left(1000 \frac{\text{kg}}{\text{m}^3}\right)\left(1 \frac{\text{m}}{\text{s}}\right)^2 = P_2 + \frac{1}{2}\left(1000 \frac{\text{kg}}{\text{m}^3}\right)\left(20 \frac{\text{m}}{\text{s}}\right)^2$$

$$334 \text{ kPa} + \frac{1}{2}\left(1000 \frac{\text{kg}}{\text{m}^3}\right)\left(1 \frac{\text{m}^2}{\text{s}^2}\right) = P_2 + \frac{1}{2}\left(1000 \frac{\text{kg}}{\text{m}^3}\right)\left(400 \frac{\text{m}^2}{\text{s}^2}\right)$$

$$334 \text{ kPa} + 500 \frac{\text{kg}}{\text{ms}^2} = P_2 + 200,000 \frac{\text{kg}}{\text{ms}^2}$$

(Sample problem continued on next page)

> **(SAMPLE PROBLEM CONTINUED)**
>
> Recall that a pascal (1,000th of a kilopascal), as a unit of pressure, is a measurement of force over area. Thus, a pascal can be written in terms of newtons over meters squared (1 Pa = 1 N/m²) or in kilograms per meter-second squared (1 Pa = 1 kg/ms²). Rewrite the terms in pascals and solve:
>
> $$334{,}000 \text{ Pa} + 500 \text{ Pa} = P_2 + 200{,}000 \text{ Pa}$$
>
> $$P_2 = 334{,}000 \text{ Pa} + 500 \text{ Pa} - 200{,}000 \text{ Pa}$$
>
> $$P_2 = 334{,}000 \text{ Pa} - 199{,}500 \text{ Pa}$$
>
> $$P_2 = 134{,}500 \text{ Pa} = 134.5 \text{ kPa}$$
>
> Thus, the pressure of the water leaving the nozzle is 134,500 pascals, or 134.5 kilopascals. This demonstrates that as the velocity of the fluid increases, the pressure decreases.

another goal. Such is the case with Bernoulli's principle and the Pitot tube. It might seem that knowledge about the pressure of air moving above and beneath a curved body would have little value to anyone, yet in the case of ocean going vessels and traditional aircraft, that is absolutely not the case.

The Pitot tube, presented at the Academy of Science in 1735 by Henri Pitot (1695– 1771), led to a discovery about water pressure that was not fully explained until Bernoulli articulated his theory about speed and pressure. Pitot was tasked with determining the speed of the water flowing past certain points. Measuring that speed was no easy task until Pitot realized he could measure both the static pressure of the water with a tube that was vertical to the water and the total pressure of the water with a tube that faced directly into the flow. With these measurements in hand, he could then calculate the velocity of the fluid flow. He also realized that flow velocity decreased with depth. Today the Pitot tube is used to measure the speed of vessels moving through a fluid.

—*Gina Hagler, MBA*

BIBLIOGRAPHY

Anderson, David F., and Scott Eberhardt. *Understanding Flight*. 2nd ed. New York: McGraw-Hill, 2009. Print.

Fernando, H. J. S. *Handbook of Environmental Fluid Dynamics*. Boca Raton: CRC, 2013. Print.

Matolyak, John, and Ajawad Haija. *Essential Physics*. Boca Raton: CRC, 2013. Print.

Ruban, A. I., and J. S. B. Gajjar. *Fluid Dynamics: Part I, Classical Fluid Dynamics*. London: Oxford University Press, 2014. Print.

Shankar, R. *Fundamentals of Physics: Mechanics, Relativity, and Thermodynamics (The Open Yale Course Series)*. New Haven: Yale University Press, 2014. Print.

Tucker, Paul G. *Unsteady Computational Fluid Dynamics in Aeronautics*. Dordrecht; New York: Springer, 2014. Print.

BETA RADIATION

FIELDS OF STUDY

Electromagnetism; Atomic Physics; Nuclear Physics

SUMMARY

Beta radiation is generated by the release of beta particles, which are electrons or positrons. Beta radiation is emitted by radioactive atomic nuclei undergoing beta decay because they have too many neutrons or protons. Ernest Rutherford first described beta radiation in 1899.

PRINCIPAL TERMS

- **antineutrino:** a subatomic particle with a neutral charge, the antiparticle to the neutrino, which is emitted during beta decay.
- **electron:** a negatively charged subatomic particle that is often bound to the positive charge of the nucleus but can also exist in a free state in an atom.
- **half-life:** the average time it takes for half of the unstable nuclei in a radioactive element to undergo radioactive decay.

- **mutation:** in biology, a change in the structure of a gene. Ionizing radiation such as beta radiation can alter the electric charge of the atoms in a gene.
- **positron:** a subatomic particle that has the same mass as an electron as well as an equal-but-opposite electric charge, that is, a positive charge.
- **radiation:** energy transmitted via electromagnetic waves (e.g., light, heat, x-rays) or subatomic particles (e.g., alpha particles, beta particles).
- **radioisotope:** a chemical element with unstable nuclei that give off radiation due to variations in the number of neutrons they contain.
- **weak interaction:** interaction between subatomic particles at a short distance that is influenced by the weak nuclear force, one of the four fundamental forces in nature.

Radiation without Waves

In many cases radiation refers to electromagnetic radiation (EMR), which includes visible light, x-rays, and ultraviolet rays. Rather than through waves like EMR, beta radiation is transmitted by subatomic particles ejected at high speed when a radioactive atom's nucleus undergoes beta decay. British physicist Ernest Rutherford (1871–1937) described beta radiation in 1899.

Beta decay and the particles it produces can be sorted into two types according to electric charge. Negative beta decay (beta-minus decay) occurs in an unstable atomic nucleus with too many neutrons. One of its excess neutrons is turned into a proton, an electron, and an antineutrino. The electron, a beta particle, is released from the nucleus along with the antineutrino. Positive beta decay (beta-plus decay) occurs in an unstable atomic nucleus with too many protons. One of its excess protons is turned into a neutron, a positron, and a neutrino. The positron, also a beta particle, and the neutrino are released from the nucleus.

In the medical field, beta radiation and beta particles are used in diagnosis, imaging, and sometimes treatment for certain conditions, especially cancer. Some beta particles can pass through skin and tissue. Beta radiation is ionizing, meaning that beta particles can alter the electrical charge of the atoms they hit. Thus, when beta particles hit cells and other living tissue, they can cause damage by altering the chemistry. When beta particles hit DNA, they can change the structure of genes and induce potentially harmful mutations.

Radioactivity and Isotopes

An atom is said to be radioactive if it gives off ionizing radiation. This radiation is the result of an unstable nucleus emitting energy in the form of subatomic particles in order to reach a more stable state. The three main forms of radiation emitted during radioactive decay are alpha, beta, and gamma. The average time it takes for half of a given sample of radioactive material to undergo decay is known as its half-life.

The number of protons in an atom's nucleus determines what element it is. Known as the atomic number, it is used to sort elements in the periodic table. The nucleus of an atom of a given element will always have the same number of protons as every other atom of that element as well as the same general chemical properties. However, the number of neutrons may vary. This variation in the number of neutrons creates isotopes. Isotopes have more or less total atomic weight because of the addition or

DECAY TYPE	SYMBOL	PARTICLE EMITTED	EXAMPLE	STOPPED BY	ENERGY RELEASED	SOURCE
Beta	β	Beta Particle (electron)	H-3 → He-3	Aluminum Shielding	1 MeV	Carbon-14 Cobalt-60 Strontium-90 Cesium-137

Beta particles are individual electrons or antineutrinos emitted when an atom such as hydrogen-3 converts a neutron to a proton and becomes a new atom. They can be stopped by aluminum shielding.

subtraction of neutrons; for example, carbon-14 is an isotope of carbon, which has an atomic number of 12. Isotopes can occur naturally or can be made by humans. Some isotopes are stable. Radioisotopes are unstable and subject to spontaneous radioactive decay. All elements above atomic number 83 are unstable and radioactive. Most below this number have at least one radioisotope.

THE WEAK INTERACTION AND BETA DECAY

Beta decay is mediated by the weak interaction, which is also known as the weak nuclear force. It was through the study of beta decay that Enrico Fermi first described the weak nuclear force in 1934 to explain this previously mysterious process of decay. The weak nuclear force is one of the four fundamental forces of physics. The other three are the strong nuclear force, the gravitational force, and the electromagnetic force. It is the balance between the long-range electrostatic repulsion and the short range strong force that limits the size of stable nuclei. It is the weak force that allows isotopes with too many neutrons to convert to isotopes of higher atomic number and fewer neutrons. For this reason the weak force has been called the *cosmic alchemist*.

BETA PARTICLES IN EVERYDAY LIFE

The penetrating power and ionizing ability of beta particles make them useful tools for targeted radiation therapy for some cancers. They can kill target cells with minimal damage to surrounding tissue. On the other hand, this also makes them dangerous when exposure is not controlled. Beta particles also have a modest penetrating power that makes them useful for testing the thickness of manufactured materials, such as metal foil or paper, by measuring how much of the radiation is stopped by the material as it passes through. It is perhaps most relevant to everyday life as a by-product of nuclear waste produced by reactors. The question of how best to dispose of this waste is of major environmental, health, and political concern.

—*Kenrick Vezina, MS*

BIBLIOGRAPHY

"Beta Particles." *Radiation Protection.* US Environmental Protection Agency, 7 Feb. 2013. Web. 26 May 2015.
Davidson, Michael W. "The Rutherford Experiment." *Molecular Expressions: Electricity & Magnetism.* Florida State U, 28 Feb. 2014. Web. 26 May 2015.
"Electromagnetic Radiation." *Encyclopaedia Britannica.* Encyclopaedia Britannica, 26 Nov. 2014. Web. 26 May 2015.
"Glossary: Beta Decay." *Jefferson Lab.* Thomas Jefferson Natl. Accelerator Facility, n.d. Web. 26 May 2015.
Lucas, Jim. "What Is the Weak Force?" *LiveScience.* Purch, 24 Dec. 2014. Web. 26 May 2015.
"Radiation Basics." US Nuclear Regulatory Commission. USNRC, n.d. Web. 26 May 2015.

BLACKBODY RADIATION

FIELDS OF STUDY

Quantum Physics; Thermodynamics

SUMMARY

Blackbodies are ideal physical objects that absorb all frequencies of electromagnetic radiation from any source. As a result of this total absorption, the only radiation a blackbody emits is thermal radiation generated by its temperature. This radiation, called blackbody radiation, is emitted at full efficiency across the entire spectrum. Discoveries about the nature of blackbody radiation ultimately led to the development of quantum mechanics in the early twentieth century.

PRINCIPAL TERMS

- **continuous energy levels:** the idea that there are limitless levels of energy between each point on a continuum.
- **Planck's law:** a mathematical description of the amount of radiation emitted at different frequencies by a blackbody at a given temperature.
- **quantum mechanics:** the branch of physics that deals with phenomena on a subatomic scale.

- **Stefan-Boltzmann law:** a mathematical description of the total radiant energy emitted by a blackbody, relating it to the temperature of the blackbody raised to the fourth power.
- **ultraviolet catastrophe:** the erroneous prediction, based on the laws of classical physics, that a blackbody would emit an infinite amount of energy at short wavelengths, starting around the ultraviolet region.

A Few Questions Remain

At the end of the nineteenth century, scientists believed that only a few unexplained phenomena stood between their current understanding of physics and a complete understanding of how the universe works. All they needed, they believed, was to tie up those loose ends. Then, they would be in a position to account for everything they could observe around them.

Scientists had good reason for this belief. Isaac Newton's (1642–1727) three laws of motion accounted for the behavior of objects at rest and in motion. James Clerk Maxwell's (1831–79) work with electromagnetism had tied together two fields of physics to explain behavior in a significant area. Once they cleared up the rest, they could use their knowledge to look at new problems as they arose.

The problem was that these scientists were looking at phenomena that took place at the readily observable level—the things they could observe with the naked eye or with optical microscopes. However, the discovery of subatomic particles and subsequent study of how those particles interact led to the realization that the laws that govern the very large do not govern the very small. In fact, the same element can behave in different ways depending on the scale on which its behavior is being observed.

Among these unexplained phenomena were questions about the nature of light, or visible electromagnetic radiation. Scientists knew that the color of the light emitted by an object changes based on the temperature of that object. Cooler fires burn with a reddish flame, while hotter fires burn blue or white; an

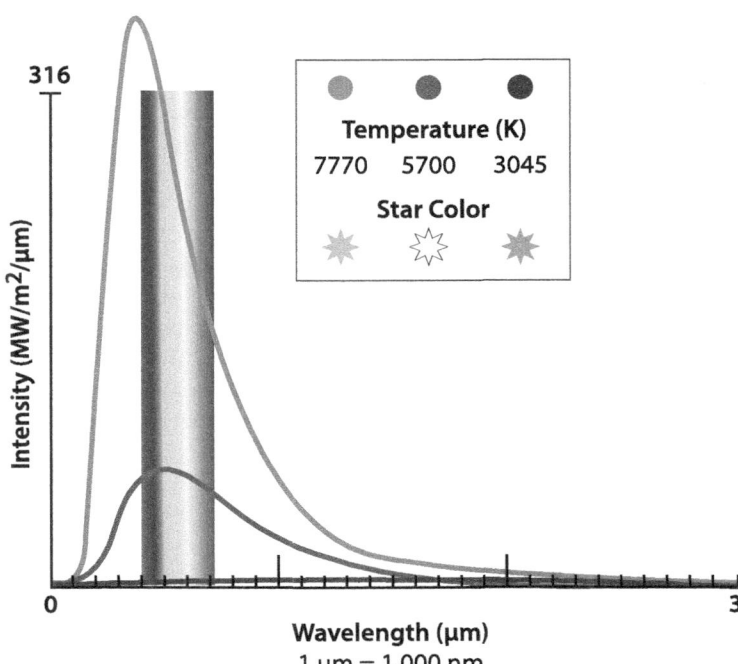

Models of the intensity of blackbody radiation emitted from three different sources: a star with a temperature of 7770 K, the sun (5700 K), and a light bulb (3045 K). The curve shows the intensity of electromagnetic radiation emitted at all wavelengths. The star symbols display the approximate color of light waves emitted from each source: blue-violet (star), all visible light (sun), and red-orange (light bulb).

iron rod glows red when heated, then changes colors as it cools. They also knew that all objects at thermodynamic equilibrium emit exactly as much, or as little, radiation as they absorb. In other words, good absorbers are good emitters, and poor absorbers are poor emitters.

German physicist Gustav Kirchhoff (1824–87) and chemist Robert Bunsen (1811–99) discovered in 1859 that different substances absorb radiation at particular frequencies and then reemit it at those same frequencies. This is the principle behind spectroscopy. Kirchhoff then considered under what conditions an object would emit radiation based solely on its temperature rather than its chemical composition. In 1860, he introduced the concept of the blackbody, an idealized body that would absorb all incident radiation, reflecting none. This would cause the interior of the blackbody to heat up, which in turn would cause it to emit thermal radiation. This thermal radiation, which came to be known as blackbody radiation, would be emitted across all wavelengths and frequencies. The peak intensity would depend solely on the temperature of the blackbody.

MEASURING BLACKBODY RADIATION
No genuine blackbody exists in nature, although some objects, such as the sun and other stars, come close. To approximate such a body, Kirchhoff proposed using an opaque box or cavity with a single small hole in the side. This cavity would be a near approximation of a perfect absorber. Any radiation that entered through the hole would be mostly absorbed by the interior walls, with very little escaping. Similarly, if the interior of the cavity were heated, the radiation emitted from the hole would approximate a perfect emitter. With these parameters in place, Kirchhoff challenged his colleagues to measure the energy curve of this temperature-based radiation.

Using experimental data, Slovenian physicist Josef Stefan (1835–93) determined in 1879 that the energy radiated by a blackbody is proportional to the temperature of the body raised to the power of four. Five years later, one of Stefan's former doctoral students, Austrian physicist Ludwig Boltzmann (1844–1906), mathematically derived the same relationship based on the laws of thermodynamics. This relationship, now known as the Stefan-Boltzmann law, can be expressed as

$$L = A\sigma T^4$$

where L is the luminosity of the blackbody in watts (W), or joules per second (J/s); A is the surface area in meters squared (m^2); T is the temperature in kelvins (K); and σ is the Stefan-Boltzmann constant, equal to 5.670373×10^{-8} W/$m^2 \cdot K^4$. It can also be written more simply as

$$E = \sigma T^4$$

where E is the total energy output in joules per second per meter squared (J/s·m^2).

THE ULTRAVIOLET CATASTROPHE
In 1905, English physicists Lord Rayleigh (1842–1919) and James Jeans (1877–1946) published the Rayleigh-Jeans law, which Rayleigh had first derived five years earlier. This law, based on classical physics equations, was meant to calculate how much radiation a blackbody would emit at a given frequency or wavelength based on its temperature. (Frequency and wavelength are inversely proportional; the higher the frequency, the shorter the wavelength.) However, although the law was accurate at lower frequencies, it predicted that the radiation output would increase exponentially at higher frequencies. That is, blackbodies would emit infinite amounts of high-frequency radiation. This was obviously not the case, or else any thermal radiation emitted by a body at thermodynamic equilibrium would reduce anything or anyone nearby to ash. The discrepancy was later dubbed the ultraviolet catastrophe, as the errors began around the ultraviolet range.

The source of the catastrophe was the assumption that electromagnetic radiation was emitted at continuous energy levels. In other words, it was assumed that light acted as a wave under all conditions and at all frequencies. Scientists had long debated whether light was a particle or a wave—in other words, whether it traveled in discrete units or as a continuous stream of energy. By the start of the twentieth century, the wave theory had largely prevailed. This was due in part to Maxwell's electromagnetic equations, which worked on the assumption that different forms of light (infrared, visible, ultraviolet) were simply different frequencies of the same electromagnetic waves. The ultraviolet catastrophe was one indication that the wave theory of light may not be the whole truth.

German physicist Max Planck (1858–1947) found a solution while addressing a similar issue raised by fellow physicist Wilhelm Wien (1864–1928). Several years before Rayleigh and Jeans, Wien had derived an equation to describe the complete thermal radiation spectrum. However, Wien's equation had the opposite problem of Rayleigh and Jeans's: it diverged at lower frequencies, starting around the infrared range. Planck revised the equation and added a constant based on experimental data. This constant had the effect of quantizing energy—that is, treating it as discrete packets rather than continuous waves. Planck published his revised equation, known as Planck's law, in 1900, shortly after Rayleigh began working on what would become the Rayleigh-Jeans law. Planck's law resolved the problems of Wien's equation and, it was later realized, of the Rayleigh-Jeans law as well. However, it raised new questions about physics as a whole.

A New Branch of Physics

Planck backed into his discovery of energy quantization while trying to reconcile Wien's equation with reality. However, he was not convinced that energy was truly quantized. He believed that the constant was merely a mathematical convenience that did not reflect reality. Shortly thereafter, in 1905, Albert Einstein (1879–1955) introduced the concept of wave-particle duality—the idea that light can behave as both a particle and a wave.

What Planck did not appreciate at the time was that he was describing the phenomenon that, when understood, would change the very nature of physics. The idea that something such as energy could behave one way at one level and another at a different level would become the basis of a new branch of physics. Quantum mechanics would answer the unresolved questions of classical physics while introducing a host of new questions about the behavior of particles at small scale. It would throw the traditional view of the universe on its head and usher in a new way of looking at the world.

—Gina Hagler, MBA

Bibliography

Al-Khalili, Jim. *Quantum: A Guide for the Perplexed.* London: Phoenix, 2012. Print.

Baggott, J. E. *The Quantum Story: A History in 40 Moments.* New York: Oxford University Press, 2011. Print.

Fayer, Michael D. *Absolutely Small: How Quantum Theory Explains Our Everyday World.* New York: AMACOM, 2010. Print.

Ford, Kenneth W. *101 Quantum Questions: What You Need to Know about the World You Can't See.* Cambridge: Harvard University Press, 2011. Print.

Fowler, Michael. "Black Body Radiation." *Modern Physics.* U of Virginia, 7 Sept. 2008. Web. 9 June 2015.

Halpern, Paul. *Einstein's Dice and Schrödinger's Cat: How Two Great Minds Battled Quantum Randomness to Create a Unified Theory of Physics.* New York: Basic, 2015. Print.

Kragh, Helge. "Max Planck: The Reluctant Revolutionary." Physicsworld.com. Inst. of Physics, 1 Dec. 2000. Web. 26 June 2015.

Mastin, Luke. "Quanta and Wave-Particle Duality." The Physics of the Universe, 2009. Web. 26 June 2015.

Susskind, Leonard, and George Hrabovsky. *The Theoretical Minimum: What You Need to Know to Start Doing Physics.* New York: Basic, 2013. Print.

BOHR ATOM

FIELDS OF STUDY

Quantum Physics; Atomic Physics

SUMMARY

Building on the observations of Thomson, Rutherford, Young, Einstein and others, Bohr's model of atomic structure was a revolutionary break from classical mechanics. It yielded excellent agreement with experiment for single electron atoms, but could not be extended in any straightforward way to many-electron atoms. Bohr assumed that the angular momentum of the orbiting electron is quantized in units of $h/2\pi$ where h is Planck's constant. Bohr's theory was an important way station on the way to the quantum mechanics of Schrödinger and Heisenberg.

PRINCIPAL TERMS

- **angular momentum:** the rotational momentum of an object around an axis. For a particle in circular orbit it equals mass times velocity times radius of the orbit.
- **conservation of energy:** a fundamental law of physics that states that the amount of energy in the universe remains constant over time. Although the energy can be transformed or transferred, it cannot be created or destroyed.
- **conservation of momentum:** in physics, the principle that the total momentum in a isolated system is always constant.
- **electron:** a negatively charged subatomic particle that is often bound to the positive charge of the nucleus but can also exist in a free state in an atom.

- **photoelectric effect:** a phenomenon that describes the emission of electrons from matter (typically metal) upon exposure to electromagnetic radiation.
- **photon:** a massless elementary particle that is the smallest possible unit, or quantum, of light and other electromagnetic radiation.
- **quantum mechanics:** the branch of physics that deals with matter interactions on a subatomic scale, based on the concepts that energy is quantized, not continuous, and that elementary particles exhibit wavelike behavior.
- **wave function:** a function that describes the quantum state of a system and represents the probability of finding the system in a given state at a given time.

ATOMIC THEORY

Before the twentieth century began, it was theorized that matter consisted of atoms that contained positive and negative components. There was, however, no good workable hypothesis of the arrangement of those components within the atoms. Physicists had also observed the black lines that appeared in the transmission and absorption spectra of compounds, but the cause of the lines was unknown. In 1897, Thomson identified the electron as a subatomic particle that could be made to exist apart from its oppositely charged, and much more massive, counterpart that accounted for essentially all of the mass of an atom.

Ernest Rutherford's gold foil experiment in 1911 demonstrated that atoms had at their core a very small, dense nucleus surrounded by what amounted to a great deal of empty space containing the few electrons. This meshed well with Thomson's observations and formed the basis of a theory of atomic structure in which the central nucleus is surrounded by a cloud of electrons. But this alone could not provide an explanation of the relationships observed in the black lines of atomic spectra.

In 1905, Einstein published a paper describing the photoelectric effect, which had been found to be dependent on the wavelength, and therefore the frequency, of the incident light and independent of the intensity of the incident light. The hypothesis that the electrons in the atoms could accept only a photon of the correct energy in order to be emitted was the basis of quantum theory. In 1913 Bohr used the quantum theory to explain the pattern of black lines in spectra of hydrogen as corresponding to the specific energy of the photons absorbed or emitted by electrons in the atoms as they transitioned between higher and lower energy levels. The mathematics of the relationships developed into the science of quantum mechanics.

THE BOHR HYDROGEN ATOM

The significant aspect of Bohr's atomic theory was the manner in which it departed from the thinking of the time, which was based on classical mechanics. The assumption of the time was that the structure of the atom and the behavior of the electrons it contained must be in accord with classical, or Newtonian, physics., In order to obey the law of conservation of energy and the law of conservation of momentum in that paradigm, an orbiting body, such as an electron orbiting a nucleus, must continually radiate energy in order to maintain stability. However, because atoms clearly did not radiate energy continuously, the idea of orbiting electrons was not well received by classical physicists.

Drawing on Young's "double slit experiment" that had demonstrated the ability of electrons to have wave characteristics, Bohr proposed the radical idea that the electrons "in orbit" about a nucleus do so either as waves or in a manner that is described by wave behavior. On the basis of the physics of objects following circular orbits, Bohr hypothesized that the angular momentum of the electrons is quantized in units of the reduced Planck's constant, $\hbar = h/2\pi$. Although he was not able to reconcile this hypothesis with classical mechanics, he was able to use his hypothesis to derive a formula that predicted the spectral lines of the hydrogen atom.

THE ONE-ELECTRON ATOM, THE RYDBERG CONSTANT AND THE BALMER SERIES

In 1885, Balmer showed that the lines observed in the visible spectrum of hydrogen is described by the relationship

$$1/\lambda = R(1/2^2 - 1/n_2^2)$$

where n_2 is an integer value greater than 2, and R is the Rydberg constant having the value $1.097 \times 10^7 m^{-1}$. By substituting a second integer value in place of 2, the Balmer series relationship could be extended to

> **SAMPLE PROBLEM**
>
> Calculate the "mass" of the electron given the following values:
>
> $$c = 2.99 \times 10^8 \text{ m.sec}^{-1}$$
>
> $$h = 6.625 \times 10^{-34} \text{ J.sec}$$
>
> $$e = 1.60 \times 10^{-19} \text{ C}$$
>
> $$R = 1.097 \times 10^7 \text{m}^{-1}$$
>
> **Answer:**
> First ensure that all values are in the appropriate units. The value of c must be converted to m.sec^{-1}, as 2.99×10^8 m.sec^{-1}. Rearrange the equation
>
> $$R = (mZ^2/2\ \hbar^3 c)(e^2/4\pi\varepsilon_0)^2$$
>
> to obtain
>
> $$m = R(2\ \hbar^3 c/Z^2)(e^2/4\pi\varepsilon_0)^{-2}$$
>
> Substitute the given values, and carry out the calculation to obtain
>
> $$9.1 \times 10^{-31} \text{kg}$$

describe other series of lines that could be identified in the spectrum of hydrogen. These are the Lyman, Paschen, Brackett and Pfund series. Their relationships are predicted by the more general form of the Balmer series equation, which is

$$1/\lambda = R(1/n_1^2 - 1/n_2^2).$$

It was then realized that the above formula would be obeyed by the emission spectrum of neutral hydrogen, singly ionized helium, doubly ionized lithium and so on, which are all single electron levels.

The quantum nature of the transitions involved is strongly suggested by the fact that n_1 and n_2 are integers. The precision with which wavelengths could be determined from spectra produces a very accurate value of the Rydberg constant. Calculating the energy of a particular level as

$$E = -(Z^2/n^2)(e^2/4\pi\varepsilon_0)^2(m/2\hbar^2).$$

Bohr found that the difference in energy between two specific energy levels can be described by the equation

$$E_2 - E_1 = -(mZ^2/2\ \hbar^2)(e^2/4\pi\varepsilon_0)^2(1/n_1^2 - 1/n_2^2),$$

which is essentially the same equation as that describing the Balmer series and yields an expression for the value of the Rydberg constant as

$$R = (mZ^2/2\ \hbar^3 c)(e^2/4\pi\varepsilon_0)^2.$$

Here for simplicity we have assumed an infinitely heavy nucleus and that the atom has z-1 units of electrical charge.

The success of this prediction, in the excellent agreement found between the observed and calculated values of the Rydberg constant, were very strong arguments in favor of Bohr's postulate that the electron in an atom has quantized angular momentum.

THE LEGACY OF THE BOHR ATOM

In practical terms, the Bohr atom has value in the present day as learning tool for understanding the basic principles of quantum mechanics. That, however, is the foundation of the standard model of the physical universe and our understanding of nuclear physics. Harvesting solar energy efficiently depends on the development of materials that exhibit the photoelectric effect as efficiently as possible, maximizing the absorption each photons that cause the release of electrons. Spectral analysis enables the identification of elements and compounds both on Earth and in distant stars. At the quantum level, atomic structure controls the movement of electrons in the transistors of computer chips, which is expected to be maximized by the development of computers that function on the quantum level using light transmission rather than just the current nanometer scale that depends on the physical movement of electrons. Understanding the Bohr atom and its role in the development of the theoretical basis of these and future technologies is the Bohr atom's essential legacy.

—*Richard M. Renneboog M.Sc.*

BIBLIOGRAPHY

Bohr, Neils *The Theory of Spectra and Atomic Constitution* Cambridge, UK: Cambridge University Press, 1924. Print.

Bohr, Niels *Atomic Theory and the Description of Nature* Cambridge, UK: Cambridge University Press, 1961. Print.

Galison, Peter, Gordin, Michael and Kaiser, David, eds. *Science and Society: Quantum Histories* New York, NY: Routledge, 2001. Print.

Kraghe, Helge *Niels Bohr and the Quantum Atom. The Bohr Model of Atomic Structure 1913 – 1925* Oxford, UK: Oxford University Press, 2012. Print.

Longair, Malcolm *Quantum Concepts in Physics* New York, NY: Cambridge University Press, 2013. Print.

BOSE CONDENSATION

FIELDS OF STUDY

Quantum Physics; Statistical Physics

SUMMARY

When a collection of particles or quasi particles obeying Bose Einstein statistics has low enough total energy so that many are in the ground translational state we say that a Bose condensation has occurred. Examples of such states are the superconducting ground state of metals and the translational states of super fluids. For a collection of atoms to exhibit such behavior they must each be in the same state of total angular momentum as well as in the same translational state and thus be at very low absolute temperature.

A Bose condensate can emit single atoms in a manner analogous to the optical laser, and has been able to reduce the speed of light as well as to store the light and emit it at a later time. Bose condensates may account for "dark matter" and "dark energy."

PRINCIPAL TERMS

- **angular momentum:** the rotational momentum of an object around an axis, defined as the product of its moment of inertia and its angular velocity.
- **quantum mechanics:** the branch of physics that deals with matter interactions on a subatomic scale, based on the concepts that energy is quantized, not continuous, and that elementary particles exhibit wavelike behavior.
- **quantum state:** the condition of a physical system as defined by its associated quantum attributes.
- **standard model:** a generally accepted unified framework of particle physics that explains electromagnetism, the weak interaction, and the strong interaction as products of interactions between different types of elementary particles.
- **uncertainty principle:** the idea, proposed by Werner Heisenberg, that one can determine with high precision either the position of a particle at a given time or its momentum, but not both.
- **wave function:** a function that describes the quantum state of a system and represents the probability of finding the system in a given state at a given time.

ATOMIC STRUCTURES AND QUANTUM MECHANICS

The modern theory of atomic structure is based on the principles of quantum mechanics, which in turn is based on the standard model of particle physics. According to quantum mechanics the overall wave function of a many particle system must change sign whenever the coordinates of two Fermi particles are interchanged. This is the most general form of the Pauli exclusion principle, so called because it requires that the probability of finding two identical particles at the same point with the same spin to be zero exactly. Whenever the space can be broken into regions such that the wave functions have negligible overlap between them this reduces to the requirement that only one electron of each spin can occupy an orbital, which was the form in which the principle was originally stated and is still presented to chemistry students today.

It is the Pauli principle that is responsible for the stability of ordinary matter. If, however, the atoms are cold enough so that a significant number will be in the lowest energy state, a Bose condensation is indeed possible

How to Make a Bose Condensate

Although the existence of Bose concentrates was theorized much earlier, the physical reality could not be demonstrated until the technology that would enable their formation could be developed. Not until the 1990s was it possible to produce temperatures sufficiently cold for a Bose concentrate to form. In 1995, Eric Cornell and Carl Wieman used the techniques of laser cooling and trapping, coupled with an advanced magnetic containment method, to cool rubidium atoms in the gas phase to a temperature of just under two hundred billionths of a degree above absolute zero (1.7×10^{-7} K). Magnetic containment is required in order to eliminate the extra mass that a material container would pose to the system, representing a prohibitively large reservoir of heat energy that would also have to be removed by cooling to the same temperature as the gas atoms used in the experiment.

In a similar way, Wolfgang Ketterle produced a Bose concentrate using sodium atoms. The significant feature of a Bose concentrate produced in this way is that the individual atoms coalesce into a "super-atom." The Bose condensate superatom produced by Cornell and Wieman consisted of some 2000 individual rubidium atoms but its quantum properties were consistent with it being just a single atom. Its size is significant, because it was large enough to observe directly by microscopy. Since 1995, Bose condensates have been observed using a number of specific isotopes, including ^{1}H, ^{7}Li, ^{23}Na, ^{39}K, ^{41}K, ^{52}Cr, ^{85}Rb, ^{87}Rb, ^{133}Cs, ^{170}Yb, ^{174}Yb and ^{4}He in an excited state. In principle, just two atomic particles should be able to form a Bose condensate. Such a condensate was observed experimentally by Deborah Jin and her colleagues in 2003.

Advanced Mathematical Physics

As might be expected from their theoretical basis in quantum mechanics, the mathematical description of Bose condensates does not depend on simple arithmetic, and is complicated by the fact that the universe does not consist of ideal matter. For example, the simplest mathematics describing an aspect of Bose condensation is restricted to an ideal three-dimensional gas whose atoms do not interact with each other. The critical temperature for the transition of this gas to a Bose condensate is given by

$$T_C = (n/\zeta(3/2))^{2/3}(2\pi\hbar^2/mk_B) \approx 3.3125(\hbar^2 n^{2/3}/mk_B)$$

where T_C is the critical temperature, n is the particle density, ζ is the Riemann "zeta function" for which $\zeta(3/2)$ has a value of approximately 2.6124, $\hbar$ is the reduced Planck's constant, m is the mass per boson, and k_B is the Boltzmann constant.

Interesting Properties of Bose Condensates

Liquid helium has long been observed to have the property of superfluidity. It has been proposed that liquid helium is a superfluid because it is a Bose condensate. Superfluidity, in which a fluid flows with no friction, has also been observed in other Bose condensates. Superconductivity, in which electrons flow through a conductor with no electrical resistance, is another property of Bose condensates. In comparison with optical lasers, which emit light as a coherent collimated beam, a Bose condensate has the ability to emit a stream of single atoms, as a so-called atom laser. This feature has numerous possible applications in its own right. But perhaps the most incredible property of Bose condensates is their interaction with light.

Light is believed to have a dual nature, being composed of photons yet behaving as a wave phenomenon. A photon is essentially a 'particle' of light. Bose condensates have the ability to not only slow down the motion of light passing through them, but to stop it completely. It can be easily imagined that Bose condensates are a medium that under the right conditions does not support the propagation of light. This is a great oversimplification, however, because the interaction of Bose condensates with light is much more complex than this, as is demonstrated by the fact that a Bose condensate can be made to store light and release it again after a period of time.

These properties all have potential applications in different fields that are currently being heavily researched, including the development of quantum computers. The fact that Bose condensates are real also offers the intriguing possibility with regard to so-called dark matter and dark energy that they may account for a very large fraction of the universe that currently cannot be detected.

—Richard M. Renneboog M.Sc.

BIBLIOGRAPHY

Calmet, Xavier, ed. *Quantum Aspects of Black Holes.* Cham: Springer, 2015. Print.

Leggett, Anthony J ,. *Quantum Liquids : Bose Condensation and Cooper Pairing in Condensed-matter Systems.* Oxford [u.a.]: Oxford U, 2015. Print.

Levin, Kathryn, Alexander L. Fetter, and Dan M. Stamper-Kurn, eds. *Ultracold Bosonic and Fermionic Gases.* Amsterdam: Elsevier, 2012. Print.

Martellucci, Sergio, Arthur N. Chester, Alain Aspect, and Massimo Inguscio. *Bose-Einstein Condensates and Atom Lasers.* Boston, MA: Springer US, 2002. Print.

Pitaevskii, Lev, and Sandro Stringari. *Bose-Einstein Condensation.* Oxford: Clarendon, 2013. Print.

Pitaevskii, Lev P. *Bose-Einstein Condensation and Superfluidity.* New York: Oxford University Press, 2016. Print.

BRA-KET NOTATION

FIELDS OF STUDY

Quantum Physics

SUMMARY

The mathematics of quantum mechanics developed from the vectors and matrices of classical mechanics and the differential calculus and integration of wave mechanics. Bra-ket notation provides an independent, simple system of calculation.

PRINCIPAL TERMS

- **dot product:** the product of the lengths of two vectors and the cosine of the angle between them; also called the inner product or the scalar product.
- **eigenfunction:** any mathematical function for which the solution yields the original function multiplied by a constant (an eigenvalue).
- **quantum mechanics:** the branch of physics that deals with matter interactions on a subatomic scale, based on the concepts that energy is quantized, not continuous, and that elementary particles exhibit wavelike behavior.
- **quantum state:** the condition of a physical system as defined by its associated quantum attributes.
- **spin:** the intrinsic angular momentum of a subatomic particle.
- **vector:** a mathematical representation of a property whose value has both magnitude and direction.
- **wave function:** a function that describes the quantum state of a system and represents the probability of finding the system in a given state at a given time.

VECTORS, MATRICES, AND DIRAC NOTATION

Many physical quantities, including velocity and force, consist of two distinct components: magnitude and direction. Such quantities are called vectors. The counterpart of the vector is the scalar, which has magnitude but no direction. For example, the velocity of an object describes not just its speed—a scalar quantity that measures how fast the object is moving—but also the direction of travel. A car driving north at sixty miles per hour has a different velocity from a car driving west at sixty miles per hour, even though their speeds are the same.

In plane, or Euclidean, geometry, it is easy to see how vectors function. A classic example is a boat crossing a flowing river. If the boat travels directly across the river under the force of its motor, it will reach the opposite shore somewhere downstream of its initial target due to the force of the current. The point at which the boat will touch the shore is determined by the overall vector of the boat's motion. This vector is the sum of the two force vectors (the motor and the current) acting on it. In such a case, the two vectors function relative to a common origin point, and the resultant vector is obtained by adding them together.

Vectors can also be multiplied. The result of vector multiplication is called the dot product, so named because of the notation used. If two vectors a and b are

Paul Dirac (1902–84) introduced bra-ket notation in an article published in 1939. He later expanded on it in the third edition of his book *The Principles of Quantum Mechanics*, originally published in 1930 and still commonly used in the study of quantum mechanics. By Nobel Foundation. Public domain via Wikimedia Commons

multiplied together, the operation is written as $a \cdot b$. The dot product is a scalar quantity. It is equal to the quantity $ab\cos\theta$, where θ is the angle between the two vectors a and b.

A series of vectors and operations can be represented as a matrix rather than as equations. This allows them to be more easily manipulated. Such manipulations are best performed using bra-ket notation, also called Dirac notation, after its developer, English physicist Paul Dirac (1902–84). In the simplest form of bra-ket notation, the operator of the equations is the *bra*, and the vectors being operated on are the *ket*.

Bra-ket notation is a specialized form of mathematical notation that separates the common operator from the vectors being operated on. In quantum mechanics, it is used to describe the quantum state of a physical system. The ket, written in the format | ket>, represents the quantum state in an vector space. The bra, written in the format <bra |, represents a function that translates the state into specific conditions. A bra-ket pair is written as <bra | ket>, representing the dot product of the bra and the ket.

Subatomic Particles and Wave Mechanics

Simple vector relationships are readily solved without bra-ket notation. However, the vectors describing the behavior of subatomic particles are much more complex. The basic premise of quantum mechanics is that energy is absorbed and emitted in discrete units, called quanta (the plural of quantum). Quantum mechanics also asserts that subatomic particles behave as both particles and waves, a concept known as wave-particle duality. The behavior of these particles is determined by their quantum states, described by specific quantum numbers. One such quantum number is spin, which represents the angular momentum of the particle. Different types of particles have different spins.

Subatomic particles are modeled using wave mechanics. In effect, they are treated as "matter waves." These waves are described by the Schrödinger equation, devised by Austrian physicist Erwin Schrödinger (1887–1961). Solutions to the Schrödinger equation are called wave functions, represented by the Greek symbol psi, Ψ. A wave function can be derived from the Schrödinger equation that will describe the behavior of any given particle, including its spin. This function is termed an eigenfunction if the Hamiltonian operator (H) acting on it is equal to the energy operator (E) of the system. This is normally written as

$$H\Psi = E\Psi,$$

which is the time-independent form of the Schrödinger equation. The energy operator corresponds to the total energy of the system. The Hamiltonian operator is a function representing the sum of the system's kinetic energy operator and its potential energy operator. (If the equation is time-dependent, the Hamiltonian operator shows how the energy of the system changes over time.) Thus, E represents the eigenvalues of the function H, such that for $n = 1, 2, 3 \ldots,$

$$H\Psi n = En\Psi n.$$

Bra-ket notation is used to simplify the calculations carried out on the wave function. Dirac developed this system in order to reconcile the different approaches to quantum theory taken by Schrödinger and German physicist Werner Heisenberg (1901–76). Schrödinger's approach relied on wave mechanics, Heisenberg's on vector and matrix calculations. Bra-ket notation simplifies the expression and solving of these complex mathematical relationships, allowing them to be addressed apart from any preexisting coordinate system.

Continuing with Bra-Ket Notation

Bra-ket notation is particularly valuable in quantum physics, as it allows very large vector statements to be manipulated in compact form. However, it is also useful in other fields, particularly those incorporating concepts from quantum mechanics. Arguably, any theoretical studies that involve both vector or matrix calculation and differential calculus can benefit from the use of bra-ket notation.

—*Richard M. Renneboog, MSc*

Bibliography

Beddard, G. S. *Applying Maths in the Chemical and Biomolecular Sciences : An Example-based Approach.* Oxford; New York: Oxford University Press, 2009. Print.

Dahl, Jens Peder,. *Introduction to the Quantum World of Atoms and Molecules.* New Jersey: World Scientific, 2009. Print.

Dick, Rainer. *Advanced Quantum Mechanics : Materials and Photons.* New York: Springer-Verlag, 2014. Print.

Finn, J. Michael. *Classical Mechanics.* Sudbury, MA: Jones and Bartlett, 2010. Print.

Pereyra, Pedro. *Fundamentals of Quantum Physics: Textbook for Students of Science and Engineering.* Berlin: Springer, 2012. Print.

BRITISH THERMAL UNIT (BTU)

FIELDS OF STUDY

Classical Mechanics; Thermodynamics

SUMMARY

The British thermal unit (BTU) is a traditional unit of energy. One BTU is equal to the energy needed to heat one pound of water by one degree Fahrenheit. It has largely been replaced by the joule.

PRINCIPAL TERMS

- **barrel of oil equivalent:** the energy output of burning one barrel (42 gallons, or 159 liters) of crude oil; equal to 5.8×10^6 BTU or 6.1×10^9 joules.
- **calorie:** the energy needed to raise the temperature of 1 gram (0.002 pounds) of water by 1 degree Celsius (1.8 degrees Fahrenheit); equal to 3.96×10^{-3} BTU or 4.19 joules.
- **heat value:** a measurement of the energy released as heat when a specific amount of a specific substance is burned; typically applied to fuels and foods in units of energy or mass.
- **International System of Units (SI):** a widely used standardized system of units for measuring natural phenomena, based on and largely identical to the metric system.
- **joule:** the SI unit of energy, equivalent to the work done by applying a force of one newton over a distance of one meter. One BTU is equal to 1,055.06 joules.
- **therm:** the energy released by burning 100 cubic feet (2.83 cubic meters) of natural gas; equal to 1.02×10^5 BTU or 1.08×10^8 joules.
- **ton of coal equivalent:** the energy released by burning 1 US short ton (0.91 metric ton) of coal; equal to 19.49×10^6 BTU or 21.28×10^{10} joules.

Measuring Energy

The British thermal unit (BTU) was developed during the Industrial Revolution as a way to measure the amount of energy contained in the coal used to power steam engines. It was defined as the energy needed to heat 1 pound (453.59 grams) of water by 1 degree Fahrenheit (0.556 degrees Celsius). The

FROM	TO		
	BTUs	Joules	Foot-Pounds
BTUs	1	1055	778
Joules	0.0009478	1	0.7376
Foot-Pounds	0.001285	1.3558	1

FROM	TO			
	Horsepower	Watts	Foot-Pounds Per Second	BTUs Per Second
Horsepower	1	746	550	0.7068
Watts	0.001341	1	0.7376	0.00095
Foot-Pounds Per Second	0.00182	1.356	1	0.001285
BTUs Per Second	1.415	1055	778	1

Conversion tables for units of energy and units of power.

joule has replaced it in science as the standard unit of energy within the International System of Units (SI). One BTU is equal to 1,055.06 joules.

The BTU is one of many traditional units of energy. Another such unit is the calorie, defined as the energy needed to raise the temperature of one gram of water by one degree Celsius. It is most commonly used in chemistry. (Note that the "calories" in food are in fact kilocalories, sometimes written as Calories; one Calorie is equal to one thousand calories.) Other traditional units were developed based on the heat values of fuels such as oil, coal, and natural gas. One barrel of oil equivalent is the energy released by burning one barrel of crude oil; the ton of coal equivalent, one short ton of coal; and the therm, one hundred cubic feet of natural gas. Despite the potential for confusion, these units are still used in their respective industries. For example, therms are commonly used in home heating.

Modern Use of the BTU

BTUs are used in a number of industries, such as the power industry, to describe energy output. BTU may also be used as shorthand for BTUs per hour, a measure of power (energy spent over time). Despite the prevalence of the joule, the BTU is a widely accepted standard in many countries that use the metric system.

—Kenrick Vezina, MS

Bibliography

"British Thermal Unit (BTU)." *Encyclopaedia Britannica*. Encyclopaedia Britannica, 3 Sept. 2013. Web. 3 Apr. 2015.

Crowell, Benjamin. "Conservation of Mass and Energy." *PhysWiki: The Dynamic Physics E-Textbook*. U of California, Davis, 6 Mar. 2014. Web. 3 Apr. 2015.

Crowell, Benjamin. "Energy." *PhysWiki: The Dynamic Physics E-Textbook*. U of California, Davis, 24 June 2014. Web. 3 Apr. 2015.

"December 1840: Joule's Summary on Converting Mechanical Power into Heat." APS News, Dec. 2009: n. pag. APS Physics. Web. 3 Apr. 2015.

"Energy Units." *American Physical Society Physics*. Amer. Physical Soc., 2015. Web. 13 May 2015.

"Energy Units and Calculators Explained." *Energy Explained: Units and Calculators*. US Energy Information Administration, 3 Oct. 2014. Web. 13 May 2015.

C

CALCULATING SYSTEM EFFICIENCY

FIELDS OF STUDY

Thermodynamics; Electronics; Classical Mechanics

SUMMARY

Calculating system efficiency is a means of monitoring system functions and identifying aspects of a system that may be improved. The concept and methods are applicable in all fields, but especially in technology and engineering. Mechanical, electrical, and thermal efficiency are discussed.

PRINCIPAL TERMS

- **electrical efficiency:** the ratio of the power applied to an electrical circuit to the power delivered by a particular device in the circuit.
- **imperfect system:** any system that functions with less than 100 percent efficiency.
- **mechanical efficiency:** the ratio of the power applied to a mechanical system relative to the power delivered by the system.
- **system input:** a force, such as voltage or torque, applied to a system.
- **system output:** the work that the system performs.
- **thermal efficiency:** the ratio of the work performed by a system relative to the heat energy that is supplied to the system.
- **work:** in mechanical systems, the result of a force operating through a distance.

SYSTEMS AND WORK

A system consists of components that work together to perform a desired function. A system input is a force that causes the system to function and perform work. Work performed by the system is the system output. In a mechanical system, components might include pulleys, gears, drive belt, and other physical devices. These devices function together to perform work such as pumping water, lifting heavy weights, or propelling a vehicle. The system input for a mechanical system is also physical in nature. It can be the force supplied by muscles, the torque supplied by an attached engine, or the thrust produced by a jet engine, for example. The mechanical efficiency of such a system is affected by factors such as friction and slippage. Friction is the resistance of one component's motion relative to another. It always consumes some of the power within a system. Slippage occurs when not all of the power that could be delivered is actually delivered between components. An imperfect system is one in which the system input's power is not delivered with 100 percent efficiency to the system output. Since these factors are present in every mechanical system, no mechanical system can be 100 percent efficient. The mechanical efficiency of a system can be calculated by comparing the system output to the system input.

An electrical circuit is also a system designed to perform work. Electrical circuits function through the movement of electrical charge between two points. The factors that affect electrical efficiency are analogous to those that affect mechanical efficiency. The system input to an electrical system is the applied voltage, or electromotive force. This is also referred to as the electrical potential difference between two points in the circuit. This is typically delivered by a battery or a rectifier circuit for direct current (DC) applications. It is delivered by a generator for alternating current (AC) applications. The voltage for a small flashlight, for example, may be provided by three small dry-cell batteries. The batteries provide a constant electrical potential of 4.5 volts. In contrast, the voltage of a television set is typically provided by the alternating electrical potential of 110 volts of a standard North American wall socket. Electrical and electronic devices typically produce heat when they are in use. This can be thought of as the by-product of the friction of electrons moving through the components of the circuitry. This is the resistance of the circuit and is the primary source of the heat generated by the circuit. A type of slippage exists in electrical systems. This slippage is the eddy currents and voltage losses that exist at junctions, such as solder joints, where different materials connect different components. As in mechanical systems, these factors combine to impair the overall efficiency of the system. The efficiency of any electrical system is typically very high and may be almost 100 percent for devices such as transformers, in which the system output is nearly equal to the system input.

Systems that rely on the transfer of heat from one part of the system to another are known as heat engines. Heat engines include steam engines and internal combustion engines. The efficiency with which such systems perform work relative to the heat that is transferred is the thermal efficiency of the system.

Calculating Efficiency

System efficiency is calculated by comparing the system output to the system input and expressing the result as a percentage. For electrical systems, load power is the system output. Total power is the system input. Electrical efficiency is calculated as

$$\text{Efficiency} = (\text{Load power} \times 100\%) / \text{Total power}.$$

This same equation can be applied to each individual component of an electrical system as well as to the overall system or any part of the system. In such cases, it defines the electrical efficiency of that specific component or structure.

The thermal efficiency of a heat engine depends on the conversion of heat energy to work output. Since all heat engines absorb some of the heat that is produced without converting it to a work output, the thermal efficiency of a heat engine is never 100 percent. In an internal combustion engine, for example, combustion of the fuel produces heat. This heat causes the combustion gases to expand and push against the pistons, converting chemical energy into physical work. Not all of the heat produced is applied in this way, however. The material from which the engine is made absorbs a portion of the heat, which is then carried away by a coolant. Some of the heat is carried out as hot gases

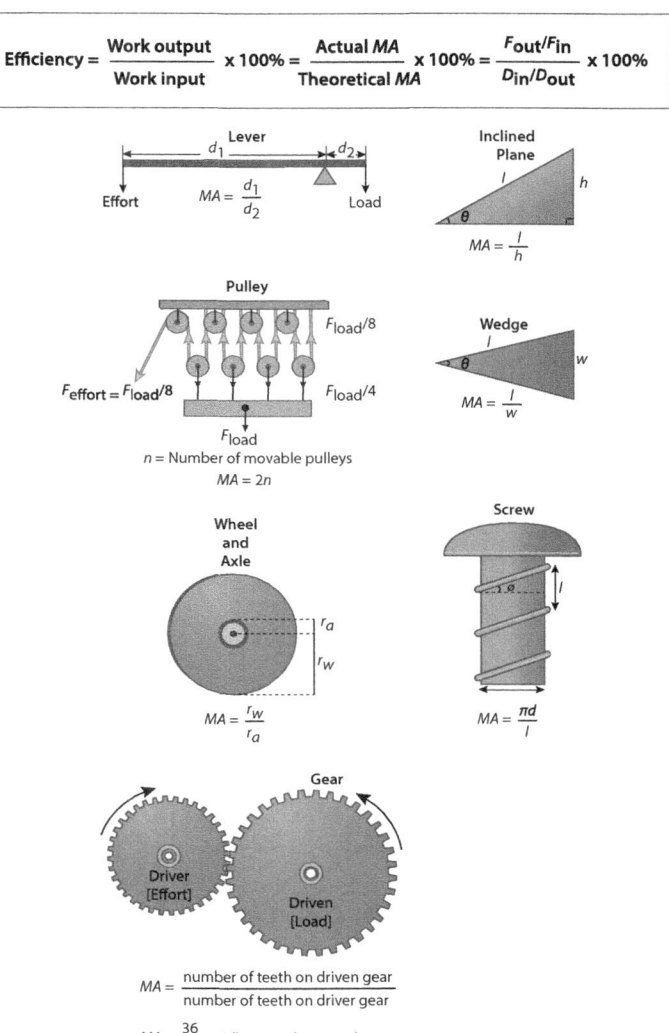

The efficiency of simple machines can be calculated by finding the ratio of the work output divided by the work input and is typically converted to a percentage. It can also be calculated by measuring the actual mechanical advantage (actual MA) divided by the ideal mechanical advantage (ideal MA). The ideal mechanical advantage (MA) for each type of simple machine is calculated differently.

are exhausted. Such heat losses can be compared to slippage in mechanical systems, in which some of the input power is not delivered to the output as work but simply lost as waste. The thermal efficiency of such a system is calculated as:

$$\text{Efficiency} = [(\text{Total heat} - \text{Lost heat}) \times 100\%] / \text{Work output}.$$

Note that both heat and work require the same units (joules).

The efficiency of a mechanical system is perhaps also the simplest to calculate. It is the ratio of the actual work output to the work input expressed as a percentage. It is calculated as

$$\text{Efficiency} = (\text{Work output} \times 100\%) / \text{Work input}.$$

The mechanical efficiency can also be calculated as the ratio of the actual mechanical advantage to the ideal mechanical advantage expressed as a percentage, or

$$\text{Efficiency} = (\text{Actual mechanical advantage} \times 100\%) / \text{Ideal mechanical advantage}.$$

The actual mechanical advantage of a machine is the ratio comparing the output force to the input force, taking into account all the limitations on the efficiency of real-world machines. In contrast, the ideal mechanical advantage is the ratio comparing the output force to the input force, ignoring those limitations.

REAL-WORLD EFFICIENCY

Mechanical and electrical systems are central to all aspects of modern human endeavor. In any enterprise or economy, the systems that function most efficiently are also the systems that provide the highest return on investment. Efficiency is related to all aspects of production and quality control in industry and, indeed, in all aspects of modern life. The goal of efficiency, in any application, is to achieve the desired system output with the least waste of system input, no matter whether it is preparing a field for planting, manufacturing the most advanced microprocessor chip, or engineering the acoustic characteristics of a concert hall for the best enjoyment of an artistic performance.

—*Richard M. Renneboog, MSc*

BIBLIOGRAPHY

Bolton, J. *Classical Physics of Matter*. Philadelphia: Inst. of Physics Pub., 2000. Print.

Chabay, Ruth W., and Bruce A. Sherwood. *Matter and Interactions*. 4th ed. Hoboken: Wiley, 2015. Print.

SAMPLE PROBLEM

A certain pulley system being used to lift a weight of 150 kilograms (330 pounds) to a height of 3 meters (9.8 feet) requires the user to exert a constant pulling force of 40 kilograms (88 pounds) through a distance of 21 meters (68.6 feet). What is the mechanical efficiency of the pulley system?

Answer:

Since the newton is defined as the force required to accelerate a mass of 1 kg at a rate of 1 m/s^2 (1 N = 1 kg·m/s^2), and the units will cancel out when expressed as a percentage, it is acceptable to use kilograms instead of newtons as the units of force in such calculations.

The mechanical efficiency of a system is the work input to bring about a corresponding work output, expressed as a percentage. Work is calculated as the product of force and distance. So,

$$\text{Work input} = (40 \text{ kg} \times 21 \text{ m}) = 840 \text{ kg·m}$$

and

$$\text{Work output} = (150 \text{ kg} \times 3 \text{ m}) = 450 \text{ kg·m}.$$

The mechanical efficiency of the pulley system is therefore

$$(450 \text{ kg·m} \times 100\%) / 840 \text{ kg·m} = 53.57\%$$

Gibbons, Patrick C,. *Physics*. 2nd ed. Hauppauge, N.Y.: Barron's, 2008. Print. Barron's EZ 101 Study Keys.

Nave, R. "Mechanics." *HyperPhysics*. Georgia State U, 2012. Web. 19 Aug. 2015.

Parasiliti, Francesco, and Paolo Bertoldi. *Energy Efficiency in Motor Driven Systems*. Berlin: Springer, 2003. Print.

Pfeiffer, Friedrich. *Mechanical System Dynamics*. Corrected 2nd ed. Berlin: Springer, 2010. Print. Lecture Notes in Applied and Computational Mechanics.

Thumann, Albert, and Harry Franz. *Efficient Electrical Systems Design Handbook*. Lilburn: Fairmont, 2009. Print.

CIRCULAR MOTION

FIELDS OF STUDY

Classical Mechanics

SUMMARY

Circular motion has captivated scientists and philosophers since antiquity. Although circular motion is described using distinct equations and terminology, these are all derived from Newton's laws of motion. In particular, the interplay between inertia and centripetal force accounts for everything from planetary motion to the feeling of being pressed against the side of a car during a sharp turn.

PRINCIPAL TERMS

- **angular momentum:** the rotational momentum of an object around an axis, defined as the product of its moment of inertia and its angular velocity.
- **centrifugal force:** a perceived force that seems to act outward on a rotating object, pushing it away from the center of its circular path; commonly confused with centripetal force.
- **centripetal force:** the force that impels a rotating object inward, acting at a right angle to its momentum.
- **Coriolis force:** the effects of the spinning motion of a planet on objects at the planet's surface, such as deflection in large-scale wind patterns.
- **inertia:** the principle that an object remains at rest or continues moving in the same direction at the same speed unless an outside force acts on it.
- **radian:** abbreviated rad, the International System of Units standard unit of angular measure, the length of the corresponding arc in a unit circle. A full circle is compose of 2π radians.

CIRCULAR MOTION AND NEWTON'S LAWS

Since antiquity, scientists and philosophers have recognized circular motion, particularly in the context of planetary orbits, as distinct from linear motion. This is somewhat misleading, however. While there are distinct equations to describe the properties of circular motion, these are all derived from the laws of motion codified by English physicist Isaac Newton (1642–1727). The same rules that govern the orbits that interested Renaissance astronomer and physicist Galileo Galilei (1564–1642) also govern a car turning around a race track or a tennis ball rotating with backspin. Angular momentum, just like linear momentum (often simply momentum), is conserved in a system. The total momentum of that system before and after any impact or exchange of energy is constant.

Objects moving in a circular path are subject to inertia, as described by Newton's first law. An object at rest will stay at rest, and an object in motion will continue moving in the same direction at the same speed, unless acted on by an outside force. In other words, objects "want" to move in a straight line. For an object to move in a curved path, a force must be acting on it. In circular motion, a centripetal force "pulls" or "pushes" the object toward the center of its circular path. Centripetal force simply describes the way a particular force acts. Gravity is the centripetal force in planetary orbits; the tug of a yo-yo string provides centripetal force during the around-the-world trick. The competing influences of inertia (straight-line tendency) and centripetal force (pull or push

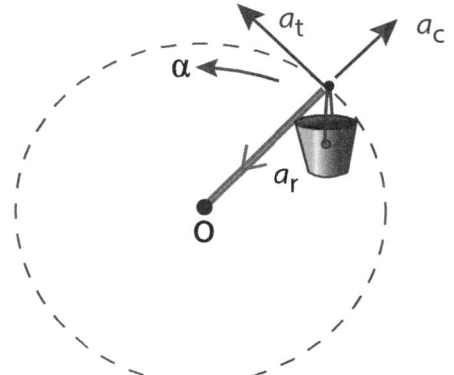

A bucket swinging by a rope in a circle from a fixed center (O) exhibits circular motion. The bucket accelerates in multiple directions simultaneously, including centrifugal acceleration (a_c) directed away from the center, centripetal acceleration (a_r) directed toward the center, and tangential acceleration (a_t) directed perpendicular to centrifugal and centripetal accelerations. These three accelerations summate to angular acceleration (a), which causes the circular motion of the bucket.

toward the center) result in the familiar arc of circular motion.

UNIFORM VERSUS NONUNIFORM CIRCULAR MOTION
Circular motion can be uniform or nonuniform. An object in uniform circular motion travels at the same speed throughout its path, with a constant rate of acceleration. It may seem strange at first to think that an object with constant speed is accelerating, but remember that acceleration is dependent on change in velocity, which in turn is dependent on speed and direction. If an object is maintaining a constant speed but constantly changing direction, it is still accelerating.

This constant acceleration is also predicted by Newton's laws. Because an object moving in a circular path must always be acted on by a centripetal force, it must always be accelerating. By definition, a mass must be accelerating if a nonzero net force is acting on it. With uniform circular motion, an object will always be accelerating toward the center of the circle.

Many of the most familiar forms of circular motion are uniform, or close enough to be treated as such. Examples include planetary orbits, the spin of a Ferris wheel, and the rotation of Earth on its axis. Even objects that vary in speed over time, such as a windmill varying with wind speed, often exhibit periods of roughly uniform motion.

Nonuniform circular motion, by contrast, involves an object moving in a circular path with varying speed and acceleration. A car accelerating into and out of a turn is one real-life example. Because the speed changes over time, so does the acceleration. In nonuniform circular motion, the vector of acceleration at any given moment may not be aimed at the center.

CENTRIFUGAL AND CENTRIPETAL FORCE
Anyone who has ever been a passenger in a car during a sharp turn can attest to the fact that there seems to be another force acting outward, away from the center of the arc. The same sensation can be felt during sharp turns in roller coasters and other amusement-park rides. This is commonly referred to as centrifugal force. However, it is not an actual force but a trick of perception.

Consider a passenger in a car driving in a straight line at a uniform speed. The car and the passenger

---

SAMPLE PROBLEM

Imagine a ride called the Rotor at a local carnival. It is a large upright metal cylinder, approximately 3 meters across, attached to a massive motor concealed below. Tim, a young man weighing about 50 kilograms, walks in and lines up with his friends along the inner walls of the cylinder. When the ride turns on, the cylinder begins to spin, increasing in speed until it is spinning at a rate of 33 revolutions per minute. The force applied to the riders is great enough that when the floor drops away, Tim and his friends remain pinned to the inside walls of the cylinder. After a few seconds at top speed, the floor comes back up, and the ride slows to a stop. Given that information, calculate the centripetal force acting on Tim while the ride was at maximum speed.

Answer:

To begin, identify which variables are needed to solve for F_c in the formula for centripetal force:

$$F_c = \frac{4\pi^2 rm}{T^2}$$

Mass (m) and the radius (r) and period (T) of the circular motion are needed. The mass value of the object in question (Tim) is 50 kilograms. The radius of the circular path of motion can be inferred from the diameter for the cylinder: half of 3 meters is 1.5 meters. The period can be derived, logically, from the revolutions per minute. If the ride completes 33 revolutions in one minute, it must complete one revolution in 1/33 of a minute, or approximately 1.82 seconds.

Next, plug these values into the equation and calculate:

$$F_c = \frac{4\pi^2 rm}{T^2}$$

$$F_c = \frac{4\pi^2 (1.5 \text{ m})(50 \text{ kg})}{(1.82 \text{ s})^2}$$

$$F_c = \frac{4\pi^2 (75 \text{ kg} \cdot \text{m})}{3.3124 \text{ s}^2}$$

$$F_c = 893.88 \frac{\text{kg} \cdot \text{m}}{\text{s}^2} = 893.88 \text{ N}$$

The centripetal force acting on Tim is about 893.88 newtons. This is enough to overwhelm his weight (the force of gravity acting on his mass) and to pin him to the inner wall of the ride while it is at maximum speed, but it is not enough to do him any harm.

are both moving forward at the same rate; the passenger perceives no force. When the car accelerates, the passenger feels as though he or she is being pressed backward into the seat. In fact, the seat, along with the rest of the car, is accelerating forward and pushing the person forward. The inertia of the passenger's body resists, creating the illusion of a backward force. Likewise, centrifugal force is a trick of perception caused by the inertia of an object in motion resisting inward acceleration caused by centripetal force.

The equations for uniform circular motion are all derived from Newton's second law of motion, which relates force (F) to the mass (m) and acceleration (a) of an object as follows:

$$F = ma.$$

For linear motion, acceleration is expressed in terms of change in velocity (v) over time. Using the relationship of a circle's radius (r) to its circumference, the acceleration of an object in circular motion (a_c) can be written as follows, where r is the radius of the circular path, shown here:

$$a_c = v^2 / r.$$

The velocity component of this equation can be further broken down. It can be thought of as the circumference of the circular path ($2\pi r$) divided by the period of rotation or revolution (T):

$$v = 2\pi r / T.$$

This works because velocity is equal to the absolute distance traveled (displacement) over the time it took to travel said distance. For circular motion, angular velocity is often more useful. Angular velocity measures the distance an object travels relative to the amount of a circle it completes in terms of radians. By definition, a full circle (360 degrees) is equal to 2π radians of rotation. Because radians, as a unit, are based on the relationship between the angle of rotation in a circle and the corresponding proportion of the circumference covered, they are the only unit of measurement for angles that works with these equations.

Plugging these expressions into the Newton's second law gives a formula for the force acting on an object in circular motion, which always act inward toward the center of the circular path:

$$F_c = ma_c$$

$$F_c = m\frac{\left(\frac{2\pi r}{T}\right)^2}{r}$$

$$F_c = \frac{4\pi^2 rm}{T^2}$$

Note that force is measured in newtons (N). One newton is the amount of force it takes to accelerate one kilogram of mass at a rate of one meter per second per second—that is, one kilogram-meter per second squared (kg·m/s^2). When using this formula for F_c, the values must be in terms of kilograms, meters, and seconds in order to produce an accurate calculation of force in newtons.

Circular Motion around the World

Examples of circular motion abound. Many motors, whether in electric toothbrushes or cars, generate circular motion from an external energy source, such as electricity or fuel combustion. Merry-go-rounds spin children on playgrounds. Indeed, the entire planet is spinning, as are the solar system, the galaxy, and even, it seems, the entire universe.

Understanding circular motion not only helps understand motion at an everyday, human scale but also influences aspects of life at the planetary level. Understanding the motion of the moon has helped humans understand tides and navigate the seas. Understanding the interaction between Earth's rotation (faster at the equator than the poles) and its atmosphere explains the Coriolis effect, which causes winds that would normally travel across the planet in a straight line to bend clockwise in the Northern Hemisphere and counterclockwise in the Southern Hemisphere.

—*Kenrick Vezina, MS*

Bibliography

Henderson, Tom. "Circular Motion and Satellite Motion." N.p.: *Physics Classroom*, 10 June 2013. Digital file.

Nave, Carl R. "Circular Motion." *HyperPhysics*. Georgia State U, 2012. Web. 12 Aug. 2015.

Riebeek, Holli. "Planetary Motion: The History of an Idea That Launched the Scientific Revolution." *Earth Observatory*. NASA, 7 July 2009. Web. 12 Aug. 2015.

Simanek, Donald E. "Mechanics." *Brief Course in Classical Mechanics*. Lock Haven U, Feb. 2005. Web. 28 Apr. 2015.

"Surface Ocean Currents: The Coriolis Effect." *NOS Education*. Natl. Oceanic and Atmospheric Administration, 22 June 2015. Web. 12 Aug. 2015.

Terr, David. "Uniform Circular Motion." *Wolfram MathWorld*. Wolfram Research, 22 July 2015. Web. 12 Aug. 2015.

CLOSED SYSTEMS AND ISOLATED SYSTEMS

FIELDS OF STUDY

Thermodynamics

SUMMARY

This article discusses two of the three types of thermodynamic system, with examples of how they behave. Closed systems cannot transfer matter across their boundaries, but they can transfer energy. Isolated systems can transfer neither matter nor energy.

PRINCIPAL TERMS

- **conservation of energy:** a law of physics that states that within an isolated system, the amount of energy remains constant; energy can be neither created nor destroyed, only transformed.
- **entropy:** a measure of how close a system is to equilibrium; also described as the amount of disorder in a system.
- **equilibrium:** the condition of a thermodynamic system in which there is no energy flow.
- **open system:** a system in which both matter and energy can be exchanged between the system and its surroundings.
- **system boundary:** the border of the system being considered.

Thermodynamic Systems

Thermodynamics is the study of the propagation of energy, usually in the form of heat, within a system boundary. Systems can be open, closed, or isolated. An open system is one that can exchange both matter and energy across its boundary. A closed system allows only energy transfer. An isolated system allows neither energy nor matter to be transferred. A perfectly isolated system is not really possible, although the universe itself is sometimes thought of as an isolated system. However, one might imagine an impossibly good thermos, capable of maintaining a beverage at one exact temperature forever, with no heat ever escaping; this would be an isolated system. Isolated systems follow the law of conservation of energy, meaning within them, energy can never be created or destroyed. It can only change its form. For example, if one were to put warm water and an ice cube in this ideal thermos, the ice cube would melt, and the water would cool down, but the total amount of heat in the thermos would stay the same.

The example of the ice cube melting inside the thermos is also an example of entropy, the tendency of systems to move toward a state of equilibrium. (Sometimes this is described as moving toward a state of greater disorder.) Equilibrium means the energy in a system is evenly distributed. In the example of the ice water, at first the water will be warmer and the

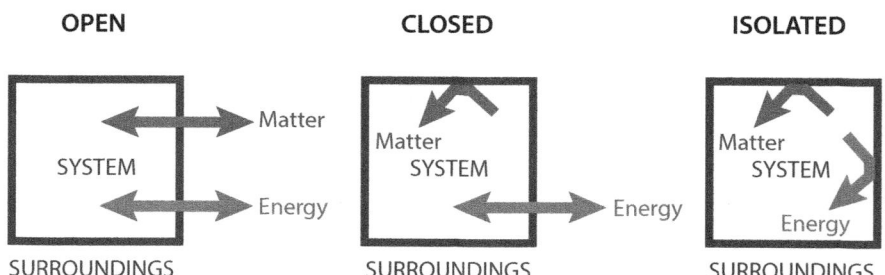

There are three classifications of systems based on whether energy and/or matter is exchanged between the system and its surroundings. Closed systems do not exchange matter with their surroundings, but do exchange energy. An isolated system does not exchange matter or energy. In reality, systems are rarely truly isolated, but isolated systems are used for theoretical calculations.

ice cube will be colder. Eventually, however, the ice cube will melt and the system will reach a state where it is all water of one temperature. In real life, thermoses are not isolated systems, but closed systems. Therefore, because of the transfer of energy through the wall of the thermos (that is, across the system boundary), the water inside the thermos will eventually reach the same temperature as the space outside the thermos. This is an example of how closed systems in contact (the thermos and the space around it) reach equilibrium between one another.

—*Gina Hagler, MBA*

BIBLIOGRAPHY

Anderson, G. M. *Thermodynamics of Natural Systems.* Cambridge: Cambridge University Press, 2005. Print.

Haynie, Donald T. *Biological Thermodynamics.* Cambridge: Cambridge University Press, 2008. Print.

Johnson, Erin R., and Axel D. Becke. *Density Functionals: Thermochemistry.* Berlin: Springer, 2015. Print.

Lemons, Don S. *Mere Thermodynamics.* Baltimore: Johns Hopkins University Press, 2009. Print.

Oliveira, Mário J. de. *Equilibrium Thermodynamics.* Berlin: Springer, 2013. Print.

Struchtrup, Henning. *Thermodynamics and Energy Conversion.* Berlin: Springer, 2014. Print.

CONCAVE AND CONVEX

FIELDS OF STUDY

Optics; Electromagnetism

SUMMARY

Mirrors and lenses have been instrumental in the development of science and society over the course of history. These devices have been used in military, scientific, and biological applications for thousands of years. This article explains the laws and properties that make these uses possible.

PRINCIPAL TERMS

- **converging:** the joining of light rays after an interaction with a mirror or a lens.
- **diverging:** the separation of light rays after an interaction with a mirror or a lens.
- **focal point:** the place where light from a source converges after it reflects off a mirror or refracts through a lens.
- **inversion:** the reversal of the way an image looks after an interaction with a mirror or lens.
- **refraction:** the bending of the direction of light when it goes from one substance to another.
- **reflection:** the bouncing of light with an angle equal to the angle at which it first comes in contact with an object.
- **Snell's law:** a mathematical description of the refraction of light as it goes from one medium to another.

THE STUDY OF OPTICS

One of the most important fields of physics is the study of light and optical instruments. These instruments have provided a vast understanding of different kinds of phenomena and objects. With his knowledge and understanding of optical instruments, Galileo Galilei (1564–1642), for example, created a telescope that led him to observe the moons of Jupiter, sunspots, and other astronomical occurrences.

Optical instruments have advanced science in the microscopic realm as well. For example, through his use of the compound microscope, Robert Hooke (1635–1703) was able to describe and name the cell for the first time. With the help of optical instruments, other scientists have been able to develop theories of genes and microbes, instrumental in the detection and control of diseases.

Furthermore, Isaac Newton (1642–1727) developed and built the first useful reflecting telescope. In turn, the Newtonian telescope led to advancements in the development of other reflecting telescopes. Newton used a mirror instead of a lens, focusing light through the use of reflection. Before the Newtonian telescope, telescopes used lenses, which are refractors. They used the property of refraction through a lens to focus light.

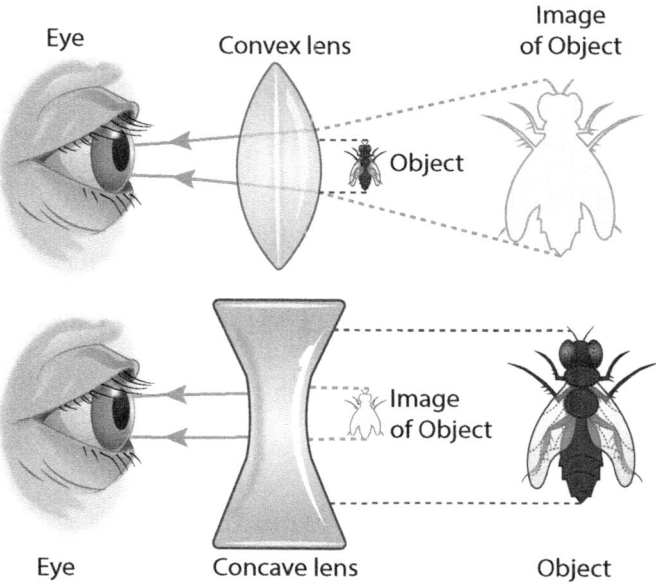

Lenses and mirrors can be curved so that the edges expand out (concave) or so that their edges bend in (convex). Light will react differently with each of these shapes, causing an object to be seen through these two types of lenses to appear different sizes. An object viewed through a convex lens will appear larger than reality; this is useful for farsightedness. An object viewed through a concave lens will appear smaller than reality; this is useful for nearsightedness.

FOCUSING LIGHT

Many different techniques have been employed to focus light into a point. Originally, opticians preferred lenses to mirrors because the process of making them was easier. Two different kinds of lenses were produced: converging, or convex lenses, and diverging, or concave lenses. Converging lenses let light through and focus it past the lens. These lenses converge the light to a point where the image forms. Most camera lenses are converging lenses. They have to focus the light into the detector (a film or sensor), the focal point in a camera.

Diverging, or concave, lenses, on the other hand, refract the light in such a way that it spreads apart on the other side of the lens. These lenses create images that appear to have come from in front of the lens. This is useful in the creation of holograms and in the correction of nearsighted vision. When a nearsighted person puts a diverging lens in front of his or her eyes, far-away images appear closer. These images are picked up by the eye. Since the images are closer, the person is able to see them clearly. In the same manner, a converging lens can be used to correct farsightedness. In this case, the lens makes objects appear to be farther away than they are. These methods can fix these two eye conditions by using basic principles of physics.

When light passes from one substance to another, its speed changes slightly, and it is bent or directed in a slightly different direction. Different substances react differently to light going through them. Some slow down light more than others do. The slowing of light compared to the speed of light in a vacuum is quantified in the constant called index of refraction (n). The bending of the path of light depends on the angle from which the light comes (θ_1), the angle at which the light emerges (θ_2), and the index of refraction of each substance (n_1 and n_2, respectively). This is known as Snell's law. It follows the mathematical formula

$$n_1 \sin \theta_1 = n_2 \sin \theta_2.$$

The angles θ_1 and θ_2 are measured with respect to the line perpendicular to the boundary between the two substances. Imagine light from the sun hits the surface of a pool at an initial (or incidence) angle (θ_1) of 30 degrees. The index of refraction of air (n_1) is 1. The index of refraction of water (n_2) is 1.33. To calculate the angle of refraction (θ_2), substitute these values into the formula for Snell's law, as below.

$$n_1 \sin\theta_1 = n_2 \sin\theta_2$$
$$n_2 \sin\theta_2 = n_1 \sin\theta_1$$
$$\sin\theta_2 = \frac{n_1}{n_2}\sin\theta_1$$
$$\theta_2 = \sin^{-1}\left(\frac{n_1}{n_2}\sin\theta_1\right)$$
$$\theta_2 = \sin^{-1}\left(\frac{1}{1.33}\sin 30°\right)$$
$$\theta_2 = 22.08°$$

FROM LENSES TO MIRRORS

After many advancements in glassmaking, mirrors became easier to make. In 1668, Newton completed the first successful reflecting telescope. Instead of using lenses in his telescope, Newton focused the light with a converging, or concave, mirror. The secondary mirror in telescopes can be diverging, as it sends the light back without inverting it. It also allows for a wider field of view. Mirrors follow a much simpler optical rule than lenses do. The angle of incidence (θ_i) equals the angle of reflection (θ_f) as shown here:

$$\theta_i = \theta_f.$$

Both plane and curved mirrors invert images through image inversion. When looking at the moon through a telescope people might notice that the image is inverted. What looks like the left side of the moon in the telescope is actually the right side. The same effect happens when one looks at oneself on a concave mirror or through a convex lens. This occurs because mirrors reflect light in the same direction they receive it, while lenses refract light that goes through them.

DAILY USE OF MIRRORS AND LENSES

The industry of glassmaking has existed for millennia. From the glass beads found on King Tut's body to the legendary glassmakers of Venice and the modern makers of telescope mirrors and lenses, mirrors and lenses have been essential parts of everyone's life. Without these developments nearsighted and farsighted people would not be able to see the world around them clearly. The technology industry also depends heavily on mirrors to make microchips. Curved televisions, monitors, and other devices are becoming popular because of the advancements in mirror- and glassmaking. Future discoveries in physics and astrophysics will rely on both large mirrors and those that can reflect infrared and other forms of light.

—*Angel G. Fuentes, MS*

BIBLIOGRAPHY

Breinig, Marianne. "Reflection and Refraction." *Elements of Physics II*. U of Tennessee, Dept. of Physics and Astronomy, n.d. Web. 6 May 2015.

Conrady, A. E. *Applied Optics and Optical Design*. Newburyport: Dover, 2013. Print.

Giambattista, Alan, and Betty McCarthy Richardson. *Physics*. 3rd ed. New York: McGraw, 2015. Print.

Nave, R. "Refraction of Light." *HyperPhysics*. Dept. of Physics and Astronomy, Georgia State U, n.d. Web. 6 May 2015.

Rasmussen, Seth C. *How Glass Changed the World: The History and Chemistry of Glass from Antiquity to the Thirteenth Century*. New York: Springer, 2012. Print.

Young, Hugh D., Philip W. Adams, and Raymond J. Chastain. *Sears and Zemansky's College Physics*. 10th ed. Hoboken: Pearson, 2016. Print.

CONSERVATION OF CHARGE

FIELDS OF STUDY

Particle Physics; Electronics

SUMMARY

Conservation of charge is perhaps the most fundamental of all universal laws, at least equal to the law of conservation of energy. The unit charge of the electron, proton and other sub-atomic charged particles is the result of combinations of elementary particles called quarks. The concept of conservation of charge is essential to the functioning and predictability of matter interactions, including the most basic processes of living systems and the most advanced electronic technologies.

PRINCIPAL TERMS

- **beta particle:** an electron that is emitted from an unstable nucleus via radioactive decay.
- **conservation of energy:** a fundamental law of physics that states that the amount of energy in a system remains constant over time. Although the energy can be transformed or transferred, it cannot be created or destroyed.
- **coulomb:** the basic unit of charge in the International System of Units (SI).
- **current:** the rate at which an electric charge, usually in the form of electrons, moves through a wire or other conductive material.

- **positron:** a subatomic particle that has the same mass as an electron as well as an equal-but-opposite electric charge, that is, a positive charge.
- **protons:** subatomic particles that, with neutrons, make up the mass of an atom's nucleus; they have functionally the same weight as neutrons but hold a positive electric charge.
- **quark:** an elementary fermion that combines with other quarks to form a baryon, such as a proton or a neutron, or with an antiquark to form a particle called a meson.
- **standard model:** a generally accepted unified framework of particle physics that explains electromagnetism, the weak interaction, and the strong interaction as products of interactions between different types of elementary particles.

THE ELEMENTARY PRINCIPLE

The law of conservation of charge is the most fundamental principle of the standard model of particle physics, and says simply that the amount of charge in any particle interaction is constant. It is a subsidiary to the law of conservation of energy, which states that the amount of energy in the universe is constant. In balancing chemical reactions involving ions, conservation of charge is first encountered in the statement that the net charge in the products must be exactly equal to the net charge in the reactants. In balancing redox chemical equations involving the transfer of one or more electrons between atoms, the number of electrons given up in the oxidation statement must be exactly equal to the number of electrons taken up in the reduction statement so that when summed together the two statements balance and charge has neither been created nor destroyed.

The principle is rather more complex in particle physics, which allows opposite charges to annihilate each other and be converted to energy, as when an electron and a positron come into contact with each other. The amount of charge in the process changes, but charge balance is maintained. That is, charge is conserved in the process. In another example, in some radioactive decay processes, a neutron can decompose into a proton and an electron or beta particle, with the emission of an "antineutrino."

The neutron has no electrical charge, and the positive charge of the proton is balanced by the negative charge of the electron. There is thus no net change of charge, and charge is therefore conserved. In another process, the proton can convert into a neutron by emission of a positron and a "neutrino." In this process the positive charge of the proton continues in the positive charge of the positron, and charge is again conserved.

CHARGE IS A QUANTUM QUANTITY

The concept of subatomic particles such as the electron and the proton having a unit charge is easy to grasp in light of the standard model and quantum mechanics as the basis of the modern theory of atomic structure. Prior to Thomson's identification of electrons and protons as charged particles, and their movement between two points as the essential feature of electrical current, electricity was thought to be a kind of fluid. This is easy to understand given that the individual nature of the moving charges cannot be detected on the macroscopic scale. For example, an electrical device such as an incandescent light bulb that consumes 100 watts of power when operating at the standard household voltage of 110 volts passes 6 X 10^{18} individual charges, or one coulomb of charge, through it every second.

The magnitude of the individual charge of an electron is not comprehensible on a scale measured in meters, and was not clearly understood until it was actually measured in Milliken's "oil drop experiment."

SAMPLE PROBLEM

The electron and the proton in a hydrogen atom are separated by a distance of approximately 5.3 X 10^{-11} meter. Calculate the magnitude of the electrical force F_e between the two particles.

Answer:

Use the Coulomb's law equation and the corresponding values of ε_0, q_1, q_2 and r to obtain

$$F_e = (1/4\pi\varepsilon_0)(q_1 q_2/r^2)$$

$$= (9.0 \times 10^9 \, Nm^2/C^2)(1.6 \times 10^{-19} \, C)^2/(5.3 \times 10^{-11} \, m)^2$$

$$= 8.1 \times 10^{-8} \, n.$$

The unit charge, whether positive or negative, now has its basis in the interaction of elementary particles called quarks. In the standard model, there are six flavors of quark, termed up, down, top, bottom, charm, and strange, along with their antiquark counterparts. The unit charge is produced in the specific combination of quarks that for particles bearing a unit charge. This, unfortunately, can make the concept rather confusing since it requires that certain quarks are designated as possessing fractional charges for the mathematics to work out in accord with observations from high-energy collision experiments.

The important feature to note is that in describing the interactions of quarks and other particles, charge is conserved and the charged particles obey the mathematics of Maxwell's and Gauss's equations. Indeed, it has been through this very predictable behavior that many of the subatomic particles have been identified.

ELECTRICAL FORCE

The force between any two charged particles that each bear a unit charge is inversely proportional to the square of the distance between them, and is stated as Coulomb's law

$$F = (1/4\pi\varepsilon_O)(q_1 q_2 / r^2)$$

where ε_O is the "permittivity constant" having the value 8.85418×10^{-12} C^2Nm^{-2}, q_1 and q_2 are the unit charges, each of which has the magnitude 1.6×10^{-19} C, and r is the distance between the two point charges (a point charge is the overall charge on a particle or object treated as though it is located only at the center of the particle rather than dispersed over the volume occupied by the particle). For particles and objects that bear more than one unit charge, the values of q_1 and q_2 are multiplied accordingly.

A FOUNDATION PRINCIPLE OF MODERN TECHNOLOGY

The constancy of charge is a vital foundation of modern technology, and even of physical existence. Matter and energy can be converted from one to the other according to Einstein's equation $e = mc^2$, which applies to subatomic and elementary particles as well as to larger particles like protons and neutrons. But these particles, and the charge they contain, consist of quark combinations. If charge was not conserved, and could disappear spontaneously, then so would the mass associated with the quarks and higher particles, including atoms and molecules that make up the physical universe.

In practical terms, it would be impossible to know or predict the, or rely on, the operation of electrical and electronic systems, including the oxidation and reduction cycles that function to maintain life. Transmission of signals between nerves, for example, involves the transfer of electrical charge between synapses. Transmission of electrical energy through the power grid involves the transfer of electrical charge between points of different electrical potential.

The conservation of charge is so fundamental that if it were otherwise the universe as we know it could not exist. In technology, calculations and control of electrical current are based on the concept of the unit charge and its conservation. The relationship between current, electrical potential (or voltage), resistance, capacitance and inductance are true and useful relationships only because the unit charges that are their basis are conserved. This is an especially important concept in the functioning of semiconductor devices, in which the currents and voltages involved are quite small. Power generation is another field in which an understanding of charge conservation is important with regard to current technology and will become even more important to the success of research into better and more efficient methods of power generation.

—*Richard M. Renneboog M.Sc.*

BIBLIOGRAPHY

Bird, John. *Electrical and Electronic Principles and Technology*. 5th ed. London [u.a.]: Routledge, 2014. Print.

Grant, I. S., and W. R. Phillips. *Electromagnetism*. 2nd ed. Chichester: John Wiley, 2011. Print.

Newman, Jay. *Physics of the Life Sciences*. New York: Springer, 2008. Print.

Purcell, Edward M., and David Morin. *Electricity and Magnetism*. 3rd ed. Cambridge [etc.]: Cambridge University Press, 2013. Print.

Trefil, James. *The Nature of Science: An A-Z Guide to the Laws and Principles Governing Our Universe*. Boston: Houghton Mifflin, 2003. Print.

CONSERVATION OF ENERGY

FIELDS OF STUDY

Classical Mechanics; Electromagnetism; Nuclear Physics

SUMMARY

The motion of an object can be described by considering the various forms of energy the object has. In the absence of dissipative forces, the energy of a system can be neither created nor destroyed. This principle can be used to find the varying amount of kinetic energy of an object, which directly relates to the speed of that object.

PRINCIPAL TERMS

- **kinetic energy:** energy due to any kind of motion, be it rotation, vibration, or translation.
- **potential energy:** energy that is stored in objects and has the potential to become other forms of energy, such as kinetic energy.
- **total mechanical energy:** the sum of all the kinetic and potential energies of an object in a closed system.

CONSERVATION OF ENERGY AND MASS

To study the motion, or kinematics, of an object, one can apply Isaac Newton's (1642–1727) laws of motion and obtain results that match what happens in the real world. However, there is one small issue with this approach: it gets a lot more complicated when dealing with variable accelerations. When a car stops at a red light, it is not accelerating at that moment. When the light turns green, the driver applies a variable amount of pressure to the gas pedal. That variable amount of pressure produces a variable acceleration of the car. In these cases, applying Newton's laws of motion produce different results depending on the acceleration value used. Calculating energy is a simple approach to solving these kinds of problems. This is due to the fact that the total amount of energy an object has never changes. Physicists call this principle the conservation of energy.

Energy is conserved by all objects in the absence of friction or any other dissipative force. If a person rubs a finger on a table for a long period of time, his or her finger will get hotter and hotter. If the person hits the table with an open palm, his or her palm will also be warmer than before. This is because some of that person's energy was turned into heat, which is a form of energy. Heat then moves away from the person's hand in the form of infrared radiation and goes back into the environment. While it may seem like the energy was not conserved, it is still part of a larger overall system and did not disappear completely.

The law of conservation of energy states that the total energy in the universe is always the same, and therefore energy can be neither created nor destroyed. In other words, it is not possible to add energy to the universe or take some energy out of the universe. This law can be applied to any system in which one can assume no dissipating forces exist. In 1905, Albert Einstein (1879–1955) recognized in his theory of relativity that mass is itself a form of energy. Thus the law of conservation of energy also addresses the conservation of mass, in that the total amount of mass and energy in the universe is constant.

Inside the sun, mass is constantly being turned into energy. This is the energy that warms Earth. In a way, fossil fuels are a form of energy from the sun, if one considers the law of conservation of mass and energy. Eons ago, the sun converted mass into energy in the form of light and heat. When that light and heat reached Earth, plants, bacteria, and algae turned the sun's energy into mass in the form of food, which provided them with energy to live and grow. That energy was converted back into mass in the form of the newly grown plants, bacteria, and algae. Animals then ate those plants, bacteria, and algae as food, which they converted to energy. Over time, the remains of flora and fauna fossilized, becoming coal and oil. Humans use coal and oil as fuel to provide energy. Conservation of energy is all around, and it affects everyone in more ways than one might think.

The discovery of the principle of conservation of energy is often attributed to Dr. Joseph E Mayer, a ship's surgeon, who noted that crewmen who needed to be bled (standard medical practice at the time) had venous blood that seemed to be brighter in color when his ship was in the tropics than when his ship was more northern latitudes. He concluded–correctly–the human body did not require as much energy when it was already warm. He submitted his paper to a journal which promptly rejected it. But the editor coached Mayer on revisions until the science met the standards of the journal and then published it.

Different Forms of Energy

In order to understand conservation of energy, one must understand the different forms and properties of energy. Everything that is in motion has a form of energy called kinetic energy. In fact, the temperature of a room is defined as the average kinetic energy of the particles in the room. The air molecules in a room are in a constant state of motion. Not only are they moving around, they are also vibrating. If on average they are moving faster, then they have more kinetic energy, which makes the whole room warmer. If on average they move slower, then less kinetic energy leads to lower temperatures. The kinetic energy (K) of an object, in joules (J), is mathematically defined by the object's mass (m) and its velocity (v), as in the following equation:

$$K = \frac{1}{2}mv^2$$

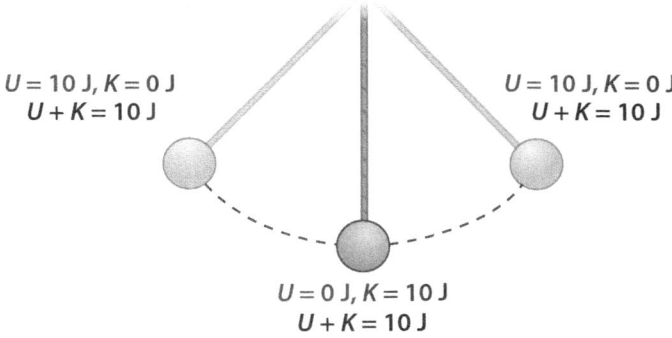

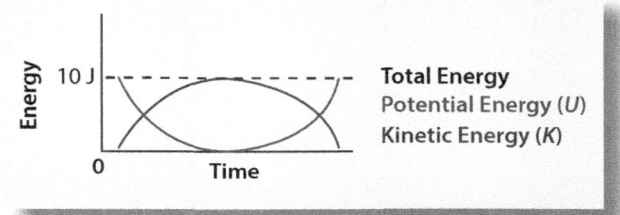

A pendulum at a maximum height will have maximum potential energy and zero kinetic energy. A pendulum at minimum height will have zero potential energy and maximum kinetic energy. At any point along the pendulum's path, the potential energy and kinetic energy will equal the total mechanical energy of the system and will remain constant. The graph shows how potential energy and kinetic energy change over time and how total energy remains constant.

But that is not the only form of energy objects can have. Objects tend to fall to Earth's surface due to the planet's gravitational pull. Another simple way to say this is that the object has extra amounts of energy when it is above the surface of Earth. This energy is known as the gravitational potential energy (U_g), and it is determined by the object's height above the surface (h), its mass (m), and the strength of Earth's gravitational field (g). When a pencil is held, it has potential energy. As the pencil falls to the ground, it starts to lose some of that potential energy, which becomes kinetic energy. The pencil's velocity increases as it continues to fall. When it hits the ground, all of its potential energy has been turned into kinetic energy, and it lands at its highest possible speed. At that point, dissipative forces cause the energy to be lost to the surrounding environment. Mathematically speaking, the gravitational potential energy, in joules, is expressed as

$$U_g = mgh$$

where g equals 9.8 meters per second per second, or meters per second squared (m/s²), near the surface of Earth. There are many other forms of stored or potential energy. There is stored energy in chemicals that are about to react in a chemical reaction. Some of this energy is released in the reaction as heat.

Another form of potential energy is found by the stretching and compression of a spring. If there is a mass (m) attached to a spring that has been fixed to a wall, and someone pulls on the mass without letting go, the mass now has potential energy. If the mass is released, it will begin to oscillate, gaining kinetic energy. At one point during this oscillation, the mass will compress the spring to the maximum possible amount and stop moving for a fraction of a second. When this happens, all the kinetic energy gained has been transformed back into potential energy. Then the spring will push back on the mass, allowing the stored energy to be transformed into kinetic energy. The entire process repeats itself for as long as the mass is allowed to oscillate. The potential energy stored in a spring is a function of the distance the spring is stretched or compressed (x) and the properties of the particular spring used, summarized as its unique spring constant (k). The spring constant is a measurement of how rigid the spring is and how it reacts to being stretched or compressed. In the International System of Units, it is measured in newtons per meter (N/m). Mathematically, the potential energy of a spring in joules, known as the elastic potential energy (U_e), is found using the following equation:

$$U_e = \frac{1}{2}kx^2$$

Elastic potential energy only exists if a spring is part of the system in question. An object can have multiple forms of energy at once. A pendulum that is oscillating is moving, therefore it has kinetic energy, and is at a distance from the surface, so it has gravitational potential energy. When the pendulum is at its highest point, all its energy is in the form of gravitational potential energy. When it is at its lowest point, it has zero potential energy and the highest amount of kinetic energy it can have. In between, it has different amounts of potential and kinetic energies. As described above, energy in a closed system is conserved. That means that all of the energy in the pendulum is always the same. Physicists have defined the total mechanical energy (E) as the sum of all the kinetic and potential energies of an object in a closed system:

$$E = K + U_g + U_e.$$

When physicists say that the energy is conserved, they mean that mechanical energy is conserved. This means that there is no change in the total mechanical energy, or that the initial mechanical energy (E_i) equals the final mechanical energy (E_f):

$$E_i = E_f.$$

Substituting the definitions of the different forms of energy into the mechanical energy conservation equation, the equation becomes

$$K_i + U_{g,i} + U_{e,i} = K_f + U_{g,f} + U_{e,f}.$$

ENERGY PRODUCTION

By using energy to solve problems about motion, one can arrive at the same result without having to deal with variable forces and accelerations. This has wide implications and applications for the real world. When hydroelectric power plants produce energy, they do so by converting potential and kinetic energy into electrical energy. As water falls down from the top of a lake behind a dam through pipes called penstocks, it loses potential energy and gains kinetic

---

SAMPLE PROBLEM

A single-car roller coaster with a mass of 500 kilograms (kg) sits at the top of the track, waiting to begin its motion. After its initial descent, it travels up a smaller hill with an altitude of 10 meters (m). When at the top of this hill, it has a speed of 15 m/s. What is the kinetic energy of the car when it reaches the ground at the bottom of the smaller hill?

Answer:

In order to calculate the final kinetic energy, the initial kinetic and potential energies must be found. There are no springs in this system, so there is no elastic potential energy. First, use the information about the speed on the smaller hill to calculate the initial kinetic energy:

$$K_i = \frac{1}{2}mv_i^2$$

$$K_i = \frac{1}{2}(500 \text{ kg})(15 \text{ m/s})^2$$

$$K_i = 56,250 \text{ J}$$

Then use the same information to calculate the initial gravitational potential energy:

$$U_{g,i} = mgh_i$$

$$U_{g,i} = (500 \text{ kg})(9.8 \text{ m/s}^2)(10 \text{ m})$$

$$U_{g,i} = 49,000 \text{ J}$$

(Sample probem continues on next page)

> **(SAMPLE PROBLEM CONTINUED)**
>
> Do the same thing for the final potential energy:
>
> $$U_{g,f} = mgh_f$$
>
> $$U_{g,f} = (500 \text{ kg}) (9.8 \text{ m/s}^2) (0 \text{ m})$$
>
> $$U_{g,f} = 0 \text{ J}$$
>
> Note that when the car is at the bottom of the hill, its height above the ground is 0 meters, so it has no final gravitational potential energy. Using the values for initial kinetic energy and initial and final gravitational potential energy, use the equation for the conservation of total mechanical energy to find the final kinetic energy:
>
> $$K_i + U_{g,i} = K_f + U_{g,f}$$
>
> $$56{,}250 \text{ J} + 49{,}000 \text{ J} = K_f + 0 \text{ J}$$
>
> $$K_f = 105{,}250 \text{ J}$$
>
> The final kinetic energy is 105,250 joules.

energy. This allows the water to move faster and faster as it falls. It then hits the blades of a turbine and transfers its energy into the turbine, causing it to spin. The turbine turns a generator, which produces electrical energy. Conservation of energy can also be seen in the production of energy by other means. Most of the electricity produced in the United States comes from the burning of coal. When coal is used to heat water to produce electricity, the power plant cannot produce more energy than is stored in the coal as chemical potential energy. The same can be said about gasoline cars. The kinetic energy obtained by the explosive reaction of gasoline in an engine cannot be greater than the chemical energy stored in that gasoline before the reaction. This means that no one will ever be able to attain infinite speeds by means of propulsion, as an ever-increasing need for kinetic energy comes from an ever-increasing amount of stored potential energy, and there is a limited, and not infinite, amount of energy in the universe. Nothing in the universe that has mass can propel itself at or faster than the speed of light.

—*Angel G. Fuentes, MS*

BIBLIOGRAPHY

"Circus Physics: Conservation of Energy." *Circus*. PBS, 2010. Web. 21 Apr. 2015.

"Conservation of Energy: Physics." *Encyclopedia Britannica*. Encyclopedia Britannica, 23 Jan. 2014. Web. 21 Apr. 2015.

"Conservation of Energy." *Khan Academy*. Khan Acad., 2015. Web. 21 Apr. 2015.

Giambattista, Alan, and Betty McCarthy Richardson. *Physics*. 3rd ed. New York: McGraw, 2015. Print.

Moskowitz, Clara. "Fact or Fiction? Energy Can Neither Be Created nor Destroyed." *Scientific American*. Scientific Amer., 5 Aug. 2014. Web. 21 Apr. 2015.

Young, Hugh D., Philip W. Adams, and Raymond J. Chastain. *Sears and Zemansky's College Physics*. 10th ed. Hoboken: Pearson, 2016. Print.

CONVECTION AND CONDUCTION

FIELDS OF STUDY

Thermodynamics

SUMMARY

Convection and conduction are thermal processes. Conduction is the movement of heat through a material via very small atomic or molecular collisions in the material. Convection is the movement of head specifically through a fluid, via larger-scale movement of the material. First observed in steam engines in the nineteenth century, these processes are used in such applications as cooling computers and nuclear reactors, and in heating homes.

PRINCIPAL TERMS

- **heat transfer:** transfer of thermal energy from one region or system to another.
- **ideal gas law:** the scientific law relating pressure, volume, and temperature in a gas where the particles do not interact except to collide in perfectly elastic collisions.
- **radiation:** heat transfer via subatomic particle emission.
- **temperature:** a value proportional to the average speed of the particles inside a material.
- **thermal conductivity:** the ease with which heat propagates through a material; measured in watts per meter kelvin.
- **thermodynamics:** the study of heat flow and related forms of energy.

Convection, Conduction, and Heat

Heat is the collective random motion (kinetic energy) of atoms in a substance. In any material, atoms move randomly, vibrating, spinning, and even changing location. Temperature indicates the average speed of the particles, while heat is about the total energy. The amount of this kinetic energy is the heat of an object.

According to the principles of thermodynamics, heat can be transferred from one material to another by three methods. The first is radiation, which requires no physical contact. The second and third methods, convection and conduction, do require contact. In essence, the difference between convection and conduction is that in convection, the matter moves, but in conduction, only the heat does.

The ideal gas law provides a good idea of what goes on when one heats something, based on the idea of heat as exciting atoms. It states that the product of absolute pressure (p) and volume (V) equals the product of the number of molecules present (n), the universal gas constant R, and the absolute temperature (T), written as

$$pV = nRT.$$

Basically, the more molecules are wedged into an area and the faster they are moving, the more they will try to spread out and the more pressure they will put on their enclosure (since they are moving faster, they collide with more force). The next area to consider is how that energy gets around, known as the process of heat transfer.

Convection and Conduction

Convection only occurs in materials where the molecules are free to move. A hot fluid or gas is less dense than a cold one because the molecules in the hot one are more energetic and spread out. In a teacup, the cool tea sinks to the bottom and the warm tea rises, but if one thinks of a pot of water on a stove, then as the warm water rises and the cool water sinks, one finds that the warm water is cooled by the contact with the air and the cold water is warmed by the stove. Soon a cycle forms as the now-cooled water at the surface begins to sink and the now-hot water near the flame begins to rise. In this cycle, the water is circulated through the pot, rising from the bottom as hot water, coming to the surface, cooling, and sinking back down to be warmed again. This is a convection cycle.

Conduction is easier to visualize. Imagine a cup of tea that is very hot. The molecules of tannin and water and even the cellulose in the leaves at the bottom are vibrating and shaking and bouncing. They strike the walls of the teacup with a lot of energy, causing them to vibrate too. Soon the atoms and molecules in the walls of the teacup are nearly as energetic as the tea inside. When one goes to drink the tea, one's skin

Principles of Physics — Convection and Conduction

Water boiling in a pot of water on a stovetop demonstrates heat transfer via conduction and via convection. Convective heat transfer occurs with hot water transferring to the top of the pot, pushing cooler water down to be heated. Conductive heat transfer occurs from the stovetop to the pot it touches and from the pot through the connected handle to the hand.

comes in contact with the energetic molecules in the cup. These collide with the skin, causing the molecules there to gain energy and move faster. Soon the warmth spreads through one's fingers.

The heat has spread through the direct collisions of the energetic cup molecules with the more lethargic molecules of the skin. The amount of heat transferred in a given period of time is determined by the materials' thermal conductivity, which is measured in watts per meter kelvin (W/m·K).

Thermal conduction is computed as the amount of heat transferred (Q) over time (t). This is equal to the product of the thermal conductivity (k), the area (A), and the difference between the higher and lower temperatures (ΔT), over a given distance (d). Its formula is written as:

$$\frac{Q}{t} = \frac{kA(\Delta T)}{d}$$

APPLICATIONS

Convection is present in many important processes in earth science. It is the driving force behind plate tectonics, where hot magma rises from the molten and radioactive core of the planet and cools as it rises, eventually hardening into a thin crust at the surface. As more magma rises, the hardened crust rides the currents across the surface, before sinking back down to the core. A current rises from Earth's core, reaching the surface in an arc of volcanoes running from Iceland, down the middle of the Atlantic Ocean, and up into East Africa and into the Red Sea. That is why Iceland is so volcanically active. It is growing a little bigger every year. Conversely, hot magma spreading the East African Rift Valley will result in East Africa breaking away and the possible creation of a new ocean. Meanwhile, off the coasts of British Columbia, Alaska, Russia's Kamchatka Peninsula, and Japan, the cooled magma beneath the surface sinks back down into the core. That is why there are deep trenches off the coast.

Equally, convection circulates the water in the oceans. Warm water from the equator cools as it reaches the

---

SAMPLE PROBLEM

Consider the case of a window into a zoo's penguin enclosure. The pane is 2 meters by 5 meters, and 0.01 meters thick. It has a thermal conductivity of 1 watt per meter-kelvin (W/m·K). If the interior of the enclosure is −3 degrees Celsius and the exterior a warm 24 degrees Celsius, how much heat is transferred through the pane in a minute?

Answer:

To solve this problem, first find the area of the window using the length and the width:

$$2 \text{ m} \times 5 \text{ m} = 10 \text{ m}^2$$

Next, find the difference in temperature by converting from degrees Celsius to kelvins, then subtracting the lower temperature from the higher one:

$$K = °C + 273.15$$

$$24 °C + 273.15 = 297.15 \text{ K}$$

$$-3 °C + 273.15 = 270.15 \text{ K}$$

$$297.15 \text{ K} - 270.15 \text{ K} = 27 \text{ K}$$

Sample problem continues on next page

(SAMPLE PROBLEM CONTINUED)

(Note that it does not matter much whether one converts from Celsius to kelvin, because degrees Celsius and kelvins have the same increment, the only difference being the zero point.)

Plug these values, along with the given time, thickness (distance between the inside and outside), and thermal conductivity, into the thermal conduction equation

$$\frac{Q}{t} = \frac{kA(\Delta T)}{d}$$

$$Q = \frac{kA(\Delta T)t}{d}$$

$$Q = \frac{(1\,\frac{W}{m \cdot K})(10\text{ m}^2)(27\text{ K})(60\text{ s})}{0.01\text{ m}}$$

$$Q = \frac{16{,}200\text{ W}\cdot\text{m}\cdot\text{s}}{0.01\text{ m}}$$

$$Q = 1{,}620{,}000\text{ W}\cdot\text{s} = 1{,}620{,}000\text{ J}$$

Thus, we find a transfer of 1,620,000 watt seconds, or 1,620,000 joules.

---

poles until it sinks. This circulation is very complex, as currents of cold water run beneath the surface until they are warmed at the equator. In essence, this is the reason the Gulf Stream runs north from the Caribbean to the Arctic Ocean. (In practice, it is a little more complicated, as salinity also greatly affects the water's density.) This whole process is called the thermohaline circulation.

It is hard to find an engineering field where heat transfer is not an important principle. Most machines must be cooled one way or another. Cooling fans make use of convection, while a computer's heat sink uses a block of easily heated material to draw heat from the circuits. In fact, it was the convection in steam engines that gave rise to the early examination of thermodynamics, by scientists like Sadi Carnot (1796–1832). Carnot wrote during the early part of the nineteenth century and outlined much of what would later become thermodynamics by careful analysis of the steam engine and trying to explain questions like why a steam engine using superheated steam was better than one using hot water or whether a better fluid than steam could be used. In so doing, Carnot created a theoretical model of an engine that is still used today. Later scientists would expand these ideas with a greater understanding of atoms and chemistry, although quantum thermodynamics remains an area of active research.

Convection and conduction are vital in any engineering project, from building a more energy-efficient house to keeping a nuclear power plant from overheating. In the case of buildings, the search for better insulation materials, as well as better layouts for passive cooling via convection, is ongoing. In a nuclear reactor, engineers try to design a fault-tolerant means of dispersing a great deal of heat. The cooling towers used in many nuclear plants are designed to work with convection, being open at the bottom. This allows air to flow into them, where it is heated by a hot water pipe. This effectively sucks water into the tower, improving efficiency.

—*Gina Hagler, MBA*

BIBLIOGRAPHY

Bejan, Adrian. *Convection Heat Transfer*. Hoboken: Wiley, 2013. Digital file.

"Conduction, Convection and Radiation." *Conduction, Convection and Radiation*. IOP Inst. of Physics. Web. 10 June 2015.

Kaviany, M. *Heat Transfer Physics*. New York: Cambridge University Press, 2014. Print.

Nave, R. "Ideal Gas Law." HyperPhysics. Dept. of Physics and Astronomy, Georgia State U, 2012. Web. 10 July 2015.

Shabany, Younes. *Heat Transfer: Thermal Management of Electronics*. Boca Raton: CRC, 2010. Print.

"Thermodynamics." *Thermodynamics*. Ed. Nancy Hall. NASA, 5 May 2015. Web. 10 June 2015.

COSMIC RADIATION

FIELDS OF STUDY

Particle Physics; Nuclear Physics

SUMMARY

Cosmic radiation, or cosmic rays, is ultra-high-energy radiation that originates beyond Earth's atmosphere. This article describes its discovery, its composition, its origins, its effect on human life, and ongoing observational experiments.

PRINCIPAL TERMS

- **active galactic nucleus:** the region in the center of certain types of galaxies that emits massive amounts of energy across most, if not all, of the electromagnetic spectrum; believed to result from a supermassive black hole.
- **alpha particle:** a particle consisting of two protons and two neutrons, identical to a helium nucleus, which is emitted from an unstable nucleus via radioactive decay.
- **background radiation:** the total amount of ionizing radiation to which Earth is constantly exposed from both natural and artificial sources.
- **beta particle:** an electron that is emitted from an unstable nucleus via radioactive decay.
- **Cherenkov radiation:** electromagnetic radiation emitted when a charged particle, such as an electron, travels through a given medium faster than light would in that same medium.
- **Compton scattering:** the collision of a high-energy photon with a lower-energy electron, causing energy to be transferred from the photon to the electron and changing the angle of the photon's trajectory.
- **electromagnetic field:** a physical field consisting of a combined electric field (generated by stationary electric charges) and magnetic field (generated by moving electric charges) that affects the behavior of charged objects in its vicinity.
- **gamma ray:** a high-energy photon that is usually emitted by an unstable nucleus via radioactive decay; it may also be produced by various other means, including particle-antiparticle annihilation and the interaction of atmospheric particles with cosmic rays.

MYSTERIOUS "EXTRA" PARTICLES

Humans are exposed to constant background radiation from both natural and artificial sources. Common sources include radon gas, foods such as bananas and carrots (which produce certain isotopes of potassium), medical imaging technology, and consumer products such as cigarettes. Background radiation is ionizing, meaning it has the potential to damage DNA and cause subsequent health problems. However, the average annual worldwide effective dose is 3.01 millisieverts (mSv); one sievert is equal to one joule of energy per kilogram of body mass. This is well below the lowest annual dose linked to a measurable increase in the incidence of cancer among a population (100 mSv).

Radioactivity was discovered in the late nineteenth century by French physicist Henri Becquerel (1852–1908). Subsequent research by him and others, including physicists Marie Curie (1867–1934), Ernest Rutherford (1871–1937) and Paul Villard (1860–1934), revealed that certain elements emit different types of radiation. The researchers identified three distinct types: positively charged alpha particles, negative beta particles, and neutral gamma rays. All three are forms of ionizing radiation.

In 1912, Austrian physicist Victor Hess (1883–1964) identified another source of ionizing radiation: cosmic radiation, or cosmic rays. Hess observed that the amount of ionizing radiation in Earth's atmosphere increased with altitude. This implied that it originated somewhere beyond the atmosphere. Later experiments determined that cosmic rays vary with latitude as well because their paths are affected by Earth's electromagnetic field, meaning that they are electrically charged.

Over the next decade, further experiments found that cosmic rays consist of about 83–90 percent protons, 9–15 percent helium nuclei (alpha particles), and 1 percent heavier nuclei such as carbon, iron, and lead. These heavy nuclei are also called HZE particles due to their high atomic number (Z) and high energy (E). Proportions vary depending on the source of the rays. Cosmic radiation accounts for

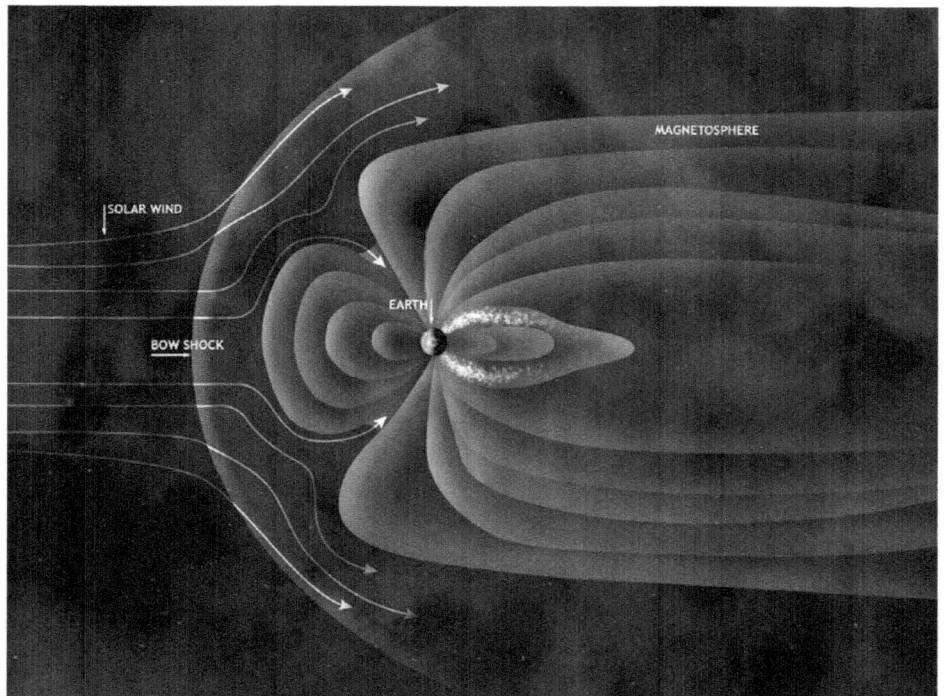

Earth's magnetic field shields it from solar winds in the same way that the solar system's heliosphere protects it from cosmic rays generated elsewhere. © NASA/CXC/M. Weiss

about 0.3–0.5 mSv of the average annual background radiation exposure. Commercial airline pilots and astronauts experience higher doses due to the amount of time they spend at high altitudes.

Stellar Origins

Cosmic rays can be divided into several categories. The main types are solar cosmic rays, or solar energetic particles (SEPs), galactic cosmic rays (GCRs), and anomalous cosmic rays (ACRs). SEPs come from the sun and are believed to result from solar flares, with a smaller amount produced by coronal mass ejections. GCRs originate outside the solar system but generally within the Milky Way galaxy. They are most likely produced in supernovas. ACRs are believed to be atoms of neutral interstellar gas that were ionized by solar ultraviolet radiation.

Cosmic rays are extremely high-energy particles. Their energies generally range from 10 megaelectronvolts (MeV; 10^7 eV), the lower range of SEPs and ACRs, to 1 petaelectronvolt (PeV; 10^{15} eV), the upper range of GCRs. Another, rarer type of cosmic radiation is ultra-high-energy cosmic rays (UHECRs), which have energies greater than 10^{15} eV. Their origin is unknown, but many astronomers believe that they come from outside the Milky Way. They may be the product of active galactic nuclei at the centers of certain types of galaxies, such as radio galaxies and Seyfert galaxies. The most energetic UHECR ever observed had an energy of 3.2×10^5 PeV (3.2×10^{20} eV)—about the same kinetic energy as a baseball traveling 100 kilometers per hour (62 mph), concentrated in a single atomic nucleus.

SEPs, GCRs, ACRs, and UHECRs are all considered primary cosmic rays because they come from outside Earth's atmosphere. Primary cosmic rays are deflected by the magnetic fields of other bodies in space, so they appear to arrive at Earth from all directions. When they enter Earth's upper atmosphere, they collide with atoms and molecules, producing a shower of secondary cosmic rays. These include lower-energy particles such as protons, electrons, neutrons, muons, pions, neutrinos, and alpha particles, as well as electromagnetic radiation such as x-rays and gamma rays. Because muons are the most resistant to energy loss and the effects of Earth's electromagnetic field, they make up more than half of the cosmic rays that reach sea level. Muons can penetrate Earth's surface down to the level of deep underground mines. Neutrinos also easily reach Earth from the upper atmosphere, but most pass through it without detection.

The collision of high-energy cosmic rays with much slower atmospheric particles can result in Compton scattering. This is when a photon collides with a charged particle, transferring some of its energy to that particle. If the charged particle gains enough energy to travel faster than the speed of light in atmosphere, it creates a shock wave that produces a flash of blue light. This light is called Cherenkov radiation. It can be used to determine the source and intensity of primary cosmic rays.

Other cosmic-ray detection methods include extensive air shower (EAS) arrays, cloud chambers, and nitrogen fluorescence. An EAS array is a collection of small ground-based devices that detect the passage of charged particles. A cloud chamber is a chamber filled with supersaturated air. When a charged particle passes through the chamber, the water vapor condenses around it. Nitrogen fluorescence occurs when charged particles excite atmospheric nitrogen, causing it to emit radiation.

HIGH ENERGY, HIGH STAKES

Although most cosmic rays are rendered harmless by Earth's atmosphere, some have enough energy to disrupt electronic circuits at ground level. A 1996 study by IBM estimated that cosmic rays cause one error per 256 megabytes of RAM per month. The problem is greatly intensified for electronics beyond Earth's atmosphere, such as satellites and spacecraft. Cosmic rays also pose a danger to humans who spend long periods in space. Thus, cosmic-ray shielding will be an important consideration for future space travel.

Cosmic rays have beneficial aspects as well. They served as the original source of high-energy particles for early physics experiments, before the widespread construction of particle accelerators. They are also responsible for the production of unstable isotopes in Earth's atmosphere, including carbon-14, which is used for carbon dating in archaeology. Ground-based and satellite research projects, such as the IceCube Neutrino Observatory and the Fermi Gamma-Ray Space Telescope, rely on cosmic rays to produce neutrinos and to identify sources of high-energy gamma rays, respectively.

—*Lindsay Brownell, MS*

BIBLIOGRAPHY

Biermann, Peter L., et al. "Active Galactic Nuclei: Sources for Ultra High Energy Cosmic Rays!" Institute for Nuclear Theory. U of Washington, 20 Feb. 2008. Web. 10 Aug. 2015.

"Cosmic Rays." *Cosmicopia*. NASA, 11 May 2012. Web. 10 Aug. 2015.

"Fermi Overview." *Fermi Gamma-Ray Space Telescope*. NASA, 5 Mar. 2013. Web. 10 Aug. 2015.

"In Search of Cosmic Rays." *ASPIRE*. U of Utah, 1997–98. Web. 10 Aug. 2015.

Mewaldt, Richard A. "Cosmic Rays." *Macmillan Encyclopedia of Physics*. Ed. John S. Rigden. Vol. 1. New York: Simon, 1996. Space Radiation Lab, Caltech. Web. 10 Aug. 2015.

Stern, David P., and Mauricio Peredo. "Cosmic Rays." *Exploration of the Earth's Magnetosphere*. Phy6.org, 24 Jan. 2005. Web. 10 Aug. 2015.

CREST AND TROUGH

FIELDS OF STUDY

Classical Mechanics; Fluid Mechanics; Electromagnetism

SUMMARY

In a transverse wave (e.g. an ocean wave), the crest is the highest point of the wave and the trough is the lowest point. Measurements based on crests and troughs reveal a wave's wavelength and amplitude. When two waves interact, the relative positions of crests and troughs determines whether they will boost one another or cancel each other out.

PRINCIPAL TERMS

- **amplitude:** the maximum amount of displacement at a point on the wave from its position of rest.
- **antiphase:** waves with a phase difference of 180 degrees relative to one another. The crests of one wave align with the troughs of the other and vice versa.
- **constructive interference:** when waves meet in the same medium and are in similar phase.
- **destructive interference:** when waves meet in the same medium and are in dissimilar phase.
- **maximum:** the point of greatest wave height, also known as a crest or peak. It is the point of greatest positive displacement from the position of rest in a wave.

- **minimum:** the point of greatest wave depth, also known as a trough or valley. It is the point of greatest negative displacement from the position of rest in a wave.
- **transverse:** type of wave that displaces its medium perpendicular to the direction of energy transfer. Ocean waves, for example, move water vertically but transmit energy horizontally across the surface.
- **wavelength:** the full length of a complete wave cycle, measured as the distance between crests of a wave.

CRESTS, TROUGHS, AND WAVE PROPERTIES

In a transverse wave—which displaces its medium perpendicular to the transfer of energy—the crests and troughs are the areas of maximum and minimum height from the resting position (equilibrium), respectively. The distance between two crests or two troughs is the wavelength of a wave. The distance between the top of a crest or bottom of a trough relative to the equilibrium position—such as the flat surface of the sea in an ocean wave—determines the wave's amplitude. A concept related to wavelength is frequency, or the number of waves per unit of time.

These properties reveal important information about the energy transmission ability of a wave. A long wavelength indicates a lower capacity for energy transfer over time. A large amplitude indicates a high energy content in each cycle of the wave. High-energy waves have short wavelengths and high frequencies. Low-energy waves have long wavelengths and low frequencies.

INTERFERENCE AND PHASE

The orientation of crests and troughs is determined by a wave's phase. Two waves of the same frequency are said to be in antiphase relative to one another if the crests of one wave line up with the troughs of the other. When two waves traveling through the same medium encounter one another, their relative phase determines the nature of their interaction. If they are in sync, with crests corresponding to crests, the waves will reinforce one another. This is constructive interference. If they are in antiphase, they will dampen each other. This is destructive interference. Careful observation of crests and troughs can reveal crucial information about transverse waves and their potential interactions.

—*Kenrick Vezina, MS*

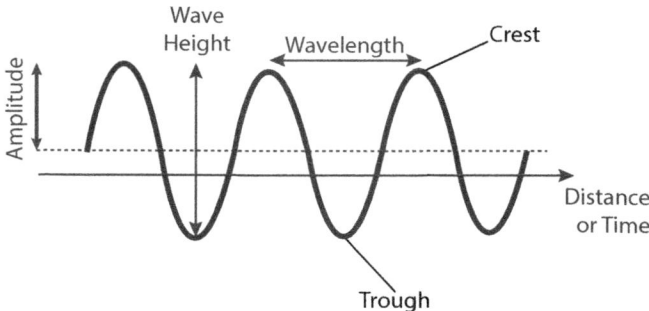

A wave has a maximum *y* value, called a crest, and a minimum *y* value, called a trough. The difference between these two values is called the wave height. A wave's amplitude is half the wave height. A wave's wavelength is measured from one crest to a consecutive crest or from one trough to a consecutive trough.

BIBLIOGRAPHY

"Anatomy of an Electromagnetic Wave." Mission: Science. NASA, n.d. Web. 19 June 2015.

Fleisch, Daniel, and Laura Kinnaman. *A Student's Guide to Waves*. New York: Cambridge University Press, 2015. Print.

Freegarde, Tim. *Introduction to the Physics of Waves*. Cambridge: Cambridge University Press, 2013. Print.

Newton, Roger G. *Waves and Particles: Two Essays on Fundamental Physics*. Hackensack: World Scientific, 2014. Print.

Pain, H. J., and Patricia Rankin. *Introduction to Vibrations and Waves*. Hoboken: Wiley, 2015. Print.

Russell, Daniel A. "Longitudinal and Transverse Wave Motion." *Acoustics and Vibration Animations*. Pennsylvania State U, 18 Feb. 2015. Web. 19 June 2015.

D

DENSITY AND SPECIFIC GRAVITY

FIELDS OF STUDY

Classical Mechanics

SUMMARY

Every object has mass, which measures the amount of matter contained in the object, and volume, which measures the three-dimensional space it takes up. The ratio of these values gives an object's density. Comparing the density values of different objects yields specific gravity, a value that provides information about whether objects will float and the concentrations of various solutions.

PRINCIPAL TERMS

- **gram:** abbreviated g, a standard unit of mass in the International System of Units; derived from the kilogram (kg, thousands of grams), which is the base unit of mass.
- **mass:** the amount of matter contained in an object; it influences gravitation, motion, and—along with volume—density and specific gravity.
- **slug:** an alternate unit of mass, defined as a mass that accelerates one foot per second squared (1 ft/s^2) when acted upon by one pound-force (lb$_F$).
- **standard temperature and pressure (STP):** a temperature of 273.15 kelvins (0 degrees Celsius or 32 degrees Fahrenheit) and pressure of 101.3 kilopascals (1 atmosphere); used in chemistry and physics to establish a standardized set of conditions for experimentation.
- **volume:** the space occupied by an object or substance, measured in cubic meters (m^3); together with mass, it determines the density of an object and its specific gravity. Alternative units include liters (L) or cubic feet (ft^3).
- **weight:** an object's heaviness; the force imparted to an object by gravity acting on its mass, often measured in kilograms or pounds, although as a force, it should be given in newtons.

WHAT IS SPECIFIC GRAVITY?

Density is a measure of how "tightly packed" the matter in a substance or object is. It is the ratio of an object's mass relative to its volume. Mass quantifies how much matter is present in an object. Volume measures its three-dimensional space. Density (d) is calculated as the mass (m) divided by volume (V).

Mass can be measured in grams (g) or kilograms (kg), but slugs are an alternative unit of mass sometimes used in the United States and imperial measurement systems.

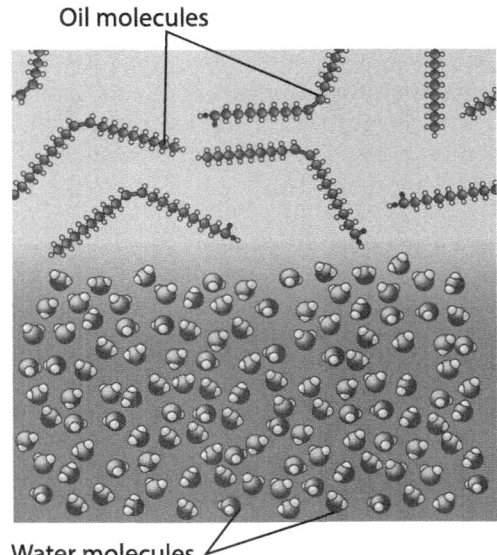

Specific gravity = (density of sample)/(density of water)
Specific gravity oil = (.91)/(1.0) = .91 g/cm^3
Specific gravity water = (1.0)/(1.0) = 1.0 g/cm^3

Molecules of oil do not pack as nicely as molecules of water, making olive oil less dense than water. Specific gravity of a substance is calculated by dividing the sample's density by the density of water (1.0 g/cm^3). The specific gravity of olive oil is 0.91 g/cm^3. Since it is less than water, it floats on top of the water.

One slug is 14,593 grams, or about 31 pounds. Volume is measured in cubic centimeters (cm^3) when dealing with density, although liters or cubic inches are sometimes used.

Specific gravity describes the relationship between the density of a substance relative to that of some other substance—very often water. It is the ratio of two density values. It can be described as determining whether one substance will float in another. Specific gravity (SG) is computed as the density of the target object or substance (d_{object}) divided by the density (d_{ref}) of the reference object or substance. A substance will float if its density is less than that of the substance it is in; in other words, if its specific gravity is less than one. For example, ice in water has a specific gravity of about 0.9—its density is slightly less than that of liquid water, so it floats.

The density of some object or substance can also be found using its specific gravity and the density of its reference substance. In this case, the density of the object would be equal to the object's specific gravity times the density of the referent.

It is possible to determine whether an object will float or sink if its mass and volume are known. Objects float when the buoyant force of the substance they are immersed in is enough to overcome their weight, that is, the force of gravity acting on their mass. The denser a substance is, the greater its buoyant force.

STANDARD TEMPERATURE AND PRESSURE

By default, the specific gravity of solids and liquids is measured relative to the density of water, and gases relative to the density of air, with both reference substances at standard temperature and pressure (STP). STP is a temperature of 273.5 kelvins (0 degrees Celsius or 32 degrees Fahrenheit; the freezing point of water) and an atmospheric pressure of 101.3 kilopascals (roughly one atmosphere, the atmospheric pressure at sea level). STP is a way of standardizing the values against which densities and other values are compared. Solids, liquids, and gases all have different densities at different temperatures and pressures, with gases having the greatest variability. This is because gases, unlike solids and liquids, can change their volume. They will expand to fill whatever container they are in.

SAMPLE PROBLEM

Consider a solid ball. It has a mass of 60 grams. Its volume, measured using displacement, is 120 cubic centimeters. What is its specific gravity relative to freshwater (density 1 g/cm^3), and will it float in it? Will it float in vegetable oil (density 0.92 g/cm^3)? Will it float in a bowl of a mystery liquid with a mass of about 102 kilograms and a volume of 7,540 cubic centimeters?

Answer:

Begin by calculating the density of the ball in grams per cubic centimeter, using the formula for density:

$$d = m / V$$

$$d = 60 \text{ g} / 120 \text{ cm}^3$$

$$d = 0.5 \text{ g/cm}^3$$

To get the specific gravity of the ball relative to freshwater, divide its density by the density of water:

$$SG_{ball-water} = d_{ball} / d_{water}$$

$$SG_{ball-water} = 0.5 \text{ g/cm}^3 \div 1 \text{ g/cm}^3$$

$$SG_{ball-water} = 0.5$$

To determine if the ball will float in water, ask whether its specific gravity is greater than or lower than a value of one—0.5 is less than 1.0, so it floats. For vegetable oil, repeat the process:

$$SG_{ball-oil} = d_{ball} / d_{oil}$$

$$SG_{ball-oil} = 0.5 \text{ g/cm}^3 \div 0.92 \text{ g/cm}^3$$

$$SG_{ball-oil} \approx 0.54$$

(Sample problem continues on next page)

PRACTICAL APPLICATIONS OF SPECIFIC GRAVITY

Beyond the classroom application of determining whether a given object or substance will float in water,

> (SAMPLE PROBLEM CONTINUED)
>
> The specific gravity relative to the oil is still less than 1, so it still floats (this makes intuitive sense, given that the density of water and vegetable oil is so similar: 1.0 g/cm³ versus 0.92 g/cm³).
>
> For the mystery liquid, begin by calculating its density (remember to convert kilograms to grams before plugging in the given values):
>
> $$d_? = m_? / V_?$$
>
> $$d_? = 102{,}000 \text{ g} / 7{,}540 \text{ cm}^3$$
>
> $$d_? = 13.53 \text{ g/cm}^3$$
>
> Then use the same specific gravity equation as above:
>
> $$SG_{\text{ball-?}} = d_{\text{ball}} / d_?$$
>
> $$SG_{\text{ball-?}} = 0.5 \text{ g/cm}^3 \div 1.353 \text{ g/cm}^3$$
>
> $$SG_{\text{ball-?}} \approx 0.369$$
>
> This very small value for specific gravity indicates that the ball would float nearly entirely on the surface of the liquid, which is extremely dense. The mystery liquid is modeled after mercury, the only metal that is a liquid at room temperature.

air, or another medium, specific gravity sees use in industry as a shorthand for the values of various solutions. This is possible due to the close relationship between density and specific gravity. With a variety of water-based solutions, specific gravity becomes an easy way to track the concentration of various substances in solution.

Consider a soft-drink factory, which uses a variety of syrups and other water-based solutions when mixing together drinks. In these situations, tables list various values of specific gravity for each substance relative to water and the corresponding concentration levels of that substance.

—*Kenrick Vezina, MS*

BIBLIOGRAPHY

"Density." *Encyclopaedia Britannica*. Encyclopaedia Britannica, 27 Jan. 2014. Web. 10 July 2015.

"How Do I Calculate Density?" *Math You Need, When You Need It*. Science Education Resource Center, Carleton Coll., 16 June 2015. Web. 26 June 2015.

"How Things Float." SeaPerch. Assn. for Unmanned Vehicle Systems Intl., 2013. Web. 10 July 2015.

Nave, R. "Mass and Weight." *HyperPhysics*. Dept. of Physics and Astronomy, Georgia State U, 2012. Web. 10 July 2015.

Pounder, Elton R. *Physics of Ice*. London: Pergamon, 1965. Print.

"Q & A: What Is Specific Gravity?" Q & A: Physics Questions? Ask the Van. Dept. of Physics, University of Illinois at Urbana-Champaign, 22 Oct. 2007. Web. 26 June 2015.

DIODE

FIELDS OF STUDY

Electronics

SUMMARY

The term diode indicates a structure having two electrodes. In operation a diode permits the flow of electricity in one direction only. Analog diode tubes have been replaced by transistor diode structures that function in either analog or digital devices.

PRINCIPAL TERMS

- **bias:** the extent to which a measurement differs from the real value of the property being measured, or an intrinsic factor in a method that consistently causes such deviation.
- **capacitor:** an electrical part consisting of two conductors separated from each other by a nonconductor, allowing it to store electrical charge temporarily.
- **circuit:** a closed path along which electricity travels.

- **continuity:** a clear path for electricity from point A to point B.
- **mechanical switches:** devices that use moving parts and direct physical force to bring contacts together to let electricity flow.
- **nonmechanical switches:** devices that use electromagnetism to open and close a circuit.
- **power rating:** the maximum electrical power a device can use without being damaged.
- **voltage:** the difference in electric potential between two points, measured in volts; electric current flows naturally from the higher-voltage point to the lower-voltage point.

ELECTRIC CIRCUITS

When a voltage, a difference in electrical potential, exists between two points in a conducting material, conventional current (that is, the flow of imaginary positive charges) is established from the point of highest potential to the point of lowest potential. If there is no work being done by this current, then this structure is known as a short circuit. In even simple circuits, however, the electrical current performs various functions to produce a desired output. Various devices are used to control and manipulate the flow of electrons. The diode is a device that permits electrical current to flow in one direction only within an electrical circuit. Two kinds of diodes have been in common use. Before the development of semiconductor-based transistor technology, a vacuum tube device containing an anode and a cathode functioned as a diode. Transistor diodes are defined as a single *pn* junction structure between two types of semiconductor materials. The *n*-type semiconductor material functions as the cathode, and the *p*-type semiconductor functions as the anode. Both the vacuum tube and transistor diodes require the application of a bias voltage in order to function and allow the flow of electricity thought the circuit.

DIODE APPLICATIONS

In an electrical circuit, current flow is normally turned off and on by the use of mechanical switches such as the common light switch and ON-OFF button, or by non-mechanical switches such as contacts operated by solenoids. The diode can also be used as a kind of electronic switch to break the continuity of a circuit or to control current flow through some portion of a circuit. The bias voltage to a diode can be applied

---

SAMPLE PROBLEM

Calculate the value of the limiting resistor required to produce a current of 0.005 amperes when the bias voltage is 9 volts.

Answer:

Using the relationship

$$I_f = V_{bias} / R_{limit}$$

rearrange to obtain the expression for R_{limit} as

$$R_{limit} = V_{bias} / I_f.$$

Use the given values of bias voltage and current to obtain

$$R_{limit} = (9V) / (0.005A)$$

$$= 1800 \, \Omega \, (\text{ohms}).$$

---

in either direction, termed forward bias and reverse bias. The diode is forward biased when the positive side of the voltage source is connected to the anode, and the negative side of the voltage source is connected to the cathode of the diode. This allows current to flow through the diode in the direction from the anode to the cathode. In the reverse bias configuration, the connections are reversed so that the positive voltage source is connected to the cathode and the negative voltage source to the anode. This configuration does not allow current to flow through the diode. Changing the direction of the voltage applied to the diode during the operation of the circuit allows the diode to be used as an effective stop switch, until the voltage is reversed again. The capacitor is another device that prevents the flow of electricity in a circuit, and is often used to condition the current flow through diodes and other devices by screening out the development of transient voltages and their associated currents. However, a capacitor is not constructed from semiconductor materials. Current flowing through a diode generates heat due to the friction of electrons moving through the material. The amount of heat that the material can tolerate before being destroyed is limited, and diode devices are given a power rating to indicate their allowable operating limits.

CURRENT CONTROL

Diodes in circuits are typically associated with a limiting resistor that determines the amount of current that can pass through a forward biased diode. The forward current is determined by the ratio of the bias voltage to the limiting resistance, as

$$I_f = V_{bias} / R_{limit}.$$

DIODES AND TRANSISTORS

Combinations of diodes are routinely used to condition, control and manipulate the waveforms of electrical current in various devices. The basic diode structure is the foundation structure of all transistors. The addition of a second *n*-type semiconductor segment on the other side of the *p*-type segment produces the transistor, which is the basis of all computer and digital electronic devices in the present day.

—*Richard M. Renneboog M.Sc.*

BIBLIOGRAPHY

Kubat, Milan. *Power Semiconductors.* Reprint of 1984 1st ed. Berlin: Springer-Verlag, 2013. Print.

Lutz, Josef, Heinrich Schlangenotto, Uwe Scheuermann, and Doncker De Rik. *Semiconductor Power Devices Physics, Characteristics, Reliability.* Berlin: Springer, 2014. Print.

Pulfrey, David L. *Understanding Modern Transistors and Diodes.* Cambridge; Madrid: Cambridge University Press, 2010. Print.

Voldman, Steven H. *ESD: Circuits and Devices.* 2nd ed. Chichester, United Kingdom; Hoboken, NJ: John Wiley & Sons, 2015. Print.

DISTANCE AND AREA

FIELDS OF STUDY

Classical mechanics

SUMMARY

Distance and area are some of the most useful measurements and calculations to make in everyday life—whether calculating the square footage of a room or measuring a desk. Starting with one-dimensional measurements of distances, one can calculate the area of various two-dimensional shapes and the total surface area of three-dimensional objects. Though the formulas used to make these calculations were devised centuries ago, they are still in wide use.

PRINCIPAL TERMS

- **accuracy:** how closely a measurement of a property matches the real value.
- **derived unit:** unit of measure that is described in terms of two or more base units; for example, meters are a base unit of distance, whereas square meters are derived units of area calculated from meters of height multiplied by meters of depth.
- **displacement:** the absolute distance between two points.
- **International System of Units (SI):** also known as the metric system; a widely used standardized system of units for measuring natural phenomena.
- **measurement:** quantifying an observation (e.g., the length of a person's foot) using discrete units (e.g. meters); alternately, the unit or system used to do so.
- **meter:** the SI base unit of distance (or length) measurement.
- **precision:** how well a measurement agrees with other measurements of the same phenomenon.
- **scale:** a description of the area under observation in broad approximation.

MEASURING SPACES

The distance between two points could be the distance between two cities, the distance between two ends of a ruler, or the distance between the bottom of one's foot and the top of one's head. In these cases, "distance" is equivalent to length, width, and height. It is always a quantitative measurement of the space between two points. In the International System of Units (SI), the meter (m) is the standard unit in which distances are measured. Americans commonly use feet (ft) instead of meters for making measurements. One meter is equal to approximately 3.28 feet.

The foot was widely used in many countries before the advent of the metric system. But it varied in its value by location. This created a problem. In their local areas where everyone was using the same standard, measurements may have resulted in high precision. However, their accuracy would be questionable—especially when translating measurements between locations that had different ideas of what a foot meant. The standardization of the SI has helped to ensure that scientists and engineers are able to make measurements of maximum accuracy and precision without worrying that the units they are using are a source of uncertainty. At larger scales of observation, units of kilometers (km) or miles (mi) may be used instead of meters or feet. At smaller scales, centimeters (cm) or inches (in) may be used.

Sometimes, it is worth making distinctions between the actual distance traveled and the distance between two points. Consider the winding road a car may take to get from home to the store. In this case, displacement describes the absolute distance between the car's starting point and its end point.

Area is calculated using measurements of distance. This results in a derived unit, such as the square meter (m^2). When dealing with three-dimensional shapes with multiple sides, the total area of each side is the surface area. This is also measured in square meters and calculable using well-known formulas.

CALCULATING SURFACE AREA

It is possible to estimate the surface area of everyday objects by choosing a similar three-dimensional geometric shape and applying the mathematical formulas for the surface area of that shape. For instance, a pencil can be considered a cylinder.

The process of determining the surface area of a shape begins with taking measurements of its length, width, and height. After taking these measurements, the values are plugged into the formula for the appropriate shape to get an approximation of the object's surface area.

For more complicated shapes, two options exist: if a shape can be broken down into component shapes, the surface area can be estimated. Consider a spire. It is, roughly speaking, equivalent to a cylinder with a cone on top. The surface area could be estimated by calculating the surface area of the cylindrical and conical portions and then subtracting the value of the connecting circular surface from each.

ESTIMATING THE SURFACE AREA OF A TIRE

Given a tire of a 0.75-meter radius (r) and a 0.2-meter tread width (w), derive the circumference in meters. This will allow for an estimate of the surface area in square meters.

To figure out the circumference, one must know the formula for the circumference of a circle. Circumference (C) equals two multiplied by pi (π) multiplied by the radius:

$$C = 2\pi r$$

Dimensions : 1 dimension : 2 dimensions		
Shape	Perimeter (distance)	Area
Circle (with radius r and diameter d)	$2 \cdot \pi \cdot r$	$\pi \cdot r^2$
Triangle (with sides a, b, c and height h)	$a + b + c$	$\frac{1}{2} \cdot h \cdot b$
Rectangle (with sides a and b)	$2(a + b)$	$a \cdot b$

Measurements of distance and area are different for each shape. Distances are one-dimensional while areas are two-dimensional and their units are squared. Calculating the perimeter, or distance around a shape, is different for a circle, triangle, and rectangle. Calculating the area covered by the shape is also different for circles, triangles, and rectangles.

$$C = 2\pi\,(0.75 \text{ m})$$

$$C \approx 4.712 \text{ m.}$$

To estimate surface area, one must first decide on a three-dimensional geometric shape that best approximates a tire. In this case, a short cylinder works best (imagine the tire is laid on one of its flat sides). The formula for the surface area (A) of a cylinder is the circumference (C, or $2\pi r$) of one of the circular ends, multiplied by the height (h) of the cylinder, plus two multiplied by the area of each of the circular ends ($2\pi r^2$):

$$A = 2\pi rh + 2\pi r^2$$

In this case, the value of h is the same as the width of the tire. Thus, h is equal to 0.2 meters, giving:

$$A = 2\pi\,(0.75 \text{ m})(0.2 \text{ m}) + 2\pi\,(0.75 \text{ m})^2$$

$$A = 2\pi\,(1.5 \text{ m}^2) + 2\pi\,(0.5625 \text{ m}^2)$$

$$A \approx 4.48 \text{ m}^2.$$

Area in the Everyday

The ability to estimate area and surface area is useful in a variety of real-world scenarios. For example, in real estate, understanding area helps one make sense of square footage, which tells how much living space there is. If one is selling, buying, or renting a house or apartment, one can calculate the square footage. If one wants to paint a room, estimating the surface area of the walls will allow one to determine how much paint to purchase.

—*Kenrick Vezina, MS*

Bibliography

Hegg, Robin. "The Surface Area Effect." *IEEE Spark*. IEEE, Apr. 2014. Web. 25 Mar. 2015.

Klein, Herbert Arthur. *Science of Measurement: A Historical Survey.* 1974. New York: Dover, 2012. Digital file.

Somervill, Barbara A. *Distance, Area, and Volume.* Chicago: Heinemann, 2011. Digital file.

Thompson, A., and B. N. Taylor. "The NIST Guide for the Use of the International System of Units." National Institute of Standards and Technology. US Dept. of Commerce, 5 Oct. 2010. Web. 25 Mar. 2015.

Wagner, Mark. *The Geometries of Visual Space.* Mahwah: Erlbaum, 2006. Print.

Weinstein, Eric W. "Surface Area." *MathWorld*. Wolfram Research, 19 Mar. 2015. Web. 25 Mar. 2015.

DISTORTION

FIELDS OF STUDY

Acoustics

SUMMARY

Whenever an audio or visual signal is transmitted, as in an amplifier or camera lens, the process can alter the signal so that it is not reproduced accurately. This results in distortion. Many types of distortion exist, and it is sometimes used for creative effect.

PRINCIPAL TERMS

- **amplifier distortion:** an imperfect reproduction of the original signal when transmitting sound electronically.
- **barrel distortion:** the optical distortion caused by a wide-angle lens that makes an image seem inflated at the center, as though stretched across a barrel.
- **interference:** the distortion of a wanted signal by an unwanted one, such as an unwanted radio station "bleeding into" another because both are broadcasting on the same wavelength.
- **pincushion distortion:** the optical distortion caused by a telephoto lens that makes an image seem bunched up in the center.

- **quantization distortion:** a distortion common in digital signal processing, such as encoding music as audio files, that results from mapping an original signal of possibly infinite values to a signal with a smaller, countable set of values.
- **signal-to-noise-and-distortion (SINAD) ratio:** the ratio of total signal power received to noise and distortion received, indicating how well a signal or waveform has been reproduced.
- **transverse wave:** a wave that displaces its medium perpendicular to the direction of energy transfer.
- **warping:** the digital manipulation of an image that mimics distortion and can be used to correct for distortion or to introduce various creative effects.

What Is Distortion?

No method of transmitting or reproducing a signal is perfect. Distortion occurs when a signal is altered by the process of transmission or reproduction. There are many types of distortion that affect a variety of signals, from the visual signals one's eyes receive through a telescope to the quality of audio signals reproduced from digital storage.

In general, the term distortion refers to the electronic transmission and reproduction of audio signals, often music. However, it also applies in a variety of other situations. All forms of distortion can be evaluated using a signal-to-noise-and-distortion (SINAD) ratio, which compares the total amount of information received—the signal, noise, and distortions combined—to just the noise and distortions. This ratio measures how well the signal was reproduced. The SINAD ratio is similar to the more familiar signal-to-noise ratio, but it accounts for distortion as well.

Optical and Image Distortions

Optical distortions occur as a result of the way lenses gather and bend light before transmitting it to the eye. One familiar type of distortion is the fish-eye effect, which occurs when looking outward through a heavily convex lens, such as a peephole in a door. This apparent inflation of the center of an image is a type of barrel distortion, so called because the image looks as though it has been wrapped around a barrel. In barrel distortion, light waves closer to the center of the lens are magnified more than those at the edges. The inverse effect, called pincushion distortion, is caused by an extreme telephoto lens. It results in an image that appears to be compressed toward the center. A pincushion distortion occurs because light waves farther from the center are magnified more than those closer in.

Some types of digital image distortion mimic these effects. The process of purposely distorting an image through digital means is called warping. It may be used to correct for unwanted distortion or to manipulate images for creative effect.

Interference and Audio Distortion

Sound may undergo many types of distortion, depending on how it is transmitted or reproduced. Sound is transmitted as a wave through air, water, or another medium. These sound waves may bounce off surfaces, causing echoes, or they may encounter interference from other sound waves of similar frequencies.

When music is being reproduced and transmitted electronically, an amplifier is often used to increase the voltage before the signal is sent to a speaker to

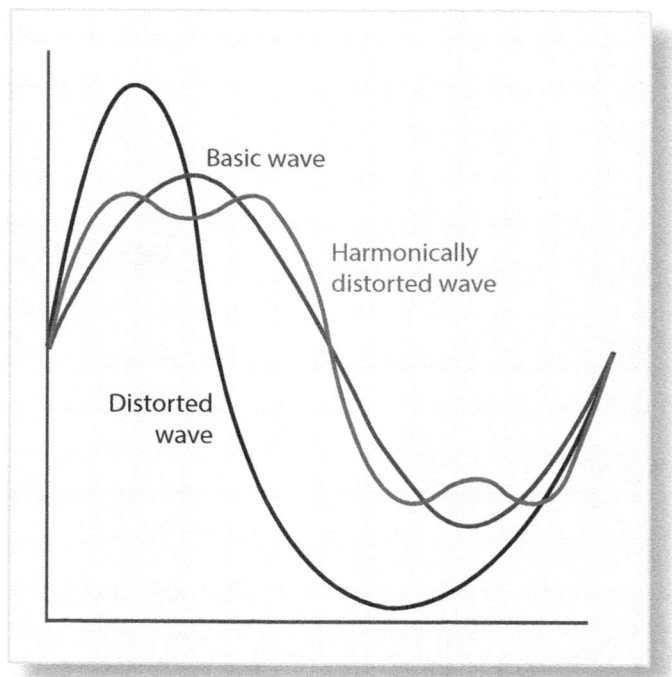

Distortion occurs when a wave is bombarded with another wave of a different frequency; the result is a distorted wave. One example is when a wave is distorted by its harmonic; the result is a harmonic distortion.

produce sound. The signal must be scaled up linearly, and the voltage raised proportionally, to maintain the initial waveform. Nonlinearity results in amplifier distortions, such as harmonic distortion (the clipping of wave peaks) and intermodulation distortion (nonharmonic mixing of different frequencies). The greater the amplification, the more noticeable amplifier distortion may become.

As with image warping, distortion may be used purposely in music. The distortion of electric guitars that originally resulted from faulty amplifiers has become integral to many types of music, including rock and blues. Special equipment is used to create this distortion.

Radio waves and other forms of electromagnetic radiation are transverse waves. These types of waves oscillate up and down as they move. Radio waves are used to transmit information through the air, including audiovisual signals for analog television broadcasts. If they run into other radio waves, the amplitudes of the colliding waves may combine to form a higher-amplitude wave, or they may cancel each other out. This phenomenon, known as interference, results in what is commonly called snow or static on analog television sets. Interference can also result when unshielded electrical wires or other sources of electromagnetic fields are too close to wires transmitting an electrical signal.

Distortion via Encoding

When sound is stored digitally, as in the common MP3 file format, it is typically compressed. A sound recording produces a large amount of data, which translates to a very large data file. An algorithm can then be used to create a much smaller MP3 file, which contains significantly less data but is still capable of recreating a very close approximation of the original sound.

All audio-encoding systems are subject to quantization distortion. This results from the mapping of the original signal of possibly infinite (continuous) values to a smaller, countable set of values. The process frequently involves rounding and truncation, leading to an imperfect reproduction. A similar problem exists for storing photographic data.

Other Forms of Distortion

Various distortion subtypes exist in every engineering or physics subfield that deals with signal transmission, storage, or reproduction. The common thread in these cases is that no method of transmission or reproduction is perfect. All such methods will introduce some form of distortion that must be compensated for if the integrity of the signal is to be preserved.

—*Kenrick Vezina, MS*

Bibliography

Barbour, Eric. "The Cool Sound of Tubes." IEEE Spectrum. IEEE, Aug. 1998. Web. 10 Sept. 2015.

Crowell, Benjamin. "Images, Quantitatively." *UC Davis PhysWiki*. University of California, Davis, 24 June 2014. Web. 10 Sept. 2015.

Henderson, Tom. "Refraction and the Ray Model of Light." *The Physics Classroom*. Physics Classroom, 1996–2015. Web. 10 Sept. 2015.

Lesurf, Jim. "Analog Techniques and Audio." *The Scots Guide to Electronics*. University of St. Andrews, n.d. Web. 15 Sept. 2015.

Nave, Carl R. "Amplifiers." *HyperPhysics*. Georgia State U, n.d. Web. 14 Sept. 2015.

van Walree, Paul. "Distortion." Toothwalker.org. 2001–15. Web. 10 Sept. 2015.

DOPPLER EFFECT

FIELDS OF STUDY
Acoustics; Harmonics; Relativity

SUMMARY

This article discusses the Doppler effect with respect to both mechanical and electromagnetic waves. The relative movement of a wave source alters the waves' length and apparent frequency from the point of view of an observer. The wavelength shortens when the wave source and receiver move toward each other and longer when they move away from each other, although the actual frequency and velocity of the waves does not change.

PRINCIPAL TERMS

- **beat frequency:** the apparent frequency at which waves from two or more sources create constructive interference.
- **blueshift:** the apparent shortening of the wavelength of electromagnetic radiation due to the movement of the radiation source toward the observer.
- **frequency:** the number of complete cycles of a phenomenon, such as a wave, that occur per unit of time.
- **general relativity:** the theory stating that gravity is a geometric property of space and time.
- **pitch:** the perceived relative frequency of audible sound waves.
- **redshift:** the apparent lengthening of the wavelength of electromagnetic radiation due to the movement of the radiation source away from the observer.
- **vector:** a geometric concept describing both the magnitude of a property and the direction in which it operates.
- **wavelength:** the distance from a point in one wave to the same point in the next wave.

THE NATURE OF WAVES

Many physical phenomena act through the propagation of waves. Wave motion is a type of harmonic motion in which energy moves independently of matter. Waves take many different forms, from ocean waves to electromagnetic radiation that is invisible to the naked eye. However, all waves share certain defining features. The distance between any point in one wave and the identical point in the next wave is called the wavelength. Wavelength is usually defined as the distance from crest to crest of each wave. The frequency of a wave is the number of times the wave occurs in a unit of time. The magnitude and direction of these concepts are expressed using vectors.

The Doppler effect, or Doppler shift, explains how the observed frequency of a wave changes depending on the observer's motion relative to the source of the wave. Austrian physicist Christian Doppler (1803–53) first proposed the effect in 1842. Knowing that wave frequency relates to motion has allowed scientists to understand the nature of easily observable wave phenomena, such as ocean or sound waves, and to make complex measurements in astronomy and other fields.

Most people have observed the Doppler effect in action in relation to sound. Sound travels in waves through a medium (such as the air) from its source (such as an ambulance siren) to a receiver (such as a listener's ear). The pitch, or perceived frequency of a sound, depends on the position and movement of the listener relative to the sound source. For example, to a stationary observer hearing an ambulance approach, the pitch of the siren sound will grow higher as the ambulance comes near. Once the ambulance passes, the pitch of the siren sound will start to drop again. Yet the sound of the siren itself never changes. To the ambulance driver, the pitch remains constant.

The Doppler effect holds whether the object in motion is the source, the observer, or the medium itself. It also applies to waves that do not need a medium, such as light waves. In all cases, the Doppler effect means that from the point of view of an observer, the frequency of a wave increases as the wave source approaches and decreases as the source recedes.

The Doppler effect happens because frequency and wavelength are inversely related. As frequency (the number of waves per unit time) increases, the wavelength decreases, and vice versa. This is why the wave frequency appears to change, even though the actual frequency remains constant. When the

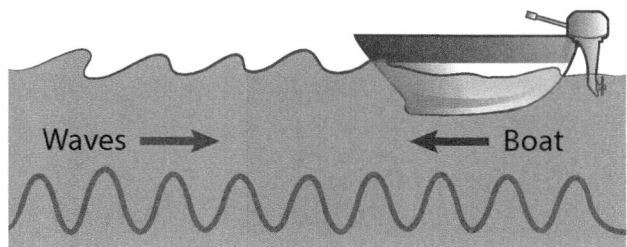

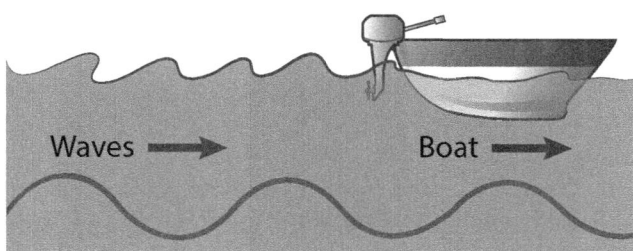

The phenomenon called the doppler effect occurs with light waves, sound waves, and physical waves. As an object approaches oncoming waves, the waves seem to speed up (increase frequency or shorten wavelength). As an object moves away from oncoming waves, the waves seem to slow down (decrease frequency or lengthen wavelength). An observer in a boat approaching waves would experience the boat hitting a new wave more often the faster it goes toward the waves. An observer in a boat heading away from waves would experience the boat hitting a new wave less often the faster it goes.

source approaches an observer, each new wave originates closer to the observer than did the previous wave, causing the distance between two successive waves—the wavelength—to shorten. Because the wavelength decreases, the waves reach the observer with greater frequency. The opposite happens when the wave source moves away from the observer. Each wave originates farther from the observer than did the previous wave, creating more space between two successive waves. This increases the wavelength and decreases the frequency.

ELECTROMAGNETIC OBSERVATIONS

While the Doppler effect is easily observed in sound waves even at relatively low speeds, it is harder to observe with faster-moving waves, such as electromagnetic radiation. Advanced sensing methods have made measuring these waves possible, leading to many useful applications. Most notably, the Doppler effect has allowed astronomers to make measurements that help determine the nature of the universe.

Spectroscopy, or the study of how radiation and matter interact, is used to analyze the frequencies of radiation emitted by objects. Applying this technique to the light of distant stars revealed that the spectral emissions of those stars are shifted to different wavelengths. In the visible light spectrum, red light has the longest waves, and blue light has the shortest. According to the Doppler effect, those stars whose wavelengths are shifted toward the longer, red end of the spectrum must be moving away from the point of observation—in this case, Earth. This phenomenon is called redshift. The wavelengths of other stars are shifted toward the shorter, blue end of the spectrum. This phenomenon, called blueshift, means that those stars are moving toward the Earth. These shifts allow astronomers to measure the distances between celestial objects and determine the relative speeds at which they are moving closer to or farther from Earth. American astronomer Edwin Hubble (1889–1953) used this information to formulate Hubble's law, which defines the relationship between redshift and velocity of movement away from Earth. Hubble's law helped establish the groundwork for the big bang theory and the theory of the expanding universe. Such phenomena also support the predictions of the theory of general relativity, such as the gravitational lensing that results from the bending of light by massive gravitational bodies.

The electromagnetic Doppler effect has been used in other applications as well, including several forms of Doppler radar. These technologies use microwave signals to examine the relative velocity of a target object. A microwave beam is emitted, and its reflection is picked up with a sensor. Comparing the frequencies of the original and the reflected signal provides a beat frequency, or pattern of constructive and destructive interference between the two waves, which can be used to calculate the target's velocity relative to the radar device. The most well-known examples of Doppler radar are the radar guns used by police and the weather-tracking systems used by meteorologists to gauge the motion of precipitation patterns.

CALCULATING DOPPLER SHIFTS

The shift of observed frequency in the Doppler effect is proportional to the relative difference in the velocity of the wave source and the observer. The general equation for both moving sources and moving observers can be given as

$$f' = f\frac{v \pm v_o}{v \mp v_s}$$

where f' is the observed frequency, f is the true frequency, v is the velocity of the waves, v_o is the velocity of the observer, and v_s is the velocity of the source. The top sign is used when the observer is moving toward the source, or vice versa. The bottom sign is used when they are moving apart. For a stationary observer of sound waves from a moving source, the apparent frequency can be calculated according to the simplified equation

$$f' = f\frac{v}{v \mp v_s}$$

This equation is derived from the general equation by making v_o equal 0. Similarly, the equation for a moving observer and a stationary source is derived by making v_s equal 0.

REAL-WORLD CALCULATIONS

The general equation for the Doppler effect is not an exact calculation in all cases. It applies only when the source or the observer (or both) is moving in a direct line toward the other. Different calculations are required when they are moving at angles to one another. Also, when the waves in question are sound waves, the calculation depends on the speed of the wave through the medium, while the calculation for light waves does not. Since the speed of light in space is considered to be a constant according to general relativity theory, calculating Doppler shifts of light frequencies is more complex.

—*Richard M. Renneboog, MSc*

SAMPLE PROBLEM

A train whistle with a frequency of 500 hertz (Hz) is approaching a stationary person at a speed of 50 kilometers per hour (km/h). Calculate the apparent frequency of the whistle to the stationary observer if the speed of sound through the air is 342 meters per second (m/s).

Answer:

Because the units of speed must be the same, it is first necessary to convert the speed at which the whistle is moving to the same units as the speed of sound:

50 km/h = 50000 m/h

1 h = 3600 s

50000 / 3600 = 13.9 m/s

The apparent frequency is calculated using

$$f' = f\frac{v}{v - v_s}$$

where f = 500 Hz, v = 342 m/s, and vs = 13.9 m/s:

$$f' = 500\frac{342}{342 - 13.9}$$

$$f' = 521.18 \text{ Hz}$$

The apparent frequency of the whistle is 521.18 hertz.

BIBLIOGRAPHY

Albin, Edward F. *Earth Science Made Simple*. New York: Broadway, 2004. Print.

Fleisher, Paul. *Doppler Radar, Satellites, and Computer Models: The Science of Weather Forecasting*. Minneapolis: Lerner, 2011. Print.

Hiebl, Ewald, and Maurizio Musso, eds. "Christian Doppler: Life and Work, Principle and Applications." Proceedings of the Commemorative Symposia in 2003, Salzburg, Prague, Vienna, Venice. Pöllauberg: Living Ed., 2007. Print.

Kirshner, Robert P. "The Extravagant Universe: Exploding Stars, Dark Energy, and the Accelerating Cosmos." Princeton: Princeton University Press, 2004. Print.

Petkov, Vesselin. *Relativity and the Nature of Spacetime*. 2nd ed. New York: Springer, 2011. Print.

Rees, W. G. *Physical Principles of Remote Sensing*. 3rd ed. New York: Cambridge University Press, 2013. Print.

Schmidt, Werner, and Asim Kurjak. Color *Doppler Sonography in Gynecology and Obstetrics*. Stuttgart: Thiem, 2005. Print.

DYNAMIC SYSTEMS THEORY

FIELDS OF STUDY

Thermodynamics; Classical Mechanics; Harmonics

SUMMARY

Dynamic systems theory examines the properties of systems that change over time, as opposed to static systems that do not undergo time-related change. Systems that change incrementally are treated with difference equations. Systems that undergo continuous change are treated with differential calculus.

PRINCIPAL TERMS

- **dynamic systems:** systems that are subject to change.
- **entropy:** the degree of randomness of the components within a system.
- **mixing:** being able to transform a dynamic system with multiple phase spaces in its initial state in more than one way over time to reach a target state that has completely overlapping phase spaces.
- **periodicity:** the extent to which a property repeats over time; regular recurrence.
- **phase space:** a space in which all of a dynamic system's possible states are represented.
- **thermodynamics:** the study of the transfer of heat energy into other forms of energy, and vice versa.

Dynamic Systems and Static Systems

Any physical system can be described either as a static system or a dynamic system. The components of a dynamic system change, while those of a static system do not. A golf ball on a tee is a static system because it undergoes no change. When the ball is struck by a golf club, it becomes part of a dynamic system that includes the golf club, the golfer, and Earth through the force of gravity and the atmosphere. These components act to change the state of the system over time. Mechanical systems are dynamic systems while they are performing functions. When they are idle, they are static. Chemical systems are dynamic systems at all times because atoms and molecules in the system are in a constant state of motion. Motion is defined as a change in location over time.

Atomic and molecular motion reflect changes in the heat energy of the system, as described by thermodynamics. A central feature of thermodynamics and of systems in general is the entropy of the system. Entropy is the extent to which the system is in a state of disorder or randomness. A crystalline solid, for example, has low entropy compared to the same material in the liquid state. In the solid form, atoms and molecules are locked into a rigid, ordered array, though they still move due to vibrations and rotations of bonds. As the temperature of the solid rises, the atoms and molecules move more energetically. At the material's melting point, these motions are energetic enough to overcome the forces that maintain the orderly array of molecules. The atoms and molecules can then move more freely instead of remaining locked in place. Because the orderliness of the system has decreased, its entropy has increased.

Entropy can be seen in a simple cup of coffee when sugar is added to it. As the coffee's water molecules interact with the sugar's sucrose molecules, the temperature of the sucrose increases to that of the water, and the sugar dissolves. Thus, the sugar becomes an integrated component of the overall system.

Mixing and Phase

The cup of coffee is a dynamic system that undergoes mixing as the coffee combines with milk and sugar. In its initial state, the system has multiple phase spaces. A phase space is a space that shows all of the possible states of a system. For the cup of coffee system, the sugar, coffee, and milk each have a separate phase space. The cup of coffee system can be disturbed in several ways. It can be stirred. It can be swirled. It can be shaken. Yet no matter which way the cup of coffee is disturbed, over time, the contents combine to reach the same target state, a drinkable cup of coffee. In this target state, the phase spaces of the coffee, sugar, and milk completely overlap one another.

Systematic Change

Dynamic systems can undergo change in different ways. The mathematics used in dynamic systems theory differs accordingly. Systems that change incrementally are described using difference equations that use specific value differences in their solution.

To explain how a system changes over time, dynamic systems theory is applied. The billiard table is a dynamic system in which the table border is the physical boundary and the billiard balls are units of matter that interact with each other over time.

Systems that change continuously require the use of differential equations that use derivative and integral calculus in their solution. Some systems that undergo continuous change do so in a cyclic manner. They may change their value between two limiting values or over a regular period of time. This periodicity can be described using differential calculus based on sine and cosine functions.

The broad study of particle dynamics examines those systematic changes that affect motion and energy, and is described using both Newtonian and relativistic mechanics. Newtonian mechanics describes the actions of the planets and other bodies that make up the solar system. Although Newtonian mechanics can also be used to describe the motion and energy of atoms and molecules, these phenomena are more precisely described by quantum mechanics and other disciplines. On the fundamental subatomic level, quantum field theories alone are adequate to describe the motion and energy of the relevant particles.

Working with Dynamic Systems

The primary feature of any dynamic system is the way in which the motion and energy of its components change over time. Complex machines, while carrying out their defined functions, are dynamic systems having strict boundaries. A robotic welding machine, for example, cannot carry out functions beyond those defined within the computer algorithm that controls the robot's movements. There are many applications, however, that do not have such boundaries. Chemical engineering uses processes that depend on mixing for their effectiveness and the facilitation of their control. An explosion, such as occurs in an internal combustion engine, is a dynamic system. The flow of water through a turbine to generate electricity is also a dynamic system. Understanding such systems enables both their control and their improvement, as well as the development of new systems and applications.

Biological systems are perhaps the ultimate examples of dynamic systems. In the human body, for example, a multitude of chemical reactions involving many different molecules take place every second. The natural world itself is an adaptive dynamic system that has undergone a continual process of dynamic change over billions of years. One example of adaptive dynamic change is the coevolution of different species due to some beneficial relationship. Bees and flowers are an example of coevolution in which both species have evolved features that benefit the other species. Bees have evolved in ways that better utilize the nectar of flowers to maintain the survival of bees, while flowers have evolved nectar production and structural features that better enable visiting bees to pollinate the flowers that they visit to obtain nectar. The mathematics of biological dynamic systems are accordingly complex, but they are also essential for understanding the ecologies of those systems.

—*Richard M. Renneboog, MSc*

BIBLIOGRAPHY

Devaney, Robert L. *An Introduction to Chaotic Dynamical Systems*. 2nd ed. Boulder: Westview, 2005. Print.

Gros, Claudius. *Complex and Adaptive Dynamical System: A Primer*. 4th ed. Cham: Springer, 2015. Print.

Meyers, Robert A. *Mathematics of Complexity and Dynamical Systems*. New York: Springer, 2012. Print.

Perko, Lawrence. *Differential Equations and Dynamical Systems*. 3rd ed. New York: Springer, 2001. Print.

Smith, Hal L. *Monotone Dynamical Systems: An Introduction to the Theory of Competitive and Cooperative Systems*. 1995. Providence: Amer. Mathematical Soc., 2008. Print.

Thirring, Walter. *Classical Mathematical Physics: Dynamical Systems and Field Theories*. 2nd rev. ed. Berlin; New York: Springer, 2010. Print.

Wainwright, J., and Ellis, G. F. R. *Dynamical Systems in Cosmology*. Cambridge: Cambridge University Press, 2005. Print.

Won, Chang-Hee, Cheryl B. Schrader, and Anthony N. Michel, eds. *Advances in Statistical Control, Algebraic Systems Theory, and Dynamic Systems Characteristics*. Boston: Birkhäuser, 2008. Print.

EDISON EFFECT

FIELDS OF STUDY
Electronics

SUMMARY
The Edison effect refers to the emission of electrons from the surface of a material that has been made sufficiently hot. Thermionic emissions are the valence electrons that have acquired sufficient energy to escape from the valence shell of a metal atom.

PRINCIPAL TERMS

- **atomic model:** a theoretical representation of the structure and behavior of an atom based on the nature and behavior of its component particles.
- **closed system:** a physical or chemical reaction system defined by certain boundary conditions that prevent any components, reactants, or products from entering or exiting the system.
- **conductivity:** the ability of a material to transfer heat (thermal conductivity) or electricity (electrical conductivity) from one point to another.
- **electrical charge:** a property of subatomic particles that causes them to exert a force on each other, either attractive (if their charges are of opposite signs) or repulsive (if they are of the same sign); by convention, a proton is assigned a charge of 1+ and an electron is assigned a charge of 1−.
- **electron:** a fundamental subatomic particle with a single negative electrical charge, found in a large, diffuse cloud around the nucleus.
- **electron configuration:** the order and arrangement of electrons within the orbitals of an atom or molecule.
- **electron shell:** a region surrounding the nucleus of an atom that contains one or more orbitals capable of holding a specific maximum number of electrons.
- **valence electron:** an electron that occupies the outermost or valence shell of an atom and participates in chemical processes such as bond formation and ionization.

ATOMIC STRUCTURE
According to the atomic model matter is believed to be composed of atoms. Each atom is ascribed an internal structure in which a very small, dense nucleus is surrounded by a much larger, but very nebulous, cloud of electrons. Each electron is restricted to a defined region of space about the nucleus according to the specific energy that is required for it to occupy that particular region. Each atom therefore has a specific electron configuration in which a specific number of electrons occupy each electron shell. The valence electrons are those normally found in the outermost, or valence, electron shell. Electrons were first identified as subatomic particles by Thomson in 1897. Thomson arbitrarily designated the electrical charge carried by electrons to be the negative charge, and that carried by the nuclei of atoms as the positive charge. The separation of positive and negative electrical charge as components of different particles is essential to the Edison effect, or thermionic emission as it is now known.

WHERE THE CHARGES GO
In the last quarter of the nineteenth century, Frederick Guthrie first reported that a hot metal ball having a negative electrical charge would slowly lose that charge, but the same hot metal ball retained a positive charge. A similar effect was observed by Edison in 1880 inside the closed system of his experimental light bulb. Using a carbonized bamboo filament in the bulb resulted in the blackening of the inside of the glass bulb. Eventually it was determined that the blackening was related to the transmission of electricity through the empty space within the bulb. As electrical current flow was later understood to be the movement of electrons through a material, the

> **SAMPLE PROBLEM**
>
> The melting points (in °C) of aluminum, tungsten, silver, copper, nickel and iron are 659 °C, 3400 °C, 960 °C, 1080 °C, 1450 °C and 1530 °C, respectively. If Edison tried to use these as a source for thermionic emissions, which ones would be successful?
>
> **Answer:**
>
> Since thermionic emission only becomes significant for metals in vacuum at temperatures above 730 °C, only aluminum would melt before emitting any ions thermally. The other metals would all function as thermal ion sources.

conductivity of the vacuum was determined to be due to the movement of electrons through empty space.

The emission of electrons from the surface of a hot material came to be known as the Edison effect. Electrons that are emitted in this way are valence electrons from the atoms of the hot material. Their emission leaves the remaining ion with a positive charge, but because it is part of the bulk matter of the material, the positively charged atoms remain as an integral part of the hot matter, in agreement with Guthrie's observations.

In principle, heating atoms increases the energy that they contain, and at a sufficiently high temperature the valence electrons may acquire enough energy to escape from the valence shell and leave the respective atom. The emission of electrons and other ions by increasing the thermal energy of atoms, as in the Edison effect, is termed "thermionic emission." This emission only becomes significant when the temperature of a metal in a vacuum is greater than 730 °C (1,340 °F). This is the temperature at which metal filaments in incandescent light bulbs and vacuum tubes, which is why devices constructed with such tubes require a warm-up period before they will function properly.

Crystal Confusion

Thermionic emission is most common from metal surfaces but is also known from other compounds. A recent fad is the use of heated blocks of salt or other mineral to release ions that are touted as healthful. Such thermal emission of ions from crystals is achieved when the combination of thermal energy and electrical potential is sufficient to overcome the lattice energy of the particular crystal, enabling some ions to escape from the surface.

—*Richard M. Renneboog M.Sc.*

Bibliography

Anders, André. *Cathodic Arcs From Fractal Spots to Energetic Condensation.* New York, NY: Springer US, 2009. Print. Springer Ser. on Atomic, Optical, and Plasma Physics.

Balkan, N., ed. *Hot Electrons in Semiconductors : Physics and Devices.* Oxford; New York: Clarendon ; Oxford University Press, 1998. Print.

Clark, Ronald William. *Edison: The Man Who Made the Future.* London: Bloomsbury Reader, 2011. Print.

Humy, Fernand Emile D'. *The Birth of the Vacuum Tube, "the Edison Effect."* New York: Newcomen Society of England, American Branch, 1949. Print.

Mansfield, Michael, and Colm O'Sullivan. *Understanding Physics.* 2nd ed. Hoboken: Wiley, 2012. Print.

Zangwill, Andrew. *Physics at Surfaces.* Cambridge [Cambridgeshire]; New York: Cambridge UP, 1988. Print.

EIGENVECTORS

FIELDS OF STUDY

Quantum Physics

SUMMARY

Eigenvectors are mathematical descriptors of a property that has both magnitude and direction. Certain mathematical operations acting on an eigenfunction yield the original function multiplied by a constant. This constant is the eigenvalue of the operation. Vectors describe position and direction using the Cartesian coordinate system. Bra-ket notation is used to simplify complex vector and wave-function equations.

PRINCIPAL TERMS

- **bra-ket notation:** a system of mathematical notation developed to more easily manipulate very large vector equations.
- **Cartesian coordinate system:** a system that uses a pair or trio of numbers to indicate the location of a point in two-dimensional or three-dimensional space relative to a set point of origin.
- **eigensystem:** the set of all eigenvectors of a matrix paired with their respective eigenvalues.
- **eigenvalue:** the mathematical constant that, when multiplied by an eigenfunction, yields the solution to the operation performed on the function.
- **matrix:** a mathematical notation in which a series of coordinates is written in an array and can then be manipulated according to set rules.
- **vector:** the mathematical statement of a property that has both magnitude and direction.

EIGEN

The German word *eigen* means "own," in the sense of belonging. When used as a prefix in mathematics and physics, it signifies that the property it represents is unique and complete within itself. In the mathematics of quantum mechanics, an eigenfunction is a mathematical function consisting of operators acting on a vector property. The solution of an eigenfunction is equal to the original function multiplied by a constant, which is the eigenvalue of the function. For example, the derivative of the function $f(x) = ce\lambda x$, where differentiation (finding the derivative) is represented by the notation $f'(x)$, is $\lambda ce\lambda x$, which can also be written as $\lambda f(x)$. Thus, the constant λ (which could also be a function) is an eigenvalue of the eigenfunction $f(x) = ce\lambda x$, and $f'(x) = \lambda f(x)$.

The quantum state of a physical system, including the energy and position of the electrons in its atoms, is best described using the Schrödinger equation. This equation is a type of wave equation, similar to the equations that describe sound waves and electromagnetic waves, because elementary particles such as electrons behave like waves on a quantum scale. Solving the Schrödinger equation for a particular system yields a wave function, represented by the symbol Ψ, which is a much more complex function than the one above. Mathematical manipulation of Ψ readily produces stable energy-state results that are eigenvalues of the eigenfunction Ψ. This can be used to describe and plot atomic orbitals.

QUANTUM MATH

The energy state and position of an electron in an atom can be described mathematically in one of two ways. One approach is based on the mathematics of classical mechanics, using vectors and matrices. The other, based on wave mechanics, relies on differential calculus. Both express the same properties and quantities and must therefore produce the same results for specific conditions. Vector calculations use the Cartesian coordinate system (x, y, and z axes), developed by French mathematician René Descartes (1596–1650), for position. They describe changes in position and energy as matrix transformations. Another coordinate system, known as spherical coordinates, is often used in wave mechanics. This system, based on the two-dimensional polar coordinate system, uses radial distance and two angles (r, θ, and φ) for position. It describes change of position and energy with differential equations and integrals using operators. An operator denotes the function or transformation that is to be carried out on the wave function. In both forms of calculation, the complex conjugates ($a - bi$ and $a + bi$) of wave functions and vector expressions are included to normalize the equations of state for the system.

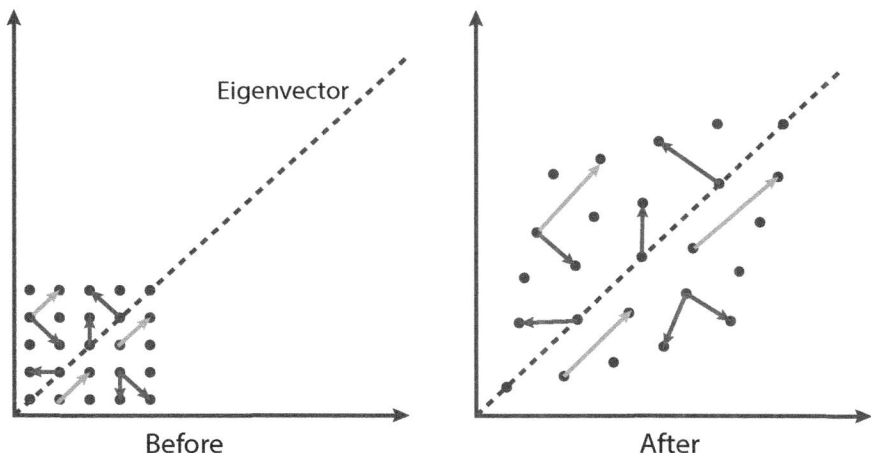

Arrows on a grid show how an image changes when transformed by an eigenvector. The dashed line extending from each graph's origin at a slope of 1 is the eigenvector along which the grid of arrows is stretched. After the arrows are transformed, those that run parallel to the eigenvector (blue arrows) maintain their angle but are stretched, those that run perpendicular to the eigenvector (purple arrows) maintain their angle and length, and those that run at any other angle to the eigenvector (red arrows) change length and angle after transformation.

The calculations are simplified using bra-ket notation, devised by British physicist Paul Dirac (1902–84). In bra-ket notation, each *bra* represents one or more operations or matrix manipulations to be performed on the state vector or wave function represented by the *ket*. The product of the two, known as the dot product or the inner product, is represented as follows:

<bra | ket>.

The bra is most often the complex conjugate of the ket. Thus, the product of the wave function Ψ and its complex conjugate Ψ^* is written as

<Ψ | Ψ>

because <Ψ | is equal to | Ψ>*, which is the complex conjugate of | Ψ>.

Bras and kets can be mixed and matched as needed throughout the calculation sequence. The bra-ket notation does not require specific coordinates, which makes it convenient for either method of calculation used in quantum mechanics.

Eigenvectors

Every vector property has magnitude and an associated direction. Think of the forces acting on a swing as an example. Gravity acts vertically, pulling the swing down toward the ground. A person pushing on the swing applies a horizontal force to it, resulting in its displacement from the neutral vertical position. Because the motion of a swing is that of a pendulum, it can be described using wave mechanics or vector mathematics. The position of the swing and its direction of motion follow a progression of vectors relative to the swing's neutral position. These two properties can be described in terms of the distance of the swing from its point of suspension (r) and the angle of the displacement of r from the neutral position (θ). The cyclic variation of the swing's motion is reflected in the cyclic variation of the vectors and angular coordinates that it follows.

For any such treatment, specific conditions exist for which the resultant vector is in the same direction as the original vector, though its magnitude may differ. For a given set of vectors and operators represented by the matrix A, there are one or more vectors x that, when multiplied by A, result in a vector equal to x multiplied by λ:

$$Ax = \lambda x.$$

The value λ is therefore an eigenvalue of A, and the vector x is an eigenvector. All possible eigenvalues of a matrix, when paired with their associated eigenvectors, form an eigensystem.

—*Richard M. Renneboog, MSc*

Bibliography

Anton, Howard. *Elementary Linear Algebra*. 11th ed. Hoboken: Wiley, 2013. Print.

Beddard, Godfrey. *Applying Maths in the Chemical & Biomolecular Sciences: An Example-Based Approach*. New York: Oxford University Press, 2009. Print.

Bowman, Gary E. *Essential Quantum Mechanics*. New York: Oxford University Press, 2008. Print.

Dahl, Jens P. *Introduction to the Quantum World of Atoms and Molecules.* Hackensack: World Scientific, 2001. Print.

Dick, Rainer. *Advanced Quantum Mechanics: Materials and Photons.* New York: Springer, 2012. Print.

Finn, John Michael. *Classical Mechanics.* Sudbury: Jones, 2010. Print.

Pereyra, Pedro. *Fundamentals of Quantum Physics.* New York: Springer, 2012. Print.

ELASTIC AND INELASTIC COLLISIONS

FIELDS OF STUDY

Classical Mechanics

SUMMARY

Whether a car hits a truck, a train hits a stalled car, or one billiard ball hits another, collisions are inevitable in the world. Sometimes collisions conserve energy, and sometimes they dissipate energy. If energy is conserved, the collision is said to be an elastic collision; if energy is lost, the collision is either inelastic or completely inelastic. Regardless of the type of collision, one thing is certain: momentum is always conserved.

PRINCIPAL TERMS

- **collision:** an interaction in which two or more bodies come into contact and briefly exert force on each other.
- **conservation of momentum:** in physics, the principle that the total momentum in a closed system is always constant.
- **dissipation:** the irreversible loss of energy from a system.
- **kinetic energy:** the energy a body contains as a result of its motion.

Crashing Objects

Every time somebody walks through a room, millions of collisions are happening every second. Not only is the person's body colliding with the surrounding air molecules, but his or her feet are colliding with the floor with every step. In each of the countless collisions that take place every day, numerous and varied forces act on the colliding objects as they make contact with each other. Some of these forces are nonconservative forces, resulting in the dissipation of energy, especially kinetic energy.

One such nonconservative force is friction. Although friction causes some loss of energy in every collision, it would be impossible to move without it. Every time a person takes a step, he or she is pushing back against the floor, creating friction. Isaac Newton's (1642–1727) third law of motion states that for every action, there is a reaction of equal magnitude and opposite direction. So as the person's foot exerts a friction force backward on the floor, the floor exerts an answering friction force on the foot that propels the person forward. Without friction, people would not be able to walk.

Nonconservative forces are almost inescapable on Earth. In most collisions between large bodies, energy is lost through these forces. When a person claps his or her hands and then keeps them together, all of the kinetic energy from the movement is lost. This energy does not disappear from the universe; it is transformed into other kinds of energy. Some of the kinetic energy becomes thermal energy, which is why clapping repeatedly for more than a few seconds causes a person's hands to feel warm. Some of it becomes sound energy, which is why clapping makes a sharp sound.

Due to the abundance of nonconservative forces, most collisions between large bodies are inelastic collisions. Inelasticity is the property that allows objects to be deformed by a collision, while elasticity is the property that allows objects to return to their original shape. In an elastic collision, the kinetic energy of the system—in this case, the colliding objects—is conserved. A good example of an elastic collision is one in which two objects of equal mass collide and interchange velocities. In an inelastic collision, the kinetic energy is not conserved, and part of it is lost to the surrounding environment. This is the most common type of collision. A good example of an inelastic collision is a car crash.

A collision in which the maximum possible kinetic energy is lost is said to be perfectly or completely inelastic. In a completely inelastic collision, the colliding objects stick together. An arrow hitting a target and a tennis ball sticking to a Velcro surface are good examples of completely inelastic collisions.

Regardless of whether a collision is elastic or inelastic, momentum is always conserved. Linear momentum (p) is defined as the product of an object's mass (m) and its velocity (v). Momentum and velocity are represented by bolded variables to indicate that they are vector quantities; that is, they have both a magnitude and a direction. (The counterpart to a vector quantity is a scalar quantity, such as mass, which has magnitude but no innate direction.) Due to this, one must consider the momentum of an object by the direction of its motion.

Conservation of momentum states that the initial overall momentum of a system (p_i) is equal to the final overall momentum of the system (p_f). These values represent the sum of all initial and final momenta, respectively, of each individual object in the system. To find the initial momentum, add up the individual initial momenta of each of the objects. The same procedure is used to find the final momentum.

Inelastic and Completely Inelastic Collisions

In an inelastic or completely inelastic collision, some kinetic energy is lost. Both types of collision are treated the same way, and momentum is conserved in both cases. The only difference between the two is that in a completely inelastic collision, the colliding objects stick together.

Imagine a car with a mass of 2,000 kilograms (kg) traveling east at 25 meters per second (m/s). The car hits a truck stopped at a red light. The truck has a mass of 4,000 kg. After the collision, the car and truck are stuck together and moving at the same velocity. The car and the truck together constitute a system, so their shared final velocity can be calculated by first determining the total momentum of the system (dealing with magnitude only and ignoring direction):

$$p = mv.$$

Initially the truck is not moving, meaning that its velocity is 0 m/s and it has no momentum. After the collision, because the car and the truck stick together, their velocities are the same. Thus, the total momentum of the system before the collision is calculated as follows, using measurements for the car:

$$p_i = m_c v_{c,i}$$

$$p_i = (2,000 \text{ kg})(25 \text{ m/s})$$

$$p_i = 50,000 \text{ kg} \cdot \text{m/s}.$$

Because momentum is conserved, p_f must also equal 50,000 kg·m/s. This information can be used to calculate the car and truck's final velocity. Both are moving at the same velocity, so their velocities can be represented by the same variable (v_f):

$$p_f = m_c v_f + m_t v_f$$

$$p_f = (m_c + m_t) v_f$$

$$50,000 \text{ kg} \cdot \text{m/s} = (2,000 \text{ kg} + 4,000 \text{ kg})(v_f)$$

$$8.33 \text{ m/s} = v_f.$$

The final velocity of both vehicles is approximately 8.33 m/s (rounded).

Because this is a completely inelastic collision, energy is lost, although momentum is not. To calculate the change in energy, find both the initial and the final kinetic energy of the system. Kinetic energy is a

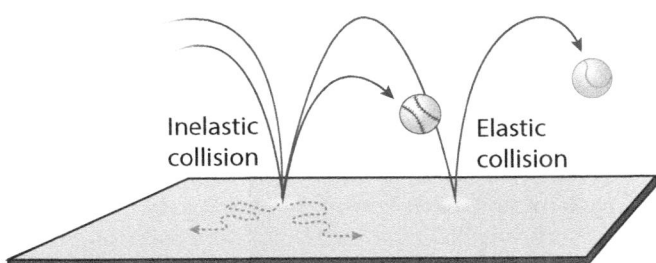

A tennis ball retains its energy and its maximum height throughout elastic collisions with the ground. A baseball loses some of its energy when colliding with the ground, making it an inelastic collision. Because of this energy loss, the maximum height the baseball can reach on its next bounce is reduced with each collision.

function of an object's mass (m) and the square of its velocity (v):

$$K = \frac{1}{2}mv^2$$

The initial and final kinetic energies of the system are calculated as follows:

$$K_i = \frac{1}{2}m_c v_{c,i}^2$$

$$K_i = \frac{1}{2}(2,000 \text{ kg})(25 \text{ m/s})^2$$

$$K_i = 625,000 \text{ J}$$

$$K_f = \frac{1}{2}(m_c + m_t)v_f^2$$

$$K_f = \frac{1}{2}(2,000 \text{ kg} + 4,000 \text{ kg})(8.33 \text{ m/s})^2$$

$$K_f = 208,166.7 \text{ J}$$

The initial kinetic energy is 625,000 joules (J). (A joule is equal to 1 kg·m²/s².) The final kinetic energy is 208,166.7 J. The energy lost in the collision is equal to the change in kinetic energy:

$$\Delta K = K_f - K_i$$

$$\Delta K = 208,166.7 \text{ J} - 625,000 \text{ J}$$

$$\Delta K = -416,833.3 \text{ J}.$$

A total of 416,833.3 J of energy is lost during this completely inelastic collision.

While the car and the truck are colliding, they are exerting forces on each other: the car hits the truck, and the truck hits back. These forces are equal in magnitude but opposite in direction. As a result, both the car and the truck experience the same change in momentum. Because momentum is a function of mass and velocity, and the truck has a greater mass than the car, this means that the change in the truck's velocity will be smaller than the change in the velocity of the car.

Now imagine that instead of the two vehicles sticking together after the collision, the truck moves forward at a velocity of 11 m/s and the car moves forward at 3 m/s. Although the collision is not completely inelastic, some energy is still lost. The initial kinetic energy is the same as in the previous example (625,000 J). The final kinetic energy is calculated as follows:

$$K_f = \frac{1}{2}m_c v_{c,f}^2 + \frac{1}{2}m_t v_{t,f}^2$$

$$K_f = \frac{1}{2}(2,000 \text{ kg})(3 \text{ m/s})^2 + \frac{1}{2}(4,000 \text{ kg})(11 \text{ m/s})^2$$

$$K_f = 251,000 \text{ J}$$

Subtract the initial kinetic energy from the final kinetic energy:

$$\Delta K = K_f - K_i$$

$$\Delta K = 251,000 \text{ J} - 625,000 \text{ J}$$

$$\Delta K = -374,000 \text{ J}$$

Only 374,000 J of energy are lost in this collision. Less energy is lost than in the previous scenario because the collision is not completely inelastic. As in the completely inelastic collision, the truck exerts a force on the car that is equal in magnitude and opposite in direction to the force of the car hitting the truck. And again, due to momentum being dependent on mass, the truck suffers the smaller change in velocity.

ELASTIC COLLISIONS

In an elastic collision, energy is conserved. This type of collision is extremely rare on Earth, because nonconservative forces such as friction almost always play a role. However, in some cases, the effects of these forces are small enough that they can be ignored.

One such case is that of balls bouncing off one another. Imagine a ball with a mass of 10 kg, moving at 5 m/s along an imaginary x axis in the positive ($+x$) direction, from left to right. This ball hits another, stationary ball of equal mass. The collision happens off-center, so that after the collision, both balls are moving diagonally: while both are now traveling in $+x$ direction at a velocity of 2.5 m/s, one ball is simultaneously moving up the y axis, in the positive ($+y$) direction, at a velocity of 2.5 m/s, and the other ball is moving down the y axis, in the negative ($-y$) direction, also at 2.5 m/s. Because momentum is a vector quantity and this case deals with multiple directions, each axis has to be considered separately.

Only one ball is moving before the collision, so the initial momentum of the system consists only of the momentum of that one ball, and only in $+x$ direction. The ball's initial momentum is calculated as follows:

$$p_{1,x} = m_1 v_{1,i,x}$$

$$p_{i,x} = (10 \text{ kg})(5 \text{ m/s})$$

$$p_{i,x} = 50 \text{ kg·m/s}.$$

After the collision, both balls are moving in $+x$ direction with a velocity of 2.5 m/s. The final momentum in this direction is:

$$p_{f,x} = m_1 v_{1,f,x} + m_2 v_{2,f,x}$$

$$p_{f,x} = (10 \text{ kg})(2.5 \text{ m/s}) + (10 \text{ kg})(2.5 \text{ m/s})$$

$$p_{f,x} = 25 \text{ kg·m/s} + 25 \text{ kg·m/s} = 50 \text{ kg·m/s}.$$

Both the initial and final momenta in $+x$ direction are 50 kg·m/s, so momentum along this axis is conserved.

Now consider the y axis. The initial momentum along this axis is 0 kg·m/s, because before the collision, neither ball is moving in either $+y$ or $-y$ direction. After the collision, one ball moves in $+y$ direction at a velocity of 2.5 m/s, and the other ball moves in $-y$ direction at the same velocity. Because momentum is a vector quantity, the momentum of the ball moving in $+y$ direction is positive, while the momentum of the ball moving in $-y$ is negative. Thus, the total combined final momentum along the y axis is:

$$p_{f,y} = m_1 v_{1,f,x} + (-m_2 v_{2,f,x})$$

$$p_{f,y} = (10 \text{ kg})(2.5 \text{ m/s}) - (10 \text{ kg})(2.5 \text{ m/s})$$

$$p_{f,y} = 25 \text{ kg·m/s} - 25 \text{ kg·m/s} = 0 \text{ kg·m/s}.$$

The final combined momentum is also 0 kg·m/s. Momentum along the y axis is also conserved.

COLLISIONS ALL AROUND

Because nonconservative forces mean that most collisions on Earth result in a loss of energy, solving problems based on conservation of energy is not a very useful technique. However, using the principle of conservation of momentum will always lead to the correct answer. Collisions happen constantly, whether they consist of a person walking and colliding with the surrounding air molecules or two cars colliding on the highway. Even in recreational

---

SAMPLE PROBLEM

In a game of pool, one player hits the cue ball, which has a mass of 0.26 kg, toward the red three ball, which has a mass of 0.17 kg. The cue ball hits the stationary red ball at a velocity of 1.5 m/s in an elastic collision. The red ball moves to the left at approximately 1.57 m/s, while the cue ball moves to the right at 0.8 m/s. How much energy is transferred to the red ball during the collision?

Answer:

This is an elastic collision, so there is no loss of kinetic energy from the system. Thus, the system's initial kinetic energy and its final kinetic energy are the same:

$$K_i = K_f$$

The total kinetic energy of the system consists of the combined individual kinetic energies of the cue ball and the red ball:

$$K_{c,i} + K_{r,i} = K_{c,f} + K_{r,f}$$

Because the red ball is stationary at first, it has no initial kinetic energy, so $K_{r,i} = 0$. Solve for $K_{r,f}$, then substitute the definition of kinetic energy for $K_{c,i}$ and $K_{c,f}$ and simplify:

$$K_{c,i} = K_{c,f} + K_{r,f}$$
$$K_{c,i} - K_{c,f} = K_{r,f}$$
$$(\tfrac{1}{2} m_c v_{c,i}^2) - (\tfrac{1}{2} m_c v_{c,f}^2) = K_{r,f}$$
$$\tfrac{1}{2} m_c (v_{c,i}^2 - v_{c,f}^2) = K_{r,f}$$

Plug in the known values for each variable and solve:

$$\tfrac{1}{2} m_c (v_{c,i}^2 - v_{c,f}^2) = K_{r,f}$$
$$\tfrac{1}{2} [0.26 \text{ kg}] \left[(1.5 \text{ m/s})^2 - (0.8 \text{ m/s})^2 \right] = K_{r,f}$$
$$K_{r,f} = 0.2093 \text{ J}$$

The final kinetic energy of the red ball is 0.2093 J. Because it had no initial kinetic energy, that is also the amount of energy that is transferred to it during the collision.

activities, proper collision calculations can be made by using conservation of momentum.

In particle physics, collisions are extremely important. Particle accelerators, such as the Large Hadron Collider, work by crashing particles into other particles. The resulting elastic collisions break the particles apart into their respective components, allowing scientists to study the makeup of matter itself.

—Angel G. Fuentes, MS

BIBLIOGRAPHY

"Conservation of Momentum." *NASA Glenn Research Center: The Beginner's Guide to Aeronautics.* NASA, 5 May 2015. Web. 21 July 2015.

Fitzpatrick, Richard. "Conservation of Momentum." *Classical Mechanics: An Introductory Course.* University of Texas at Austin, 2 Feb. 2006. Web. 21 July 2015.

Giambattista, Alan, and Betty McCarthy Richardson. *Physics.* 3rd ed. New York: McGraw, 2015. Print.

"Impacts and Linear Momentum." *Khan Academy.* Khan Acad., n.d. Web. 21 July 2015.

Nave, Carl R. "Elastic and Inelastic Collisions." *HyperPhysics.* Georgia State U, n.d. Web. 21 July 2015.

Young, Hugh D., Roger A. Freedman, A. Lewis Ford, and Francis Weston Sears. *Sears & Zemansky's College Physics.* 14th ed. Boston, Mass: Pearson, 2016. Print.

ELECTRIC CIRCUITS: PARALLEL VS. SERIES, DIAGRAMS AND COMPONENTS

FIELDS OF STUDY

Electronics; Electromagnetism

SUMMARY

This article describes four basic parts and two basic configurations of elements used in constructing electrical circuits. The elements of an electrical circuit use the current flowing through them to produce a desired result. The basic method of calculating the total resistance of series and parallel resistors is described.

PRINCIPAL TERMS

- **capacitor:** an electrical part consisting of two conductors separated from each other by a nonconductor, allowing it to store electrical charge temporarily.
- **diodes:** devices in which an anode and a cathode, or the transistor equivalent, control the direction of current flow.
- **elements:** the parts of an electrical circuit, each of which has a specific function.
- **ground:** a direct connection to a larger body by which excess current is carried away, preventing errant electrical potentials from being generated in the circuit.
- **hot conductor:** in an electrical circuit, the conductor that brings current into a component element (the conductor having the higher electrical potential).
- **inductor:** a conducting coil in which the current generates a proportional magnetic field that, in turn, impedes the flow.
- **neutral conductor:** in an electrical circuit, the conductor that accepts the current coming out of the component elements (the conductor having the lower electrical potential).
- **parallel:** circuits or segments of circuits in which any two different elements are joined at two common points side by side.
- **resistor:** a device or material that resists the flow of electrons through it.
- **series:** circuits or segments of circuits in which any two different elements are joined only at one common point end to end.

THE BASIC ELECTRIC CIRCUIT

A circuit is a structure in which some property moves through an interconnected series of stages from a starting point to a finishing point. In an electric circuit, that property is the flow of electrons through a conductor under the influence of an electromotive force (emf). The emf results from the difference in electrical potential between two points, such as the terminals of a battery or the connection slots of a wall

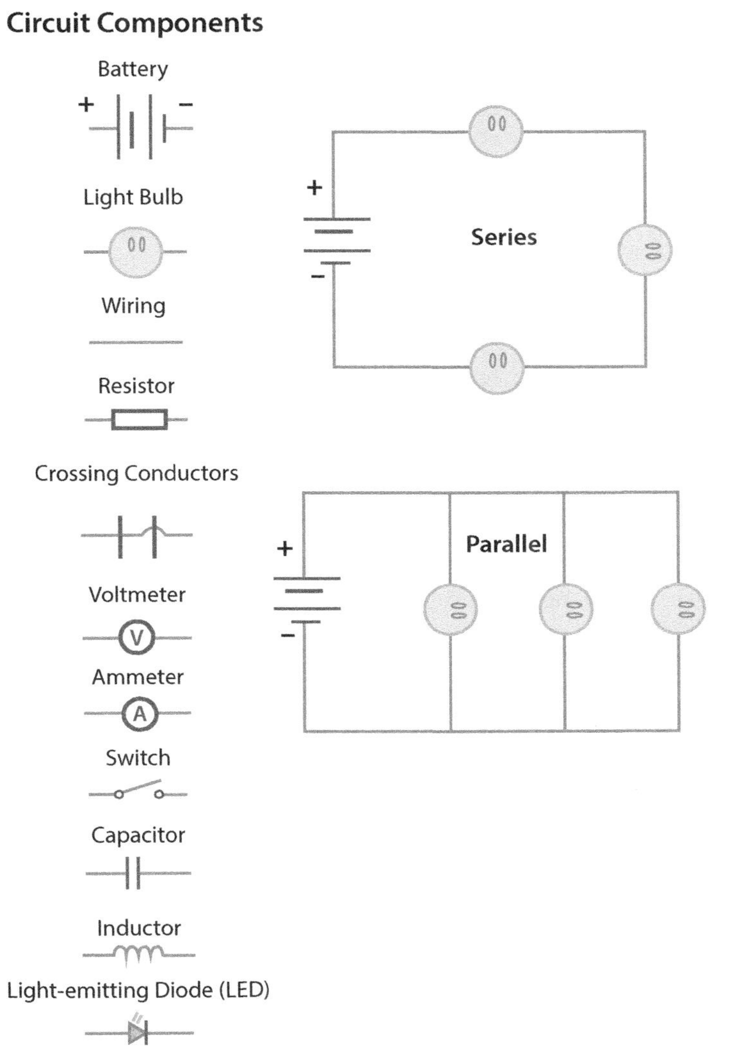

Electric Circuits: Parallel vs. Series, Diagrams and Components

Circuit Components

- Battery
- Light Bulb
- Wiring
- Resistor
- Crossing Conductors
- Voltmeter
- Ammeter
- Switch
- Capacitor
- Inductor
- Light-emitting Diode (LED)

Circuit diagrams use standard symbols for each component so that diagrams can be interpreted easily without a key. Common circuit components and their symbols are shown at left. The two simple circuit diagrams on the right show three light bulbs connected to a battery in series and in parallel.

outlet. The difference in electrical potential between two points is the voltage (V). The flow of electrons through the circuit is called the current. It always moves in the direction from a point of higher electrical potential to one of lower potential. The current is measured in units called amperes (A or amp).

Electrical current is either direct current (DC) or alternating current (AC). Direct current is produced by a constant voltage that drives current flow in one direction only. Alternating current is produced by a potential that varies in a cyclic manner between a maximum positive potential and an equivalent negative potential. The current flow thus switches direction as the applied potentials alternate between positive and negative. Switches, wires, and conductors are nonpolarized, meaning that they work equally well regardless which way the current flows. Devices that function differently based on the flow are polarized.

An electrical circuit is useful only when it does a desired function. Simply connecting two points together with a conductor creates a "short" circuit that will drain electrical current from one point to the other freely. A functional circuit consists of various elements that act with the current to produce a desired result. The basic component elements are the resistor, the capacitor, and the inductor. Devices called diodes also carry out important functions in directing the current flow. Each different element has a standard symbol to identify it easily in a circuit diagram. Each wire is represented by a straight line. The circuit diagram is essentially a map of the circuit that shows how each element is connected in relation to the other elements and to the voltage source. The important functional characteristic of each element, such as its resistance or capacitance, is typically also shown to be the respective element. Conventional current diagrams indicate current flowing from positive to negative, while electron flow diagrams show it flowing from negative to positive. Because the direction of flow affects how polarized devices work, it is important to know the type of diagram one is using.

Resistors resist current flow. They also change the electrical potential from one end of the resistor to the other as a "voltage drop." In complex circuits, resistors are generally used to produce specific electrical potentials or voltages. This permits the use of a wide variety of devices that require different voltages in the overall circuit function.

An inductor has a similar role. An inductor is a simple coil of wire. Current flow through the coil generates a magnetic field. This in turn induces a voltage that impedes the current flow and is directly

95

proportional to the current flowing through the inductor. Inductors are generally used to stabilize an electrical signal within a complex circuit.

Because of its construction, a capacitor is a nonconductive element in any circuit. Capacitors consist of two conducting surfaces separated by a nonconductor. A capacitor can store electrical charge, releasing it when the driving potential is removed. The amount of charge that a capacitor can store depends primarily on the applied voltage and the size of the conductive surfaces inside the capacitor. The discharge from even a small capacitor can be very damaging or even fatal. The electrical spark of a Taser weapon, for example, is produced by the discharge of capacitors. In complex circuits, capacitors are often connected between the main circuit and a ground to stabilize or "condition" the electrical signal in the circuit.

Diodes were originally vacuum-tube devices that used a cold anode and a heated cathode to drive current flow in one direction only. Transistor-based devices use semiconductor structures to achieve the same function. Diodes are used in both DC and AC circuits. In a DC circuit, diodes prevent errant currents from flowing in the wrong direction. In AC circuits, diodes "rectify" the alternating current by allowing the flow of current in one direction and blocking the current flow in the opposite direction. By combining different elements, diodes in an AC circuit can produce the equivalent of a DC current.

Circuit Structures

The elements in an electrical circuit can be combined in many different ways, but all circuits consist of just two basic structures. Elements may be connected in series or in parallel. In a series circuit, elements are connected head to tail so that the same current flows through each element. Electrical work is done at each element and decreases the voltage across the element according to its resistance (R). This "work" is the power consumed by the element, expressed as watts. In a parallel circuit, various elements are connected side by side at common points. The voltage measured across all of the parallel elements is the same. The total current entering a parallel circuit is the same as the current leaving, but a different current can flow through each parallel element. Complex circuits are constructed using a combination of series and parallel sections, forming a series-parallel circuit.

In all circuits, the hot conductor represents the high potential by which current flows from the voltage source. The neutral conductor represents the lower potential by which current flows back toward the voltage source. In a three-conductor system, the third conductor leads to the ground. To prevent short circuits from forming, the different conductors are isolated from each other using insulators, devices or materials that are very poor conductors of electrical current.

Calculating the Load

The electrical service in buildings is primarily a parallel-circuit system. This ensures that the voltage supplying all of the outlets is the same throughout the system. This cannot be done with a series-circuit system as the voltage at each outlet would decrease according to the number of devices that are drawing current. The load on a circuit is the sum of the power being drawn by the devices operating in the circuit. A parallel circuit distributes the load so that all of the devices receive the correct voltage supply. A series circuit can properly supply only one functioning device at a time.

The resistances in a series circuit add together according to the expression

$$RT = R1 + R2 + R3 + \text{etc.}$$

where each R value is expressed in ohms (Ω). In a parallel circuit, the total resistance is distributed as the reciprocal values, according to the expression

$$1/RT = 1/R1 + 1/R2 + 1/R3 + \text{etc.}$$

Circuit Complexity

Electrical circuits range from extremely simple to extremely complex. Analog circuits, through which current flows constantly, are generally much simpler in structure. Digital circuits, such as those in integrated circuit and computer-processor chips, are extremely complex. Many different elements besides resistors, capacitors, inductors, and diodes are typically used in common circuits, and each additional element makes the analysis and calculation of circuit properties that much more complex.

—*Richard M. Renneboog, MSc*

BIBLIOGRAPHY

Alexander, Charles, and Matthew Saddiku. *Fundamentals of Electric Circuits*. 5th ed. New York: McGraw, 2012. Print.

Brookes, A. M. P. *Basic Electric Circuits*. 2nd ed. Elmsford: Pergamon, 2014. Print.

Dorf, Richard C., and James A. Svoboda. *Introduction to Electric Circuits*. 9th ed. Hoboken: Wiley, 2014. Print.

Nilsson, James William, and Susan A. Riedel. *Electric Circuits*. New York: Prentice, 2008. Print.

Pulfrey, David L. *Understanding Modern Transistors and Diodes*. New York: Cambridge University Press, 2010. Print.

Schulz, Alexander L. *Capacitors: Theory, Types, and Applications*. Hauppauge: Nova Science, 2010. Print.

Yorke, R. *Electric Circuit Theory*. Elmsford: Pergamon, 2013. Print.

ELECTRIC POTENTIAL

FIELDS OF STUDY

Classical Mechanics; Electronics; Electromagnetism

SUMMARY

Electric charge is a property of atomic particles. As charged particles move, they generate electric currents. By properly storing these currents, electric potential energy can be stored in devices such as batteries and capacitors later to be released to serve a purpose. The relations that affect electric charges are described in this article.

PRINCIPAL TERMS

- **conductor:** a material that has a low resistance to electric charges, allowing them to move through it easily.
- **coulomb:** the basic unit of charge in the International System of Units (SI).
- **current:** the movement of electric charges from one place to another.
- **insulator:** a material that has a high resistance to electric charges, preventing them from moving through it easily.
- **joule:** the SI derived unit of energy, equal to one kilogram–square meter per second squared ($kg \cdot m^2/s^2$).
- **voltage:** the work done per unit charge when moving a charge against an electric field.
- **work:** the use of energy to move an object over a distance by means of the application of force.

Electric Potential Energy

As a rainstorm gathers in the atmosphere, a dance and motion of charges begins to take place. As the storm gets stronger, it creates faster updraft winds. These updraft winds pick up water droplets and raise them high in the clouds. At these high altitudes, the temperatures are extremely low, and the water freezes to form ice. As more water droplets are carried up by the updrafts, they start to collide with some of these ice particles. During these collisions, electrons are taken away from the ice. The electrons remain lower in the cloud. Eventually enough electrons build up to produce a lightning strike. These electrons travel through the air, which is typically an insulator. However, when enough electrons build up, they can break through the insulator, turning it into a conductor. During the buildup of charges at the base of the cloud, an interesting effect takes place. Electrons at the base of the cloud start attracting positive particles on the ground. This interaction builds up the electric potential energy in the system.

The electric potential energy between the clouds and ground is due to the configuration of electrons and protons, which are treated as point charges, or idealized dimensionless charged particles. Electric potential is defined as electric potential energy per unit charge, or the electric potential energy of a single point charge at any point in an electric field. It is equal to the amount of work that would be necessary to carry the charge to that point when moving against the electric field.

Calculating Electric Potential

The electric potential energy (U_e) between two point charges is a function of the electrostatic constant (k), also called Coloumb's constant; the values of the individual point charges (Q_1 and Q_2); and the distance between the two charges (d), measured in meters (m):

$$U_e = k\frac{Q_1 Q_2}{d}$$

The electrostatic constant is measured in newton–square meters per coulomb squared (N·m²/C²) and is equal to 8.99×10^9 N·m²/C². Any physical energy is measured in the International System of Units (SI) unit of energy, the joule (J). The charges are measured in SI units of coulombs (C). In a system of more than two charges, the total electric potential energy is the sum of the electric potential energy of each pair of charges. The equation to calculate the electric potential (V) of one of the two point charges is similar to the equation for electric potential energy. It, too, is a function of the charge amount (Q), the electrostatic constant (k), and the distance to the charge (d):

$$V = k\frac{Q}{d}$$

Electric potential is measured in volts (V), an SI derived unit named in honor of the Italian physicist Alessandro Volta (1745–1827). This equation calculates the electric potential generated by one point charge. Electric potential is simply electric potential energy per unit charge. Therefore, it can also be defined in as electric potential energy (U_e) per unit of charge (Q):

$$V = \frac{U_e}{Q}$$

When dealing with accumulations of charges, a different technique must be used. Imagine two horizontal charged plates separated by a certain distance. The charges in these plates create a uniform electric field between them. Finding the electric potential using the equation above for point charges would take a great deal of time, because the number of individual charges can be very large. Instead, we turn to the electric field. The electric potential at any point in a uniform electric field with a strength of E newtons per coulomb (N/C), generated by two plates separated by a given distance (d), is

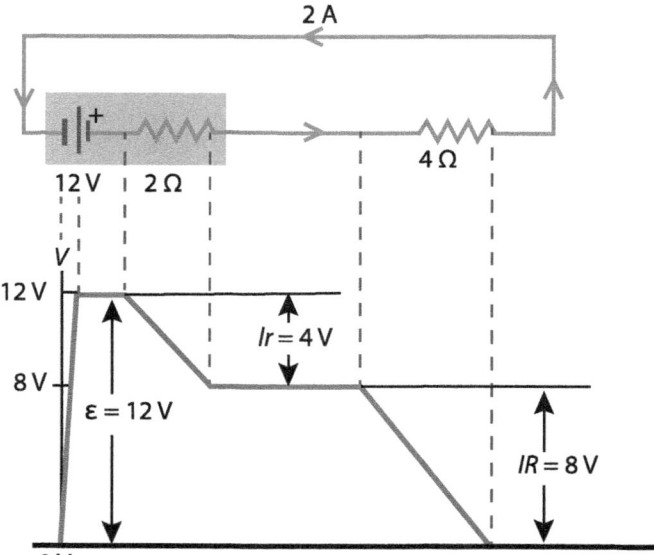

A circuit diagram of two resistors in series connected to a battery and drawing an electric current of 2 amps. A voltage diagram shows the change in voltage, or electric potential, is along the circuit. The voltage on the negative side of the battery is 0, and jumps to 12 volts on the positive side of the 12-volt battery. It drops to 8 volts after passing by the first 2-ohm resistor, then drops to 0 volts after passing by the 4-ohm resistor. The voltage change across each resistor is calculated by multiplying the current by the resistance.

$$V = Ed.$$

On its own, the electric potential of a single point charge is not a meaningful quantity. However, the difference in electric potential between two points (ΔV) is a very useful quantity. This quantity, called potential difference, is more commonly known as voltage. It is the amount of work, measured in units of electric potential (i.e., volts), that would be necessary to move a charge between the two points in the opposite direction of the electric field.

Electric potential due to a uniform electric field generated by parallel plates is something that affects the world daily. Electronic devices are made with a collection of smaller circuit parts. One of these parts is a capacitor, or a pair of parallel conductive plates that store electric potential energy between them. The electric potential energy (U_e) stored by a single capacitor in a circuit is a function of the capacitance (C) of the capacitor, measured in farads (F), and the potential difference or voltage (ΔV) between the two plates:

SAMPLE PROBLEM

A 1.0 nanocoulomb (nC) point charge is located at a point 2 m away from another charge, Q. The electric potential at this location is 4.5 V. If the 1.0 nC charge is moved to 4 m from charge Q, what is its new electric potential?

Answer:

In order to calculate the electric potential (V_2) at a distance (d_2) of 4 m from charge Q, begin by defining the electric potential at the original point (V_1), at a distance ($d1$) of 2 m from charge Q. This value was already given as 4.5 V.

$$V_1 = k\frac{Q}{d_1} = 4.5 \text{ V}$$

Note that the second point is twice as far from charge Q as the original point:

$$d_2 = 2d_1$$

Now, to calculate V2, use the same equation as for V1:

$$V_2 = k\frac{Q}{d_2}$$

Replace d_2 with $2d_1$:

$$V_2 = k\frac{Q}{2d_1}$$

This new equation can be rewritten as

$$V_2 = \frac{1}{2}(k\frac{Q}{d_1})$$

$$V_2 = \frac{1}{2}V_1$$

Solve:

$$V_2 = \frac{1}{2}V_1$$

$$V_2 = \frac{1}{2}(4.5 \text{ V})$$

$$V_2 = 2.25 \text{ V}$$

At twice the distance from charge Q, the electric potential is 2.25 V, one-half of the electric potential at the original point.

$$U_e = \frac{1}{2}C\Delta V^2$$

Imagine a capacitor of 6 microfarads (μF) is placed in a circuit and achieves a potential difference of 5 V. The electric potential energy stored in that capacitor can be calculated thus:

$$U_e = \frac{1}{2}C\Delta V^2$$

$$U_e = \frac{1}{2}(6\times10^{-6} \text{ F})(5 \text{ V})^2$$

$$U_e = 0.000075 \text{ J} = 7.5\times10^{-5} \text{ J}$$

CURRENTS AND CIRCUITS

Although charged particles run through them, circuits do not work using individual charges. They run on a current, a series of charges moving as a function of time. In a circuit there is a voltage that is supplied by a power source, for example a battery. The voltage supplied by the battery depends on the amount of resistance and the current allowed by the resistance. This relation is given by Ohm's law, which states that the voltage (ΔV) in a circuit is a function of current (I), measured in amperes (A), and the resistance (R), measured in ohms (Ω):

$$\Delta V = IR.$$

Solving this equation for current shows that it is equal to voltage divided by resistance:

$$I = \frac{\Delta V}{R}$$

Resistors work by slowing down or stopping the electrons from moving through a circuit. That is, the more resistance, the less current will flow. A good example of how resistors dissipate potential energy is an incandescent lightbulb. They are made out of a filament that has very high resistance properties. As electrons flow through the filament, they are slowed down or stopped by the filament. The electric potential energy the electrons carry cannot just disappear. It changes form into thermal energy, which is emitted as light and heat.

The energy in an electrical circuit is directly related to the work done by the charges in the circuit, the current. Work is the energy used by a force to move an object over a distance. A simple circuit can be made using a battery, which provides the voltage, and one resistor. The work done by the system (W) is a function of the voltage supplied by the battery (ΔV), the current in the circuit (I), and the amount of time the current stays flowing (t):

$$W = \Delta V I t.$$

In this same circuit energy is being dissipated by the resistor. There is a limited amount of energy in the battery. Eventually the battery will stop working. The resistor has slowly used up all the energy. The power (P) at which the resistor works and dissipates energy is a function of the voltage (ΔV), the current (I), the resistance (R):

$$P = \Delta V I = I^2 R = \frac{\Delta V^2}{R}$$

All of these versions of the power equation can be used to calculate the power dissipated by the resistor.

Charging Circuits

Hearts beat because of an electrical discharge that makes the heart contract, thereby pushing blood into the arteries. Unfortunately, like most circuits, the circuit that makes the heart beat sometimes fails. It can fail in many ways. Sometimes it gets out of control. Sometimes it stops working altogether. In each of these cases, scientists have developed a device that can help the heart beat normally again. Pacemakers and defibrillators make use of capacitors. They supply a charge or current that gets stored in a capacitor. When the capacitor stores as much charge as it can, no more charge flows through this circuit. As soon as the pacemaker detects that the heart has stopped beating or is beating at different pattern, it releases the charge. At this point it begins to charge the capacitor again. In many ways these capacitors work like batteries that supply the shock needed to keep the heart beating.

—*Angel G. Fuentes, MS*

Bibliography

"Electric Potential." Encyclopaedia Britannica. Encyclopaedia Britannica, 3 Apr. 2014. Web. 19 June 2015.

"Electric Potential Energy." Khan Academy. Khan Acad., n.d. Web. 19 June 2015.

"Introduction to Circuits and Ohm's Law." Khan Academy. Khan Acad., n.d. Web. 19 June 2015.

Giambattista, Alan, and Betty McCarthy Richardson. *Physics*. 3rd ed. New York: McGraw, 2015. Print.

Nave, Carl R. "Electric Potential Energy." HyperPhysics. Georgia State U, n.d. Web. 19 June 2015.

Young, Hugh D., Roger A. Freedman, A. Lewis Ford, and Francis Weston Sears. *Sears & Zemansky's College Physics*. 14th ed. Boston, Mass: Pearson, 2016. Print.

ENERGY AND POWER

FIELDS OF STUDY

Classical Mechanics; Electronics; Electromagnetism

SUMMARY

There are many different ways to measure the amount of energy used to perform a task. Some involve knowing the force used and the resulting velocity. Others use the work done and the time it took to do that work. When dealing with circuits, there is a third way of calculating the energy used and the power expended.

PRINCIPAL TERMS

- **displacement:** the difference between the initial position of an object and its final position, regardless of its path.
- **force:** the result of an interaction between two objects that changes the pattern of motion of the objects. A force can be a pull or a push.
- **joule:** the base unit of energy, equal to one kilogram-meter squared per second squared (kg·m^2/s^2).
- **kilowatt-hour:** a unit for measuring electricity consumption, equal to one thousand watts of power consumed over one hour, or 3.6×10^6 joules.

- **newton:** the base unit of force, equal to one kilogram-meter per second squared (kg·m/s^2).
- **potential energy:** the energy that is stored in objects and can be converted to other forms of energy, such as kinetic energy.
- **watt:** the base unit of power, equal to one joule per second (J/s).
- **work:** the energy used by a force to move an object over a given distance.

From Energy to Power

As a person lifts an object, he or she is increasing the object's gravitational potential energy. That energy is being transferred to the object at different amounts per second, as the force a person expends when lifting an object is not constant. When the object is lifted, work is being done on the object. The work done on the object, like the force expended, is also not constant. The amount of work done per unit of time is what physicists call "power." Whether it takes a person three seconds or ten minutes to lift a box one meter above the ground, the amount of work done is the same; it is the power that is different. In terms of the amount of energy spent per second in each case, the person who took ten minutes to lift the box spent energy at a much lower pace. The person developed less power, which is why it took the person longer to lift the box. The power spent by an object is equal to the total work done divided by the time it took to do that work.

Because work is a measurement of the difference in energy, in the International System of Units (SI), it is measured in joules (J). This means that power is measured in units of joules per second (J/s), or watts (W). Named after Scottish engineer James Watt (1736–1819), one watt represents the amount of work done in joules over a period of time measured in seconds. Power is also the amount of radiant energy produced by a light source. For instance, a hundred-watt light bulb produces one hundred joules of energy each second.

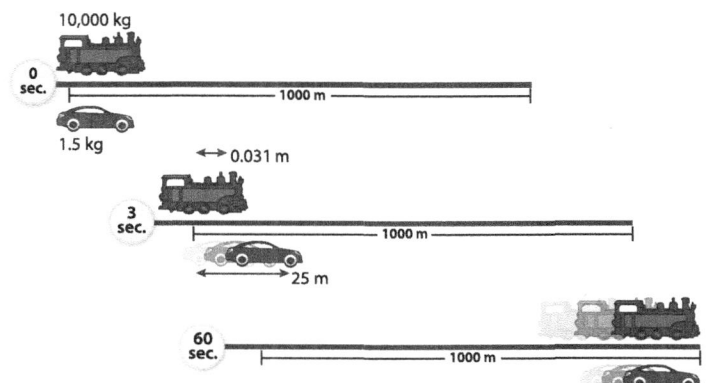

	Time	Distance	v_f	a	m	F	ΔK	P
Train	3	0.031 m	1.033 x 10^{-2} m/s	3.444 x 10^{-3} m/s^2	10,000 kg	34.44 N	0.534 J	0.356 W
Car	3	25 m	16.67 m/s	3.858 x 10^{-6} m/s^2	1.5 kg	5.787 x 10^{-6} N	208.417 J	9.647 x 10^{-5} W
Train	60 s	1000 m	16.67 m/s	7.716 m/s^2	10,000 kg	77160 N	1389444.5 J	128657.2 W
Car	60 s	1000 m	16.67 m/s	0.278 m/s^2	1.5 kg	0.417 N	208.417 J	6.951 W

Change in Kinetic Energy = ΔK = ½ mv^2
Power = P = F · v
F = m · a

A train and a car will display different changes in kinetic energy and produce different magnitudes of power. The added mass of the train strongly affects its kinetic energy and power. A train that reaches 60 km/h after 60 seconds travels alongside a car that reaches 60 km/h after 3 seconds. However, since the car remains at the same velocity after 3 seconds they both reach 1000 m at the 60 second mark. The car's kinetic energy remains constant from 3 seconds to 60 seconds as its power increases, but the train increases both kinetic energy and power from 3 seconds to 60 seconds.

Calculating Power

Power (P) is a function of work done (W) over a given period of time (t):

$$P = \frac{W}{t}$$

Therefore, if a washing machine does 9×10^5 joules of work per load, and one load takes 1,800 seconds to complete, then the power spent by the washing machine in one load is calculated as follows:

$$P = \frac{9 \times 10^5 \text{ J}}{1,800 \text{ s}}$$

$$P = 500 \text{ J/s} = 500 \text{ W}$$

This definition can be mathematically expanded by considering the definitions of work and energy themselves. Work is the difference in energy between a starting point and an end point. It depends on the force (F) acting on an object and the displacement (d) of the object. This relationship is given by the equation

$$W = Fd.$$

Substituting this value into the original equation for power produces the equation

$$P = \frac{Fd}{t}$$

Remembering that velocity (v) is equal to distance over time, this equation can be simplified:

$$P = Fv.$$

The SI unit of force is the newton (N). The energy spent to exert one newton of force over a distance of one meter is equal to one joule.

There are many ways to calculate power; which method to use depends on the type of problem. For example, imagine a hundred-kilogram man walking into an elevator. The elevator takes the man up to the second floor, a displacement of ten meters. It takes the elevator five seconds to do this. The force the elevator has to overcome is the weight of the man himself. Force is equal to the product of mass (m) and acceleration (a):

$$F = ma$$

Therefore, the weight of an object is also equal to its mass times its acceleration— specifically, the acceleration due to gravity (g), which on Earth is 9.8 meters per second squared (m/s^2). This is the rate at which the velocity of an object in freefall will increase. Given this information, the power required for the elevator to lift the man is calculated as follows:

$$P = \frac{Fd}{t}$$

$$P = \frac{(100\,\text{kg})\left(9.8\,\frac{\text{m}}{\text{s}^2}\right)(10\,\text{m})}{5\,\text{s}}$$

$$P = \frac{9{,}800\,\frac{\text{kg}\cdot\text{m}^2}{\text{s}^2}}{5\,\text{s}}$$

$$P = 1{,}960\,\text{J/s} = 1{,}960\,\text{W}$$

As seen in this example, power depends directly on the distance the object moves, the time it takes to do so, and the force, which itself depends on the mass of the object. Thus, accurate measurements of these quantities must be taken in order to obtain correct measurements of power.

These examples deal with the power spent against a specific kind of potential energy, known as gravitational potential energy. Other forms of potential energy exist, such as electric potential energy. To calculate electric potential energy, one must use a different equation of power. The power provided by batteries for circuits is the product of the current (I) that flows from the batteries, measured in amperes (A), and the voltage (V) supplied by the batteries:

$$P = IV.$$

Note that one volt (V), the base unit of electric potential difference, is equal to one watt per ampere. If a six-volt battery supplies a current of 0.002 ampere into a circuit, then the power provided by the battery is calculated as follows:

$$P = (0.002\,\text{A})(6\,\text{V})$$

$$P = (0.002\,\text{A})(6\,\text{W/A})$$

$$P = 0.012\,\text{W}$$

Measurements of Power

Knowing the amount of energy used by appliances, machines, and other devices in homes is of great importance to people and companies. Power companies install wattmeters in their customers' homes to measure the amount of energy consumed by a household. This amount can then be quoted to the

SAMPLE PROBLEM

A 1,500 kg plane starts moving down the runway in order to take off. The engines are running at a power of 3.5×10^5 W. It takes the plane 35 s to travel from the starting position to the takeoff point, a distance of 2 km. What is the speed of the plane as it takes off? To calculate acceleration (a), use the distance formula

$$d = \frac{at^2}{2}$$

Answer:

The equation for power is calculated as the force producing the motion multiplied by the velocity of the object. Because the acceleration of the plane is proportional to the force applied, as stated in Newton's second law of motion, the force can be calculated. First, find the acceleration of the plane by rearranging the distance formula to solve for a:

$$d = \frac{at^2}{2}$$

$$a = \frac{2d}{t^2}$$

$$a = \frac{2(2{,}000 \text{ m})}{(35 \text{ s})^2}$$

$$a = 3.27 \text{ m/s}^2$$

Next, use the rate of acceleration to calculate the force the engines must produce in order to achieve that acceleration, based on the mass of the plane:

$$F = ma$$

$$F = (1{,}500 \text{ kg})(3.27 \text{ m/s}^2)$$

$$F = 4{,}905 \text{ kg} \cdot \text{m/s}^2 = 4{,}905 \text{ N}$$

Finally, solve the power equation for the speed of the plane, using the force exerted on the plane and the given engine power:

$$P = Fv$$

$$v = \frac{P}{F}$$

$$v = \frac{3.5 \times 10^5 \text{ W}}{4{,}905 \text{ N}}$$

$$v = \frac{3.5 \times 10^5 \frac{\text{J}}{\text{s}}}{4{,}905 \frac{\text{kg} \cdot \text{m}}{\text{s}^2}}$$

$$v = 3.5 \times 10^5 \frac{\text{kg} \cdot \text{m}^2}{\text{s}^3} \times \frac{1}{4{,}905} \frac{\text{s}^2}{\text{kg} \cdot \text{m}}$$

$$v = 71.4 \text{ m/s}$$

The plane is traveling at 71.4 m/s at takeoff.

---

consumer in units of kilowatt-hours (kWh). A kilowatt-hour is used to measure energy consumption per unit time, such as the total energy used during a last billing cycle. One kilowatt-hour is equal to 3.6×10^6 joules.

Other types of measuring devices are employed when measuring the power produced by machines. Dynamometers are used to measure the power produced by engines. These devices measure the rotational speed and the torque produced by the engine to calculate the power the engine can achieve.

ENERGY VERSUS POWER

The words "energy" and "power" are commonly used to mean the same thing. People quote the energy of light bulbs they buy by incorrectly quoting the power. These quantities are very much related, yet they represent different things. While energy is the ability of an object to perform work, power is the rate of energy use. Energy consumption is a subject that affects every society. When buying compact fluorescent light bulbs (CFLs), for instance, people have noticed that they have lower power ratings. A CFL with a power rating of sixty watts can be as bright as a hundred-watt incandescent light bulb because the CFL uses most of this power to produce visible light rather than heat. Hundred-watt incandescent bulbs use a smaller proportion of that power to produce visible light; the rest is used to heat the element. Accurately measuring the energy spent and how fast it is being spent affects everything from the amount of power produced by utility companies to horsepower in cars.

—*Angel G. Fuentes, MS*

Bibliography

"Electrical Power." BBC Bitesize. BBC, n.d. Web. 5 May 2015.

Giambattista, Alan, and Betty McCarthy Richardson. *Physics*. 3rd ed. New York: McGraw, 2015. Print.

Khan, Sal. "Work and Energy (Part 2)." Khan Academy. Khan Acad., 2015. Web. 30 Apr. 2015.

Nave, Carl R. "Work." HyperPhysics. Georgia State U, 2012. Web. 5 May 2015.

Santo Pietro, David. "Power." Khan Academy. Khan Acad., 2015. Web. 5 May 2015.

Young, Hugh D., Roger A. Freedman, A. Lewis Ford, and Francis Weston Sears. *Sears & Zemansky's College Physics*. 14th ed. Boston, Mass: Pearson, 2016. Print.

ENTHALPY

FIELDS OF STUDY

Thermodynamics

SUMMARY

Enthalpy is the total amount of energy bound within a body or system by its structure. It is equal to the system's internal energy plus the product of its volume and pressure. Enthalpy can only be measured in terms of its change. When a system undergoes a process or reaction, its enthalpy increases or decreases depending on whether energy is absorbed or released.

PRINCIPAL TERMS

- **activation energy:** the minimum energy required for a chemical reaction to take place.
- **bond energy:** the energy required to break the chemical bonds in one mole of molecules and separate them into their component atoms.
- **endothermic:** describes a chemical reaction or process that requires the input of energy from an external source in order to proceed.
- **exothermic:** describes a chemical reaction or process that results in an output of energy from the system.
- **Hess's law:** the principle that the overall enthalpy change of a given reaction remains the same regardless of the number of reaction steps involved and is equal to the sum of the enthalpy changes of any and all intermediate steps; also called Hess's law of constant heat summation.
- **standard enthalpy of formation:** the change in enthalpy that results from the formation of one mole of a given substance, with the reactants and the products in their standard states.
- **standard state:** a set of standard thermodynamic reference conditions for a given substance, generally calculated at a temperature of 298.15 kelvins (25 degrees Celsius or 77 degrees Fahrenheit) and a pressure of 105 pascals (14.5 pounds per square inch).
- **thermodynamics:** the study of the relationships between heat, energy, and work in a system.

Enthalpy and Energy

All physical systems contain energy. This energy takes different forms as the system undergoes change. It may be expressed as kinetic energy, potential energy, chemical energy, or some other form of energy. The enthalpy of a system is equal to the sum of all its internal energy (U) plus the product of its volume (V) and pressure (p). The internal energy is the amount of energy that would be necessary to create the system, and the quantity pV is the amount of energy that would necessary to make room for the system by displacing the environment in the space it occupies.

Enthalpy represents the ability of the system to do nonmechanical work and release energy in the form of heat. The enthalpy of a system cannot be measured directly. Instead, it is usually expressed in terms of enthalpy change, or how much energy the system loses or gains during a chemical reaction or other process.

When two atoms or molecules interact in a chemical reaction, the energy changes that occur determine the outcome of the reaction process. In any reaction, the two reacting materials, or reactants, first combine to form an intermediate activated complex. The amount of energy contained in this activated complex is called the activation energy of the reaction. This is the minimum energy level that is required for the reaction to take place. If the activation energy is greater than the sum of the energies of the individual reactants, then an input of energy is required for the reaction to proceed.

The activation energy of a chemical reaction includes the energy necessary to break the bonds of any molecules taking part in the reaction. The atoms in a molecule form covalent bonds by sharing electrons between them. Ionic bonds, or electrovalent bonds, are formed through an exchange of electrons between oppositely charged atoms. To break such a bond requires a specific amount of energy, called the bond energy. The bond energy of a given molecule is the amount of energy required to break all the bonds in one mole (6.022×10^{23} particles) of that molecule.

The freed atoms then bond in new ways to form the product or products of the reaction. If the energy contained in the products is less than that of the reactants, the reaction is exothermic, and the excess energy is released into the surrounding environment as heat. If it is greater, the reaction is endothermic, and the system has experienced a net gain of energy from its surroundings. Endothermic reactions normally require the input of energy as heat or light in order to proceed. Some endothermic processes occur readily just by absorbing heat energy from their environment. In such a case, the vessel in which the reaction takes place will become colder.

All atoms and molecules are in constant motion. They vibrate, rotate, and, in the case of fluids, move through space. Even the atoms that make up molecules are constantly moving, although their movements are restricted by the bonds between them. Thus, all of these particles have kinetic energy. This kinetic energy contributes to the total enthalpy of a system.

The enthalpy of a system is different from the heat of a system. Heat is an expression of the physical motions of the atoms and molecules. According to kinetic theory, the temperature of a substance is a function of the kinetic energy contained in its constituent particles. The more kinetic energy the particles have, the more heat is expressed.

Chemical Reactions and Hess's Law

The thermodynamics of chemical reactions measures changes in the energy of a system, particularly the changes in heat energy. Exothermic reactions reduce the amount of energy in the system, causing a decrease in enthalpy. The energy released is given off as heat. Endothermic reactions increase the enthalpy of the system, because energy must be added for the system to complete the reaction. The rate and magnitude of an enthalpy change depend on the amount and physical states of the materials involved in the reaction, the temperature at which the reaction occurs, and the pressure and volume of the system.

A chemical equation typically expresses a reaction as a single step from reactants to products. This is an oversimplification, showing only the difference between the start and end points of the reaction. It does not account for any of the many steps that may occur during the process. Many chemical processes can be described in terms of a series of distinct steps with identifiable reactants and products. Each step is associated with specific changes in thermodynamic properties. The sum total of all of these changes must be equal to the overall change in thermodynamic properties associated with the reaction. This is the basic principle of Hess's law, named for Swiss Russian chemist Germain Hess (1802–50). Although

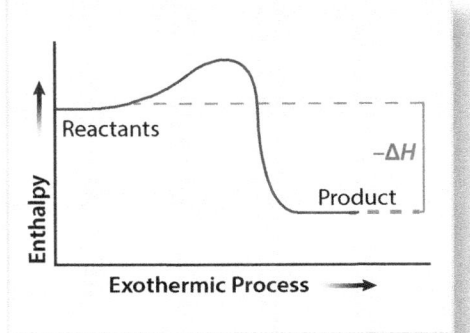 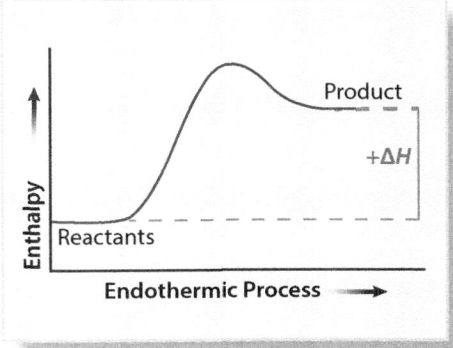

The enthalpy of a system is measured in terms of its change. A model of an exothermic process shows that the enthalpy change is negative, meaning that the enthalpy of the product(s) of the process is lower than that of the reactant(s). A model of an endothermic process shows that the enthalpy change is positive, meaning that the enthalpy of the product(s) is higher than that of the reactant(s).

it is practically impossible in most cases to measure directly thermodynamic changes that take place during intermediate reaction steps, Hess's law can be used to calculate their values quite accurately.

Like all forms of measurement, Hess's law calculations rely on standardized values of properties. These values are determined based on the standard state of the materials in a reaction. The standard state of a material is generally its most stable common form under standard conditions of temperature and pressure. The conditions typically used are known as standard ambient temperature and pressure (SATP): a temperature of 298.15 kelvins (K), or 25 degrees Celsius (°C) or 77 degrees Fahrenheit (°F), and a pressure of 100,000 pascals (Pa), or 14.5 pounds per square inch (psi). The thermodynamic properties of materials under these conditions allow for the calculation of a reference value called the standard enthalpy of formation. This is the enthalpy change that would result from the formation of one mole of a given substance in its standard state. Typically, it is measured in units of kilojoule per mole (kJ/mol). The standard enthalpies of formation for common reactions and substances can be found in many reference books. Generally, a change in some property, such as enthalpy, is indicated by the Greek letter delta (Δ).

ENTHALPY AND ENTROPY

If enthalpy can be thought of as the energy that holds materials together, entropy can be thought of as the energy that drives materials apart. It is a measure of the degree of disorder in a system. Entropy limits the amount of work that the energy within a system can perform. The energy that is available for a system to do work is called its Gibbs energy. The Gibbs energy (G) of a system is calculated as the difference between its enthalpy (H) and the product of its temperature (T) and its entropy (S):

$$G = H - TS.$$

SAMPLE PROBLEM

Calculate the enthalpy change (ΔH) for the reaction of one mole of pure carbon (C) with one mole of pure oxygen (O_2) to produce one mole of pure carbon monoxide (CO), assuming that all materials are in their standard states. Use the following values for standard enthalpies of formation of the intermediate steps:

$$C + O_2 \rightarrow CO_2 : \Delta H = -393.5 \text{ kJ/mol}$$
$$CO + \tfrac{1}{2}O_2 \rightarrow CO_2 : \Delta H = -283 \text{ kJ/mol}$$

Answer:

First, determine an overall chemical equation for the process:

$$C + O_2 \rightarrow CO$$

This equation is unbalanced, because it leaves one oxygen atom unaccounted for. Balance the equation:

$$C + O_2 \rightarrow CO + \tfrac{1}{2}O_2$$

Because O2 is the standard state of oxygen, its standard enthalpy of formation is 0 kJ/mol. Thus, its production will not affect the enthalpy change of the reaction, and it can be ignored. Note that although fractional coefficients (e.g., 1/2) are usually not used when balancing chemical equations, they are acceptable when making calculations based on Hess's law, because they make it easier to keep track of the steps in the process.

Next, use the information provided to determine the order of steps in the reaction and the overall enthalpy change. The products of one or more steps will become the starting materials for the next steps. Remember to reverse the sign of ΔH for any reaction steps that are reversed.

$$C + O_2 \rightarrow CO_2 : \Delta H = -393.5 \text{ kJ/mol}$$
$$CO_2 \rightarrow CO + \tfrac{1}{2}O_2 : \Delta H = +283 \text{ kJ/mol}$$

Because the product of the first step (C O_2) is the reactant of the second step, these two terms "cancel" each other. Combine the steps into a single reaction and calculate the overall change in enthalpy:

$$C + O_2 \rightarrow CO + \tfrac{1}{2}O_2 :$$
$$\Delta H = -393.5 \text{ kJ/mol} + 283 \text{ kJ/mol} = -110.5 \text{ kJ/mol}$$

The overall change in enthalpy is -110.5 kJ/mol. The negative sign indicates that the system lost enthalpy, meaning that the reaction was exothermic. One-half mole of O_2 is left over at the end of the reaction; again, this can be ignored.

A change in Gibbs energy can be calculated using a very similar equation:

$$\Delta G = \Delta H - T\Delta S.$$

This equation can be used to determine how much energy a system has available to perform work following any changes in enthalpy, entropy, or both.

—*Richard M. Renneboog, MSc*

Bibliography

Atkins, Peter. *The Laws of Thermodynamics: A Very Short Introduction.* New York: Oxford University Press, 2010. Print.

Atkins, Peter, and Julio de Paula. *Atkins' Physical Chemistry.* 10th ed. Oxford: Oxford University Press, 2014. Print.

Borgnakke, Claus, and Richard E. Sonntag. *Fundamentals of Thermodynamics.* 8th ed. Hoboken: Wiley, 2013. Print.

Foulkes, Frank R. *Physical Chemistry for Engineering and Applied Sciences.* Boca Raton: CRC, 2013. Print.

Linder, Bruno. *Elementary Physical Chemistry.* Hackensack: World Scientific, 2011. Print.

Mortimer, Robert G. *Physical Chemistry.* 3rd ed. Burlington: Elsevier, 2008. Print.

ENTROPY

FIELDS OF STUDY

Thermodynamics; Cryophysics

SUMMARY

Entropy is a difficult concept to explain, but is easily demonstrated in everyday events. All systems contain energy as a combination of energy that is available to do useful work and energy that is lost. Entropy is related to temperature, and is most commonly described as the degree of disorder within a system. The concept of entropy is widely used in both a technical and a non-technical sense, indicating that it is a universal concept.

PRINCIPAL TERMS

- **activation energy:** the minimum energy required for a chemical reaction to take place.
- **conservation of energy:** a fundamental law of physics that states that the amount of energy in a system remains constant over time. Although the energy can be transformed or transferred, it cannot be created or destroyed.
- **degrees of freedom:** the number of physical parameters required to specify the position and configuration of a particle or other body.
- **enthalpy:** a measure of the total internal energy (thermal energy) of a system, the product of its volume and pressure.
- **equilibrium:** the condition of a thermodynamic system in which there is no energy flow.
- **kinetic theory of gases:** the theory that atomic and molecular motion determines the behavior of gases.
- **standard temperature and pressure (STP):** standard reference conditions when dealing with gases, defined by the International Union of Pure and Applied Chemistry as a temperature of 273.15 kelvins (0 degrees Celsius or 32 degrees Fahrenheit) and pressure of 101.3 kilopascals (1 atmosphere); used in chemistry and physics to establish a standardized set of conditions for experimentation.
- **state variables:** external factors, such as temperature and pressure, that determine the physical state of matter.

THE SECOND LAW OF THERMODYNAMICS

The first law of thermodynamics is the law of conservation of energy; energy can neither be created nor destroyed, only changed from one form to another. In theory, any number of scenarios can be devised, however, that do no violate this principle but nevertheless do not, and cannot, occur. The second law of thermodynamics considers changes to the enthalpy or heat content of a system. In essence, it states that any change to the enthalpy of a system requires the expenditure or investment of energy in the form of work done by or on the system. Accordingly, it is always observed that heat moves toward a place where there is less heat and not away from it. In a system that is at equilibrium there is no net movement of heat within the system. Effecting a change to the system, typically by changing one or more of the state variables, effects a change in the heat contained in the system. The overall enthalpy of the system remains the same, but the ability of the system to do work changes in a way that depends on the temperature of the system. The enthalpy of the system consists of two components. One component is not temperature-dependent, and is termed Gibbs' free energy (G). The other component is the entropy (S) of the system and is temperature-dependent. Thus the enthalpy of a system at equilibrium is given by the equation

$$H = G + TS$$

and any change to the system that alters the enthalpy also alters both the Gibbs' free energy and the entropy, as

$$\Delta H = \Delta G + T\Delta S.$$

Entropy is a difficult concept to explain, but is easily demonstrated in everyday events, and non-events. For example, while it would not violate the first law of law of thermodynamics for water to freeze instead of boil as the temperature is raised because the energy would still be conserved, the second law of thermodynamics – the entropy law – forbids such a transformation, and indeed such a transformation is never observed in reality. Another way of stating the principle is that the second law of thermodynamics makes it impossible to "uncook" an egg. The effect of entropy is commonly demonstrated in principle by considering systems based on the kinetic theory of gases, primarily because the effect is most apparent in such systems, usually in relation to the conditions of standard temperature and pressure (STP). The number of degrees of freedom of the particles in such a system is finite, which facilitates the demonstration of the principle and allows its extension to larger systems such as the universe.

ORDER VS. CHAOS

One simple way of stating the concept of entropy is to assign entropy as the amount of disorder or chaos that exists in a system. Using the gas example, begin with the gas contained in a specific volume at the same constant temperature. Such a system is at equilibrium, and will remain in that state until something acts on the system to change its temperature, volume and pressure. Note that there is only one

---

SAMPLE PROBLEM

Use the Arrhenius equation and the enthalpy equation to obtain an equation for the entropy in terms of activation energy.

Answer:
The Arrhenius equation is

$$k = Ae^{-E_a/RT}$$

and the enthalpy equation is

$$H = G - TS.$$

Therefore,

$$S = -(H - G)/T$$

$$k = A_e^{-E_a/(R(H-G)/-S)}$$

$$= A_e^{SE_a/(R(H-G))}$$

$$ln(k) = ln(A) + Se_a/(R(H-G))e^{SE_a/(R(H-G))}$$

$$ln(k) - ln(A) = Se_a/(R(H-G))e^{SE_a/(R(H-G))}.$$

Let $S(R(H-G)) = b$. Therefore,

$$ln(k) - ln(A) = ln(k/A) = bE_a e^{bE_a} = S/(S(H-G))E_a e^{bE_a}.$$

Now let $ln(k/A) = x$. Therefore,

$$x(R(H-G)) = SE_a e^{bE_a}$$

$$S = x(E_a(H-G))/(E_a e^{bE_a})$$

state in which the system has no entropy contribution to the enthalpy. At absolute zero the entropy term disappears. At all other temperatures, the system will have an entropy component. In this example, the kinetic theory of gases states that the pressure of the gas within the system is due to the motion and collisions of the gas molecules with the boundaries of the container. If the temperature is then raised, the enthalpy of the system increases (ΔH), as does the Gibbs' free energy (ΔG) that describes the amount of energy in the system that is available to do useful work. However, as the temperature increases, so too does the negative contribution of the entropy component ($-T\Delta S$). The increasing entropy of the system is attributed to the higher energy and hence more random motions of the gas molecules. This can be understood as a portion of the system's ability to do useful work being redirected to or consumed by the non-useful work of driving the more random motions of the gas molecules. This principle is easily extended to all physical systems. Phase changes are a very easily understood application of entropy, and the effect is especially easy to see in the phase change from solid ice at 0°C (32°F) to liquid water at the same temperature. A plot of temperature versus time of this transition will be seen to increase steadily until the temperature of 0°C is attained. At that point it will remain constant while both solid ice and liquid water exist at the same temperature. When no more solid ice remains the temperature will then continue to rise steadily. The energy that is required to enter the system to overcome the crystal structure of ice to allow it to become liquid water is termed the "latent heat of fusion", and can be thought of as the energy required to counter the increase in entropy of the water molecules as they go from being bound into the highly ordered solid state to being free to move about in the completely random liquid state. The concept has been extended into essentially all fields of study and application, including creationist theory, although the physical sense of the term is meaningful only in technical fields that rely on thermodynamics.

CHEMICAL APPLICATIONS

A significant physical aspect of chemical reactions is a quantity termed the activation energy, E_a. This is the energy that two components must have in the transition state in order to be able to carry on through to form products. A transition state is formed when the two reacting components combine to produce a structure that can either revert back to the original components or continue the reaction progression to form the products. The energy of the transition state is the activation energy of the reaction, and depending on the particular nature of the reaction it may range from a value of zero to no upper limit. The activation energy of a reaction is temperature-dependent, and is found in the Arrhenius equation describing the rate of a reaction

$$k = Ae^{-E_a/(RT)}$$

where k is the reaction rate, T is the absolute temperature, R is the gas constant from the ideal gas law ($PV = nRT$), and A is a pre-exponential factor for the specific reaction. Because the equation for the enthalpy of the system is also temperature dependent, it can be rearranged and used directly in the Arrhenius equation. This is a valuable aspect of research into many different systems, and especially biochemistry and pharmacology. Pharmaceutical compounds are designed to work within the most complex physical system and must do so without causing that system to fail. Thus it is vitally important that the manner in which the energy balances within the system are modeled and understood in fine detail. The particular discipline is termed pharmacokinetics. In another field, chemical engineers use entropy calculations in designing the functional parameters of a process.

ENTROPY AND THERMAL SYSTEMS

All machines and physical systems that function in a cyclic manner for whatever purpose are affected by entropy. Many aspects of the operation of such systems must be considered in their design and construction, because no system is perfect. Even the most minor amount of friction removes energy available for useful work from the system by producing heat. Accordingly, physical system design in all fields of engineering consider entropy as an important factor in designing to maximize the efficiency of the system. Combustion engines, regardless of their application, are effectively heat engines that convert the enthalpy of fuel into useful work. Heating and cooling systems are another set of physical systems that function exclusively by the transfer of heat. The concept of entropy is so universal that astrophysicists employ it in studying the function of the Sun and other stars, and have even theorized entropy as the ultimate destroyer of the universe, slowly depleting the enthalpy of the universe until everything in it is at a temperature of absolute zero.

—Richard M. Renneboog M.Sc.

BIBLIOGRAPHY

Atkins, P. W., and P. W. Atkins. *The Laws of Thermodynamics: A Very Short Introduction.* Oxford; New York: Oxford University Press, 2010. Print.

Ben-Naim, Arieh. *A Farewell to Entropy: Statistical Thermodynamics Based on Information: S=logW.* Singapore: World Scientific, 2011. Print.

Dugdale, J. S. *Entropy and Its Physical Meaning.* London; Bristol, PA: Taylor & Francis, 1996. Print.

Kreuzer, H. J., and Isaac Tamblyn. *Thermodynamics.* Singapore; Hackensack, N.J.: World Scientific, 2010. Print.

Singh, Vijay P. *Entropy Theory and Its Application in Environmental and Water Engineering.* Chichester, Hoboken, NJ: Wiley-Blackwell, 2013. Print.

Starzak, Michael E. *Energy and Entropy.* New York: Springer, 2014. Print.

Thess, André. *The Entropy Principle Thermodynamics for the Unsatisfied.* Berlin: Springer-Verlag, 2014. Print.

Wilson, A. G. *Entropy in Urban and Regional Modelling.* New York: Routledge, 2013. Print.

EULER'S LAWS OF MOTION

FIELDS OF STUDY

Classical Mechanics

SUMMARY

Fifty years after Isaac Newton published his laws of motion in the *Principia,* Swiss mathematician and physicist Leonhard Euler added to them with his own laws of motion. Euler's laws take Newton's laws, which apply only to a singular point of mass, and extend them to an entire rigid body and to rotational motion.

PRINCIPAL TERMS

- **angular momentum:** the momentum of a rotating object, equal to the object's moment of inertia (its rotational inertia) times its angular velocity; also called rotational momentum.
- **center of mass:** the point in an object or system around which the mass of said object or system is evenly distributed.
- **fixed reference frame:** a frame of reference that is fixed to the environment and not to the subject being observed; sometimes specified as "Earth-fixed" or "space-fixed."
- **linear momentum:** an object's mass times its velocity; often called simply "momentum," as the basic concept is defined in terms of an object moving in a straight line.
- **Newton's laws of motion:** three laws devised by physicist and mathematician Isaac Newton to describe the motion of objects in relation to the forces acting on them.
- **rigid body:** an idealization of a solid object that assumes that it cannot be deformed by the forces acting on it.
- **torque:** the tendency of a force to cause an object to rotate, defined mathematically as the rate of change of the object's angular momentum; also called moment of force.

EXTENDING NEWTON'S LAWS OF MOTION

In 1686, English physicist and mathematician Isaac Newton (1643–1727) published his *Philosophiae Naturalis Principia Mathematica.* In it, he laid out three physical laws that govern interactions between objects and the forces acting on them. These laws are now known as Newton's laws of motion. The first law states that an object at rest tends to stay at rest, and an object in motion stays in motion, unless acted on by an outside force. The second law states that the net force applied to an object is equal to the resulting change in the object's momentum per unit of time—that is, its mass times its acceleration. The third law states that every action produces an equal and opposite reaction. These laws form the foundation of classical mechanics and, by extension, of all of modern physics.

Newton's laws are generalizations, devised with reference to a singular point of mass that takes up zero hypothetical space. Fifty years after they were published, Swiss physicist Leonhard Euler (1707–83) extended them by applying them to continuous bodies made up of these Newtonian point masses.

EULER'S FIRST LAW OF MOTION
Mathematically, Newton's second law is stated as

$$F = ma$$

where F is the net force applied to an object (specifically, a point mass), m is its mass, and a is its acceleration. Euler's first law extends this law to apply to an entire rigid body. A rigid body is a solid object that does not bend, twist, compress, or otherwise deform when acted on by an outside force. When assuming a rigid body, Euler's first law says, the total force acting on the body is equal to the sum of the forces acting on each individual particle in the body.

Euler's first law depends on the concept of center of mass. A point mass located at a body's center of mass and having the same mass as the body will follow the same trajectory as said body when acted on by the same force. For the human body, for example, the center of mass is typically located beneath the belly button. In practical terms, this means that Newton's first law ($F = ma$) can be rewritten for a rigid body as

$$\Sigma F = ma_{cm}$$

where ΣF indicates the sum of all external forces acting on all particles of the body (as the symbol represents summation) and a_{cm} is the acceleration of the body's center of mass.

LINEAR AND ANGULAR MOMENTUM
While the term "momentum" is usually used to mean linear momentum, there are in fact two types of momentum. Linear momentum is the momentum of an object moving in a straight line. For a point mass, it is defined as

$$p = mv$$

where p is momentum, m is mass, and v is velocity.

In contrast, angular momentum is the momentum of an object that is rotating or traveling in a circle. Angular momentum (L) is defined as

$$L = I\omega$$

where I is the moment of inertia and ω is the angular velocity. Moment of inertia, or rotational inertia, is the inertia of a rotating body. It represents an object's tendency to resist angular acceleration, just as an object tends to resist linear acceleration. Moment of inertia can be further broken down in terms of mass (m) and the distance of the object from the point of rotation (r), that is, the radius of the circular path the object is traveling:

$$I = r^2 m$$

Angular velocity, meanwhile, can be written in terms of the radius of the path of circular motion (r) and the object's tangential linear velocity (v) at a given instant:

$$\omega = v/r.$$

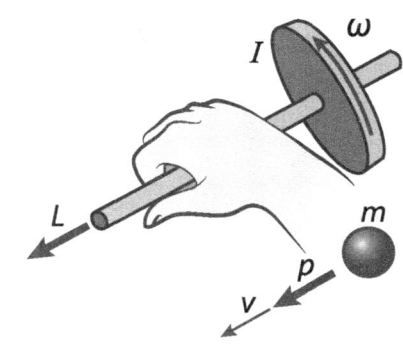

Angular Momentum	=	Moment of Inertia	•	Angular Velocity
L	=	I	•	ω
Linear Momentum	=	Mass	•	Velocity
p	=	m	•	v

Euler's first law is used to calculate linear momentum (p) by multiplying mass (m) of an object by its velocity (v). Euler's second law is used to calculate angular momentum (L) by multiplying moment of inertia (I) by angular velocity (ω). Drawing of a right hand holding an axle with a wheel attached; the fingers curve in the same direction as the angular velocity and the thumb points in the direction of the angular momentum.

SAMPLE PROBLEM

A child, Sarah, is riding the merry-go-round at a local playground. The merry-go-round is 3 meters (m) across, and Sarah, who has a mass of 30 kilograms (kg), is sitting on the outside edge. If the merry-go-round completes one full rotation every two seconds (s), calculate Sarah's angular momentum (L) in units of newton-meter-seconds (n·m·s). Note that one newton-meter-second is equal to one kilogram–square meter per second (kg·m^2/s).

Answer:
Because Sarah is so small relative to the circular path of her motion, she can be treated as a point mass, so the simpler equations for circular motion can be used:

$$L = I\omega = r2m\omega$$

Her mass is given as 30 kg. She is sitting on the outside edge of the merry-go-round, so the circular path of her motion is the same as the circle formed by the ride. The radius of the ride is simply half the diameter:

$$r = d/2$$

$$r = 3 \text{ m} / 2$$

$$r = 1.5 \text{ m}$$

Angular velocity, which measures revolutions per second, is given in units of radians per second (rad/s). One full circle is made up of 2π radians. If Sarah completes one full circle every two seconds, calculate her angular velocity in rad/s:

$$\omega = 2\pi/2 \text{ s}$$

$$\omega = \pi/\text{s}$$

Plug these values into the equation:

$$L = r^2 m \omega$$

$$L = (1.5 \text{ m})^2 (30 \text{ kg})(\pi/\text{s})$$

$$L = (2.25 \text{ m}^2)(30 \text{ kg})(\pi/\text{s})$$

$$L = 212.058 \text{ kg·m}^2/\text{s} = 212.0575 \text{ n·m·s}$$

Sarah's angular momentum is approximately 212.058 newton-meter-seconds.

Thus, in terms of radius, mass, and linear velocity, the equation for angular momentum is as follows:

$$L = (r^2 m)(v/r) = rmv.$$

These equations apply to any type of circular motion. They work for spinning motion about an internal axis, such as the rotation of a planet. They also work for motion in a circular path around an external axis, such as the orbit of a planet around the sun.

EULER'S SECOND LAW OF MOTION
Euler's second law of motion extends Newton's laws to apply to rigid bodies in circular motion. It states that for an object rotating about a given point, whether that point is an external axis in a fixed reference frame or the object's own center of mass, the rate of change of angular momentum is equal to the sum of all external torques, or moments of force, acting about that point. Solving this equation involves differential calculus, specifically finding the derivative of L with respect to time (t). However, if the object is only moving in two dimensions (e.g., a flat, spinning disk), the equation can be rewritten in terms of the center of mass:

$$\Sigma M = r_{cm} \times a_{cm} m + I\alpha.$$

Here, M is moment of force, r_{cm} is the distance of the object's center of mass from the axis of rotation, a_{cm} is the tangential acceleration of the center of mass relative to the axis, and α is the object's angular acceleration. As before, m is the mass of the object, and I is its moment of inertia.

WHAT EULER'S LAWS MEAN
Euler's laws can be more difficult to calculate than the equations derived from Newton's laws. In particular, applying Euler's second law in three dimensions requires differential calculus. However, the fundamental idea behind Euler's math is easy to understand. Newton's laws assume a single, infinitely small point mass. Euler's laws assume rigid bodies made up of a continuous collection of these point masses all stuck together. This is a reasonable, if very basic, description of the actual composition of matter: countless tiny atoms bonded together, themselves consisting of subatomic particles linked by fundamental forces.

—*Kenrick Vezina, MS*

BIBLIOGRAPHY
Coddington, Richard C. "Inertial Frame, Euler's First Law." Department of Agricultural and Biological Engineering. University of Illinois at Urbana-Champaign, 2015. Web. 27 Aug. 2015.

Hall, Nancy. "Newton's Laws of Motion." *The Beginner's Guide to Aeronautics*. NASA, 5 May 2015. Web. 27 Aug. 2015.

Henderson, Tom. "Motion in Two Dimensions." *The Physics Classroom*: Physics Classroom, 2012. Digital file.

Negahban, Mehrdad. "Equations of Motion for a Rigid Body (Euler's Laws)." Department of Engineering Mechanics. University of Nebraska, 1999–2002. Web. 27 Aug. 2015.

Ruina, Andy, and Rudra Pratap. *Introduction to Statics and Dynamics*. N.p.: Oxford University Press, 2010. PDF file.

Simanek, Donald. "Kinematics." A Brief Course in Classical Mechanics. Lock Haven U, Feb. 2005. Web. 27 Aug. 2015.

FALLING BODIES AND PHYSICS

FIELDS OF STUDY

Classical Mechanics

SUMMARY

How and why things fall—whether they are falling down, arcing through the air, or falling into orbit around one another—are simultaneously two of the most familiar and most Summary topics in physics. Isaac Newton was the first to write useful formulas to explain gravity and its effects on objects more accurately. In the early twentieth century, Albert Einstein's theory of general relativity replaced Newton's theory of gravity. However, Newton's equations are still used often due to their relative simplicity and accuracy regarding everyday scales of measurement.

PRINCIPAL TERMS

- **centripetal force:** for an object moving in a uniform circle, the force directed toward the center of the circle.
- **free fall:** falling only under the influence of gravity.
- **gravitational force:** any two objects in the universe attract one another through a force proportionate to their mass. This attractive force is gravity.
- **instantaneous velocity:** the velocity (speed and direction of travel) of an object in motion at any one instant of time.
- **reference frame:** the velocity of an object relative to the objects around it and the point of observation.
- **terminal velocity:** the maximum velocity of an object in free fall, determined by the drag and buoyancy of the object relative to the force of gravity.
- **trajectory:** the path of a thrown or falling object, such as a baseball.

Gravity's Influence: Falling Objects

Every object in the universe, from a Ping-Pong ball to a planet, exerts a gravitational force on objects around it. This gravitational force is proportional to the size (mass) of the object and the distance between it and the object it is acting on. Standing on Earth's surface, an individual experiences a stronger gravitational force from Earth than Jupiter, which is nearly 590 million kilometers (370 million miles) away. Gravity is the source of an object's weight. A scale is simply measuring the force with which an object is being pulled toward the gravitational source. In essence, objects (such as houses, people, even planes) are always falling toward the center of the earth, but the resistance of the crust stops them. This is why astronauts weigh so much less on the surface of the moon. The moon has much less mass than Earth and exerts less gravitational pull on the astronauts there. They are also so far from Earth that its gravity has a small effect on their weight as compared to the moon's. A ball thrown from one person to another will have a vertical displacement and a horizontal displacement. The velocity (m/s) of the baseball can be broken down into vertical and horizontal components. The ball maintains a constant horizontal velocity, but the vertical velocity changes as the ball rises to its maximum vertical displacement and then falls to the catcher.

When an object falls under the force of Earth's gravity, it accelerates at a uniform rate. This rate varies according to distance from Earth and resistance from air or another medium an object is moving through. It also changes under the gravitational pull of other bodies, such as the moon. However, for most purposes, a standard gravitational acceleration (written as g) of 9.8 meters per second per second (m/s^2)—which represents an object in free fall at sea level—is sufficient.

A common free-fall myth insists that a penny dropped from the top of the Empire State Building will become a lethal projectile. This is because it is accelerating at 9.8 m/s² for a distance of several hundred feet. Terminal velocity, however, limits the penny to a nonlethal (if painful) velocity. The faster an object falls, the more powerful the effect of drag from the air around it becomes. Eventually, the object reaches a point where the acceleration due to gravity is balanced by the deceleration due to drag. The acceleration becomes zero, and no matter how much longer the object falls, it will not speed up any further. The object's instantaneous velocity (vf) is the speed and direction of its motion at a given moment. This is calculated as the product of the time spent (t) and the acceleration due to gravity (g).

An object dropped straight down is not, in absolute terms, moving in a straight line. The earth is rotating about its axis, and the entire planet is orbiting the sun. Therefore, the reference frame chosen for calculation is very important. English physicist Isaac Newton (1642–1727) assumed the stars in the sky were fixed points and made his calculations of motion relative to them. This method was good enough to lead to generally accurate predictions in most cases. However, scientists have since learned that the stars are not stationary at all. Similarly, for most calculations, it is safe to ignore Earth's rotation and movement through space. If someone wants to determine the velocity of a penny dropped from the Empire State Building, he or she typically is not interested in its velocity relative to the rest of the universe—just relative to the building, the observer, and the street below.

The trajectory of a falling or thrown object is different depending on the frame of reference used to define it. If the frame of reference were set by two children playing catch on Earth's surface, a thrown ball's trajectory would be a simple arc. If the sun were set as the key reference point, then its trajectory would be complicated by the movement of the thrown ball around Earth and around the sun.

THE HISTORY OF GRAVITATIONAL THEORY

Theories to explain why and how objects fall are as old as antiquity. The ancient Greek philosopher Aristotle (ca. 384–ca. 322 BCE) theorized that more massive objects were pulled toward the center of the universe as a result of their innate heaviness. He predicted that heavy objects should fall faster than lighter ones.

In the seventeenth century CE, Galileo Galilei (1564–1642) found that objects fall at the same rate

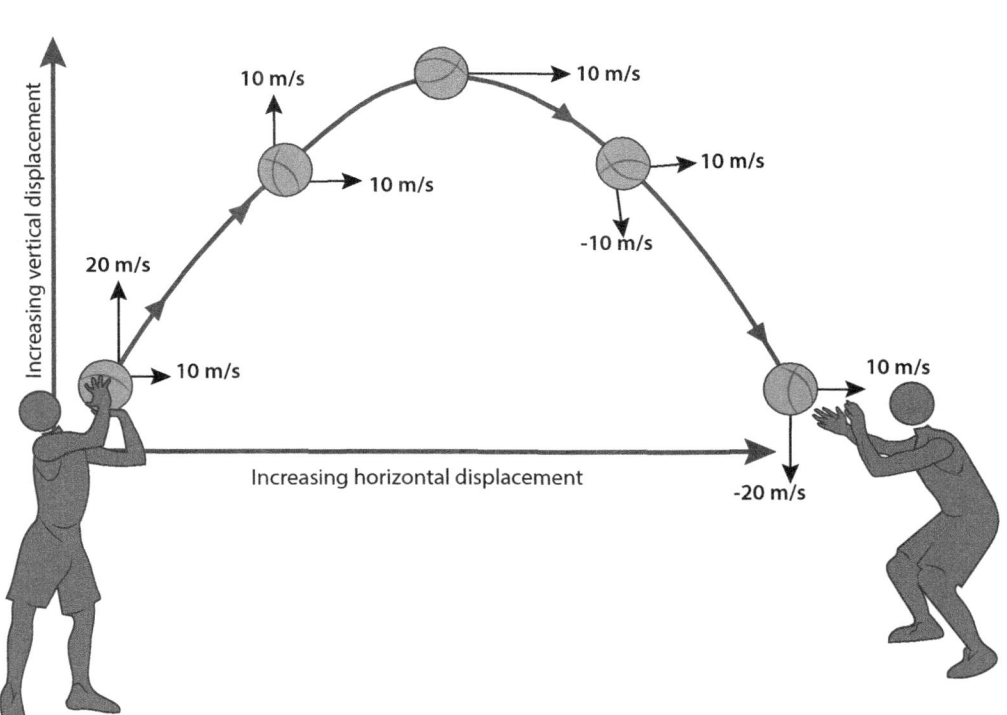

A ball thrown from one person to another has a vertical and horizontal displacement. The ball's velocity can be broken down into vertical and horizontal components. Horizontal velocity remains constant, but vertical velocity changes as ball rises and then falls.

regardless of their mass, contrary to Aristotle's ideas. Galileo is famously said to have dropped objects of various masses off the Leaning Tower of Pisa and recorded the time it took them to hit the ground, demonstrating that objects fell at the same rate regardless of their mass. Historians have since decided this was more likely a theoretical experiment rather than a physical one.

Later that same century, Newton codified the inverse-square law of gravitation. This law stated that the gravitational force exerted by one object on another is directly proportional to the object's mass and inversely proportional to the square of the distance between the centers of the two objects. Newton's equations were remarkably effective under most circumstances. However, when dealing with extremes (such as the orbit of Mercury), small errors in their predictions were found.

In 1915 Albert Einstein (1879–1955) introduced the theory of general relativity, which resolved these errors in prediction by proposing that gravity is the effect of mass and energy warping space-time. A massive object like the sun creates a large depression in space-time, causing smaller objects like the earth to move around it. So far, Einstein's predictions have been very accurate. Newton's theory has been replaced, but Newtonian gravity is still in common use for everyday calculations. Einstein's theory is used mainly in cases of extreme gravity or when great accuracy is needed. When dealing with falling or thrown objects, it is generally safe to assume Newton's laws apply.

Circular Motion and Centripetal Force

For objects moving in a circle—such as a ball being spun on a string or the moon orbiting the earth—centripetal force describes the force acting against the object's momentum and pulling it toward the center of the circular path. In the case of orbits, gravity is a centripetal force.

This concept underpins the most practical idea for how to create artificial gravity in space. Consider a Tilt-a-Whirl at a fair: as the ride spins, the inside passenger feels heavier and pinned against the inside of the cart. A spacecraft spinning at a sufficient speed would generate a centripetal force that would mimic a planet's gravity, even in the void of space. If the floor of the spacecraft were oriented toward the center of the spinning ring, it would be the astronaut's feet that would be pinned, creating a feeling similar to gravity. However, the spacecraft would need to be larger than anything conceived of so far if it were to spin at a sustainable speed. It would also need to prevent the astronauts inside from experiencing

---

SAMPLE PROBLEM

Consider a baseball weighing 0.15 kilograms dropped from the top of a building. The baseball strikes the ground with a velocity of 20 meters per second. With this information, calculate the height of the building it was dropped from and the time it took to land.

Answer:

First, make note of what is known. The problem made no mention of extreme wind or other complicating factors. From this, it is safe to assume that simple Newtonian equations for acceleration due to gravity will suffice. Neither the object's mass nor wind resistance will make a significant difference in the rate of its descent. Because the ball is dropped from a stationary point, it is safe to assume the trajectory it travels is a straight line down to the ground. Since everyday conditions are a valid assumption, the value of acceleration due to gravity (g) can be set to 9.8 m/s².

Start by calculating (t). Using the formula for the instantaneous velocity (v_f), plug in the velocity for the ball at impact and the acceleration due to gravity:

$$v_f = gt$$

$$20 \text{ m/s} = 9.8 \text{ m/s}^2 \times t$$

$$t = \frac{20 \text{ m/s}}{9.8 \text{ m/s}^2}$$

$$t = 2.04 \text{ s}$$

With a value for time (t), distance (d) fallen can be calculated using the following equation:

$$d = \tfrac{1}{2} g t^2$$

$$d = \frac{(9.8 \text{ m/s}^2)(2.04 \text{ s})^2}{2}$$

$$d = \frac{(9.8 \text{ m/s}^2)(4.162 \text{ s}^2)}{2}$$

$$d = 20.39 \text{ m}$$

a dizzying, potentially dangerous difference in the force acting on their feet versus their heads.

THE FUTURE OF GRAVITATIONAL PHYSICS

Although its predictions have been proven by many experiments, the gravitational theory provided by general relativity is not perfect. In particular, it is incompatible with modern quantum mechanics. Attempts to translate general relativity into a quantum framework have proven incomplete at best. Thus, quantum gravity—the field of physics that attempts to describe gravity using the principles of quantum mechanics—is an area of much contention in modern theoretical physics. If the trajectory of gravitational theory across history is any indication, scientists will continue to refine gravitational theory so that it makes accurate predictions under increasingly extreme conditions.

—*Kenrick Vezina, MS*

BIBLIOGRAPHY

"Falling Bodies and Uniformly Accelerated Motion." *Encyclopaedia Britannica*. Encyclopaedia Britannica, 29 Jan. 2015. Web. 27 May 2015.

Feltman, Rachel. "Why Don't We Have Artificial Gravity?" *Popular Mechanics*. Hearst Communications, 3 May 2013. Web. 27 May 2015.

MacDougal, Douglas W. *Newton's Gravity: An Introductory Guide to the Mechanics of the Universe*. New York: Springer, 2012. Print.

"Reference Frame." *Encyclopaedia Britannica*. Encyclopaedia Britannica, 1 Feb. 2015. Web. 27 May 2015.

Taylor, Nola. "Einstein's Theory of General Relativity." *Space*. Purch, 10 Apr. 2015. Web. 27 May 2015.

"Terminal Velocity (Gravity and Drag)." NASA. NASA, 12 June 2014. Web. 27 May 2015.

Wolchover, Natalie. "Could a Penny Dropped off a Skyscraper Actually Kill You?" *Scientific American*. Scientific Amer., 5 Mar. 2012. Web. 27 May 2015.

FARADAY'S LAW

FIELDS OF STUDY

Electromagnetism

SUMMARY

Electricity and magnetism are directly related through Faraday's law of magnetic induction. Along with Lenz's, Ampère's, and Gauss's laws, Faraday's law is expressed as one of Maxwell's equations. It is fundamental to how electrical devices function.

PRINCIPAL TERMS

- **electromagnetic field:** a region of space in which an electrically charged mass experiences an applied force.
- **electromotive force:** the energy supplied by an electric source such as a battery, measured in volts.
- **Lenz's law:** the principle that the current induced in a conductor by a moving magnetic field flows in the direction opposite to the motion of the magnetic field.
- **magnetic flux:** the measure of the strength of the magnetic field surrounding an active electrical conductor or a magnet, equal to the product of the external magnetic field (B), the area of the affected surface (A), and the cosine of the angle between them ($\cos \theta$).
- **magnetic induction:** the generation of magnetism within a magnetizable material by proximity to a magnetic field.
- **Maxwell's equations:** the mathematical relationships governing electromagnetism, as formulated by James Clerk Maxwell.
- **voltage:** the difference in electric potential between two points, measured in volts.

MAGNETS AND ELECTROMAGNETISM

Electricity and magnetism are familiar to most people, but the way in which they are related is not. In physics, both electricity and magnetism are aspects of electromagnetism and are characterized by the existence of an electromagnetic field. Such a field can be a magnetic field, an electric field, or a combination of both. An electric field exists because of electric charge, which can be either positive or negative. Negative electric charge is associated with the electron, while positive charge is associated with the proton.

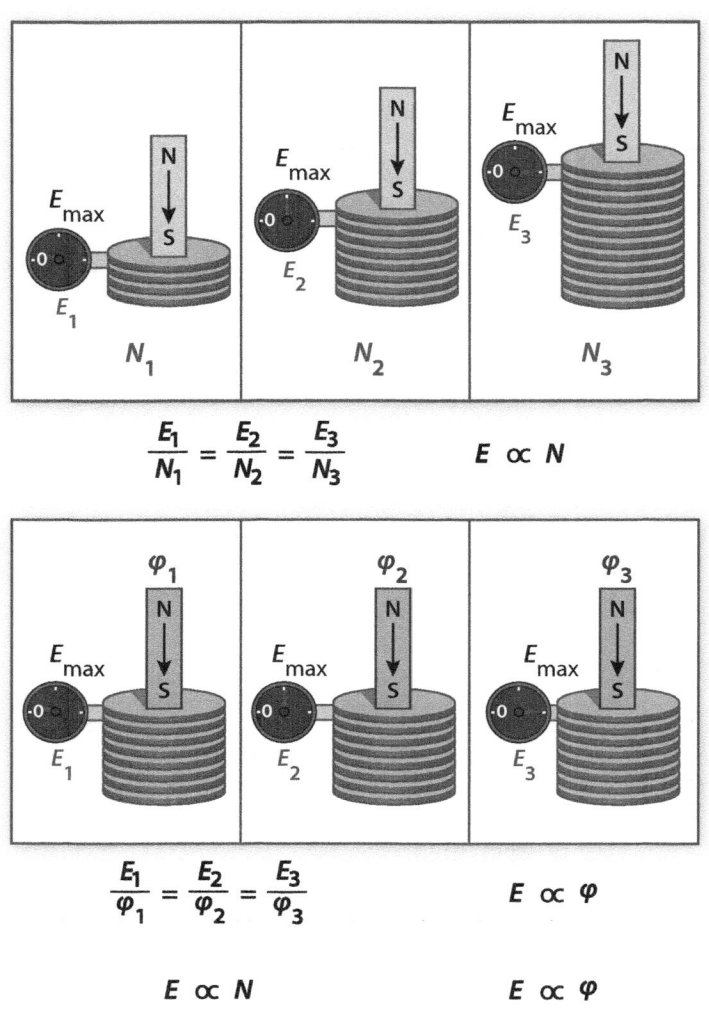

Diagrams showing the setup for Faraday's experiments to demonstrate the equation for Faraday's law. The electromotive force (emf) is directly proportional to the number of turns in a coil. When a magnet passes through a coiled conductor, the emf will increase as the number of turns increases. The emf is directly proportional to the flux of a magnet. When magnets of different sizes, or fluxes, pass through a coiled conductor, the emf will increase as the flux of the magnet increases. The emf is therefore directly proportional to the number of turns (N) times the flux (φ).

Magnetism, meanwhile, has been known since ancient times, as it occurs naturally in a type of iron ore. Seafarers used slivers of such ore to construct a simple compass called a lodestone, which always pointed north, just as modern compasses do. Earth's molten nickel-iron core generates a magnetic field around the planet, and the magnetic needle indicator of a compass aligns with this field such that one end always points toward Earth's magnetic north pole. The degree to which a compass needle aligns to a magnetic field depends on the magnetic flux within the field. If the magnetic flux is stronger than that of Earth's field, the compass needle will point toward the stronger magnet. Magnetism can be produced by magnetic induction in any object made of a magnetizable material. For instance, one can pick up an iron nail with a magnet and then use that nail to pick up other nails.

If an electric conductor is moving relative to a magnetic field, the magnetic field will generate an electric current within the conductor. This current will then generate another magnetic field about the conductor. Lenz's law states that the direction of this induced magnetic field opposes that of the magnetic field that produced the current, creating a repulsion between the two fields. The electric current that flows in the conductor is defined as the movement of electrons from atom to atom. One electron carries a single electrical charge, and so it experiences a force when in a moving magnetic field, as do all electrically charged bodies. Similarly, a magnet in a moving electric field experiences a force acting upon it. This principle, described by Faraday's law, is the basis for how generators and electric motors function.

Law of Induction

English physicist and chemist Michael Faraday (1791–1867) studied the phenomena of electricity and magnetism. In 1821, he demonstrated that the magnetic field produced by an electric current can induce movement of another magnet. This is the basic principle of the electric motor. Ten years later, in 1831, Faraday showed that a moving magnet induces an electric current in a conductor. He realized that a specific relationship exists between electricity and magnetism, which he described using his law of induction.

Faraday's law of induction is based on three principles. First, a changing magnetic field induces an electromotive force, or emf, within a conductor.

(The abbreviation "emf" is typically lowercased to distinguish it from EMF, for electromagnetic field.) This is simply demonstrated by passing a magnet across a conductor and measuring the resulting voltage between the ends of the conductor. Second, the magnitude of the emf is proportional to the rate at which the magnetic field changes. That is, the faster the magnet is moved past the conductor, the greater the emf produced. Third, the direction of the induced emf depends on the direction in which the magnet moves. This reflects Lenz's law.

Simply stated, Faraday's law says that when a magnetic field moves relative to a conductor, or vice versa, a voltage is induced in the conductor. The magnitude of the voltage produced when the conductor is in motion, called the motional emf, depends directly on the relative velocity of the magnetic field and the conductor. Slow movement produces a low voltage, and fast movement produces a higher voltage. Motional emf is calculated as the product of the magnetic field (B), the length of the conductor (l), and the velocity (v).

Faraday also theorized that there are "magnetic lines of force"—now called magnetic field lines—that make up the magnetic flux. Accordingly, each time a line crosses a conductor, it induces the corresponding emf. The more times that a line can cross the same conductor, the more times that it can induce the emf. Thus, magnetic induction is more effective for a coil of wire than for a straight wire. Each loop of wire within the coil adds to the number of times any magnetic field line can cross that conductor and induce an emf. The total voltage produced is therefore be related directly to the number of turns in the coil. The relative speed of the conductor and the magnetic field is expressed as the continuous rate of change of the magnetic flux. Faraday's law is written as

$$\xi = -N(\Delta\varphi/\Delta t)$$

where ξ is the induced emf, measured in volts; N is the number of turns in the coil; and $\Delta\varphi/\Delta t$ is the change in magnetic flux (φ) over time (t), measured in webers (Wb). One weber is equal to one volt-second (V·s). The negative sign shows that the induced emf will oppose the change in flux, per Lenz's law.

Maxwell's Equations

As Faraday and other scientists began to understand electromagnetism better through their observations and experiments, the essential question of how electricity and magnetism are transmitted through space was considered. Scottish physicist James Clerk Maxwell (1831–79) theorized that they must travel as electromagnetic waves. On this principle, he consolidated Ampère's law of magnetism, Gauss's laws of electricity and magnetism, and Faraday's law of

---

SAMPLE PROBLEM

A coil has 250 turns of wire on it. The magnetic flux passing through it changes from 0.5 webers (Wb) to 4.6 Wb over a period of 3 seconds. What voltage is induced in the coil? Calculate using Faraday's law.

Answer:

First, to determine the rate of change of the magnetic flux over time ($\Delta\varphi/\Delta t$), subtract the final flux from the initial flux. Next, divide by the change in time (the final time minus the starting time). In this case, the start time is zero, so the change in time is equal to the final time.

$$\Delta\varphi/\Delta t = (4.6 \text{ Wb} - 0.5 \text{ Wb}) / 3 \text{ s}$$

$$\Delta\varphi/\Delta t = 4.1 \text{ Wb} / 3 \text{ s} = 1.367 \text{ Wb/s}$$

Next, plug this value into the equation and calculate the induced voltage:

$$\xi = -N(\Delta\varphi/\Delta t)$$

$$\xi = -(250)(1.367 \text{ Wb/s})$$

$$\xi = -341.75 \text{ Wb/s} = -341.75 \text{ V·s/s}$$

$$\xi = -341.75 \text{ V}$$

The induced voltage is −341.75 volts.

magnetic induction into precise, related mathematical descriptions. These descriptions are known as Maxwell's equations.

The relationships described by Maxwell's equations are central to circuits, almost all of which contain some combination of transformers and inductors. A typical electronic device is the RLC circuit, consisting of a resistor (R), an inductor (L), and a capacitor (C). When designing such circuits, precise values must be calculated for the induced emf of an inductor so that the circuit operates on the input electromagnetic signal as it should. Transformers particularly depend on the principles of magnetic induction for their operation. A transformer consists of two separate coils wrapped around a single iron core and connected to each other through a magnetic field. The output coil of the transformer increases or decreases the voltage supplied by the input coil, depending on the number of turns of wire on each coil.

—*Richard M. Renneboog, MSc*

BIBLIOGRAPHY

Calle, Carlos I. *Superstrings and Other Things: A Guide to Physics*. 2nd ed. Boca Raton: CRC, 2010. Print.

Fitzpatrick, Richard. *Maxwell's Equations and the Principles of Electromagnetism*. Hingham: Infinity Science, 2008. Print.

Fleisch, Daniel. *A Student's Guide to Maxwell's Equations*. New York: Cambridge University Press, 2008. Print.

Gross, Charles A. *Electric Machines*. Boca Raton: CRC, 2007. Print.

Kelly, P. F. *Electricity and Magnetism*. Boca Raton: CRC, 2015. Print.

Nave, R. "Faraday's Law." *HyperPhysics*. Georgia State U, n.d. Web. 13 Aug. 2015.

Robbins, Allan H., and Wilhelm C. Miller. *Circuit Analysis: Theory and Practice*. 5th ed. Clifton Park: Delmar, 2013. Print.

FEYNMAN DIAGRAMS

FIELDS OF STUDY

Electrodynamics, Particle Physics

SUMMARY

Feynman diagrams are a systematic method for representing the physics underlying the interaction of elementary particles. They were introduced by Richard P. Feynman as an aid to visualizing terms in a perturbation solution for the time evolution operator of a quantum system. Each diagram represents one term in a power series expansion of the time evolution operator describing a scattering process, so the diagrams are quite amenable to physical interpretation. A set of "Feynman rules" allows the diagrams to be constructed directly without reference to the underlying mathematics.

PRINCIPAL TERMS

- **agent:** A boson, one of the carriers of the fundamental forces: a graviton, photon, gluon, or W or Z particle.
- **boson:** an elementary particle with (x component of) spin an even multiple of $h/4\pi$, possibly zero.
- **fermion:** an elementary particle with (z component of) spin an odd multiple of $h/4\pi$.
- **perturbation method:** A way of finding an approximate solution of a mathematical problem as a sum of solutions to a simpler problem, first developed in celestial mechanics.
- **propagator:** A formal solution of the quantum equations of motion; applies to the wave function at time 0, it yields the wave function at a later time.
- **scattering process:** an interaction of particles and agents, producing particles in well defined final states.

THE PROCESS OF SOLVING PHYSICS EQUATIONS

The process of solving many of the equations of physics as applied to different systems is more difficult than first appears to be true. In most cases, one cannot find an exact solution in closed form. In those cases one may have to express the solution as an infinite series. Sometimes the series can be summed to give a precise result. As an example consider the sum:

Feynman diagrams

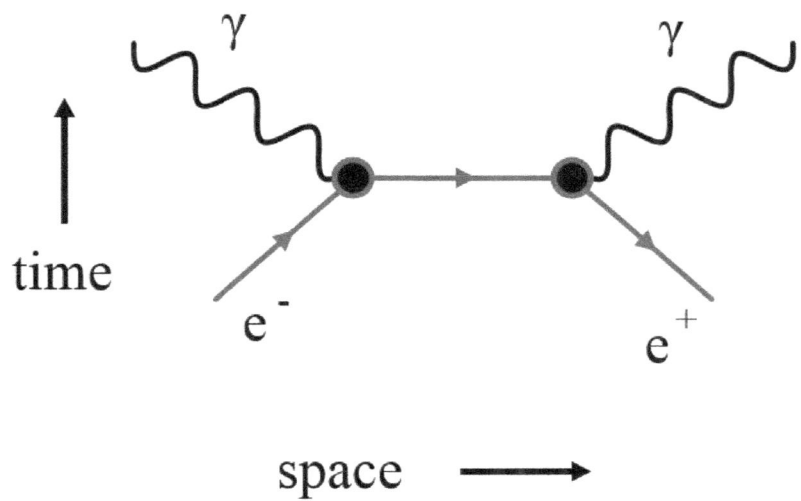

Feynman diagram of electron/positron annihilation.

$$1/2 + 1/4 + 1/8 + 1/16 + \ldots$$

That the sum of the infinite series of factions =1. To convince yourself that this is true, draw a square, then shade half of it, then half of the remaining half and you quickly realize that the series approaches one as the number of terms becomes infinite.

Consider two electrons repelling each other by the electrical force. The Feynman diagram would include the world lines of the two particles uninterrupted, then exchanging one photon then two and so on. In many cases the diagrams can be written directly, without explicit reference the mathematical processes they describe.

THE MAGICIAN

Feynman has been called the "magician" by his fellow physicists because of his ability to get mathematically meaningful results by drawing a few pictures. The figure shows a very simple process. A gamma ray is absorbed by a heavy nucleus (not shown) and its energy is converted into an electron and a positron. The electron and positron go their separate ways until the positron bumps into another electron their mass energy being turned into one or more gamma rays. Feynman's interpretation was a little different. Reverse the direction of the arrow head representing the positron and now you can read the picture as an electron evolving normally in time until it interacts with a gamma ray and decides to travel backwards in time until it bumps into a second gamma ray and behave like a normal electron again.

Feynman's exceptional creativity became apparent even to his high school physics teacher, Mr. Ralph Bader, who took steps to make sure his most promising student found challenge in his classroom. As noted in the *Feynman Lectures on Physics*, it was Ralph Bader who first drew his attention to the Principle of Least Action in classical physics. He based his doctoral thesis at Princeton University on the quantum version of this principle, which would ultimately become one of the foundations of quantum electrodynamics.

The lines in contemporary Feynman diagrams are wiggly for the bosons that carry the interaction forces from one fermion to another and smooth for the fermions themselves. They represent the world lines of the particles in the relativistic sense. They can be grouped in twos and threes (representing quarks bound into for mesons and hadrons, the two families of particles related to the proton and neutron.

A FEW WORDS ABOUT RICHARD FEYNMAN

Richard Feynman, one of the most creative individuals in twentieth century physics. Feynman worked at Los Alamos during the Second World War while completing his PhD thesis at Princeton University under John A. Wheeler. In the post World War II years he was principally involved in developing quantum electrodynamics, for which he shared the Nobel Prize in Physics for 1965 with Sin-Itero Tomonaga and Julian Schwinger. In 1963 he assumed responsibility for the set of introductory physics lectures at Caltech, which were published as a three-volume set. He became well known to the public as a member of the commission appointed to investigate the Challenger disaster. He devoted much of his later years to the writing of books for a popular audience as well as exploring the frontiers of computer science.

—*Donald R. Franceschetti, PhD*

BIBLIOGRAPHY

Brown, Laurie M. *Feynman's Thesis: A New Approach to Quantum Theory.* Singapore: World Scientific, 2010. Print.

Feynman, Richard P., and Ralph Leighton. *What Do YOU Care What Other People Think?: Further Adventures of a Curious Character.* New York: Norton, 1988. Print.

Feynman, Richard P., Ralph Leighton, and Edward Hutchings. *"Surely You're Joking, Mr. Feynman!": Adventures of a Curious Character.* New York: W.W. Norton, 1985. Print.

Feynman, Richard P., Robert B. Leighton, and Matthew L. Sands. *The Feynman Lectures on Physics.* Vol. 1-3. Reading, Mass.: Addison-Wesley Pub., 1963. Print.

Gribbin, John, and Mary Gribbin. *Richard Feynman: A Life in Science.* New York: Dutton, 1997. Print.

Mehra, Jagdish. *The Beat of a Different Drum: The Life and Science of Richard Feynman.* Oxford [England]; New York: Clarendon ; Oxford University Press, 1994. Print.

FLYWHEEL

FIELDS OF STUDY

Classical Mechanics

SUMMARY

This article examines the uses and the physics of flywheels. Flywheels were essential to the Industrial Revolution and are still used in a variety of machines, from roller coasters to spaceships. The physics behind them is classical rotational dynamics. This article will examine the kinematics of rotation and how to calculate the energy stored.

PRINCIPAL TERMS

- **inertia:** a body's resistance to change in its motion.
- **kinetic energy:** the energy a body possesses due to its motion
- **potential energy:** the energy held in a system due to the positions of its elements in relation to various forces.
- **rotational energy:** the kinetic energy of a spinning object.
- **rotational speed:** the rate at which an object spins.
- **torque:** a force that causes a rotation, or the measure of such a force.

WHAT IS A FLYWHEEL?

A flywheel is a device that keeps a machine running without interruption. In a car engine, flywheels keep the engine running even when it is in neutral or is not receiving gasoline. Flywheels are able to keep machines running smoothly by storing potential energy, which is energy that depends on the position of a device's parts within a system or force field. Potential energy can be stored in chemical bonds, an electric field, or a gravitational field. The flywheel can then convert this stored energy to kinetic energy, which is the energy of motion. Specifically, a flywheel's kinetic energy is rotational energy (E_r), generated by the spinning of the wheel. The International System of Units (SI) unit of energy is the joule (J), equal to one kilogram–square meter per second squared ($kg \cdot m^2 / s^2$).

A potter's wheel is a type of flywheel in which a stepping on a foot pedal causes a stone to spin. The spinning stone causes a table to rotate, allowing a potter to shape and mold clay atop the table as it spins. The force is applied to the pedal is at intervals, but the action of the flywheel delivers a constant flow of energy to the rotating table.

Another type of flywheel might be a length of metal shaped into a coil and housed within an apparatus that keeps it from springing out. As energy is needed, the coil inside the housing is allowed to release its stored potential energy, resulting in a smooth flow of energy to the machine or objects involved.

The faster a flywheel spins, the more energy is involved. According to Isaac Newton's (1642–1727) first law of motion, an object in motion will remain in motion until a force acts on it. It is for this reason that even if a spaceship traveling through a vacuum suddenly loses all propulsion, it will keep moving in the same direction at the same speed indefinitely.

Rotational Dynamics

A top is a sort of flywheel. In order to make a top spin, an initial impulse must be applied to it such that it moves. This sort of force is called torque. It works against an object's inertia, which is the amount of resistance it has to a change in its motion.

In a spinning object such as a flywheel, the effects of inertia are more complex because everything occurs with reference to the axis of rotation. Torque is applied at a distance from this axis. This distance works as a lever does, trading the force required at any given moment for distance. The further the force is applied from the center part of the top, the more torque is required to make it spin.

The top's rotational speed is the rate at which it spins and is proportional to its rotational energy. It is measured in cycles or revolutions per second. The SI unit of rotational speed is the hertz (Hz), equal to one revolution per second. Meanwhile, rotational velocity, also called angular velocity, is a vector quantity. It is the object's rotational speed in a defined direction of motion, given in SI units of radians per second (rad/s). One radian is the section of a circle described by an arclength that is equal to the circle's radius (r), and the circumference of a circle is $2\pi r$, so the number of radians in a circle is $2\pi r/r$, or 2π. Thus, to convert from rotational speed to angular velocity, multiply the speed by 2π to convert from hertz to radians per second.

Angular velocity (ω) can also be calculated according to the linear speed (v) in meters per second and the radius (r) of the rotational path in meters:

$$\omega = v/r.$$

Flywheels have a large moment of inertia (I), also called rotational inertia, measured in SI units of kilogram–square meters (kg·m²). It can be calculated with the equation

$$I = kmr^2$$

where k is the inertial constant, determined by the shape of the flywheel; m is the mass of the flywheel; and r is its radius. If the flywheel is shaped like a bicycle wheel, with a tire on the outer edge, k is equal to 1; if it is shaped like a flat, solid disk, k equals 0.606; and if it is shaped like a disk with a hole in the center, k equals about 0.3. The flywheel's moment of inertia and its rotational speed can then be used to calculate its rotational energy (E_r), according to the following equation:

$$E_r = \frac{1}{2} I \omega^2$$

Uses of Flywheels

One of the primary advantages of flywheels over chemical batteries is that flywheels can be used over and over again. While chemical batteries wear out over time and lose their ability to hold charge, a flywheel does not have this problem. Flywheels do lose energy to friction, but scientists have been studying ways to design new, lower-friction flywheels. Eventually, very large flywheels could store and stabilize power grids based on solar and wind power, both of which have a high variance in their output.

Having a large moment of inertia allows flywheels to resist sudden changes to their rotation. As a result, they can be used to maintain a continuous power supply from an

Angular velocity = $\omega = \dfrac{\theta}{t}$

Energy = $\dfrac{1}{2} \cdot m \cdot R_F^2 \cdot \omega^2$

Force exerted on the arm (F_L) transfers to the wheel and causes rotational force (F_R) or torque to turn the flywheel. The energy stored by the flywheel is calculated by taking half of the flywheel's mass (m) multiplied by the radius squared (R_F^2) and by the angular velocity of the flywheel squared (ω^2). Angular velocity (ω) can be calculated using the angle of rotation of the wheel (θ) over time (t) or by using the linear velocity of the arm (v) and its angle (φ).

intermittent current. If current cuts out, the wheel will continue to rotate. One of the most familiar uses of this feature is in the piston engine, where the flywheel on the crankshaft continues spinning through the upstroke. A car engine's piston is powered by a small explosion in a gas-air mixture. This explosion forces the shaft down the piston, powering the wheel. The momentum of the rotating mass carries the piston into the upstroke. Diesel engines use the energy stored in the wheel to ignite the next gas mix, ensuring maximal fuel use, which is why they are more efficient than gasoline engines.

Flywheels are also used to govern rotational motion, as in the steam engine, which has little balls that spin on top of the rotational governor. The energy required to accelerate the governor keeps the engine accelerating evenly. In complex engine design, the rate at which the governor spins can be used to regulate fuel flow into the engine, basically allowing the machine to self-regulate and maintain a constant output. Such systems have been critical both in large engines, such as steam locomotives, and in power plants.

The rotational momentum of the flywheel resists sudden changes in motion and can be used to stabilize objects as well. This is commonly seen in spacecraft, often in the form of a reaction wheel, which uses the force of a wheel trying to right itself to turn the spacecraft about its axis.

—*Gina Hagler, MBA*

SAMPLE PROBLEM

If a flywheel has a moment of inertia (I) of 300 kg·m² and an angular velocity (ω) of 5 rad/s, what is its rotational energy (E_r) in joules?

Answer:

Use the equation for rotational energy, given above:

$$E_r = \frac{1}{2} I \omega^2$$

Plug in the given values for moment of inertia and angular velocity and solve, paying close attention to units:

$$E_r = \frac{1}{2} I \omega^2$$

$$E_r = \frac{1}{2}(300 \text{ kg} \cdot \text{m}^2)(5 \text{ rad/sec})^2$$

$$E_r = \frac{(300 \text{ kg} \cdot \text{m}^2)(5 \text{ rad/sec})^2}{2}$$

$$E_r = \frac{7{,}500 \text{ kg} \cdot \text{m}^2 \cdot \text{rad/sec}^2}{2}$$

$$E_r = 3{,}750 \text{ kg} \cdot \text{m}^2 \cdot \text{rad/sec}^2$$

The radian is a dimensionless unit, equal to 1, so it can be discarded:

$$E_r = 3{,}750 \text{ kg} \cdot \text{m}^2/\text{s}^2 = 3{,}750 \text{ J}$$

The flywheel generates 3,750 J of rotational energy.

Bibliography

Dresig, Hans, and Franz Holzweißig. *Dynamics of Machinery: Theory and Applications.* Berlin: Springer, 2010. Print.

Gregory, R. Douglas. *Classical Mechanics: An Undergraduate Text.* New York: Cambridge University Press, 2006. Print.

Gross, Dietmar, et al. *Engineering Mechanics 3: Dynamics.* 2nd ed. Berlin: Springer, 2014. Print.

Levi, Mark. *The Mathematical Mechanic: Using Physical Reasoning to Solve Problems.* Princeton: Princeton University Press, 2009. Print.

Nave, Carl R. "Angular Momentum of a Particle." *HyperPhysics.* Georgia State U, n.d. Web. 26 June 2015.

Nave, Carl R. "Rotational-Linear Parallels." *HyperPhysics.* Georgia State U, n.d. Web. 26 June 2015.

FOCAL POINT

FIELDS OF STUDY
Optics

SUMMARY

The properties of lenses are determined by the curvature of their surfaces and the changes in the behavior of light as it passes through the lens medium. In most applications, the rays of light passing through a lens are bent so that they comes together in a single point, called the focal point. At this point, an object viewed through the lens is said to be "in focus," usually as an enlarged image for easier viewing.

PRINCIPAL TERMS

- **aperture:** an opening of a specific dimension designed to admit light into a lens or other optical system.
- **circle of confusion:** an imperfect image produced by a lens due to the rays of light passing through it not converging at a perfect point.
- **concave:** having surfaces that curve inward, like a bowl.
- **convex:** having surfaces that curve outward, like a ball.
- **focal length:** the distance from the center of a lens or curved mirror to the focal point.
- **optics:** the science of the interaction of light with lenses and mirrors.
- **refraction:** the change in direction of a light ray when it passes from one medium to another, such as from air to water.

Light and Matter

Most solid matter is opaque and prevents the passage of light through it. Materials such as glass are transparent and allow light to pass through. The transparency of a material may vary depending on the wavelength of the light. Crystalline sodium chloride (table salt), for example, is completely transparent to infrared but not entirely transparent to all wavelengths of visible light. The study of the behavior of light as it interacts with various forms of matter, such as glass formed into lenses, is called optics.

Because glass is transparent to visible light, it is an ideal material for optical lenses. Typically, light enters an optical device through an aperture and passes through the lens or lenses before being observed, as in a telescope, or creating a lasting image, as in a camera. On the other side of the lens, the light rays bend so that they all converge at one point, called the focal point.

Properties of Lenses

When light travels from one medium to another, its direction of motion changes. This is called refraction. This effect can be observed by placing a straw in a glass of water. The portion of the straw that is above the water appears to point in a different direction than the portion below the water. This is because air and water have different refractive indexes. A refractive index is a dimensionless measure of how much light is bent by a particular material.

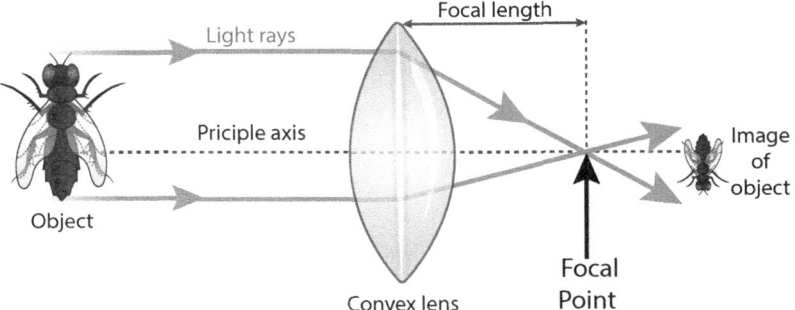

Object Image of object Convex lens Principle axis Light rays Focal Point Focal length

Light passes through a convex lens and refracts inward to a focal point along the principal axis. The focal length is the distance from the center of the lens to the focal point. Objects viewed through the lens will be in focus at the focal point. Lenses in binoculars are designed to have the focal point in your eye.

The same effect operates when light passes from air into glass. The rays of light travel in perfectly straight lines until they are refracted by the glass, causing them to travel at a slightly different angle. When the rays pass through the opposite surface of the glass, they are refracted again. If the sides of the piece of glass are parallel to one another, as in a window pane, the light emerges from the far side of the glass at the same angle at which it entered, having shifted only slightly to one side. If the light rays enter the glass perpendicular to the surface, however, the angle does not change at all.

A lens is a piece of glass that has been shaped with a very specific curve on each side. Light rays that enter the lens on one side emerge from the other side bent so that they converge at the focal point, some distance away from the center of the lens. That distance is the focal length of the lens, and it is dependent on the degree of curvature of both sides of the lens.

The focal length (f) of a thin lens can be determined experimentally by using the lens to focus an image of an object or light source on a screen. Measure the distance from the object to the lens (d_o) and the distance from the lens to the image on the screen (d_i), then calculate:

$$\frac{1}{f} = \frac{1}{d_o} + \frac{1}{d_i}$$

This is an approximation rather than an exact formula, but it is reasonably accurate for a lens whose thickness is insignificant compared to its radius of curvature. It does not work for lenses of a non-negligible thickness.

FOCAL POINT

If one or both surfaces of a lens curve outward, as in an ordinary magnifying glass, the lens is called convex or biconvex, respectively. A sharp image will be observed at the focal point of a convex lens. If one or both of the surfaces curve inward, the lens is concave or biconcave. Light rays passing through a concave lens diverge, or spread apart, instead of converging. The focal point of a concave lens is on the near side of the lens, where the diverging rays would come together if they were traced backward, and the focal length is given as a negative value. There is a virtual image at the focal point, rather than the real, clear image at the focal point of a convex lens. By combining convex and concave lenses, very large degrees of magnification can be achieved. This technique is used in powerful telescopes and optical microscopes.

SAMPLE PROBLEM

A thin biconvex lens is placed 50.0 centimeters (cm) in front of an object. It projects a sharp image of the object on a screen 35.6 cm away. What is the focal length of the lens?

Answer:
The focal length of a thin lens can be calculated according to the following formula:

$$\frac{1}{f} = \frac{1}{d_o} + \frac{1}{d_i}$$

Plug in 50.0 cm for the distance from the object (d_o) and 35.6 cm for the distance from the image (d_i):

$$\frac{1}{f} = \frac{1}{d_o} + \frac{1}{d_i}$$

$$\frac{1}{f} = \frac{1}{50.0} + \frac{1}{35.6}$$

Rearrange the formula to solve for f, then calculate:

$$\frac{1}{f} = \frac{1}{50.0} + \frac{1}{35.6}$$

$$1 = \left(\frac{1}{50.0} + \frac{1}{35.6}\right) f$$

$$\frac{1}{\frac{1}{50.0} + \frac{1}{35.6}} = f$$

$$20.8 = f$$

The focal length is approximately 20.8 cm.

CIRCLE OF CONFUSION

Lenses are typically imperfect in various ways. Thus, the focal point is not so much an actual point as it is a small circle, because the light rays coming through the lens do not all converge in exactly the same place. This actual imperfection of the theoretical focal point is called the circle of confusion, or sometimes the "blur spot." Because of the limitations of human visual acuity, however, the circle of confusion is usually not a problem. In fact, there is a range on either side of the focal point where the image through a lens appears sharp to the human eye. This range is sometimes also called the "circle of confusion," or sometimes the "circle of least confusion." It is important in photography for determining depth of field, the nearest and farthest points in an image that appear acceptably sharp.

—*Richard M. Renneboog, MSc*

BIBLIOGRAPHY

Darrigol, Olivier. *A History of Optics: From Greek Antiquity to the Nineteenth Century.* New York: Oxford University Press, 2012. Print.

Ersoy, Okan K. *Diffraction, Fourier Optics, and Imaging.* Hoboken: Wiley, 2007. Print.

Ghatak, Ajoy. *Optics.* 5th ed. New Delhi: McGraw, 2012. Print.

Laikin, Milton. *Lens Design.* 4th ed. Boca Raton: CRC, 2007. Print.

"Reflection, Refraction, and Diffraction." *The Physics Classroom.* Physics Classroom, 2015. Web. 28 July 2015.

Sharma, K. K. *Optics Principles and Applications.* Burlington: Academic, 2006. Print.

FREE BODY DIAGRAM

FIELDS OF STUDY

Electrodynamics, Particle Physics

SUMMARY

Applying Newton's second and third laws can rapidly become confusing as the force one object exerts on another always implies a force of equal magnitude, oppositely directed, acting on the original body. Free body diagrams simplify the process of applying Newton's laws to the situation at hand.

PRINCIPAL TERMS

- **acceleration:** The rate at which the velocity of an object changes with time, usually given as m/s^2.
- **force:** A push or pull on an object, generally ascribable to a single interaction.
- **mass:** The quantity of substance in an object, also equal to the ratio of net force to acceleration.
- **net force:** The sum of all the forces acting on one object.
- **pulley:** A grooved wheel, free to rotate about an effectively immobile axis, usually with ropes or cables draped over it so as too change the direction of force in each cable.
- **tension:** The force exerted at either end of a rope or cable.

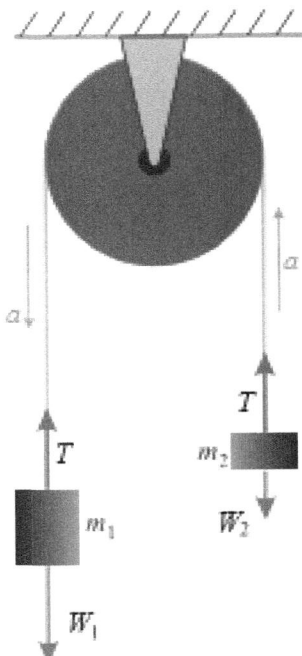

The free body diagrams of the two hanging masses of the Atwood machine. Our sign convention, depicted by the acceleration vectors is that m_1 accelerates downward and that m_2 accelerates upward, as would be the case if $m_1 > m_2$.

127

Applying Newton's Laws to Free Bodies

To take a simple example consider a rescue worker suspended from a stationary helicopter. The object of interest in the worker suspended from the cable. There are two forces acting on him. One is his weight, the gravitational attraction between him and the earth. The second is the tension in the cable. If he is stationary or moving at a constant speed, this force is zero: otherwise the force produces an acceleration satisfying Newton's second law, the net $F=ma$.

Even in this simple situation, there are several points that might confuse the beginning student. Some find it hard to believe that the Earth is an object that other forces are capable of acting upon. Others tend to forget the weight of the worker and instead, focus simply on his mass times acceleration. A simple free body diagram for the worker resolves the issue. The worker is represented by a dot and there are two arrows originating at the dot one for each force.

Coupling of Free Body Diagrams: The Atwood's Machine

Cambridge University Professor George Atwood (1745–1807) developed a pedagogical aid known as the Atwood's machine in 1784. It consists of two masses connected by a string hung over a pulley of negligible mass. Here $m_1 > m_2$ by assumption. The bodies are connected by the string tension (T) and the free body diagrams are superimposed on each mass. One can easily see in the following:

$$m_1 g - T = m_1 a$$

$$T - mg = m_2 a,$$

a system which is readily solved for the common tension, T, and acceleration, a.

Slightly more challenging problems are obtained by letting m_1 and/or m_2 slide down inclines of the same or different angles to the horizontal and perhaps adding friction. In each case, drawing and labeling the free body diagrams takes one halfway to the solution.

—*Donald R. Franceschetti, PhD*

Bibliography

Cutnell, John D., Kenneth W, Johnson, Shane Stadler, and David Young. *Physics*. 10th ed. Hoboken, NJ: Wiley, 2015. Print.

Hewitt, Paul G. *Conceptual Physics*. 12th ed. Boston, Mass: Pearson, 2015. Print.

FREQUENCY

FIELDS OF STUDY

Acoustics, Harmonics, Optics

SUMMARY

This article discusses aspects of physical and electromagnetic wave phenomena, with regard to their frequency. The frequency of a wave is expressed as the number of cycles occurring in a specific amount of time. The time required for one cycle is the period of the wavelength. Phase relationships and interference are also discussed.

PRINCIPAL TERMS

- **frequency:** the number of complete waves or cycles that occur in one unit of time.
- **harmonics:** the study of the interaction of wave phenomena.
- **hertz:** a unit of frequency defined as one cycle per second.
- **period:** the length of time for one complete cycle of a wave or other cyclic property to occur.
- **reciprocal:** the inverse of a value, calculated as 1 divided by the value.
- **sinusoidal:** having a shape or pattern of behavior that can be described by a sine wave function.
- **speed:** the distance traveled per unit of time.
- **wavelength:** the distance from any point in a wave to the identical point in the next wave, usually measured from crest to crest.

Cyclic Phenomena

The term "cycle" generally indicates something that goes around in a circle. In physics, "cycle" indicates

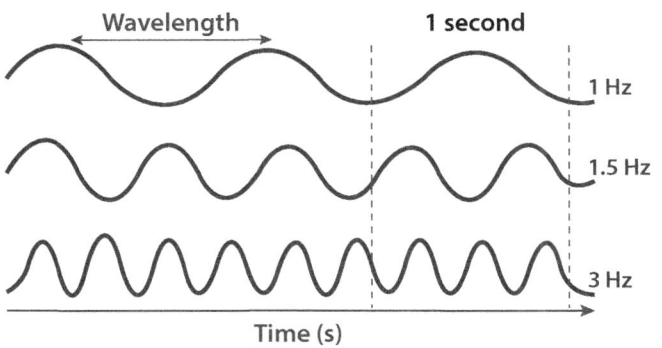

A wave's frequency is a measure of the number of waves per period of time. When calculating waves per second the unit is a hertz (Hz). A wave with a frequency of 1 Hz has one full wave (1 wavelength) in the period of one second. A wave with a frequency of 3 Hz has three full waves within the period of one second.

that a specific property or function has a value that progresses through a succession of other values and returns to the starting value in a precise manner that repeats. Phenomena that exhibit this behavior are associated with either circular or sinusoidal wave motions and properties. Such motions can be described by the same math functions, the sine and cosine.

The sine and cosine functions are themselves simple ratios of the lengths of the two sides of a right triangle at one vertex. The radius (plural: radii) of the circle can be rotated about the center by any amount to form the corresponding angle. A vertical line to the point on the circumference where it meets the displaced radius forms a right triangle with a base that is proportionately shorter than the length of the radius. In this right triangle, the displaced radius forms the hypotenuse and the vertical height of the triangle is the opposite. (The "opposite" is the side of the right triangle that is opposite the angle formed at the center of the circle.) The base of the triangle is called the "adjacent." The sine of the angle formed by the base and the hypotenuse is just the ratio of the length of the opposite to that of the hypotenuse (i.e., the radius). Likewise, its cosine is the ratio of the length of the adjacent to that of the hypotenuse. The reciprocal values of the sine and cosine are called the "secant" and "cosecant," respectively.

As the radius rotates, the angle that it forms at the center changes continuously. The value of the sine also changes accordingly. A graph of this variation produces the sideways S-shaped curve that is recognized as a sine wave. The value of the cosine follows the same pattern but is shifted from the sine values.

The cosine at any angle has the value of the sine of an angle that is greater by 90 degrees.

There are two methods of describing the amount of rotation about the center, or axis of rotation. In one, the amount is stated in degrees of rotation, with one full revolution totaling 360 degrees. The other measurement of angles is in radians (rad). One radian is the angle formed by two radii when the length of the circumference they mark off is equal to the radius of the circle. There are 2π radians in one complete revolution.

PROPERTIES OF CYCLIC PHENOMENA

All cyclic phenomena, whether wavelike or circular, share several characteristics. The primary feature of all of them is that their behaviors or values repeat in the same regular way. The number of times that the cycle of any particular phenomenon repeats in a specific amount of time is its frequency. The most common of these is revolutions per minute (rpm) for rotational movements of physical objects, and cycles per second (cps) for wavelike properties. The conventional unit for cps is hertz (Hz), in honor of Heinrich Hertz (1857–94), for his contributions to the physics of electromagnetism (EM). The term is most often used to refer to EM waves.

The duration of just one cycle of the phenomenon is the period of the cycle. The period is calculated simply as the reciprocal value of the frequency. For example, a wave frequency of 100 cps has a corresponding period of 1/100, or 0.01, seconds per cycle (spc). If that same wave is progressing, or propagating, at a speed of 100 meters per second (m/s), each cycle will have moved it through a distance of 1 meter. Since this corresponds to the distance covered by just one complete cycle of the wave, it is the specific wavelength. Wavelength is only used to describe phenomena that travel through space or time. It is not applied to rotational motions.

FREQUENCY, WAVELENGTH, AND SPEED

EM waves such as visible light all travel at the same speed—the speed of light. Physical waves, such as sound, travel at different speeds determined by the density of the medium. The speed of sound in air, for example, is 331 meters per second (about 1,086 feet per second) at 0 degrees Celsius (32 degrees Fahrenheit), and 342 meters per second (about 1,122 feet per second) at 18 degrees Celsius (65 degrees Fahrenheit). In water, the speed of sound is about 1,140 meters per second (about 3,740 feet per second). The greater the density of the medium is, the more it can transmit sound waves. Another factor that affects the transmission of both physical and EM waves is the relative motion of the wave source and the observer or receiver of the emitted waves. When the two move toward each other, the apparent frequency of the waves increases and the apparent wavelength decreases, but if they move apart, the apparent frequency decreases and the wavelength increases. This is known as the "Doppler effect." It accounts for the apparent changes to the sound of a passing train, as well as the redshift and blueshift in the light observed from distant stars.

For physical waves, speed, frequency, and wavelength are all related. For EM waves, however, the constant speed of light requires that the frequency and wavelength are related in a manner that maintains the constancy of the speed of light.

Frequency and Phase

The fundamental character of a wave is determined by its frequency. All wave phenomena have neutral points called "nodes" at which the value of the amplitude is zero. When two identical waves are synchronized such that their nodes and amplitudes coincide precisely, they are in phase. Their amplitudes do not matter. At all other times, they are out of phase by a particular amount. For example, two waves might be said to be 10 degrees out of phase. The amplitudes of waves that are in phase add together as constructive interference to produce a wave with greater amplitude. When they are out of phase, such that their positive and negative amplitudes overlap, they add together as destructive interference to produce a wave with less amplitude and loss of harmonic characteristics.

Waves of different frequencies can never be in phase. Instead, nodes and amplitudes may coincide occasionally in such a way that a harmonic beat frequency may arise. This is most apparent when one frequency is a whole-number multiple of the other. In such cases, nodes occur consistently when a number of wavelengths of one frequency span the same period of time as a different number of wavelengths of the other frequency.

—*Richard M. Renneboog, MSc*

Bibliography

Amer. Industrial Hygiene Assn. *Radio-Frequency and Microwave Radiation*. 3rd ed. Fairfax: Amer. Industrial Hygiene Assn., 2004. Print.

Chaichian, Masud, Hugo Perez Rojas, and Anca Tureanu. *Basic Concepts in Physics: From the Cosmos to Quarks*. New York: Springer, 2014. Print.

SAMPLE PROBLEM

The crests of two successive waves passing a dock were observed to be 6 meters (about 20 feet) apart. It took 7.5 seconds for the crests to travel the 100 meter (about 328 feet) length of the dock. Calculate the frequency and period of the waves.

Answer:

First, define the wavelength (λ) as the distance between any two successive wave crests. This is given as 6 m (20 ft). Next, determine the speed (s) by relating the distance that the waves travel in a certain amount of time. This is given as 100 m (about 328 ft) in a time of 7.5 s. The speed of the waves is therefore

$$s = 100 \text{ m} \div 7.5 \text{ s} = 13.33 \text{ m/s}$$

The frequency (f) can then be calculated by dividing the waves' speed by the wavelength, as

$$f = s/\lambda$$

$$f = 13.33 \text{ m/s} \div 6 \text{ m}$$

$$= 2.22 \text{ cps}$$

(The calculation using feet per second for speed and feet for wavelength must yield the same answer for the frequency of the waves.)

The period (T) of the waves is calculated as the reciprocal of the frequency, or

$$T = 1/f = 1/2.22 \text{ cps}$$

$$= 0.45 \text{ s}$$

De Pree, Christopher Gordon, and Ira Maximilian Freeman. *Physics Made Simple*. New York: Broadway, 2004. Print.

Gilbert, P.U.P.A., and W. Haeberli. *Physics in the Arts*. Burlington: Elsevier, 2008. Print.

Gunther, Leon. *The Physics of Music and Color*. New York: Springer, 2012. Print.

Kumar, B.N. *Basic Physics for All*. Lanham: University Press of America, 2009. Print.

Serway, Raymond A., John W. Jewett, and Vahé Peroomian. *Physics for Scientists and Engineers with Modern Physics*. 9th ed. Boston: Brooks, 2014. Print.

G

GAMMA RADIATION

FIELDS OF STUDY

Electromagnetism; Nuclear Physics

SUMMARY

Gamma radiation is the most energetic form of electromagnetic radiation. This article looks at the nature of gamma rays and how they are generated, touching on detection methods and their role in cosmology. Known sources of gamma radiation include high-energy collisions and nuclear decay.

PRINCIPAL TERMS

- **bremsstrahlung:** radiation generated when a particle is slowed via a collision with another particle; from the German word for "braking radiation."
- **cosmic rays:** extremely high-energy subatomic particles, mainly protons and atomic nuclei, that originate outside of Earth's atmosphere from largely unknown sources.
- **gamma-ray burst:** a high-energy flash of gamma radiation produced by a violent explosion in a distant galaxy; believed to be the brightest and most energetic electromagnetic event in the universe since the big bang.
- **inverse Compton scattering:** a collision between a high-energy electron and a low-energy photon that results in energy being transferred to the photon.
- **photon:** a massless elementary particle that is the smallest possible unit, or quantum, of light and other electromagnetic radiation.
- **radioactive decay:** the loss of energy and matter from an unstable nucleus in the form of ionizing radiation.
- **radioisotope:** an isotope of an element with an unstable nucleus.
- **synchrotron radiation:** radiation emitted by the acceleration of a charged particle traveling in a magnetic field at near light speed.
- **x-rays:** high-energy electromagnetic radiation emitted when an excited electron drops back down to a lower energy level.

Gamma Rays

Light is a form of electromagnetic radiation. The word "light" usually refers to the portion of the electromagnetic spectrum that is visible to the human eye. Sometimes it is also used to describe radiation in the infrared and ultraviolet ranges. However, there is much more to electromagnetic radiation than just light.

Electromagnetic radiation is classified according to its wavelength and frequency. The two are inversely related; the shorter the wavelength is, the higher the frequency. Energy is also inversely related to wavelength, meaning that radiation with a shorter wavelength has greater energy. The main types of electromagnetic radiation, in order of decreasing wavelength and increasing frequency and energy, are radio waves, microwaves, infrared, visible light, ultraviolet, x-rays, and gamma radiation, or gamma rays.

Gamma radiation is typically the most energetic, with the shortest wavelengths. Historically, the term referred to any electromagnetic radiation below a certain wavelength (about 10^{-11} meters) or above a certain energy level (about 100 kiloelectronvolts, or keV), while high-energy radiation that fell outside those bounds was considered x-rays. However, the distinction between the two is not quite as clear as was once thought.

Another way of distinguishing gamma rays from x-rays is by their source. While electromagnetic radiation is a wave, it is also composed of small, massless particles called photons. This is possible due to the concept of wave-particle duality, which states that all elementary particles also exhibit properties of waves. When an atom emits gamma rays or x-rays, it is in fact releasing highly energetic photons. Gamma-ray photons are generally released by the nucleus as a result of excess energy or nuclear decay. X-ray photons are

DECAY TYPE	SYMBOL	PARTICLE EMITTED	EXAMPLE	STOPPED BY	ENERGY RELEASED	SOURCE
Gamma	γ	Gamma Particle (photon)	He-3 → He-3	Thick Lead	< 10 MeV	Cosmic ray interacting with the atmosphere; Radium; Lightning strikes

Gamma rays are highly energetic photons emitted from the nucleus of an atom, generally as the result of nuclear decay. After tritium (hydrogen-3) undergoes beta decay to become helium-3, the excess energy remaining in the nucleus is emitted as gamma-ray photons. Gamma rays can also be produced through interactions with lightning strikes and cosmic rays. They can be stopped by thick lead.

released by excited electrons as they drop back down to a lower energy level.

For the most part, these two definitions—energy-based and source-based—correspond with one another. However, as scientists have developed techniques for generating x-rays with energies as high as 25 megaelectronvolts (MeV), the line has become increasingly blurred. As a result, in physics, gamma radiation is defined primarily as radiation produced in nuclear events. In astronomy, this would not be practical, as sources of high-energy radiation in outer space are often unknown. Thus, astronomers define gamma radiation based on its energy level rather than its source.

Radioactive Decay

Radioactive decay is the process by which a radioisotope emits energy from its unstable nucleus in the form of ionizing radiation. Gamma rays are produced by gamma decay, one of the three most common forms of radioactive decay. The other two forms are alpha decay and beta decay, which produce alpha particles and beta particles, respectively. An alpha particle consists of two protons and two neutrons bound together, identical to a helium nucleus. A beta particle is a high-speed, extremely energetic electron or positron (the antiparticle of the electron).

Gamma decay almost always occurs alongside or immediately after another form of decay. When a gamma-ray photon collides with an atom, its energy is transferred to one of the atom's electrons. Often, the electron absorbs so much energy that it breaks away from the atom, leaving behind a positively charged ion—hence the name "ionizing radiation." Alpha and beta particles are more ionizing, but gamma rays can penetrate more deeply. They are so energetic that they can only be blocked by thick layers of a dense material such as lead. This is what makes gamma rays so dangerous to life-forms; the ionization of molecules in the body inhibits proper cellular function, often resulting in radiation poisoning or cancer.

Generating Gamma Rays

Gamma rays are generated via several mechanisms, most involving the acceleration or deceleration of charged particles. To understand these processes, it is important to remember that photons are the particles that carry the electromagnetic force. As a result, when the motion of a charged particle is affected by electromagnetism, photons are either released or absorbed. When an electron or proton is traveling at near light speed, the amount of energy released in a collision with another particle would be enormous.

This is one way of generating bremsstrahlung, which is electromagnetic radiation produced by the sudden deceleration of a particle. When a particle such as an electron collides with another particle, such as an atom in a sheet of lead, the electron may go from near light speed to a near stop. This collision requires the electron to bleed off a great deal of energy, which it does by converting its kinetic energy into electromagnetic radiation. The more energy the electron must get rid of, the more energetic the photon and the shorter the wavelength.

A similar process is behind synchrotron radiation, which is produced by the radial acceleration of charged particles. It is named after the synchrotron, a type of circular particle accelerator. Any object that travels in a circle is constantly changing its velocity, because velocity has both magnitude and direction. Even if the object's speed remains the same, the constant change in direction means that the velocity changes, and any change in velocity is a form of acceleration. In the case of an object traveling in a circle, the acceleration is toward the circle's center. This constant acceleration results in a loss of energy, which the charged particles emit in the form of photons.

IT CAME FROM SPACE
Another source of gamma rays is the highly energetic emissions from black holes. There, high-speed electrons being sucked into the black hole's accretion disk collide with lower-energy photons generated by heat in the disk. The excess energy transfers from the electrons to the photons. This process is called inverse Compton scattering because the photons "scatter" to higher energy levels. In ordinary Compton scattering, a low-energy charged particle collides with a high-energy photon and absorbs its energy; inverse Compton scattering is, appropriately, the opposite.

Gamma-ray bursts, on the other hand, are still very mysterious. Astronomers know that in deep space, occasional brilliant flashes of electromagnetic energy produce gamma-ray spikes lasting from a fraction of a second to hours in length. These have been observed to occur only at great distances from the Milky Way galaxy, and it is thought that most of them are caused by supernovas.

Compton scattering is actually one of the primary tools for detecting gamma radiation. A detector may consist of massive tanks of water or arrays of sensors laid down on the seabed. When cosmic rays collide with this dense material, they create a flash of less-energetic radiation that can be observed as a water molecule becomes excited and releases the excess energy as a photon. Detection of gamma rays is one way astronomers and physicists can form a more complete picture of the universe, as well as try to look back at the highly energetic events that followed the big bang.

—*Gina Hagler, MBA*

BIBLIOGRAPHY
Cooper, Malcolm, et al. *X-Ray Compton Scattering*. Oxford: Oxford University Press, 2004. Print. Oxford Ser. on Synchrotron Radiation 5.
Fleishman, Gregory D., and Igor N. *Toptygin. Cosmic Electrodynamics: Electrodynamics and Magnetic Hydrodynamics of Cosmic Plasmas*. New York: Springer, 2013. Print.
Greiner, Walter, ed. *Nuclear Physics: Present and Future*. Cham: Springer, 2015. Print.
Huber, Martin C. E., et al., eds. Observing Photons in Space: A Guide to Experimental Space Astronomy. 2nd ed. New York: Springer, 2013. Print.
Kamal, Anwar. *Nuclear Physics*. Berlin: Springer, 2014. Print.
Kouveliotou, Chryssa, Ralph A. M. J. Wijers, and Stanford E. Woosley, eds. *Gamma-Ray Bursts*. New York: Cambridge University Press, 2012. Print. Cambridge Astrophysics Ser. 51.

GAUSS'S LAW

FIELDS OF STUDY

Electromagnetism

SUMMARY

Gauss's law is fundamental to how the capacitor, a key component of modern electronic circuits, works. It relates the electric flux in an electromagnetic field to the amount of electrical charge present within a defined space. Gauss's law is related to Faraday's law of magnetic induction.

PRINCIPAL TERMS

- **conservation of charge:** the principle that the net electrical charge in an electric circuit does not change.
- **Coulomb's law:** a scientific law stating that the electric flux at any defined surface is the vector

sum of the electric field strength at every point on that surface.
- **electric flux:** the measure of the strength of an electric field across an area with different electrical potentials.
- **Faraday's law:** the scientific law stating that when a magnetic field and a conductor move relative to each other, a voltage is induced in the conductor, the magnitude of which depends directly on the relative speed of the field and conductor.
- **Gaussian surface:** any closed or finite hypothetical surface.
- **Maxwell's equations:** a set of mathematical descriptions of electromagnetism, formulated by James Clerk Maxwell.
- **permittivity:** a measure of the ability of an electric field to penetrate a medium.

ELECTRIC FIELDS AND ELECTRIC FLUX

Just as a magnet has a magnetic field, produced by the presence of a north magnetic pole and a south magnetic pole, an electric field is produced by the presence of two regions with different electrical potential. These regions are typically designated as positive and negative, like the positive and negative ends of a battery. For an electric field to exist, the two regions cannot be physically connected to each other, because this would cancel out their electrical potentials. An electrical potential exists because of the electric charge that is present at that location. When two points with different electric charges are connected to each other, charge moves from the area of higher charge to that of lower charge until both have equal charges. For example, lightning occurs when the electric field caused by charge separation between the ground and the clouds is short-circuited by negative ions from the clouds connecting with positive ions from the ground.

On a much smaller scale, an electric field can be generated by connecting two plates that can act as Gaussian surfaces to an electrical source but not to each other. The source, which may be either direct current (DC) or alternating current (AC), produces a different electrical potential on the two surfaces. In the case of a battery, one surface will be positive and the other negative, depending on which terminal it is connected to. Because the surfaces are not in direct contact with each other, current cannot flow between them. Current is the movement of electric charge through a conducting medium. Certain mediums, such as air, are dielectric. This means that their molecules can be polarized by an electric field but do not normally conduct charge.

Current is usually thought of as electrons moving from atom to atom through a conducting material—that is, as negative charge flowing through a circuit. Electric current can also be thought of as the movement of the location that can accept an electron in adjacent atoms. In either case, a dielectric field prevents the movement of charge from one surface to the other. As a result, one surface becomes enriched with electric charge, while the other is depleted. This lasts until the polarity of the field is reversed or neutralized. The law of conservation of charge states that the amount of charge gained by one surface must be equal to the charge lost by the other surface. Overall, the system is electrically neutral, even though a large difference in electrical potential may exist between the surfaces and an electric field exists between them.

Gauss's law states that the electric flux is directly related to the net electric charge enclosed by a Gaussian surface. It is equal to the net electric charge divided by the permittivity of the medium. This relationship is named for German theorist Karl Friedrich Gauss (1777–1855).

FLUX AND PERMITTIVITY

Electric flux (Φ) is calculated as

$$\Phi = q/\varepsilon$$

where q is the net electric charge, given in coulombs (C), and ε is the permittivity

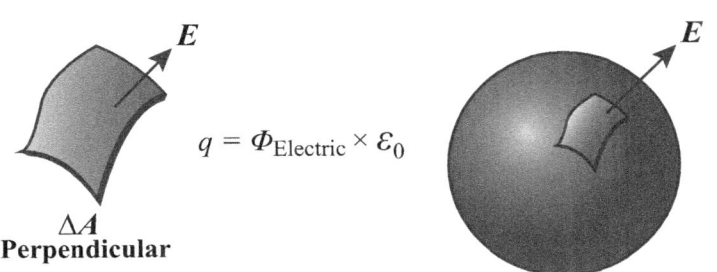

$$q = \Phi_{\text{Electric}} \times \varepsilon_0$$

The electric field (E) emitted perpendicular at any point on a surface multiplied by the summation of the area (ΔA) gives the electric flux (Φ) projecting from the surface of a volume. The electric flux multiplied by the electric constant (ε_0) will give the total charge of the enclosed space (Q).

of the medium, given in farads per meter (F/m) or coulombs per volt-meter (C/V·m). The flux is usually given in volt-meters (V·m). Permittivity refers to a medium's ability to prevent penetration by an electric field. An electric field is less able to penetrate a medium with high permittivity than one with low permittivity.

Electric field strength, a vector quantity, can be thought of as the amount of electric force that is exerted on a positive test charge, either repelling it from a positive source or attracting it toward a negative one. The greater this force, the higher the electric flux of the field for any particular charge. Mathematically, the electric flux (Φ) is the product of the electric field strength (E), the area affected (A), and the cosine of the angle between them (θ):

$$\Phi = EA\cos\theta.$$

Thus, the electric flux at any defined surface within the field can be visualized, according to Coulomb's law, as the sum of all the forces acting on each charge at each point on the surface.

Gauss's law relates electric flux to electric charge much as Faraday's law relates a magnetic field to magnetic flux. As Faraday noted, a conductor moving within a magnetic field experiences an induced voltage. That voltage is directly proportional to the rate of change of magnetic flux within the field. It also defines two areas of different electrical potential and therefore creates an electric field between those areas. By this relationship, electric and magnetic fields are shown to have a common nature as electromagnetic phenomena. Scottish physicist James Clerk Maxwell (1831–79) brought together the properties of electric and magnetic fields in a set of precise mathematical descriptions of electromagnetism, now known as Maxwell's equations.

CHARGE AND CAPACITANCE

Gauss's law is most commonly used in electronic components called capacitors. A capacitor consists of a dielectric material sandwiched between two conductors. The dielectric material can become polarized but will typically not conduct electric charge at any voltage below its particular threshold voltage. Above the threshold voltage, the electric potential between the conductors is sufficient to force electric charge to bridge the gap between them and create a short circuit. In normal use, a capacitor stores charge in the conductor with the higher electrical potential and loses charge from the conductor with lower potential, though current does not flow between them. In an electric circuit, this provides a buffer to compensate for slight fluctuations in current flow that would otherwise degrade the signal passing through the circuit.

SAMPLE PROBLEM

The permittivity of air is 8.85 picofarads (pF) per meter. Calculate the electric flux for a net charge of 3 microcoulombs (μC) between two plates of equal size separated by air. Recall that one farad equals one coulomb of charge per volt.

Answer:

Convert the picofarads to coulombs per volt:

$$8.85 \text{ pF} = 8.85 \times 10^{-12} \text{ F} = 8.85 \times 10^{-12} \text{ C/V}$$

Then, convert the microcoulombs to coulombs:

$$3 \, \mu\text{C} = 3 \times 10^{-6} \text{ C}$$

The electric flux is calculated as

$$\Phi = q/\varepsilon$$

Substitute in the known values for the charge (q) and permittivity (ε) and solve:

$$\Phi = \frac{3 \times 10^{-6} \text{ C}}{8.85 \times 10^{-12} \frac{\text{C}}{\text{V}\cdot\text{m}}}$$

$$\Phi = \left(\frac{3}{8.85}\right)\left(\frac{\frac{1}{10^6}\text{C}}{\frac{1}{10^{12}}\frac{\text{C}}{\text{V}\cdot\text{m}}}\right)$$

$$\Phi = 0.339\left(\frac{1}{10^6} \times \frac{10^{12}}{1}\right)\left(\text{C} \times \frac{\text{V}\cdot\text{m}}{\text{C}}\right)$$

$$\Phi = 0.339\left(\frac{10^{12}}{10^6}\right) \text{V}\cdot\text{m}$$

$$\Phi = 0.339 \times 10^{12-6} \text{ V}\cdot\text{m}$$

$$\Phi = 0.339 \times 10^6 \text{ V}\cdot\text{m}$$

The electric flux between the plates is 0.339 × 10⁶ volt-meters. In proper scientific notation, this is written as 3.39 × 10⁵ volt-meters.

Capacitors are designed to store a specific amount of charge when a circuit is operating. They are typically rated in units of microfarads (μF). One microfarad is one-millionth of a farad, and one farad is equal to one coulomb of charge per volt. One coulomb of charge is the amount of charge carried by an electric current of one ampere in one second, or 6.242×10^{18} electrons. Thus, one microfarad is equivalent to 6.242×10^{12} electrons when the potential difference between two conductors in a capacitor is just one volt. This is a significant amount of electric charge. Cattle prods and stun guns use capacitors to store electric charge from small batteries that produce a total voltage of as little as 9 volts. This amount of charge provides enough of an electric shock to incapacitate an adult when the capacitors are discharged.

—*Richard M. Renneboog, MSc*

Bibliography

Fleisch, Daniel. *A Student's Guide to Maxwell's Equations.* New York: Cambridge University Press, 2008. Print.

Kelly, P. F. *Electricity and Magnetism.* Boca Raton: CRC, 2015. Print.

Fitzpatrick, Richard. *Maxwell's Equations and the Principles of Electromagnetism.* Hingham: Infinity Science, 2008. Print.

Mansfield, Michael, and Colm O'Sullivan. *Understanding Physics.* 2nd ed. Hoboken: Wiley, 2012. Print.

Matsushita, Teruo. *Electricity and Magnetism: New Formulation by Introduction of Superconductivity.* Tokyo: Springer, 2014. Print.

Robbins, Allan H., and Wilhelm C. Miller. *Circuit Analysis: Theory and Practice.* 5th ed. Clifton Park: Delmar, 2013. Print.

GLUON

FIELD OF STUDY

Particle Physics

SUMMARY

This article describes the characteristics, discovery, and significance of gluons. Gluons are subatomic particles that maintain the structure of atomic nuclei and possess unique qualities useful in particle physics research.

PRINCIPAL TERMS

- **boson:** a particle that has integral spin; usually carries force and acts as the "glue" that holds matter together.
- **color charge:** a property of quarks that distinguishes quarks and gluons from each other.
- **hadron:** a subatomic particle that is made of quarks and held together by the strong force.
- **nucleon:** a type of baryon, either a proton or a neutron, that is found in atom's nucleus.
- **quark:** an elementary subatomic particle.
- **spin:** an intrinsic form of angular momentum carried by elementary particles, composite particles (hadrons), and atomic nuclei.
- **strong force:** the force that binds quarks together.

Holding It All Together

All matter encountered in daily life is composed of elementary particles. Elementary particles are particles that cannot be broken down into smaller constituent parts. There are two families of elementary particles: bosons, which carry force, and fermions, which carry matter. Fermions interact with each other by exchanging bosons. This exchange is what creates the four fundamental forces of nature: gravitation, electromagnetism, the weak force, and the strong force. Each of the four forces is carried by a gauge boson. Photons carry electromagnetism, and W and Z bosons carry the weak force. (The gauge boson of gravitation, called the graviton, is hypothesized but not yet proved to exist.) Gluons are the gauge bosons that mediate the strong force between quarks, which are the elementary particles that make up protons and neutrons in atomic nuclei.

The existence of quarks was first posited in 1964. Two American physicists, Murray Gell-Mann (b. 1929) and George Zweig (b. 1937), independently proposed them as particles that respond to the strong force. This implied that a gauge boson must exist to mediate the strong force between quarks, similar to how photons mediate electromagnetism between electrically charged particles. In 1968, British theoretical physicist Christopher Llewellyn Smith (b. 1942) found the first evidence for the existence of gluons. He noted that quarks account for only about half of a proton's momentum and theorized that electrically neutral particles must account for the other half

Gluons were directly observed for the first time in 1979 at the Positron-Electron Tandem Ring Accelerator (PETRA) in Hamburg, Germany. Since their discovery, gluons have been used as a tool to study other phenomena in particle physics. Scientists used them in the Large Hadron Collider (LHC) experiments that helped confirm the existence of the Higgs boson at the European Organization for Nuclear Research (CERN) in 2013.

What Makes a Gluon a Gluon?

All bosons have integral spin values (e.g., 0, 1, 2). Fermions have half-integral spin values (e.g., 1/2, 3/2, 5/2). Gluons have a spin of 1. Typically, particles with a spin of 1 have three polarization states. However, because gluons, like photons and unlike the W and Z bosons, are massless, they have only two polarization states: 1 and −1.

Similar to how protons and electrons possess electrical charge, quarks possess a quality called color charge. Just as electrically charged particles interact by exchanging photons, thus creating electromagnetism, color-charged particles (quarks) interact by exchanging gluons, thus creating the strong force. Individual quarks come in three "colors": red, green, and blue. (These qualities have nothing to do with the actual color of the particles, but are simply labels.) The antiparticles (antimatter particles) of quarks, called antiquarks, come in three "anti-colors": anti-red, anti-green, and anti-blue. Gluons themselves possess one color charge and one anti-color charge.

Gluons can interact with each other as well as with quarks. There are eight different types of gluons, each corresponding to a different combination of color and anti-color. The attraction between quarks of opposite color charge creates the strong force. Gluons and quarks combine to make all of the composite particles in the hadron family: mesons (one quark and one antiquark) and baryons (three quarks), the

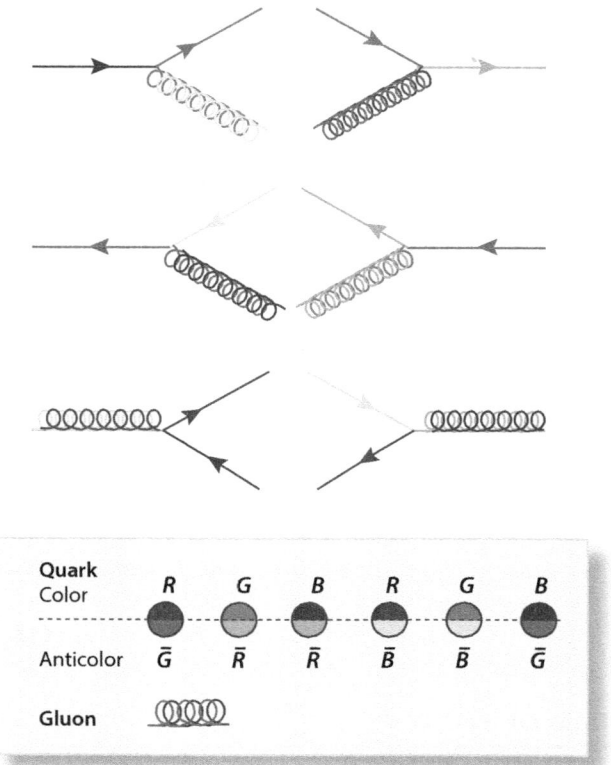

Gluons are carriers of color charge (red, green, blue) and anticolor (antired, antigreen, antiblue), or can be white. When color/anticolor gluons interact with quarks and antiquarks they exchange color in predictable patterns. The theory explaining these interactions is called quantum chromodynamics.

latter of which include the class of particles known as nucleons. The high-energy condition of many gluons interacting, known as a gluon condensate, could also be partly responsible for giving mass to some hadrons. Nucleons (protons and neutrons) are of particular importance, as they are the building blocks for most observable matter in the universe. The extreme strength of the strong force between gluons and quarks holds atomic nuclei together. This is why fission-powered nuclear weapons produce such huge amounts of energy: the fission, or breaking apart, of nuclei causes that force to be released.

Gluons' Quirky Behavior

While they form the basis for most "normal" matter humans observe, gluons also have some unusual traits that scientists are exploiting to enhance humankind's knowledge of the cosmos. Gluons and quarks have never been observed individually because of a quality of the strong force known as asymptotic freedom. Unlike the other forces, the strong force gets stronger as the distance between two quarks increases, making it difficult to pull them apart. Conversely, the strong force gets weaker as quarks are pressed more closely together. This process effectively frees the quarks to move individually. Asymptotic freedom could be responsible for a special condition known as quark-gluon plasma (QGP)—a nearly "perfect" or frictionless liquid. QGP has been created by high-speed particle collisions inside particle accelerators, which break protons and neutrons apart into gluons and quarks at 4–6 trillion degrees Celsius (7–10 trillion degrees Fahrenheit), about 100,000 times hotter than the center of the sun. The resulting hot "soup" is thought to resemble the state of the universe just fractions of a second after the big bang, about 13.8 billion years ago. Insights gained from studying QGP can help scientists learn more about how the universe's initial conditions evolved to produce galaxies, stars, planets, and, ultimately, life.

Because gluons can carry color charge, it would be theoretically possible to have a particle composed entirely of gluons. Known as "glueballs," such particles are expected to consist of either two or three gluons, have mass, and decay quickly after synthesis into pairs of several possible particles: pions (pi mesons), kaons (K mesons), or eta mesons. The observation of these emitted particles would provide further evidence in support of the standard model of particle physics. Many previously detected particles are candidates for glueballs. They have been the subject of active research for nearly two decades. The GlueX experiment, which began taking data at the Thomas Jefferson National Accelerator Facility in Newport News, Virginia, in 2014, is specifically designed to produce more definitive experimental evidence of glueballs. This will open up further avenues for the advancement of particle physics.

—*Lindsay K. Brownell, MS*

BIBLIOGRAPHY

"Color Charge." The Particle Adventure. Particle Data Group, n.d. Web. 21 Aug. 2015.

"Color Force." *HyperPhysics*. Georgia State U, n.d. Web. 21 Aug. 2015.

Dauncey, Paul D. "Nuclear and Particle Physics—Lecture 11: Parity and Charge Conjugation Conservation." HEP Group Research Pages. Imperial Coll. London, 10 June 2009. Web. 21 Aug. 2015.

Ellis, John. "The Discovery of the Gluon." *50 Years of Quarks*. Ed. Harald Fritzsch and Murray Gell-Mann. Hackensack: World Scientific, 2015. 189–98. Print.

Griffiths, David. *Introduction to Elementary Particles*. 2nd rev. ed. Weinheim: Wiley, 2008. Print.

Kyberd, Paul. "Explainer: What Are Fundamental Particles?" *The Conversation*. Conversation US, 20 Mar. 2015. Web. 21 Aug. 2015.

Moskowitz, Clara. "Hottest Particle Soup May Reveal Secrets of Primordial Universe." *LiveScience*. Purch, 13 Aug. 2012. Web. 21 Aug. 2015.

Sutton, Christine. "Strong Force." *Encyclopaedia Britannica*. Encyclopaedia Britannica, 2 Apr. 2014. Web. 21 Aug. 2015.

GRAVITATIONAL POTENTIAL ENERGY

FIELDS OF STUDY

Classical Mechanics

SUMMARY

One of the many forms of energy that an object can possess is gravitational potential energy. This energy has the ability to do work and move objects that are separated by a distance. When objects fall down to Earth, it is due to the effect of gravitational forces and gravitational potential energies.

PRINCIPAL TERMS

- **center of mass:** the weighted average position of the distribution of the mass that makes up an object.
- **displacement:** the shortest distance between the initial position of an object and its final position.

- **gravitational acceleration:** the change in velocity of a falling object caused by Earth's gravitational pull.
- **kinetic energy:** the energy an object gains due to its motion.
- **mass:** the amount of matter that makes up an object.
- **potential energy:** the energy stored in an object due to its position.
- **weight:** the force exerted on an object due to gravity.

EXCHANGE OF ENERGY

Everything on Earth's surface is being pulled down toward its center. The planet exerts a gravitational pull on all objects on around it, near and far. This property is not unique to Earth; every person on the planet is exerting a gravitational influence on all surrounding objects as well. This is due to the fact that Earth and everyone on it have mass. The concept of mass is difficult to measure, and physicists have long strived to develop a good system for doing so. In the International System of Units (SI), mass is measured in kilograms, where one kilogram is equal to the mass of the international prototype kilogram, also known as the Big K.

Mass is directly related to the weight of an object. An object's weight is equal to the force of the gravitational pull it experiences. On or near Earth's surface, the mass of the planet pulls on objects with a strength of approximately 9.81 meters per second per second, or meters per second squared, for every unit of mass. This rate, known as the gravitational acceleration, is the rate at which objects accelerate as they fall toward Earth.

FROM FORCES TO ENERGIES

When a person holds his or her phone at head height in order to speak to somebody, that phone has stored energy due to the fact that it is away from Earth's surface. This stored energy is called gravitational potential energy. Potential energy is energy that can be converted into other forms of energy in order to do work. Near Earth's surface, the gravitational potential energy (U_g) can be defined as the product of the mass of an object (m), the gravitational acceleration (g), and the object's displacement from its resting position—that is, its height above the ground (h):

$$U_g = mgh.$$

The value of U_g is given in joules (J), an SI derived unit of energy or work. One joule is equal to the energy used to apply a force of one newton (N) over a distance of one meter (m). The newton is also an SI derived unit; it is defined as the force required to accelerate a mass of one kilogram (kg) at a rate of one meter per second squared (m/s²). Thus, one joule is equal to one newton-meter (N·m), and one newton is equal to one kilogram-meter per second squared (kg· m/s²).

Now imagine the person holding the phone accidentally drops it. As the phone starts to fall, it starts to lose gravitational potential energy as its displacement from the ground decreases. That gravitational potential energy is being turned into kinetic energy (K), causing the phone to fall faster and faster. An object's kinetic energy is determined by its mass (m) and its velocity (v):

$$K = \frac{1}{2}mv^2$$

When a pendulum oscillates, it loses kinetic energy and gains potential energy as it swings up, which causes it to slow. When all its kinetic energy has turned into potential energy, the pendulum stops for an instant at the top of its swing. Then

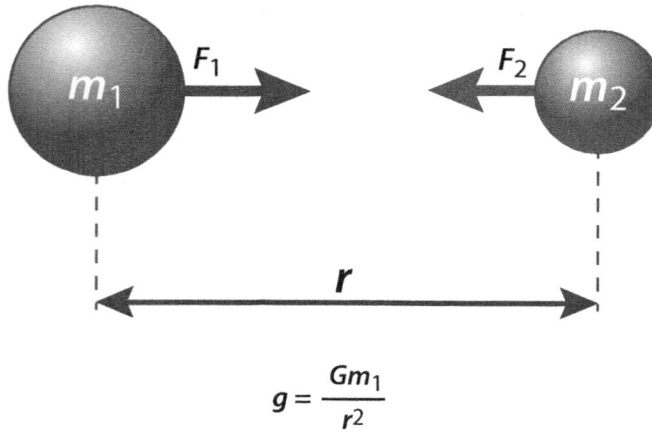

$$g = \frac{Gm_1}{r^2}$$

Any two objects with some amount of mass and some distance between them will exert gravitational force on each other. The magnitude of the force is dependent on the masses of the objects and the distance between them. The gravitational acceleration (g) on the surface of a planet is equal to the product of the universal gravitational constant (G) and the mass of the planet (m) divided by the radius of the planet (r) squared.

> **SAMPLE PROBLEM**
>
> Earth has a mass of approximately 5.97×10^{24} kilograms. Its moon has a mass of approximately 7.34×10^{22} kilograms. The average distance between Earth's center of mass and the moon's is about 3.84×10^8 meters. What is the gravitational force in newtons that the moon exerts on Earth?
>
> **Answer:**
> Calculate the gravitational force (F_g) using the universal law of gravitation:
>
> $$F_g = \frac{Gm_1 m_2}{r^2}$$
>
> $$F_g = \frac{(6.67384 \times 10^{-11} \, \tfrac{m^3}{kg \cdot s^2})(5.97 \times 10^{24} \, kg)(7.34 \times 10^{22} \, kg)}{(3.84 \times 10^8 \, m)^2}$$
>
> Use the commutative property of multiplication to rearrange the terms.
>
> $$F_g = \frac{\left(6.67384 \times 5.97 \times 7.34 \, \tfrac{kg \cdot m^3}{s^2}\right)\left(10^{-11+24+22}\right)}{(3.84 \, m)^2 \left(10^8\right)^2}$$
>
> $$F_g = \frac{292.446334032 \times 10^{35} \, \tfrac{kg \cdot m^3}{s^2}}{14.7456 \times 10^{16} \, m^2}$$
>
> Simplify the exponents and divide.
>
> $$F_g = \left(\frac{292.446334032}{14.7456}\right)\left(10^{35-16}\right)\left(\tfrac{kg \cdot m^{3-2}}{s^2}\right)$$
>
> $$F_g = 19.83 \times 10^{19} \, \tfrac{kg \cdot m}{s^2}$$
>
> Recall that 1 N is equal to 1 kg· m/s². Therefore, the gravitational force exerted on Earth by the moon is approximately 19.83×10^{19} N (rounded), or, in proper scientific notation, 1.983×10^{20} N.

for U_g is a simplification of the equation for the gravitational potential energy between any two objects. The gravitational potential energy possessed by any object due to the influence of any other object can be calculated, given the universal gravitational constant (G), the masses of both objects (m_1 and m_2), and the distance between their centers of mass (r):

$$U_g = -\frac{Gm_1 m_2}{r}$$

The value of G is 6.67384×10^{-11} m³/kg·s². Because G is a constant, its value does not change.

From this equation, physicists have obtained the universal law of gravitation. This law defines the gravitational force (F_g) exerted in newtons by one object with a mass of m_1 kilograms on a second object that has a mass of m_2 kilograms and is a distance of r meters away:

$$F_g = \frac{Gm_1 m_2}{r^2}$$

STRENGTH IN GRAVITATION

In order to calculate the force Earth exerts on the moon, one would use the same equation with all the same values as when calculating the force the moon exerts on Earth. The values of m_1 and m_2 would be switched, but because they are being multiplied and multiplication is commutative, the answer would be the same. As it turns out, Earth pulls on the moon with the same force that the moon pulls on Earth.

While this may seem unlikely, it is a direct result of Isaac Newton's (1642–1727) third law of motion. Newton's third law states that for every action, there is an equal and opposite reaction. This applies to any two objects, regardless of the difference in their masses. A single person pulls on Earth with the same force that Earth pulls on the person. So why do objects (and people) fall down to Earth but Earth does not "fall up"? This is where the difference in mass comes in. An object's acceleration is equal to the accelerating force divided by the mass of the object. Because the mass of a person is so much smaller than

it drops back down, losing potential energy and gaining kinetic energy.

The same thing can be said of a ball kicked into the air. The force of the initial kick gives the ball kinetic energy. As the ball rises, it loses some of that kinetic energy and gains potential energy. Eventually it achieves its highest point, having gained all the potential energy it can. Then it drops down, converting that potential energy back to kinetic energy and moving faster and faster as it falls.

A UNIVERSAL LAW

Gravitational potential energy does not apply only to Earth or other planets. The equation given above that of Earth, the same amount of force will cause the person to accelerate at a much, much greater rate than the planet.

—*Angel G. Fuentes, MS*

Bibliography

"The Energy of Games and Theme Rides." *BBC Bitesize*. BBC, 2014. Web. 26 May 2015.

Giambattista, Alan, and Betty McCarthy Richardson. *Physics*. 3rd ed. New York: McGraw, 2015. Print.

Khan, Sal. "Introduction to Newton's Law of Gravitation." *Khan Academy*. Khan Acad., n.d. Web. 26 May 2015.

Khan, Sal. "Work and Energy (Part 2)." *Khan Academy*. Khan Acad., n.d. Web. 26 May 2015.

Nave, Carl R. "Gravitational Potential Energy." *HyperPhysics*. Georgia State U, n.d. Web. 30 Apr. 2015.

Ohanian, Hans C., and John T. Markert. *Physics for Engineers and Scientists*. 3rd extended ed. New York: Norton, 2007. Print.

Weisstein, Eric W. "Gravitational Potential Energy." *Eric Weisstein's World of Physics*. Wolfram Research, 2007. Web. 26 May 2015.

Young, Hugh D., Roger A. Freedman, A. Lewis Ford, and Francis Weston Sears. *Sears & Zemansky's College Physics*. 14th ed. Boston, Mass: Pearson, 2016. Print.

HARMONIC OSCILLATOR

FIELDS OF STUDY

Harmonics; Classical Mechanics; Electronics

SUMMARY

Harmonic oscillations are a fundamental feature of all mass-spring systems. They are modeled by the motion of pendulums. They can be mathematically described using the sine and cosine functions. Newton's laws of motion describe the physical behaviors of pendulums. Hooke's law describes the force that restores such systems to their equilibrium positions. Communication, power generation, motor control, and all digital electronic devices depend on harmonic oscillation to function.

PRINCIPAL TERMS

- **damped oscillator:** an oscillator that is subject to friction or other braking forces.
- **Hooke's law:** the law stating that the deformation of an elastic object, such as a spring, is directly proportional to the force acting on the object, as long as the object's elastic limit is not exceeded.
- **mass-spring system:** a system consisting of an elastic object connected to an object with mass.
- **net force:** the overall force acting on a system, calculated as the vector sum of all forces acting on and within the system.
- **pendulum:** a suspended mass that can undergo regular oscillations.
- **resonance:** the response of an elastic body to a force acting on the body at its natural frequency.
- **spring constant:** a characteristic factor of a particular spring that determines the expansion or contraction of the spring when displaced by a specific force.
- **torque:** a turning force acting radially about an axis or point.

OSCILLATIONS

An oscillation is the variance of a physical property or its magnitude from one value to another and back again in a regular or cyclic manner. A harmonic oscillator is simply an object or system whose oscillation is caused by displacement that results in a restoring force. For example, imagine a marble sitting in the bottom of a circular bowl. The center of the bowl represents the marble's equilibrium point, because it is the lowest point to which the marble can roll. If the marble is held at the edge of the bowl, it has been displaced from its equilibrium point. When the marble is released, gravity acts as a restoring force, causing the marble to roll back down toward its equilibrium point. However, the marble will not stop at the bottom of the bowl. Instead, it will roll partway up the opposite side of the bowl, then back down to the bottom again and partway up the other side. This motion will repeat over and over until the marble eventually comes to rest at the bottom of the bowl. This repeated motion is oscillation, and the marble is a harmonic oscillator.

Oscillations can be described mathematically using sine and cosine waves. A sine wave begins at a value of zero (or some middle point represented by zero), increases to its maximum positive value, decreases below zero to its minimum negative value, and then increases to zero again. This is one cycle of the oscillation of a sine wave. The phrase "sine wave" describes a smooth, repetitive oscillation, such as the swing of a pendulum or the vibration of a guitar string. A cosine wave is similar to a sine wave, except that its cycle begins and ends at its maximum positive value rather than at zero. Thus, a cosine wave is also a type of sine wave. Electromagnetic waves are also sinusoidal in nature, as are waves rippling across a pond.

Oscillations are characterized by their amplitude, frequency, period, and sometimes wavelength. All oscillations have an equilibrium or neutral point in each cycle that is exactly midway between the extremes of their values. Displacement is the extent to which the value of the oscillating property varies from the neutral point at any given stage in its cycle. The amplitude of an oscillation is its maximum displacement from the neutral point. For example, the amplitude of a sine wave would be the distance of the maximum positive (or minimum negative) value from zero. The frequency of an oscillation is the number of cycles that occur in a given time. The period of the oscillation is the time required for one complete cycle. "Wavelength" is a term used to describe oscillations that travel in a linear fashion, such as electromagnetic waves. The wavelength is the distance traveled by the wave during one complete cycle of the oscillation.

Strings, Springs, and Resonance

A string on a musical instrument, such as a guitar or a piano, produces a sound when it vibrates. The frequency of the sound depends on the length of the vibrating string, and the intensity or loudness of the sound depends on the displacement of the string from its neutral position. The physical displacement of the string translates directly to the amplitude of the sound waves that are produced by the oscillation of the string as it vibrates. The sound fades and disappears as the string gradually ceases to vibrate and the amplitude decreases to zero. The frequency and the wavelength, of the sound remain the same, however, since these characteristics are not dependent on the amplitude or displacement of the oscillating string. In contrast, an ideal harmonic oscillator would produce the same values of frequency, wavelength, and amplitude over time.

A classic type of harmonic oscillator is the spring. Every spring has its own characteristic spring constant, which describes its stiffness and strength. The spring constant shows how much energy is required to displace the spring from equilibrium. It also shows how much restoring force the spring exerts to return to equilibrium. A coil spring can be either expanded or compressed. After either action, when the spring is released, it exerts a force that returns it to its neutral resting position, about which it then oscillates. This restoring force (F) is described by Hooke's law as the product of the spring's displacement (x) and its particular spring constant (k):

$$F = -kx$$

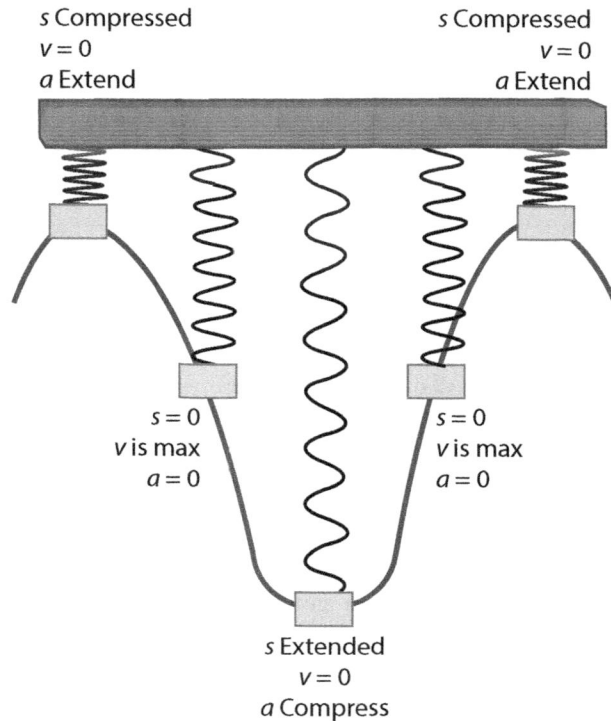

An oscillating spring and weight system shows the relationship among displacement (s), velocity (v), and acceleration (a). When the spring is fully compressed the velocity of the weight is zero and the acceleration will cause the spring to extend. When the spring is at a neutral displacement ($s = 0$) its velocity is maximum and the acceleration is zero, the spring will begin to decelerate. When the spring is fully extended the velocity of the weight is zero and the acceleration will cause the spring to compress.

The negative sign means that the restoring force is acting against the force that displaced the spring from its equilibrium position.

A spring does not have to be coiled in shape. Any structure that exerts a restoring force when displaced from its characteristic equilibrium shape can act as a spring system that obeys Hooke's law. The displacement can be lateral. It can also be caused by torque, as in some kinds of clocks or when a cable twists and untwists while supporting a suspended mass.

Hooke's law applies to all types of mass-spring systems. Every mass-spring system has its own characteristic resonance frequency (ω), which is related to its mass (m) and its particular spring constant (k), according to the following equation:

$$\omega = \sqrt{\frac{k}{m}}$$

When a system encounters another system or force oscillating at its resonance frequency, the first system resonates and vibrates at this frequency as well.

When resonant waveforms combine, their amplitudes also combine and increase, occasionally with disastrous results. Large groups of soldiers marching in lockstep are advised to break step when crossing a bridge so that they do not cause the bridge to resonate and collapse. On November 7, 1940, wind caused a bridge over the Tacoma Narrows strait in Washington State to oscillate at its resonance frequency. The amplitude of the vibration exceeded the limitations of the bridge's structural integrity, causing it to collapse. Similarly, Canadian truck drivers in the far north often take heavy loads over the "ice highway" on the frozen surfaces of lakes. More than one has been lost when the resonance frequency of the ice layer was met, causing the ice to break beneath the weight of the truck.

PENDULUMS

According to Isaac Newton's (1642–1727) second law of motion, a net force (F_{net}) acting on a body will cause the body to accelerate in the direction of the force. The acceleration (a) of the body will be directly proportional to the net force and inversely proportional to the mass (m) of the body:

$$F_{net} = ma$$

$$a = \frac{F_{net}}{m}$$

When an initial net force displaces a pendulum from its equilibrium position, it provides the pendulum with potential energy due to gravity. When the pendulum is released, gravity draws the pendulum back toward its equilibrium position. As it passes through equilibrium, the pendulum has kinetic energy. The momentum that the pendulum has acquired due to its motion, in accord with Newton's first law, requires it to continue on its path until gravity, acting as a damping force, stops the motion at a certain distance and makes the pendulum once again fall back toward its equilibrium position. This oscillating motion continues until the kinetic energy and momentum of the pendulum have been lost to friction, air resistance, and other energy-consuming factors. The net force acting on the pendulum at any point is the vector sum of the force of gravity (its weight) and any frictional or other braking forces acting on it. The force

---

SAMPLE PROBLEM

The displacement of an oscillating string is described by the equation

$$x = A\cos(\omega t)$$

where x is the displacement at time t, A is the initial displacement of the string, and ω is the frequency of the oscillation. Calculate the displacement of a string oscillating at 440 hertz (Hz), or cycles per second (1/s), at a time of 100 milliseconds (ms) after release from an initial displacement of 5 millimeters (mm).

Answer:

Assign the variable values as

$$A = 5 \text{ mm}$$

$$\omega = 440 \text{ 1/s}$$

$$t = 100 \text{ ms} = 0.1 \text{ s}$$

Calculate:

$$x = A\cos(\omega t)$$

$$x = 5 \text{ mm} \times \cos(440 \text{ 1/s} \times 0.1 \text{ s})$$

$$x = 5 \text{ mm} \times \cos(44)$$

$$x = 3.6 \text{ mm}$$

The displacement of the string 100 milliseconds after release is 3.6 millimeters.

of gravity imparts the motion, while braking forces detract from it.

A harmonic oscillator that is not subject to any braking forces is called a simple harmonic oscillator. An oscillator whose amplitude is decreased toward its equilibrium value by a braking force is called a damped oscillator. Almost all real physical oscillators are damped oscillators, because in the real world, braking forces are almost always present. To maintain the amplitude of a damped oscillation requires the constant input of energy into the system from an external source.

Torsional pendulums obey the same physical laws as linear pendulums. With a torsional pendulum, the displacement from the equilibrium position is brought about by torque about the central axis of the system. The restoring force of a torsional pendulum is not the force of gravity but the torque applied by the material that makes up the axis of the pendulum. A simple example of such a system is a metal wire or rod fixed to a weight. The radial length of the weight about the central axis determines the period and frequency at which the system can oscillate, while the radial displacement from its equilibrium position determines the spring force that acts on the system.

Oscillations in Other Systems

Electromagnetic waves and electronic signals are oscillating systems that obey the same mathematical principles as pendulums and vibrating strings. Describing such waves requires more complex calculus involving sine and cosine functions than is required for the description of simple harmonic oscillators and pendulums. Electromagnetic waves are characterized by their frequency, period, and amplitude. These are also related to circular motion through the sine and cosine functions. Electromagnetic waves combine as the vector sum of their sinusoidal waveforms and resonate at specific frequency combinations to produce purely sinusoidal waveforms. This is an essential feature of "phase shift" in electrical power generation and motor control. Digital electronic devices also depend on harmonic oscillation. They use the specific vibrational frequency of quartz or other materials to control the timing of transistor switches.

—*Richard Renneboog, MSc*

Bibliography

Beech, Martin. *The Pendulum Paradigm: Variations on a Theme and the Measure of Heaven and Earth.* Boca Raton: Brown, 2014. Print.

Chen, Y. T., and Alan Cook. *Gravitational Experiments in the Laboratory.* New York: Cambridge University Press, 2005. Print.

Giordano, Nicholas J. *College Physics. Reasoning and Relationships.* Belmont: Brooks, 2010. Print.

King, George C. Vibrations and Waves New York: Wiley, 2013. Print.

Lerner, Lawrence S. *Physics for Scientists and Engineers.* Sudbury: Jones, 1996. Print.

Matthews, Michael R., Colin F. Gauld, and Arthur Stinner, eds. *The Pendulum: Scientific, Historical, Philosophical & Educational Perspectives.* Dordrecht: Springer, 2005. Print.

HARMONICS

FIELDS OF STUDY

Harmonics; Acoustics

SUMMARY

All objects that are capable of vibrating have one or more natural frequencies. When they resonate at those frequencies, the vibrations form standing wave patterns characterized by a series of nodes and antinodes. Physical systems that resonate at their natural frequencies produce music, but they can also experience destructive consequences.

PRINCIPAL TERMS

- **antinode:** a point of maximum amplitude in a standing wave.
- **fundamental frequency:** the lowest frequency of a resonating medium; also called the first harmonic.
- **mechanical resonance:** the increase in the amplitude of motion by a medium in response to a peri-

odic applied force or a sympathetic vibration close to the fundamental frequency.
- **node:** a point of minimum amplitude (typically zero) in a standing wave.
- **octave:** a progression of eight harmonic tones.
- **overtone:** a frequency that is a whole-number multiple of the fundamental frequency of a resonating medium.
- **standing wave:** a wave pattern that maintains a static series of nodes and antinodes within a specific wavelength.

WAVES AND FREQUENCIES IN HARMONICS

Many natural phenomena are characterized as waves by the regularity of their behavior. Electromagnetic and sound waves obey mathematical relationships relating frequency and wavelength. Physical matter can also move as waves, especially fluid matter such as air, oil, and water.

Waves are characterized by their frequency, wavelength, and amplitude. Sound waves are produced by the vibration of an object, such as the string of a musical instrument. All vibrating objects have certain frequencies at which they naturally vibrate. The lowest of these is called the object's fundamental frequency. If a force is applied to such an object at a rate that is equal or very close to its natural frequency, the object will undergo mechanical resonance, which increases the amplitude of its vibrations. The frequencies at which resonance occurs are called resonance frequencies and are often the same as an object's natural frequencies. Resonance frequencies higher than the fundamental frequency are called overtones. In many cases, a note that seems to vibrate at a single frequency in fact includes overtones of that frequency as well. Such sounds are called complex tones. Most if not all naturally produced sounds are complex tones. A harmonic is an overtone that is a whole-number multiple of the fundamental frequency.

A musical string resonates when it vibrates as a standing wave. A standing wave is a pattern that forms when a wave reaches a boundary of its medium, such as the end of a string, and is reflected back in such a way that the crest (highest point) of the reflected wave meets the trough (lowest point) of the incident wave, or vice versa. (Alternately, a half-crest may meet a half-trough, a quarter-crest may meet a quarter-trough, and so on.) The two waves combine to form a resulting wave pattern whose amplitude at any given point is the sum of the amplitudes of the individual waves at that point. This effect is called interference. Thus, where a crest meets a trough (or a half-crest a half-trough, etc.), the two amplitudes add up to zero and cancel each other out. The points in a standing wave that maintain a constant amplitude of zero are called nodes. The ends of a vibrating string are also nodes, because their position is fixed. Between each pair of nodes is an antinode. These are points at which the combined amplitude of the two waves is at its maximum. For a standing wave to form, the medium must be exactly half the length of the wavelength, or some multiple of that length.

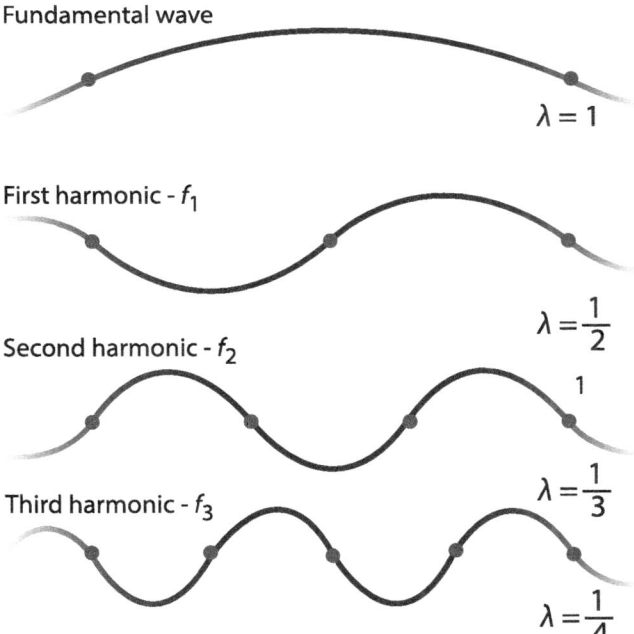

Frequency is the key property of harmonic waves. Harmonics have a frequency that is a multiple of the fundamental wave (the base of the harmonics). If a wave's wavelength (λ) is half that of the fundamental wave, it is the first harmonic of that fundamental wave. In other words, two full waves fall in the same length as the fundamental wave's wavelength. The second harmonic has a wavelength equal to 1/3 the fundamental wavelength. The third harmonic has a wavelength equal to 1/4 the fundamental wavelength.

HARMONIC SERIES

The fundamental frequency of a musical string corresponds to the wavelength that is twice the length of the string. At this frequency, the wave completes only one-half of one wave cycle before it is reflected back. A string's fundamental frequency, therefore, is

SAMPLE PROBLEM

Guitars are normally tuned to the A440 pitch standard, meaning that the A note above middle C, designated A_4, vibrates at a frequency of 440 hertz (Hz), or cycles per second (1/s). Using this standard, the fundamental frequency of the low E string is 82.41 Hz. Calculate the frequency and wavelength of the fourth harmonic of the low E string if its effective length is 0.648 meters (m).

Answer:

To calculate a harmonic frequency, the wave velocity must first be determined. The fundamental frequency (f_1) of the string is given as 82.41 Hz. Although no wavelength is provided, the wavelength that corresponds to a string's fundamental frequency (λ_1) is equal to twice the length (l) of that string:

$$\lambda_1 = 2l$$

$$\lambda_1 = 2(0.648 \text{ m})$$

$$\lambda_f = 1.296 \text{ m}$$

Use the fundamental frequency and wavelength to find the wave velocity:

$$v = f_1 \lambda_1$$

$$v = (82.41 \text{ 1/s})(1.296 \text{ m})$$

$$v = 106.80336 \text{ m/s}$$

The harmonic frequency is calculated as

$$f_n = \frac{vn}{2l}$$

The problem asks for the fourth harmonic frequency, so $n = 4$. Plug in this and the other appropriate values, and solve:

$$f_4 = \frac{(106.80336 \text{ m/s})(4)}{2(0.648 \text{ m})}$$

$$f_4 = \frac{427.21344 \text{ m/s}}{1.296 \text{ m}}$$

$$f_4 = 329.64 \text{ 1/s} = 329.64 \text{ Hz}$$

To calculate the wavelength of the fourth harmonic, substitute in the appropriate values for the wavelength equation:

$$\lambda_n = \frac{2l}{n}$$

$$\lambda_4 = \frac{2(0.648 \text{ m})}{4}$$

$$\lambda_4 = \frac{1.296 \text{ m}}{4}$$

$$\lambda_4 = 0.324 \text{ m}$$

The frequency of the fourth harmonic is 329.64 hertz, and its wavelength is 0.324 meters the lowest frequency at which it can form a standing wave, and its harmonics are the higher frequencies at which it can form a standing wave. These harmonic frequencies are determined by the equation

$$f_n = v_n / 2l$$

where f_n is the frequency of the nth harmonic, v is the velocity of the wave, and l is the length of the string. The wavelength of a harmonic can similarly be calculated using the equation

$$\lambda_n = 2l / n$$

where λ_n is the wavelength of the nth harmonic.

The velocity of a wave is calculated using the following equation:

$$v = f_n \lambda_n$$

If the same string is used and its tension remains constant, the wave's velocity will remain constant as well, regardless of its frequency or wavelength. Thus, as the frequency increases, the wavelength will decrease, and vice versa. This inverse relationship holds true for any type of wave.

The various harmonics of a particular fundamental frequency form a harmonic series. Tones that are not part of this series are considered inharmonic and fall into the category of noise. The overtones in a complex tone may be either harmonic or inharmonic. Cymbals and gongs are examples of instruments that produce sounds containing inharmonic tones.

In music, tones are arranged in scales according to their respective frequencies. The interval between a note at a certain frequency and another note at double that frequency is called an octave.

Harmonic Relationships

Every physical system has a characteristic fundamental frequency at which it will resonate. Bridges have been known to collapse because marching soldiers or high winds caused them to vibrate too close to their natural frequencies, and the resulting increase in amplitude tore the structures apart. Fluid-flow dynamics are affected by harmonics, and resonance between water waves and wind can produce large and very destructive waves. In some situations, such as the movement of fluids through a series of pipes, harmonic vibrations affect the energy economy of the system. While music is the most familiar application of harmonics, all wave and wave-like phenomena obey the fundamental physical laws of harmonic vibrations. This is especially important in electrical and electromagnetic applications such as wireless communication.

—*Richard M. Renneboog MSc*

Bibliography

Beament, James. *How We Hear Music: The Relationship between Music and the Hearing Mechanism.* Rochester: Boydell, 2001. Print.

Hartmann, William M. *Principles of Musical Acoustics.* New York: Springer, 2013. Print.

Prestini, Elena. *The Evolution of Applied Harmonic Analysis.* Boston: Birkhauser, 2004. Print.

Smith, Walter Fox. *Waves and Oscillations: A Prelude to Quantum Mechanics.* New York: Oxford University Press, 2010. Print.

Thompson, Daniel M. *Understanding Audio.* Boston: Berklee, 2005. Print.

Walker, James S., and Gary W. Don. *Mathematics and Music: Composition, Perception, and Performance.* Boca Raton: Taylor & Francis Group, 2013. Print.

Wolfe, Joe. "Strings, Standing Waves and Harmonics." *Music Acoustics.* University of New South Wales, n.d. Web. 25 Aug. 2015.

HEISENBERG UNCERTAINTY PRINCIPLE

FIELDS OF STUDY

Particle Physics; Quantum Physics; Classical Mechanics

SUMMARY

This article describes the Heisenberg uncertainty principle and examines the history of quantum mechanics leading up to it. It states that the energy and the location of an electron in an atom cannot both be determined with great accuracy. It enabled different theories to be combined into a single model for the behavior of electrons in atoms.

PRINCIPAL TERMS

- **accuracy:** the extent to which measurements of a property differ from its actual value.
- **eigenstate:** the state for which the value of a measurable, observable operator (change agent) has one exact mathematical solution, with no uncertainty.momentum: an intrinsic property of matter, the product of mass and velocity.
- **position:** in quantum mechanics, an electron's location in space relative to the nucleus of an atom.
- **precision:** the extent to which different measurements of the same property differ from one another.
- **probability density function:** the math function that describes the probability of an electron being found in a defined region of space about an atomic nucleus.
- **quantum mechanics:** a branch of physics based on the theory that energy is not continuous but rather is composed of discrete particle-like packets called "quanta"; "quantum" and "photon" are synonyms.
- **quantum state:** the energy level and various specific attributes of an electron.

ATOMIC STRUCTURE

The uncertainty principle, proposed by Werner Heisenberg (1901–76) in 1927, came at a time when physicists the world over were working to reconcile the new concepts of atomic structure and quantum theory with earlier ideas and observations. The principle was poorly accepted at first. However, it soon proved to be one of the most important concepts in the transition from classical mechanical theory to the modern quantum theory of atomic structure.

In 1897, J. J. Thomson (1856–1940) showed that the cathode rays observed in a cathode-ray tube were in fact streams of negatively charged particles with little mass. Those particles were later called "electrons." Not long after, Ernest Rutherford (1871–1937), Hans Geiger (1882–1945), and Ernest Marsden (1889–1970) identified protons as particles with more mass than electrons and the opposite (positive) charge. In 1911, they conducted the famous "alpha scattering" experiment. In this experiment, they directed a stream of alpha particles at a target of very thin gold foil. They noted that most of the particles passed directly through the foil as though it were not there. However, some particles seemed to have struck something very dense inside the foil and bounced off it. These observations were the first real evidence that atoms are composed of a very small, dense nucleus with a much larger, very diffuse cloud of electrons around it.

To describe the atom more completely, a third, neutral particle, termed the "neutron," had to exist. Because neutrons have no electrical charge, it was not possible to observe them directly by electrical means. It was not until 1932, when James Chadwick (1891–1974) demonstrated indirectly that they exist, that the basic principles of atomic structure were resolved. The resulting model was based on classical mechanics. It showed an atom with a very small, dense nucleus surrounded by electrons in specific orbits, much like planets orbiting the sun. In this model, the electrons would radiate energy constantly and would eventually fall into the nucleus as the orbit decayed.

By the early 1900s, many experiments had shown how atoms absorb and emit light. Studies also showed that only very specific wavelengths and energies along the electromagnetic spectrum were absorbed and emitted by atoms. Electrons can move into higher or lower energy states. In 1900, Max Planck (1858–1947) proposed that the energy involved in those transitions could only be transmitted or absorbed in discrete units, or "quanta." In 1913, Niels Bohr (1885–1962) amended the planetary model. He assumed that atoms could exist in static states where their electrons do not constantly radiate energy. He also assumed that electrons follow unique waveforms about the nucleus, not circular orbits. From this, Bohr developed a theory that successfully explained the observed spectrum of the hydrogen atom.

Another key observation from spectral data was that streams of electrons exhibit "wave-particle duality." Electrons can behave as particles under some conditions and as electromagnetic waves under others. Many physicists worked to reconcile these behaviors in terms of the motion of electrons about the nucleus. One important breakthrough came from Albert Einstein (1879–1955), who realized that the energy of an electron is equal to the product of its mass and the speed of light squared. These were key steps in the founding of quantum mechanics. Quantum mechanics seeks to describe the behavior of electrons in atoms using math.

One problem with early quantum theory was that wave-particle duality does not lend itself to precision and accuracy. Math solutions based on the electromagnetic properties of photons did not agree with solutions based on the particle behavior of electrons. This was mainly because scientists did not realize that electrons and photons could be the same thing. In 1924, Louis de Broglie (1892–1987) suggested that the wavelength of a photon is inversely proportional to the product of its momentum and the mass of an electron. Planck's constant, h, is the constant of proportionality.

Another problem with early quantum theory was that the math could only be resolved for simple atomic structures with one electron and one proton, and only at one energy level. Later calculations based on an initial calculation became more and more inaccurate with respect to observations. In 1927, Heisenberg explained this failure with what is now known as the uncertainty principle.

The Uncertainty Principle

In essence, the Heisenberg uncertainty principle states that it is not possible to determine both the exact energy level and the exact position of an electron in an atom at once. A single electron is so small

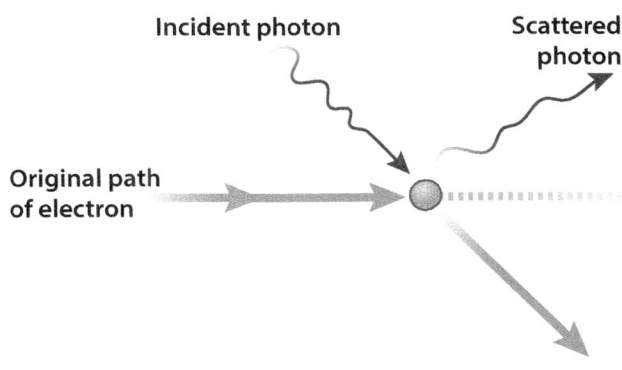

A photon of light must reflect off an electron to identify its location, but when it collides with the electron it inflicts a force on the electron that changes its velocity. Therefore, either the location or the momentum of an electron can be determined, but not both. This is the basis of Heisenberg's Uncertainty Principle.

that any measurement of its energy affects its location. Likewise, any measurement of its location affects its energy. In systems that can be described fully by classical mechanics, such effects are not relevant. For example, measuring the speed of a baseball using radar affects neither the speed nor the position of the baseball when it is measured. However, if the "baseball" were a particle of similar size to the wavelength of the electromagnetic waves from the radar gun, it would experience significant effects on its position and energy. The motion of an electron in an atom is affected by any interaction with electromagnetic radiation of a wavelength appropriate to carry out measurements. This is known as the Compton effect. It was named for Arthur Compton (1892–1962), who observed that x-rays reflected from a surface have lower energy, by consistent discrete amounts, than the incoming x-rays. Electromagnetic waves consist of photons, and photons can be the same thing as electrons. The interaction between light and an electron can therefore be thought of as two physical entities colliding, and the laws of conservation of energy and momentum apply.

Many physicists disliked the uncertainty principle because it means that precise and accurate solutions, such as those obtainable in classical mechanics, are not possible. Further progress in quantum mechanics had to rely on statistics and probabilities rather than exact solutions.

THE SCHRÖDINGER EQUATION

Erwin Schrödinger (1887–1961) worked on the same problem at about the same time as Heisenberg. Instead of treating the electron as a particle and making matrices of when, where, and how fast it was moving, he looked at the electron as a wave. He also related the matrix-based approach to his wave-mechanics-based approach. The math, while quite complex, elegantly described the behavior of electrons in atoms.

Thus, according to the uncertainty principle, an electron in a specific quantum state (with a certain energy and specific, unique values for certain attributes) exists somewhere in a range of places rather than at one specific place. Each of those regions is described by an "eigenfunction," a math function for which there can be one or more real solutions. Each of those real solutions, or "eigenvalues," corresponds to a specific eigenstate. The probability distribution function for a quantum state describes the probability of finding an electron in a specific region of space about the nucleus. These regions, and their 3-D shapes, define what are often termed "atomic orbitals." In turn, the orbitals and their electrons determine all normal chemical and molecular behavior.

—*Richard M. Renneboog, MSc*

BIBLIOGRAPHY

Aczel, Amir D. *Uranium Wars: The Scientific Rivalry That Created the Nuclear Age.* New York: St. Martin's, 2009. Print.

Aruldhas, G. *Quantum Mechanics.* 2nd ed. New Delhi: PHI, 2009. Print.

Griggs, Jessica. "To Be Quantum Is to Be Uncertain." New Scientist 23 June 2012: 8. Science Reference Center. Web. 23 Feb. 2015.

Cassidy, David C. "Quantum Mechanics, 1925–1927: The Uncertainty Principle." American Institute of Physics. Amer. Inst. of Physics, May 2002. Web. 23 Feb. 2015.

Heisenberg, Werner. *The Physical Principles of the Quantum Theory.* Trans. Carl Eckart and F. C. Hoyt. 1930. Mineola: Dover, 1949. Print.

Johnson, Rebecca L. *Atomic Structure.* Minneapolis: Twenty-First Century, 2008. Print.

Kakalios, James. *The Amazing Story of Quantum Mechanics: A Math-Free Exploration of the Science That Made Our World.* New York: Penguin, 2011. Print.

Reed, Bruce Cameron. *Quantum Mechanics.* Sudbury: Jones, 2008. Print.

Rosenblum, Bruce, and Fred Kuttner. *Quantum Enigma: Physics Encounters Consciousness.* 2nd ed. New York: Oxford University Press, 2011. Print.

HIGGS BOSON

FIELDS OF STUDY
Quantum Field Theory; Atomic Physics; Quantum Physics

SUMMARY

This essay describes the discovery and importance of the Higgs boson. The Higgs boson is the elementary particle that accounts for gravitation in the standard model of physics. The existence of such a particle was first proposed by Dr. Peter Higgs in 1964. The existence of the particle was proven in 2013 through work performed at CERN (the European Organization for Nuclear Research).

PRINCIPAL TERMS

- **boson:** carrier of one of the four fundamental forces.
- **electroweak force:** the force responsible for nuclear decay.
- **elementary particle:** one of the fundamental constituents of matter.
- **Higgs field:** a theorized field that, through interactions with matter, causes that matter to have mass.
- **lepton:** an elementary particle that has no color and thus cannot form nuclei.
- **standard model:** the contemporary framework for understanding particle physics.

ELEMENTARY PARTICLES

The standard model of particle physics is the primary theoretical model used to explain and predict

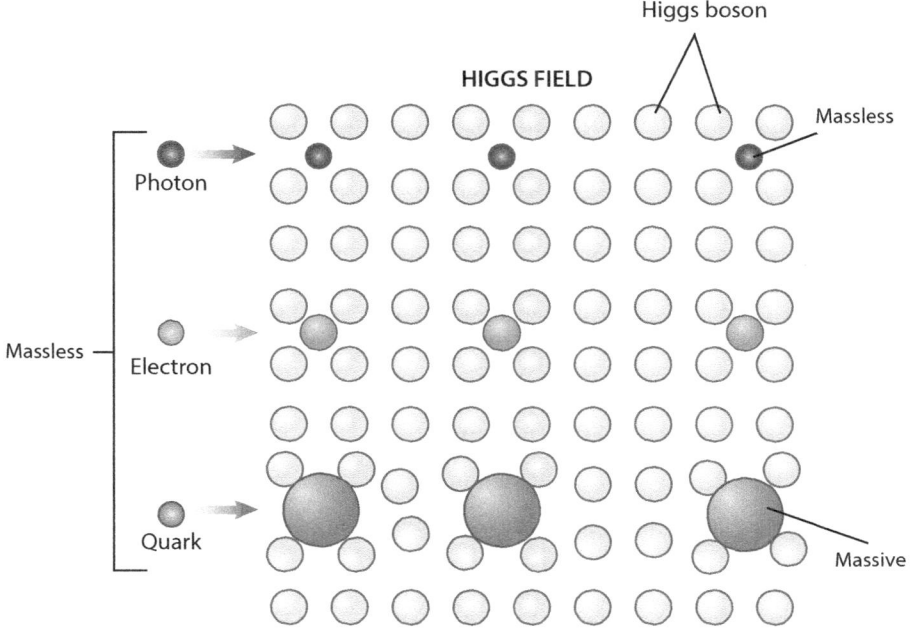

Excitations of the Higgs field, called Higgs bosons, interact with other particles, such as photons, electrons, and quarks. Particles that should be massless will have mass in the presence of the Higgs field.

particle interaction at the smallest scales and to describe the known subatomic particles. It has limitations, notably not working well with the full theory of gravitation as laid out in general relativity, but in its domain, it has been hugely successful.

At the subatomic level, everything is made of the elementary particles. The Higgs boson is the first step in reconciling gravitation and particle physics. The interactions between these particles are mediated by the fundamental forces, which are interactions among a set of distinct properties. These forces are electromagnetism, the strong nuclear force, the weak nuclear force, and gravity.

Electromagnetism is the force that causes things with opposite charges (positive and negative) to attract each other, while things with the same charge repel each other. This can easily be observed with magnets. On the atomic level, electromagnetism is also responsible for many other phenomena, such as friction. Electromagnetism is also what binds atoms together into molecules. This force is carried by the photon, which is also the particle responsible for light. A particle that carries a force is a boson and interacts with only that one force

The strong nuclear force is the force that sticks quarks together into more complex units like protons and neutrons. Strong nuclear force is linked to the property of color, which has nothing to do with visible color, but instead, is a way of tracking the property metaphorically. Color is said to come in three primary types (red, blue, and green), each with an opposite (anti-red, anti-blue, and anti-green). When the three types are combined in a stable way, they cancel out in a way similar to that in which red, blue, and green light combine to make white in a human eye. The strong nuclear force is transmitted in bosons called gluons, so called because they glue things together. It has color but no charge or mass, similar to the way in which a photon has charge but no mass or color. The reason it is called the strong nuclear force is that the attraction between differently color-charged quarks within protons is what holds the nuclei of atoms together. The attraction of the strong force is enough to overcome the electromagnetic repulsion between the protons.

The weak nuclear force changes the flavor of quarks. There are six flavors of quarks: up, down, strange, charm, top, and bottom. Each flavor of quark has a different charge, mass, and spin. If one can change the flavor, one can change those things. Weak nuclear force is carried by W and Z bosons. These bosons have mass, although bosons are not supposed to. This breaks electroweak symmetry. In physics, a symmetry is an unchanging mathematical or physical property that is necessary for a system, such as the general model, to function. Although the fuss over symmetry-breaking sounds like nitpicking and overreliance on theory, symmetries are the guiding principles of particle physics and are deeply ingrained in the math that guides the research. Electromagnetism is usually treated in concert with the weak nuclear force as electroweak force. This is because they have essentially equal strength.

Leptons are particles that do not have color and thus do not experience the strong force. They do not bond in nuclei. Quarks can be thought of as pack animals, like dogs, usually found in groups, while leptons are solitary, like cats. Electrons are the best-known kind of lepton.

The fundamental forces can be conceived of as being stored in a field, with bosons manifesting from the field to interact with other particles. A field suffuses space and gives it that property.

THE GRAVITATIONAL OUTLIER
The last fundamental force is gravity, which is observable on the largest scales and has the longest range. It has long been puzzling how gravity works, since general relativity does not mesh well with quantum mechanics. Scientists have also wondered why W and Z bosons have mass. As bosons, they should be interacting with only one force. This discrepancy led to theorizing that there could be certain conditions where the symmetries do not apply. Because gravity happens everywhere, the field responsible for it must exist everywhere. This theoretical field was dubbed the Higgs field, after the physicist Peter Higgs, one of the scientists who proposed it in 1964. The Higgs field breaks the symmetry of the electroweak interaction.

In theorizing this, they later realized that this would also explain why other fundamental particles had mass. The Higgs boson would thus be an excitation from the underlying and pervading Higgs field that would then interact with different particles in the same way that the other bosons do to give them greater or lesser mass. The theory provided a plausible explanation of the observable conditions, but there was no proof at first. Equally, it required assuming the existence of scalar fields. Up to this point, fields had been described as vector, meaning that they have a strength and direction, polarity, flavor, and color. A scalar field has only the strength. The issue was proving it.

FINDING THE HIGGS BOSON
Higgs bosons are very short-lived and require high-energy states to be observed. In order to find the Higgs boson, a new generation of technology was needed. Because these particles are so short-lived, the most effective way to find them was to search for the products of their decay. To create them, more powerful colliders were required. Colliders work by slamming particles together at very high energies. The higher energy the collision, the more interesting things scatter. It is akin to trying to learn about how a car works by ramming it into a wall and seeing what parts come flying out. Though scientists might theorize that spark plugs existed, in order to obtain proof, they would need a crash powerful enough to break the engine block open and a net fine enough to catch something as small as the plug.

This sort of collision required energy orders of magnitude beyond what the colliders of the 1960s could generate. To that end, it took the construction of CERN's Large Hadron Collider (LHC) to generate the power and provide the detectors, and the invention of a massive data sharing and computing space across the range of international researchers. Likewise, in order to sift the data, distributed computing systems were devised to allow calculations to be spread out over many computers, an early example of cloud computing. Scientists found evidence of a particle that had the properties of the Higgs boson and appeared in the proper places on July 4, 2012. This suggested that the Higgs theory was correct, but while it was clear that a new boson had been discovered, it was unclear if that boson was the Higgs boson. As more data emerged, scientists found more and more proof that this particle was indeed the Higgs boson. On March 14, 2013, it was tentatively confirmed as the Higgs boson, though there may be multiple Higgs bosons. A second run of the LHC, upgraded for greater collision energy, was planned for 2015. Further exploring the properties of the Higgs boson was one of CERN's main goals for the upgraded LHC.

—*Gina Hagler, MBA*

BIBLIOGRAPHY
Al-Khalil, Jim. *Quantum: A Guide for the Perplexed.* London: Weidenfeld, 2004. Print.
Fayer, Michael D. *Absolutely Small: How Quantum Theory Explains Our Everyday World.* New York: AMACOM, 2010. Print.
Ford, Kenneth William. *101 Quantum Questions: What You Need to Know about the World You Can't See.* Cambridge: Harvard University Press, 2011. Print.

Halpern, Paul. *Einstein's Dice and Schrödinger's Cat: How Two Great Minds Battled Quantum Randomness to Create a Unified Theory of Physics.* New York: Basic, 2015. Print.

Susskind, Leonard, and George Hrabovsky. *The Theoretical Minimum: What You Need to Know to Start Doing Physics.* New York: Basic, 2013. Print.

HORSEPOWER

FIELDS OF STUDY

Classical Mechanics; Thermodynamics

SUMMARY

Horsepower (hp) is a unit of measurement for power (work over time). It was developed by Scottish engineer James Watt to help quantify the usefulness of various machines by relating them to the power generated by horses, which were often used for heavy labor at the time.

PRINCIPAL TERMS

- **dynamometer:** a device that measures mechanical power or force, often used for the output of engines and other similar devices.
- **International System of Units (SI):** a standardized system of units and measures based on the metric system, used worldwide to enable clear and precise communication in the sciences and other disciplines.
- **power:** the rate of work performed, defined as energy consumed per unit time.
- **rotational speed (rpm):** the number of times an object rotates about a fixed axis in a set amount of time, typically measured in revolutions per minute (rpm).
- **watt:** the SI unit for measuring power (work over time), defined as one joule per second.
- **work:** the movement of an object due to the application of force, the SI unit of which is the joule.

QUANTIFYING WORK OVER TIME

Horsepower (hp) is a measure of the output power of a device, defined as work performed over time. The concept of horsepower was first developed by Scottish engineer James Watt (1736–1819). Watt had invented a type of steam engine based on rotary motion (the rotation of a wheel about an axis). He wanted to market the engine as a replacement for horses, which were then used to power mills and water pumps. To do so, Watt had to compare the performance of his engine to that of the horses.

Watt observed horses turning a mill wheel. Based on the speed and the force with which they turned the wheel, he calculated that one horse could exert enough force to move 32,572 pounds a distance of one foot per minute. He rounded this number up to 33,000. Thus, one unit of "horsepower" was defined as 33,000 foot-pounds per minute (ft·lb/min).

HORSEPOWER AND THE SI

Horsepower is not part of the International System of Units (SI), which is based on the metric system. The SI unit of power is the watt, named after Watt himself. One watt is equal to one joule of work performed (or energy expended) per second.

FROM	TO				
	BTU/hr	Joule/s	Horsepower	Kilowatt	Watt
BTU/hr	1	0.293	3.928×10^{-4}	2.93×10^{-4}	0.293
Joule/s	3.413	1	1.340×10^{-3}	1×10^{-3}	1
Horsepower	2.546×10^3	746	1	0.746	746
Kilowatt	3.413×10^3	1000	1.34	1	1000
Watt	3.413	1	1.340×10^{-3}	1×10^{-3}	1

Conversion table for units of power, including BTUs per hour, joules per second, horsepower, kilowatts, and watts.

There are several different types of horsepower. Mechanical horsepower uses Watt's measurement (33,000 ft·lb/min), which is equal to about 745.69 watts (W). Electrical horsepower is very similar, but it rounds up this amount to 746 watts. There is also a metric unit of horsepower, which is equal to 75 kilogram-meters per second (kg·m/s), or about 735.5 watts.

BEYOND THE HORSE

Although the use of horses has waned, horsepower has remained a common method of measuring the power of everyday devices. It is most often used for automobiles, but also for other forms of machinery powered by rotary motion, such as fans and washing machines. Horsepower is measured using a dynamometer, which calculates a machine's power output based on its torque (rotational force) in foot-pounds and its rotational speed in revolutions per minute (rpm).

—*Kenrick Vezina*

BIBLIOGRAPHY

Cleveland, Cutler J. "Horsepower." Encyclopedia of Earth. Environmental Information Coalition, 8 Oct. 2007. Web. 13 Mar. 2015.

Gibbon, Abby. "James Watt and the Revolution of Horsepower." *Chronicle of the Horse.* Chronicle of the Horse, 18 Aug. 2011. Web. 13 Mar. 2015.3 Science Reference Center™ Hors epower

"Horsepower." *Encyclopaedia Britannica.* Encyclopaedia Britannica, 7 July 2013. Web. 12 Mar. 2015.

"International System of Units (SI)." *NIST Reference on Constants, Units, and Uncertainty.* Natl. Inst. of Standards and Technology, Oct. 2000. Web. 11 Mar. 2015.

Kingsford, Peter W. "James Watt." *Encyclopaedia Britannica.* Encyclopaedia Britannica, 20 Nov. 2014. Web. 12 Mar. 2015.

Madureira, Nuno Luis. *Key Concepts in Energy.* Cham: Springer, 2014. Print.

Motavalli, Jim. "MIT Professor: High Weight and Horsepower Nullify Gains in Efficiency." *Wheels.* New York Times, 4 Jan. 2012. Web. 11 Mar. 2015.

IDEAL GAS LAW

FIELDS OF STUDY

Thermodynamics; Fluid Mechanics

SUMMARY

This article describes the relationship between the temperature, pressure, volume, and number of atoms or molecules of a gas. These four properties are the state variables in the ideal gas law, $PV = nRT$. Any change in one of these variables causes a change in the others. The relationship depends indirectly on the enthalpy and entropy within the gas-phase system.

PRINCIPAL TERMS

- **enthalpy:** the energy of matter expressed as heat.
- **entropy:** the property of a system that relates the degree of disorder in that system.
- **internal energy:** for an ideal gas, the energy of its atoms and molecules that is directly proportional to the absolute temperature.
- **kinetic energy:** energy that matter has due to motion.
- **kinetic theory of gases:** the theory that atomic and molecular motion determines the behavior of gases.
- **mole:** a quantity of a pure substance made up of exactly $6.02214129 \times 10^{23}$ atoms or molecules, the mass of which in grams has the same value as the mass of one atom or molecule of the substance in unified atomic mass units.
- **standard temperature and pressure (STP):** standard reference conditions when dealing with gases, defined by the International Union of Pure and Applied Chemistry as a temperature of 0 degrees Celsius (32 degrees Fahrenheit) and a pressure of 100 kilopascals (14.5 pounds per square inch).
- **state variables:** external factors, such as temperature and pressure, that determine the physical state of matter.
- **thermodynamic processes:** processes of change within a system that are related to its entropy and enthalpy.

STATES OF MATTER

All matter generally exists in one of three states at any time: solid, liquid, and gas. Solids have a fixed shape, while liquids and gases can conform to the shape of the structure that contains them. Liquids and gases are both classed as fluids because they can flow. Gases, however, can be compressed, while liquids cannot.

Most matter can change from one physical state to another according to the state variables affecting them. The state variables of pressure, temperature, volume, and amount of matter are all related in their effects on the entropy and enthalpy of the particular system. The ideal gas law states that for a hypothetical ideal gas, the product of the gas's volume (V) and the pressure (P) acting on it is equal to the product of the ideal gas constant (R), the temperature (T), and the amount of the gas in moles (n).

The atoms and molecules that make up the matter in a system have certain amounts of energy related to the absolute temperature of the system. Matter has kinetic energy when it is in motion, so the atoms and molecules that make up matter each have kinetic energy due to their motions. Atoms and molecules experience three main kinds of motion: vibration, rotation, and translation. The kinetic energy of each individual atom or molecule plus the energy contained in each chemical bond together make up the total internal energy of the system. The internal energy of a system changes when the system does "work" or when "work" is done on it.

Work, entropy, and enthalpy are all examples of thermodynamic processes because they are related to the intrinsic energy characteristics of a system. The enthalpy of a system is directly related to the absolute temperature of the system. Raising or lowering the temperature raises or lowers the system's enthalpy. Raising the enthalpy of a system, according to the kinetic theory, also raises the randomness of the atomic and molecular motion in the system, and so raises the entropy of the system. The energy that is available to do work in a system of matter is called the Gibbs free energy. This is calculated as the difference between the enthalpy and the temperature-dependent entropy of the system.

Raising the temperature of a solid material increases the movement of the atoms or molecules of that material. At some point, the attractive forces between atoms and molecules can no longer maintain the rigid connections required for the solid state, and the material becomes a liquid. The change of state from solid to liquid is called "melting" or "fusion." The change from liquid to solid is called "freezing" or "solidification." As the temperature of the liquid rises, the movement of the atoms or molecules increases until they can no longer keep in contact at all, and they separate from each other. At this point, the material becomes a gas. The change of state from liquid to gas is called "boiling" or "vaporization." The change of state from gas to liquid is called "condensation."

Another change of state is the conversion from solid directly to gas, or vice versa, without passing through a liquid state in between. This is called "sublimation."

THE STATE VARIABLES OF AN IDEAL GAS

The kinetic theory of gases states that the motions of gas atoms or molecules determine the properties and behavior of the gas. The physical state of an ideal gas is specified by the four state variables of amount, volume, temperature, and pressure. In practice, these are also sufficient for real gases.

The amount is the number of moles of gas that is contained within the system. One mole of any single substance contains exactly $6.02214129 \times 10^{23}$ atoms or molecules and has a mass in grams equal to the atomic or molecular mass of the substance. For example, the mass of one hydrogen molecule (H_2) is about 2.01 unified atomic mass units (u), so one mole of molecular hydrogen has a mass of 2.01 grams.

Volume is the space occupied by matter. The volumes of the solid and the liquid forms of matter are very similar, though not identical. However, the volume occupied by the same amount of matter as a gas is much greater. For example, one mole, or 18 grams, of pure water in the liquid state occupies a volume of 18 milliliters (mL), equal to 18 cubic centimeters (cm^3) or 1.098 cubic inches (in^3). At standard temperature and pressure (STP), however, one mole of pure water in the gaseous state occupies a volume of 22.413 liters (22,413 cm^3 or 1,367.2 in^3).

Temperature is the thermal energy of the matter. Heat energy determines the extent of atomic and molecular motions in matter. It flows from a region of higher energy to a region of lower energy until both

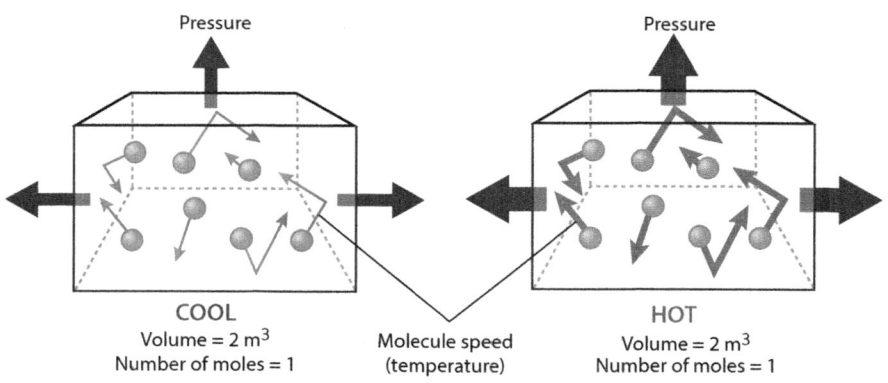

$$\text{Pressure} = \frac{\text{(Number of moles) (Gas constant) (Temperature)}}{\text{Volume}}$$

$$P = \frac{nRT}{V}$$

Pressure exerted by a gas is directly proportional to the number of molecules, or moles, of gas and the speed of the molecules, or temperature, and indirectly proportional to the volume of the gas. If the volume and moles of gas are kept constant but the temperature is increased, the pressure exerted by the gas on its container will increase.

PRINCIPLES OF PHYSICS　　　　　　　　　　　　　　　　　　　　　　　　　　　　　　　Ideal Gas Law

---

SAMPLE PROBLEM

A helium-filled weather balloon has a volume of 150 L at ground level, where the temperature is 22 °C and the pressure is 749 Torr. Calculate the mass in grams of the helium that it contains. Use $R = 0.082057 \frac{L \cdot atm}{mol \cdot K}$

Answer:

All such calculations are made simpler by first ensuring that all quantities are in the correct units. In this case, the temperature must be converted from degrees Celsius to kelvins, and the pressure must be converted from torrs to atmospheres.

To convert the temperature from the Celsius to the Kelvin scale, add 273.15:

$$K = °C + 273.15$$

$$K = 22 + 273.15$$

$$K = 295.15$$

To convert the pressure from torrs to atmospheres, divide by 760 Torr/atm:

$$\frac{749 \text{ Torr}}{760 \text{ Torr/atm}} = 0.986 \text{ atm}$$

Now, rearrange the gas law equation to obtain an expression for n:

$$PV = nRT$$

$$n = \frac{PV}{RT}$$

Insert the corresponding values of P, V, T, and R from above. Then multiply and divide.

$$n = \frac{(0.986 \text{ atm})(150 \text{ L})}{(0.082057 \frac{L \cdot atm}{mol \cdot K})(295.15 \text{ K})}$$

$$n = \frac{147.9 \text{ L} \cdot \text{atm}}{24.21912355 \frac{L \cdot atm}{mol}}$$

$$n = 6.106745 \text{ mol}$$

Convert the moles of helium to grams. Use the atomic mass of helium, which is approximately 4 u. Remember that the mass of one atom in unified atomic mass units is equal to the mass of one mole of that atom in grams.

$$m = 6.106745 \text{ mol} \times 4 \text{ g/mol}$$

$$m = 24.42698 \text{ g}$$

The mass of the helium in the weather balloon at ground level is approximately 24.427 grams.

---

regions are the same temperature. Because the heat energy of matter determines how much the atoms or molecules can separate from each other, gases occupy more volume at higher temperatures than they do at lower ones. Raising the temperature of a gas also raises its volume. Four standard scales are still used to measure temperature. Two of these, the Fahrenheit and the Celsius scales, use the freezing and boiling points of water as their reference temperatures. The other two, the Kelvin and the Rankine scales, start at absolute zero. The energy difference of each degree is the same for the Fahrenheit and Rankine scales, so one degree Fahrenheit is the same "size" as one degree Rankine. Similarly, one degree Celsius is the same size as one kelvin. For ideal gas calculations, the temperature must be stated using either the Rankine or the Kelvin scale, with the correct corresponding units for the other state variables. All temperatures are typically stated in kelvins.

The pressure, according to the kinetic theory of gases, is the force that the atoms or molecules of the gas exert when they collide with their container. Lowering the volume of matter tends to raise its pressure. Similarly, raising the temperature of the gas also tends to raise the pressure. A complete change of state can be brought about by changing the pressure of a system. Gases can be compressed sufficiently to transform into the liquid state and, in some cases, even into the solid state. Conversely, reducing the pressure in a system can drive a change of state from solid or liquid to gas. Pressure can be stated in different units. These include atmospheres (atm), pounds per square inch (psi), torrs (Torr), millimeters of mercury (mmHg), pascals (Pa), or kilopascals (kPa). As with temperature, the correct system of measurements for pressure must be used in ideal gas law calculations.

The Gas Law Equation

The relationship between pressure, volume, and temperature is an equilibrium that depends on the amount of matter. Changing any one or more of the state variables causes the others to also change, defining a new physical state. The relationship between these variables is quite precise and is used to define the gas constant R as

$$R = PV/nT$$

where P is pressure, V is volume, n is amount in moles, and T is temperature. Through careful observation, physicists such as Robert Boyle (1627–91), Joseph Louis Gay-Lussac (1778–1850), Amedeo Avogadro (1776–1856), and Jacques Charles (1746–1832) determined that the ratio of the pressure and volume of a gas to its amount and temperature is a constant value. This held true for every gas they examined. A gas always has the same volume when at the same temperature and pressure. In all cases, one mole of gas has a volume of 22.413 liters when the temperature is 0 degrees Celsius (273.15 K or 32 °F) and the pressure is 101.333 kilopascals (760 Torr, 1 atm, or 14.697 psi).

The constant relationship between the four state variables is equalized in the value of the gas constant R, which has the value of 0.082057 liter atmospheres per kelvin per mole (L·atm/mol·K), or the equivalent depending on the units being used. The correct units for each amount must be used when doing calculations with the ideal gas law. Accordingly, the ideal gas law is stated as

$$PV = nRT$$

and the units of the constant R must agree with those of the state variable being computed. Again, the ideal gas law is entirely independent of the chemical identity of the gas. Only the number of moles of gas is relevant to the calculations.

Proportional Values

In many cases, one state variable remains the same while the others change. By using some basic rules of algebra, one equation can be used for the same system in different states. This often allows new state properties to be calculated as simple ratios rather than as complete equations. The simplest example is when the amount of matter remains constant. In such cases, the product nR has the same value for both states and can be used to equate the two state expressions

$$P_1 V_1 = nRT_1$$

and

$$P_2 V_2 = nRT_2$$

as follows:

$$\frac{P_1 V_1}{T_1} = nR$$

$$\frac{P_2 V_2}{T_2} = nR$$

$$\frac{P_1 V_1}{T_1} = \frac{P_2 V_2}{T_2}$$

This allows an unknown state to be calculated directly by simple ratios when the appropriate values of T, P, and V are known, even if n is not known. This is even simpler when P, V, or T is also a constant value through the change of state.

—*Richard M. Renneboog, MSc*

BIBLIOGRAPHY

Ackerman, Steven A., and John A. Knox. *Meteorology: Understanding the Atmosphere.* 3rd ed. Sudbury: Jones, 2012. Print.

Allen, James P. *Biophysical Chemistry.* Hoboken: Wiley, 2008. Print.

Andrews, David G. *An Introduction to Atmospheric Physics.* 2nd ed. New York: Cambridge University Press, 2010. Print.

Lide, David R., ed. *CRC Handbook of Chemistry and Physics.* 95th ed. Internet vers. N.p.: Taylor, 2015. Web. 23 Feb. 2015.

Reger, Daniel L., Scott R. Goode, and David W. Ball. *Chemistry: Principles and Practice.* 3rd ed. Belmont: Brooks, 2009. Print.

Rogers, Donald W. *Concise Physical Chemistry.* Hoboken: Wiley, 2011. Print.

Zumdahl, Steven S., and Susan A. Zumdahl. *Chemistry.* 9th ed. Belmont: Brooks, 2014. Print.

J

JOULE

FIELDS OF STUDY

Classical mechanics; Thermodynamics

SUMMARY

The joule (J) is the standard unit of energy in the International System of Units. It measures the ability of a system to do work. This can include the chemical energy in the food that one's body digests to power itself or the thermal and electromagnetic energy radiated by the sun. Various nonstandard units are used in certain contexts, but all can be expressed in terms of joules.

PRINCIPAL TERMS

- **British thermal unit (BTU):** a nonstandard unit of energy measurement supposedly equivalent to the heat needed to raise the temperature of one pound of water by one degree Fahrenheit.
- **electron volts:** units of energy equal to the energy carried by electrons from an electric potential to one higher by one volt.
- **energy:** the ability of a system to perform work.
- **horsepower hour:** a nonstandard unit of power supposedly equivalent to the amount of work a horse does over an hour.
- **International System of Units (SI):** also known as the "metric system"; a widely used standardized system of units for measuring natural phenomena.
- **kilocalorie:** a nonstandard unit commonly used to measure energy content in food.
- **watt:** the SI unit of power (i.e., energy transfer over time).

MEASURING ENERGY

Questions such as how to quantify the amount of heat given off by a candle led to the first measurements of heat. Heat is recognized as one of the many forms of energy. The joule is the standard unit within the International System of Units (SI) for measuring energy. It is defined as the energy transferred (or "work done") by a force of one newton over a distance of one meter. In everyday terms, this is approximately the amount energy require to lift an apple one meter.

JAMES PRESCOTT JOULE AND BTUS

James Prescott Joule was a brewer and a physicist in England in the nineteenth century. He studied heat and its relationship to energy and work. He measured

FROM	TO		
	BTUs	Joules	Foot-Pounds
BTUs	1	1055	778
Joules	0.0009478	1	0.7376
Foot-Pounds	0.001285	1.3558	1

FROM	TO			
	Horsepower	Watts	Foot-Pounds Per Second	BTUs Per Second
Horsepower	1	746	550	0.7068
Watts	0.001341	1	0.7376	0.00095
Foot-Pounds Per Second	0.00182	1.356	1	0.001285
BTUs Per Second	1.415	1055	778	1

Conversion table for units of energy (top), including BTUs, joules, and foot-pounds. Conversion table for units of power (bottom), including horsepower, watts, foot-pounds per second, and BTUs per second.

the energy output of various systems using the British thermal unit (BTU). His work helped lay the groundwork for the first law of thermodynamics: energy in a system is neither created nor destroyed. In recognition, the SI unit of energy that eventually replaced the BTU was named after him. One BTU is equal to approximately 1,055 joules.

The Joule Today

The joule is an important component of the more familiar unit, the watt. The watt is equivalent to one joule per second. Although the joule is the standard unit, many nonstandard units find usage in certain contexts. The electron volt (abbreviated eV) is popular among physicists working at the atomic scale. It is equal to approximately 1.6×10^{-19} joules. The kilocalorie ("food calorie") is the unit used on nutrition labels to quantify the amount of energy in a given food item. It is equal to approximately 4,184 joules. Just as horsepower is still used to describe the power of large everyday machines such as cars, the horsepower hour is sometimes used to measure the power output of large machines over time. It is equal to 2,684,519.54 joules. However, each of these nonstandard measures can be expressed using the joule.

—Kenrick Vezina, MS

Bibliography

Ceraolo, Massimo, and Davide Poli. *Fundamentals of Electric Power Engineering: From Electromagnetics to Power Systems.* Hoboken: Wiley, 2014. Print.

Cooper, Christopher. *The Basics of Electric Current.* New York: Rosen, 2015. Print.

"Joule." *Encyclopaedia Britannica.* Encyclopaedia Britannica, 2 Mar. 2014. Web. 3 Apr. 2015.

Ohanian, Hans C., and John T. Markert. *Physics for Engineers and Scientists.* 3rd extended ed. New York: Norton, 2007. Print.

Pickover, Clifford A. *The Physics Book: From the Big Bang to Quantum Resurrection, 250 Milestones in the History of Physics.* New York: Sterling, 2011. Print.

"This Month Physics History: December 1840—Joule's Summary on Converting Mechanical Power into Heat." *APS News.* American Physical Soc., Dec. 2009. Web. 3 Apr. 2015.

K

KEPLER'S LAWS

FIELDS OF STUDY

Classical Mechanics

SUMMARY

This article defines Johannes Kepler's three laws of orbital motion, which describe the motions of planets around a sun-centered solar system. Isaac Newton later developed the universal law of gravitation that gave a more general explanation of planetary motion. These laws can be applied to the motion of any satellite orbiting a massive central body.

PRINCIPAL TERMS

- **Cartesian coordinates:** a pair or triplet of numbers that indicate the location of a point in a plane or in a three-dimensional space and are the signed distances from the origin.
- **centripetal force:** the force acting on an object moving along a circular path that is directed toward the center of the path or axis of rotation.
- **ellipse:** an oval surrounding two focal point points where the sum of the distances from any point on the oval to the two focal points is constant.
- **Lagrangian points:** positions where the gravitational pull of two large masses equals the centripetal force needed for a small object to maintain a stable position in orbit with the two large masses.
- **perturbation:** a disturbance in the motion or orbit of a massive body due to the gravitational pull of or impact with another object.
- **polar coordinates:** a pair of numbers indicating a point's length (radius) from a fixed center, called the pole or origin, and the angle between the radial and the polar axes.
- **two-body problem:** a mathematical description of the motion in space of two rigid, point-like objects interacting with each other.

KEPLER'S LAWS

German astronomer Johannes Kepler (1571–1630) used observations collected by Danish astronomer Tycho Brahe (1546–1601) to develop his three laws of orbital motion. Kepler's first law states that planets move in elliptical orbits with the sun at one focus point. More generally, the law says that one object moves around another object in an elliptical orbit with the system's center of mass located at one focus. An orbit can be described by its semimajor axis and eccentricity. In an ellipse, the long axis is referred to as the "major axis" and the short is referred to as the "minor axis." The semimajor axis is half the length of the major axis. Eccentricity describes the shape of an object's orbit, with an eccentricity of zero referring to a circular orbit and an eccentricity of 1 being very flat orbit. The eccentricities of the orbits of the planets in the sun's solar system are small (almost circular).

Kepler's second law states that the line connecting a planet and the sun sweeps out equal areas in equal amounts of time. A planet moves slower as it travels farther from the sun and faster as it travels closer to the sun. The point where the planet is closest to the sun is called the perihelion, and the point where it is farthest, the aphelion. The aphelion and perihelion distances (R_a and R_p) can be related to the semimajor axis (a) and the eccentricity (e) of an orbit by the following equations:

$$R_a = a(1 + e)$$

$$R_p = a(1 - e).$$

Kepler's third law states that the square of a planet's orbital period (T) is proportional to the cube of the orbit's semimajor axis (a), written as

$$T^2 \propto a^3.$$

164

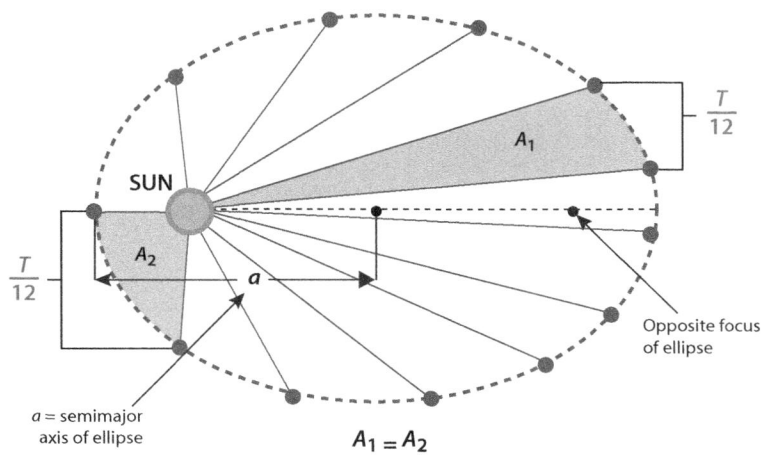

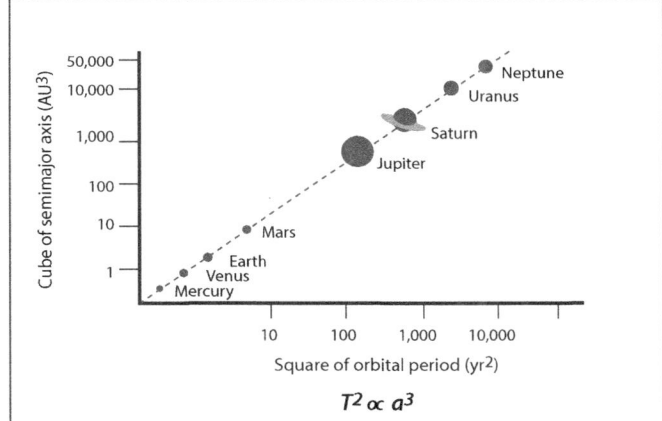

All planets orbit the sun in an ellipse with the sun at one focus point along the semimajor axis (a) of the ellipse. For equal segments of the planet's period (time it takes for a planet's full orbit, T), the areas of the ellipse (A) that it sweeps across are equal. A graph of the planets' orbital period squared and semimajor axis cubed show all planets follow the same line. The period (T) squared of all planets orbiting the sun are directly proportional to their semimajor axis (a) cubed by the same ratio.

A planet's orbital period is the time it takes for a planet to complete a full orbit around the sun. Thus, as the semimajor axis of an orbit increases, the more time it takes the planet to orbit the sun. Also, the ratio of the square of a planet's orbital period to the cube of the orbit's semimajor axis has the same value for every planet in the solar system.

NEWTON'S VERSION

Newton combined gravitation and circular acceleration to obtain a relation that included period, distance, and a central body's mass. The attractive gravitational force (F_g) between two objects with masses (m_1) and (m_2) that have distance (r) between them is:

$$F_g = \frac{Gm_1m_2}{r^2}$$

This equation is referred to as the universal law of gravitation. The constant G is the gravitational constant and is equal to 6.67384×10^{-11} m³/kg·s². The centripetal force (F_{cp}) needed for a planet to orbit the sun is the product of the velocity (v) and mass of the planet (m_2), divided by its orbital radius (r).

$$F_{cp} = \frac{m_2 v^2}{r}$$

The velocity of the planet (v) is equal to the circumference of the orbit ($2\pi r$) divided by the period (T).

$$v = \frac{2\pi r}{T}$$

Substituting this value for v in the equation for centripetal force, one arrives at:

$$F_{cp} = \frac{4\pi^2 r m_2}{T^2}$$

The centripetal force is then set equal to the gravitational force, with m_1 as the mass of the sun and m_2 as the mass of the planet.

$$F_{cp} = F_g$$

$$\frac{4\pi^2 r m_2}{T^2} = \frac{Gm_1 m_2}{r^2}$$

Rearranging terms and eliminating m_2, one arrives at the following:

$$T^2 = \frac{4\pi^2 r^3}{Gm_1}$$

For circular motion, the semimajor axis (a) replaces the distance (r) and the expression can be written simply as $T^2 = a^3$.

This relationship between period and semimajor axis was predicted by Kepler's laws. However, Kepler's laws apply only approximately to the solar system, since the mass of the sun is so much greater than the

165

mass of the planets. A more precise version of the law can be written as follows:

$$T^2 = \frac{4\pi^2 a^3}{G(m_1 + m_2)}$$

Newton's model better fits real observations and two-body problems, such as the orbit of a moon around a planet or a binary star system. The units for the semimajor axis of an orbit are normally expressed in kilometers (km) or astronomical units (AU), with the units for period expressed in days. An astronomical unit is about the average distance from Earth to the sun and is equal to about 149.6×10^6 kilometers (about 93 million miles).

Perturbations are differences in the motion of a planet from Kepler's laws that result from gravitational attraction by other planets or, sometimes, collisions. If another body is added to a system of two bodies, there are positions where the satellite can maintain a stable position. For example, if a satellite is added to the system of Earth and sun, the gravitational pull of the sun and Earth produces the centripetal force needed for the satellite to orbit with them. There are five Lagrangian points in the orbital plane, referred to as L1, L2, L3, L4, and L5. The points L1, L2, and L3 are positioned on a line that connects Earth and sun. L4 and L5 are located at the tip of two equilateral triangles with the sun and Earth at the other points. L1, L2, and L3 are unstable points, while L4 and L5 are very stable. A satellite positioned at L1, L2, and L3 will quickly be thrown off orbit.

There are different coordinate systems and methods for labeling points in a plane, including Cartesian coordinates and polar coordinates. Polar coordinates describe planetary motion best. A planet's position can be described by the radial distance between the planet and sun (r) and the angle between the planet and sun (θ).

Impact of Kepler's Laws

Kepler developed laws of orbital motion that made it apparent that the sun and the other planets do not orbit Earth, as was claimed by the Greek Egyptian astronomer Ptolemy (ca. 100–ca. 170 CE). Instead, Earth and other planets orbit the sun. Kepler's laws described the motion of the planets but not why the planets moved in this manner. Isaac Newton (1642–1727) improved upon Kepler's laws by showing that those laws are directly connected to the law of gravitation. These laws are now applied to not only planets orbiting the sun, but also to

---

SAMPLE PROBLEM

Using Kepler's laws, calculate the Earth's farthest (aphelion) and closest (perihelion) distances from the sun if Earth's semimajor axis is 149.6×10^6 km and its eccentricity is 0.0167.

Answer:

To calculate the farthest and closest that Earth travels to the sun, use the equations below that relate aphelion and perihelion distances, semimajor axis, and eccentricity.

$$R_a = a(1 + e)$$

$$R_p = a(1 - e)$$

Plug in the values for Earth's semimajor axis (a) and eccentricity (e). Then compute.

$$R_a = a(1 + e)$$

$$R_a = 149.6 \times 10^6 \text{ km } (1 + 0.0167)$$

$$R_a = 152.1 \times 10^6 \text{ km}$$

$$R_p = a(1 - e)$$

$$R_a = 149.6 \times 10^6 \text{ km } (1 - 0.0167)$$

$$R_a = 147.1 \times 10^6 \text{ km}$$

The farthest Earth gets from the sun is 152.1×10^6 km, and the closest Earth gets to the sun is 147.1×10^6 km. The slightly elliptical nature of Earth's orbit changes its distance from the sun at certain points along its travel.

Note that these distances do not coincide with warm- or cold-weather periods in either the Northern or the Southern Hemisphere. Seasons are determined by Earth's tilt on its axis, not by its distance from the sun.

natural and artificial satellites orbiting Earth or other celestial bodies.

—*Casey M. Schwarz, PhD*

BIBLIOGRAPHY

Bennett, Jeffrey O., et al. *The Essential Cosmic Perspective.* 7th ed. Boston: Pearson, 2015. Print.

Cutnell, John D., Kenneth W. Johnson, Shane Stadler, and David Young. *Physics.* 10th ed. Hoboken, NJ: Wiley, 2015. Print.

Halliday, David, Robert Resnick, and Jearl Walker. *Fundamentals of Physics.* 10th ed. Hoboken, NJ: Wiley, 2014. Print.

Henderson, Tom. "Kepler's Three Laws." *The Physics Classroom.* Physics Classroom, 1996–2015. Web. 22 Apr. 2015.

Tipler, Paul Allen, and Gene Mosca. *Physics for Scientists and Engineers.* 6th ed. Vol. 1-2. New York, NY: W.H. Freeman, 2008. Print.

Williams, David, R. "Earth Fact Sheet." NASA Space Science Data Coordinated Archive. NASA, 1 July 2013. Web. 10 July 2013.

KILOWATT-HOUR

FIELDS OF STUDY

Thermodynamics; Electronics

SUMMARY

The kilowatt-hour (sometimes "kilowatt hour") is a unit of measure for electricity consumption. It is equal to the consumption of 1,000 watts of power over the course of one hour.

PRINCIPAL TERMS

- **calorie:** the amount of energy needed to raise the temperature of one gram of water by one degree Celsius (1.8 degrees Fahrenheit) at one atmosphere of pressure, equivalent to approximately 4.2 joules.
- **consumed energy:** the amount of energy used or transferred over a period of time, often measured in kilowatt-hours.
- **electron volt:** the amount of energy carried by a single electron moved across an electric potential difference of one volt; equal to 1.6×10^{-19} joules.
- **energy:** the ability of a system to perform work.
- **International System of Units (SI):** a standardized system of units and measures based on the metric system.
- **joule:** the SI unit of energy, work, or heat; equivalent to the work done by one newton of force moving one meter.
- **kilowatt:** a unit measuring work over time, equivalent to 1,000 watts, or joules per second
- **power:** energy over time; describes rate of work and measured in joules per second.

WATTS TO KILOWATT-HOURS

The International System of Units (SI) is a global measurement system used in science and other disciplines. The watt (W) is an SI derived unit of power, or energy used over time. One watt is equal to one joule per second. The joule is another SI derived unit equal to the work of one newton of force moving a distance of one meter. Household appliances have labels that state how many watts each device uses. For example, a 60 W light bulb uses 60 joules of electrical energy per second and turns most of that energy into light and heat. The kilowatt-hour (kWh) measures the rate of energy consumption in wattage over time, in terms of kilowatts (kW) per hour. Unlike the watt and the joule, the kilowatt and kilowatt-hour are not SI units.

The kilowatt-hour is a familiar method of measuring the rate of energy used over time, but other measures exist. The measurement of time is always in units of seconds, minutes, and hours, but the units of energy vary by area of study and convention. In chemistry, the calorie (c) is a popular alternative for measuring energy in fuel sources like coal. In physics, the electron volt (eV) may be used instead. A single electron carries exactly one electron volt of energy when it moves across an electric potential difference of one volt.

FROM	TO			
	Kilowatt-hours	Watt-seconds	Terawatt-hours	BTUs
Kilowatt-hours	1	3.6×10^6	1×10^{-6}	3,412
Watt-seconds	2.778×10^{-7}	1	3.6×10^{12}	1,055
Terawatt-hours	1×10^6	2.778×10^{-13}	1	2.931×10^{-10}
BTUs	2.931×10^{-4}	9.478×10^{-4}	3.412×10^9	1

Conversion table for units of energy show the conversions among BTU, kilowatt-hours, watt-seconds, and terawatt-hours.

How Much Energy Does Your Toaster Use?

A 1,000 W toaster that is used nonstop for one hour would use exactly one kilowatt-hour. The kilowatt-hour is a way to measure the consumed energy of a home over the course of time. The monthly electrical energy consumption of most homes ranges from hundreds to thousands of kilowatt-hours. Because a kilowatt-hour expresses the rate of energy used over time, it is a useful way to compare the efficiency of various homes or the same home over months or years.

—*Kenrick Vezina, MS*

Bibliography

Dodge, David. "So What Is a Kilowatt-Hour Anyway?" *HuffPost Blog*. Huffington Post Media Group, 17 Mar. 2015. Web. 25 Mar. 2015.

"Electricity Explained: Measuring Electricity." U.S. Energy Information Administration. US Dept. of Energy, 16 Oct. 2014. Web. 27 Mar. 2015.

"Estimating Appliance and Home Electronic Energy Use." Energy.gov. US Dept. of Energy, 5 Nov. 2014. Web. 23 Mar. 2015.

"International System of Units (SI)." *Encyclopaedia Britannica*. Encyclopaedia Britannica, 14 Jul. 2013. Web. 25 Mar. 2015.

Nakagami, Hidetoshi, Murakoshi, Chiharu, and Yumiko Iwafune. "International Comparison of Household Energy Consumption and Its Indicator." *ACEEE Summer Study on Energy Efficiency in Buildings*. Amer. Council for an Energy-Efficient Economy, 2008. Web. 25 Mar. 2015.

"Residential Sector Energy Consumption." *Buildings Energy Data Book*. US Dept. of Energy, Mar. 2012. Web. 25 Mar. 2015.

KINECTIC AND POTENTIAL ENERGY

FIELDS OF STUDY

Classical Mechanics; Relativity

SUMMARY

This article describes different types of kinetic and potential energy and how they are related to the total energy of a system. Kinetic energy is expressed in terms of classical mechanics and relativistic mechanics in order to investigate a particle as it approaches the speed of the light.

PRINCIPAL TERMS

- **center of mass:** the point of a system that moves as though all of the mass of the system were located or concentrated there and all of the forces were applied there.
- **center of momentum:** the inertial frame of a system where the vector sum of the moments of all of the particles in that system is zero.
- **frame of reference:** a set of coordinate axes that serve to describe position or movement of an object with reference to that coordinate system.
- **special relativity:** the theory that states that for all inertial nonaccelerating reference frames, the laws of motion remain the same, and that speed of

light in a vacuum is the same for all observers, regardless of the observer's movement relative to the source of light or the movement of the source itself.

KINETIC AND POTENTIAL ENERGY EXPLAINED
Energy comes in many different forms. Some common forms of energy are thermal or heat energy, chemical energy, gravitational energy, and electrical energy. Kinetic energy is the energy an object possesses because of its motion. Any kind of motion produces kinetic energy, including translation (movement along a straight line), rotation, and vibration. Objects at rest have zero kinetic energy. The faster an object moves, the greater its kinetic energy.

Work is used to describe the energy transferred to or from an object by a force acting on that object. When energy is transferred to an object, the work is said to be positive. When energy is transferred away from the object, the work is negative. "Work" is defined as the displacement of the object being acted on multiplied by the force acting on it. The change in the kinetic energy of an object is equal to the net work done on that object. Also, the kinetic energy after the work is done is equal to the sum of the initial kinetic energy and the net work done. This is known as the work–kinetic energy theorem. It is true for both negative and positive work.

Potential energy is the energy a system possesses due to the configuration or positioning of objects within the system that exert forces on each other. The potential energy of a system can change if the configuration of the system changes. Potential energy can be thought of as a stored energy of an object, or as the difference between the energy of an object at a certain position and its energy at a reference position. There are different types of potential energy that relate to various types of forces, including gravitational, elastic, and electric potential energy.

A conservative force is a force for which the work done to move an object from one location to another does not depend on the path taken to get there. The work done by a conservative force is equal to the negative change in potential energy.

MEASURING KINETIC ENERGY
To correctly measure kinetic energy, one must keep in mind the frame of reference in which it is being

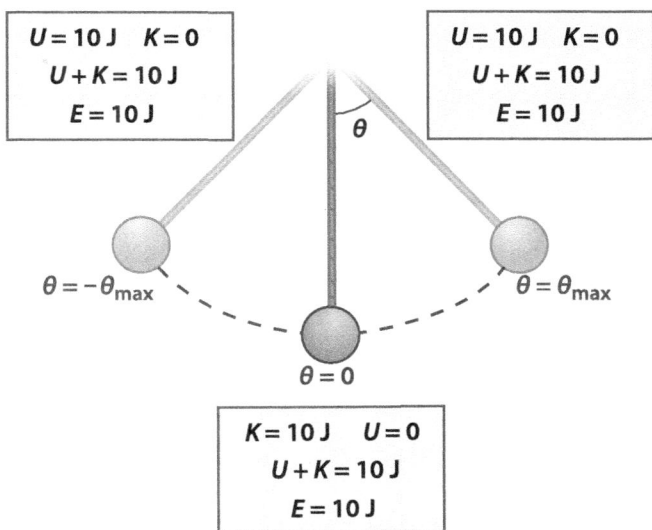

A pendulum at a maximum angle (θ) will have a minimum velocity, maximum potential energy (U), and a kinetic energy (K) of zero. A pendulum at an angle equal to zero will have a maximum velocity, a potential energy of zero, and a mechanical energy equal to the maximum kinetic energy. At any point along the pendulum's path, the potential energy and kinetic energy will equal the total mechanical energy (E) of the system and will remain constant.

measured. If a bird were to fly over the head of a stationary observer, that bird would have velocity and, thus, kinetic energy. However, if another observer were moving at the same speed as the bird, then the bird would be stationary relative to that observer, with no velocity and therefore no kinetic energy. Two observers who are moving in two different reference frames measure two different values. Even so, the total energy of an isolated system never changes within the measured reference frame.

A system's total kinetic energy is the sum of the total kinetic energy in its center-of-momentum frame and the kinetic energy the total mass would have if it were entirely concentrated in its center of mass. The center-of-momentum frame, or center of momentum for short, is the reference frame in which the sum of the momentum of a particular system is equal to zero. This frame gives the minimum values of a system's energy. The kinetic energy of a system in its center of momentum is a quantity that is conserved, meaning that its value never changes, and is the same to all observers. In other reference frames, one must take into account additional kinetic energy resulting from the total mass moving at the speed of the center of mass.

Kinectic and Potential energy

The two main forms of kinetic energy are translational kinetic energy, produced by an object moving in a straight line, and rotational kinetic energy, produced by an object rotating about its center of mass. The kinetic energy (K) of an object moving in a straight line at nonrelativistic speeds (that is, much slower than the speed of light) is described by the equation

$$K = \frac{1}{2}mv^2$$

where m is the mass of the object and v is its velocity. The equation for rotational kinetic energy is very similar, only with moment of inertia (I) instead of mass and angular velocity (ω) instead of linear velocity.

TYPES OF POTENTIAL ENERGY
The energy stored in elastic materials such as bungee cords, rubber bands, and springs due to their compressing or stretching is called elastic potential energy. The force required to compress or stretch a spring is equal to the force that will restore the spring back to its initial relaxed state. Hooke's law states that this force, called the spring force (F), is proportional to the distance from equilibrium that the spring is extended or compressed (x), as shown in the following equation:

$$F = -kx.$$

In this equation, k the spring constant, a measure of spring stiffness that is specific to the particular spring. The stiffer the spring, the greater the value of k. In the International System of Units (SI), the value of k is given in newtons per meter (N/m). The negative sign indicates that the spring force always acts in the direction opposite the displacement of the spring.

The same variables are used to calculate the elastic potential energy (U_e) for a spring, as shown in the following equation:

$$U_e = \frac{1}{2}kx^2$$

When a spring is not being compressed or stretched, it is in its equilibrium position and has no elastic potential energy stored in it.

SAMPLE PROBLEM

A 0.50 kg crate resting on a horizontal, frictionless surface is pushed into a spring and then released. The crate compresses the spring a distance of 11.0 cm. The spring has a force constant of $k = 250$ N/m. What is the velocity and kinetic energy of the crate just after it leaves the spring?

Answer:

Recall that the initial total energy (E_i) must be equal to the final total energy (E_f), and that the total energy of a system is equal to the sum of its potential energy (U) and its kinetic energy (K):

$$E_i = E_f$$

$$E = U + K$$

$$U_i + K_i = U_f + K_f$$

First, convert the compression distance from centimeters to meters. Then use the equation for elastic potential energy to calculate the initial potential energy (U_i) of the compressed spring. The initial kinetic energy (K_i) is 0 J because the crate is at rest. To keep units consistent throughout, recall that 1 J = 1 N·m = 1 kg·m²/s².

$$U_i = \frac{1}{2}kx^2$$

$$U_i = \frac{1}{2}(250 \text{ N/m})(0.11 \text{ m})^2$$

$$U_i = 1.5125 \text{ J}$$

$$K_i = 0 \text{ J}$$

(Sample problem continued on next page)

If an object moves along the y axis relative to the ground, the gravitational force does work on the object. Gravity exerts a downward force on the center of mass of an object near Earth's surface. Gravitational potential energy (U_g) is the stored energy associated with an object due to its vertical position y, or the height of the object relative to its reference position. The gravitational potential energy of an object depends on the object's mass (m), the height (h) to which it is raised, and the rate of acceleration due

> **(SAMPLE PROBLEM CONTINUED)**
>
> Use these values to calculate the final kinetic energy (K_f). This can be done because the final total energy consists solely of the kinetic energy of the released crate, just like the initial total energy consisted solely of the potential energy of the compressed spring. The final potential energy (U_f) is 0 J because the spring is no longer compressed.
>
> $$U_i + K_i = U_f + K_f$$
>
> $$(1.5125 \text{ J}) + (0 \text{ J}) = (0 \text{ J}) + K_f$$
>
> $$K_f = 1.5125 \text{ J}$$
>
> Now use the equation for kinetic energy to calculate the crate's velocity after it leaves the spring:
>
> $$K_f = \frac{1}{2}mv^2$$
>
> $$1.5125 \, \frac{\text{kg} \cdot \text{m}^2}{\text{s}^2} = \frac{1}{2}(0.50 \text{ kg})v^2$$
>
> $$\frac{(2)(1.5125 \, \frac{\text{kg} \cdot \text{m}^2}{\text{s}^2})}{0.50 \text{ kg}} = v^2$$
>
> $$\sqrt{6.05 \, \frac{\text{m}^2}{\text{s}^2}} = v$$
>
> $$v = 2.46 \text{ m/s}$$
>
> The crate's velocity is 2.46 m/s, and its final kinetic energy is 1.5125 J.

to gravity (g), which on Earth is approximately 9.8 meters per second per second, or meters per second squared (m/s²). This relationship is shown in the following equation:

$$U_g = mgh$$

It is important to establish a reference point or zero-height position. Depending on the specific situation, the ground or a tabletop can be considered the zero position. The potential energy of an object then depends on the height relative to the ground or tabletop.

RELATIVISTIC MECHANICS
Special relativity applies the principle of relativity to inertial reference frames. Inertial reference frames are frames of reference that move with a constant velocity relative to each other and in which the laws of physics hold. Newton's laws of motion for mechanical systems do not work for systems that accelerate relative to an inertial frame. When an object's speed approaches the speed of light, one must use relativistic mechanics to calculate kinetic energy. Albert Einstein's theory of relativity defines relativistic kinetic energy (K) as

$$K = \frac{mc^2}{\sqrt{1 - \left(\frac{v}{c}\right)^2}} - mc^2$$

where c is the speed of light in a vacuum, equal to 299,792,458 meters per second. As the velocity of a particle approaches the speed of light, the energy of the particle approaches infinity. In the center of momentum, the system's total energy is the rest energy (E), which can be related to mass and the speed of light by Einstein's famous mass-energy equation:

$$E = mc^2$$

The relativistic total energy is the sum of the kinetic energy and the rest energy. When a particle is at rest relative to an observer, K is equal to zero, and the relativistic total energy equation is equal to the mass-energy equation.

CONSERVATION OF MECHANICAL ENERGY
Energy can be converted from kinetic energy to potential energy and back again. The SI unit of kinetic energy is the joule (J), equal to 1 newton-meter (N·m), or 1 kg·m²/s². The total mechanical energy (E) of a system is the sum of its potential energy (U) and its kinetic energy (K):

$$E = U + K$$

Because the total energy of a system is constant, any decrease in the system's potential energy must result in a corresponding increase in its kinetic energy, and vice versa. Thus, this equation can also be written as

$$E_i = E_f$$

where E_i is initial energy and E_f is final energy. When written in terms of potential and kinetic energy, the equation becomes

$$U_i + K_i = U_f + K_f$$

where U_i and K_i are the initial potential and kinetic energies, respectively, and U_f and K_f are the final potential and kinetic energies.

A pendulum consists of a string connected to a pivot point at one end and a mass at the other end. The pendulum moves back and forth periodically. As the pendulum mass swings from its maximum height to its minimum height, its speed increases. Every time the mass falls, it loses potential energy, but it gains speed and thus kinetic energy. The mechanical energy of this system is thereby conserved.

KINETIC AND POTENTIAL ENERGY

The total amount of energy in the universe is always constant. It can change from kinetic to potential and back again, as seen when a high jumper leaps into the air and then comes back down, picking up speed along the way. Relativistic mechanics are used for objects approaching the speed of light and for transforming measurements between reference frames that move relative to each other. Einstein's special theory of relativity is necessary for modern long-range navigation, in which the precise location and speed of a moving craft are constantly monitored and updated.

—*Casey M. Schwarz, PhD*

BIBLIOGRAPHY

"Analysis of Situations in Which Mechanical Energy Is Conserved." *Physics Classroom.* The Physics Classroom, n.d. Web. 18 Apr. 2015.

Cutnell, John D., Kenneth W, Johnson, Shane Stadler, and David Young. *Physics.* 10th ed. Hoboken, NJ: Wiley, 2015. Print.

Halliday, David, Robert Resnick, and Jearl Walker. *Fundamentals of Physics.* 10th ed. Hoboken, NJ: Wiley, 2014. Print.

Schobert, Harold H. *Energy: The Basics.* New York: Routledge, 2014. Print.

Tipler, Paul Allen, and Gene Mosca. *Physics for Scientists and Engineers.* 6th ed. Vol. 1-2. New York, NY: W.H. Freeman, 2008. Print.

L

LENZ'S LAW

FIELDS OF STUDY

Classical Mechanics; Electromagnetism; Electronics

SUMMARY

Coils react to any change in the magnetic environment around them. Induced currents are created in the coils to work against any change in magnetic field. Lenz's law states that in order to conserve energy, the induced current in a loop produces a magnetic field, which then opposes any changes in the existing magnetic field through the loop. By using coils and changing magnetic fields, humans have been producing electricity for decades.

PRINCIPAL TERMS

- **conservation of energy:** total amount of energy in a closed system is always the same, therefore energy can neither be created nor destroyed.
- **current:** the motion of charged particles as a function of time.
- **electromagnetic field:** a field created by the motion of charged particles.
- **Faraday's law:** a physical law that explains how changes in the strength of magnetic fields through coils induce a current in the coil to counteract the change in the magnetic field.
- **inductance:** the property by which a current is created in conductors to resist a change in the magnetic field through the conductor.
- **right-hand rule:** a technique used to find the direction of magnetic forces and induced currents by using the right hand. The direction of the current is expressed by the fingers, and the direction of the magnetic field is expressed by the extended thumb.

INDUCED CURRENTS

Michael Faraday (1791–1867) was a British scientist who experimented extensively with currents. Although he had no formal training in physics, he was able to discover an important relation between electricity and magnetism. Faraday found that currents are induced by changing magnetic fields and currents create magnetic fields. The motion of charged particles creates electromagnetic fields. The two physical concepts are related. This is the basis of electric generators. By spinning coils around magnets or magnets around coils, electricity can be produced. Because Faraday did not have any formal training, he could not explain these effects with mathematics. It took many years and the work of Scottish physicist James C. Maxwell (1831–79) to discover the mathematical law that explains Faraday's discoveries. In order to correctly predict the behavior of induced currents, Maxwell had to include Lenz's law. The law states that the induced current in a loop produces a magnetic field, which then opposes any changes in the existing magnetic field through the loop.

THE MAGNETIC FIELD OF A CAR

When a motor vehicle stops at a traffic signal, a sensor is activated in order to detect the waiting vehicle. Many people assume that a pressure sensor is used, but if that were the case, smaller vehicles such as motorcycles would have a hard time activating the sensor. Instead, an inductive loop sensor is used, and it is designed to detect changes in the magnetic field.

Cars and other motor vehicles are made of parts that contain ferromagnetic materials, which create magnetic fields. When a car stops at a red light, the magnetic field produced by the car induces a current in an inductive loop sensor. The size of the current created is relative to the magnetic field's strength and the inductance of the coil. This, in part, is analyzed by the electronics near the traffic light, which then trigger the light to change. These inductive loop sensors work on two basic laws of electromagnetism: Faraday's law and Lenz's law.

Faraday's law states that changes in the strength of magnetic fields through coils induce a current in the coil to counteract the change in the magnetic field. An inductive loop sensor is basically a loop of wire. The voltage induced (V_{ind}) in the inductive loop or any other coil of wire is equal to the magnetic flux ($\Delta\Phi$) divided by the amount of time it takes for the field to change (t), expressed as

$$V_{ind} = -\frac{\Delta\Phi}{\Delta t}$$

The magnetic flux is a function of the magnetic field strength (B), the cross-sectional area of the loop exposed to the magnetic field (A), the number of turns in the loop (N), and the amount of time it takes for the field to change (t), or

$$V_{ind} = -N\left(\frac{\Delta(BA)}{\Delta t}\right)$$

In the example above, the area of the loop permeated by the magnetic field is constant. Therefore, the equation can be simplified to

$$V_{ind} = -N\left(A\frac{\Delta B}{\Delta t}\right)$$

In order to find the current through the loop, one must apply Ohm's law in combination with Faraday's law. Ohm's law states that the voltage (V) is the product of the current (I) and the resistance (R), or

$$V = IR.$$

By substituting this into Faraday's law, the induced current in the inductive loop sensor can be obtained. Mathematically this is expressed as

$$V_{ind} = -N\left(A\frac{\Delta B}{\Delta t}\right)$$
$$IR = -N\left(A\frac{\Delta B}{\Delta t}\right)$$
$$I_{ind} = -\left(\frac{NA}{R}\right)\left(\frac{\Delta B}{\Delta t}\right)$$

Lenz's law explains why a negative sign precedes the equation in Faraday's law. If the negative sign were not included, it would mean the induced current could create a magnetic field that enhances the existing magnetic field through the coil. In turn, this magnetic field would produce more current that would then produce even more magnetic field. The system would produce a feedback that would take the current into infinity. The law of conservation of energy states that in a closed system, the total amount of energy is always the same; energy in a closed system can neither be created nor destroyed. If an infinite current is induced, the implication is an infinite amount of energy, which is impossible. The negative sign in Faraday's law carries this second important property. It balances the magnetic field and makes the system conserve energy by keeping the induced currents opposing any change.

In order to correctly find the directions of the induced currents, a simple technique was devised. This technique, known as the right-hand rule, uses an individual's right hand. First, point the fingers of the right hand in the direction of the current. If it is a coil, curl the fingers so that they are pointing in the direction of the current. Next, extend the thumb, which will then be pointing in the direction of the induced magnetic field. This technique is also effective if the thumb of the right hand is first pointed in

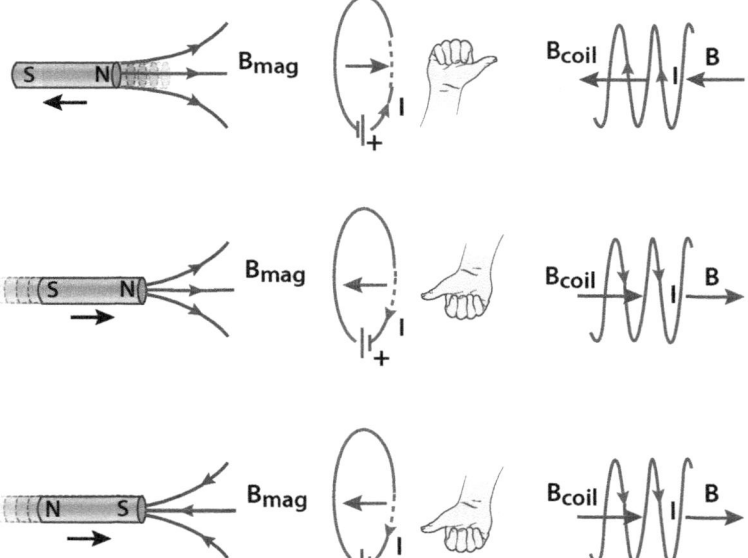

Induction opposes the change in flux of a magnetic field (B). When a magnet passes through a coil, the change in magnetic field (B_{mag}) induces a current (I) in the coil according to the right-hand rule, with a magnetic field (B_{coil}) that moves in the opposite direction.

> **SAMPLE PROBLEM**
>
> A coil of copper wire is being used by high school physics students in order to energize a light bulb. The students bring the north pole of a bar magnet closer to the coil of wire from the left side of the coil. What is the direction of the induced current in the coil?
>
> **Answer:**
>
> $$\Delta \Phi > 0$$
>
> $$\Delta \Phi_{ind} < 0$$
>
> I_{ind} is counterclockwise.
>
> Using Lenz's law, it is understood that a coil's induced current opposes any change in the magnetic environment around it. By bringing a magnet's north pole closer to the coil and from the left, the magnetic field is increasing in strength. Therefore, there are more magnetic field lines going through the coil to the right. The coil opposes this extra field, so it produces a field that is to the left. To find the direction of the current that produces the magnetic field, use the right-hand rule. Place the thumb of the right hand in the direction of the magnetic field, left. Now curl the fingers, and the direction of the fingers is the direction of the induced current, which in this case is counterclockwise.

the direction of the induced magnetic field, with the fingers then naturally curling in the direction of the induced current.

Generating Electricity

Alternating current (AC) is produced by generators at power plants that use the principles described by Faraday's and Lenz's laws. These generators, whether powered by falling water, steam, or wind, consist of coils of wire and magnets. As the water, steam, or wind makes a turbine spin, the main shaft, which has magnets attached to it, also turns. When a magnet crosses in front of one of the coils, it increases the magnetic field through the coil. This induces a current in the coil, which creates a magnetic field to cancel the extra magnetic field through the coil. As the turbine keeps spinning, the magnet moves away from the coil. Then, the coil has less magnetic field through it, and an induced current is created on the coil to increase the magnetic field through the coil. In other words, coils resist change and fight against any change in the magnetic field through them. If enough time passes, the coil gets used to the change and the induced currents disappear. But by that time, the shaft has spun more, and another magnet has moved in front of the coil.

—*Angel G. Fuentes, MS*

Bibliography

Giambattista, Alan, and Betty McCarthy Richardson. *Physics*. 3rd ed. New York: McGraw, 2015. Print.

Khan, Sal. "Conservation of Energy." Khan Academy. Khan Academy, 17 Feb. 2008. Web. 17 July 2015.

Khan, Sal. "Induced Current in a Wire." Khan Academy. Khan Academy, 3 Aug. 2008. Web. 17 July 2015.

Nave, Carl R. "Faraday's Law." HyperPhysics. Georgia State U, 2012. Web. 16 July 2015.

Shamos, Morris H., ed. "Lenz's Law." *Great Experiments in Physics: Firsthand Accounts from Galileo to Einstein*. 1959. New York: Dover, 1987. 159–65. Print.

Young, Hugh D., Roger A. Freedman, A. Lewis Ford, and Francis Weston Sears. *Sears & Zemansky's College Physics*. 14th ed. Boston, Mass: Pearson, 2016. Print.

LIGHT WAVES

FIELDS OF STUDY

Optics; Electromagnetism

SUMMARY

The nature of light as both particles and electromagnetic waves is presented. The interaction of light with matter is the basis of the science of optics and has many applications. Visible light is just a small portion of the electromagnetic spectrum, which is the range of wavelengths that make up electromagnetic radiation.

PRINCIPAL TERMS

- **diffraction:** a change in direction (bending) of a light ray as it passes around an obstruction or through an opening.
- **interference:** the meeting of two waves traveling in the same medium, causing their properties to interact.
- **photoelectric effect:** a phenomenon that describes the emission of electrons from matter (typically metal) upon exposure to electromagnetic radiation.
- **photon:** an elementary particle of light that moves and has energy but lacks mass and electrical charge.
- **polarization:** the process by which the motion of electromagnetic waves, which is perpendicular to the direction of energy transfer, is brought into alignment in the same plane.
- **reflection:** the bouncing back of waves from surfaces that do not absorb those particular wavelengths.
- **refraction:** the change in direction of a light ray as it passes from one medium to another.

THE NATURE OF LIGHT

Nuclear fusion reactions within the sun produce vast amounts of energy that are emitted into space from the sun's surface. Most of the energy emitted is in the form of electromagnetic waves. These waves cover the entire range of the electromagnetic spectrum, from extremely long, low-frequency waves such as radio through extremely short, high-frequency waves such as x-rays. Visible light is a very narrow range of frequencies within this overall spectrum. Electromagnetic energy is transmitted in the form of photons that have characteristic behaviors of both particles and waves. A photon can be thought of as a particle of energy moving through space like a wave moves through water.

Light is typically referred to by either frequency or wavelength. The frequency is defined as the number of complete wavelengths that occur, typically in one second of time. The wavelength is the distance between two peaks or two troughs of the wave. White light, such as that from the sun or an electric light, is composed of electromagnetic waves of many different wavelengths. These waves are detected by the eye as different colors when they become separated by diffraction, as when sunlight passes through a prism. The colors change continuously from red, with the longest visible wavelength, to violet, with the shortest visible wavelength. Each wavelength corresponds to a different specific energy of the photons that make up the light of a particular wavelength. Wavelengths longer than visible red are called infrared, and wavelengths longer than visible violet are called ultraviolet.

Matter that allows light to pass through it is said to be transparent to that wavelength. The speed of light is affected by the density of the medium, causing the light rays to change speed and direction when they pass from one medium to another. This is refraction, and it is readily seen when an object that is partly in air and partly in water appears to change direction. When light strikes a surface, it can either be absorbed or reflected. When light is absorbed, the material accepts the energy of the photons into its atoms and molecules. In reflection, light waves bounce off of the surface and the energy of the photons is not absorbed. When only light of certain frequencies is absorbed, the other frequencies are reflected and may be seen as colors. Leaves and grass appear green because their material absorbs light in the red, orange, yellow, blue, and violet parts of the visible spectrum and reflects the green waves.

When electromagnetic waves overlap, their amplitudes add together to produce either a greater amplitude or a lesser amplitude. Constructive interference produces a waveform with a greater amplitude than the sum of the two component amplitudes. Destructive interference produces a waveform in which the amplitude is reduced.

WAVE-PARTICLE DUALITY
Light behaves as though it is composed of particles (photons). At the same time, it behaves as though it is composed of waves. This is the essential feature of wave-particle duality. The "double slit" experiment carried out by Thomas Young (1773–1829) in 1801 demonstrated the behavior of light as a wave. When light passes through two closely-spaced narrow slits, the light that emerges exhibits a pattern of light and dark bands that are the result of constructive and destructive interference. This is a characteristic behavior of waves that is demonstrated by the interference of waves in the water of a ripple tank. In 2015, for the first time, scientists were able to take an image that showed light behaving as both a wave and a particle at the same time.

Wave behavior cannot account for the photoelectric effect, however. The photoelectric effect was identified in 1887 by Heinrich Hertz (1857–94) and was studied extensively by Robert Millikan (1868–1953). Millikan, and others, observed that when light fell on a metal surface, it could induce the metal to generate an electric current through space in a specially constructed vacuum apparatus. Eventually, it was determined that the metal was in fact emitting electrons and that monochromatic light (light having just one wavelength) produced the effect rather than the entire range of wavelengths in white light. Albert Einstein (1879–1955) explained the photoelectric effect by positing photons having specific energies corresponding to specific frequencies. The energy of the photon (E) is equal to the product of the frequency (f) and the Planck constant (h), written as:

$$E = hf.$$

Light intensity and distance are inversely related by the inverse square law. The intensity of light (I) decreases as the square of the distance (d) between the light source and the detector of the light. This can be described by the following equation:

$$I = 1/d^2.$$

Therefore, when the distance between the light source and the detector is doubled, the intensity of the light that is detected is just one-quarter of its original value.

OTHER PROPERTIES OF LIGHT
Normally, light waves do not have a uniform plane of vibration. Instead, the vibrations are radially distributed about the direction of motion. Photons of light can be thought of as pulsating balls of energy traveling through space. When reflected from a flat surface such as a paved highway or passed through an appropriate filter, the light can be altered in such a way that only vibrations parallel to one plane go on. Such polarization is used to restrict the intensity of light without altering its wavelength. Polarizing sunglasses and window coatings are perhaps the most common applications of this phenomenon.

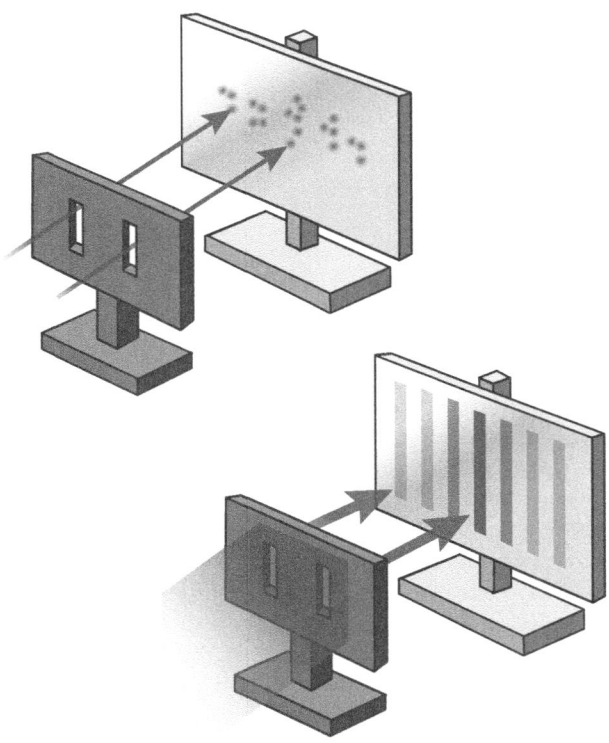

The double-slit experiment provided evidence that light acts like a wave. The interference pattern displayed on a wall from light passing through two slits in the foreground occurs because photons passing through the slits follow a wavelike motion and therefore hit certain spots on the wall more frequently.

In addition, astronomers can use the light waves of distant stars to measure distances between celestial objects. This is possible through the Doppler effect, or Doppler shift. (The Doppler shift refers to a perceived change in frequency as a wave source moves farther from or closer to an observer.) Scientists can use spectroscopy to study the movement of s bodies according to the shifting of their wavelengths of light.

—*Richard M. Renneboog, MSc*

BIBLIOGRAPHY

Goldstein, Dennis H. *Polarized Light*. 3rd ed. Boca Raton: CRC, 2011. Print.

Hillesheim, Heather E. *Sound and Light*. New York: Chelsea House, 2012. Print.

Keller, Ole. *Light: The Physics of the Photon*. Boca Raton: CRC, 2014. Print.

Lynch, David K., and William Livingston. *Color and Light in Nature*. 2nd ed. New York: Cambridge UP, 2001. Print.

Roychoudhuri, Chandrasekhar, A. F. Kracklauer, and Katherine Creath, eds. *The Nature of Light: What Is a Photon?* Boca Raton: CRC, 2008. Print.

Walmsley, Ian. *Light: A Very Short Introduction*. New York: Oxford UP, 2015. Print. Wolfe, William L. *Optics Made Clear: The Nature of Light and How We Use It*. Bellingham: SPIE, 2007. Print.

Woo, Marcus. "The Weird Quantum Behavior of Light, Captured in a Lab." *Wired*. Condé Nast, 4 Mar. 2015. Web. 29 July 2015.

LINEAR MOTION

FIELDS OF STUDY

Classical mechanics

SUMMARY

This article discusses kinematics, the study of the motion of objects, with a focus on linear motion. The motions of small objects, such as marbles, and large objects, such as planets, are all described by basic equations. With the basic equations described in this article, the position, velocity, and acceleration of objects can be calculated.

PRINCIPAL TERMS

- **acceleration:** a vector quantity that describes the rate of change in velocity over time.
- **dimension:** a direction in which an object can move. In spatial physics, the three dimensions are traditionally represented by the symbols x, y, and z.
- **displacement:** the difference between the initial position of an object and its final position, regardless of the path taken.
- **kinematics:** the study of the motion of objects.
- **magnitude:** the size, or numerical measurement, of a vector.
- **Newton's laws of motion:** three laws, defined by Isaac Newton, that describe the motion and forces acting on large objects moving at speeds much smaller than the speed of light.
- **vector:** a quantity that has both magnitude and direction.
- **velocity:** a vector quantity that describes the rate of displacement over time.

SIMPLIFYING MOTION

For many centuries, the model for the structure of the solar system stated that Earth was the center of the universe and that every celestial object moved around Earth in perfect circles. This model was established and backed up by many ancient Greek philosophers, including Aristotle (ca. 384–ca. 322 BCE). He posited that if Earth were moving, then dropped objects would be left behind in space. Because Aristotle did not see objects being left behind, he concluded that Earth must not be moving.

Many centuries later, Galileo Galilei (1564–1642) provided the key evidence that disproved this and many other Aristotelian views of the structure and motion of celestial bodies. Based on experimentation, Galileo posited that in the absence of friction, an object in motion will remain in motion. This was the beginning of a revolution in science. After Galileo's death, Sir Isaac Newton (1643–1727) improved on his work. He derived mathematical laws and equations to describe the motions of objects not only on Earth but also in space. We now know these

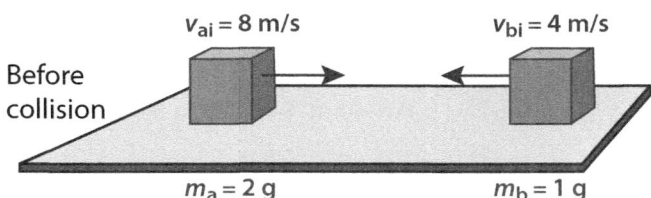

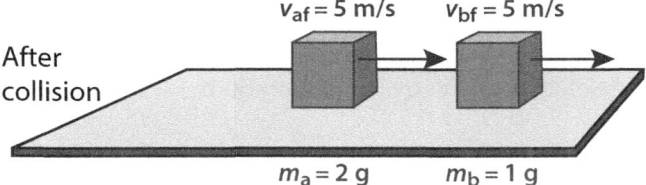

$2\,g \cdot 8\,m/s + 1\,g \cdot (4\,m/s) = 2\,g \cdot 5\,m/s + 1\,g \cdot 5\,m/s$

Model of a system where two blocks with masses of 2 g and 1 g are moving toward each other at a velocity of 8 m/s and 4 m/s respectively. Within this system, mass and momentum are conserved; only the velocity of the two blocks changes after the head-on collision. The velocity of the first block reduces in magnitude to 5 m/s and the velocity of the second block changes direction and increases to 5 m/s.

laws as Newton's laws of motion. Newton expanded on Galileo's experiments and defined his first law of motion, the law of inertia. This law states that any object at rest will remain at rest, and any object in motion will remain in the same state of motion, unless acted on by an external force. This can happen in any of the three spatial dimensions. Newton published his laws of motion and other such principles in his masterpiece *Philosophiae Naturalis Principia Mathematica* (1687), also known as the *Principia*.

THE MOTION OF OBJECTS
While Newton's first law of motion describes the state of motion of an object, it also describes how that state can be changed by external forces. An external force can change the speed of an object at rest, forcing it to move. An external force can also change the speed or direction of motion of an object already in motion.

In everyday usage, velocity and speed have similar meanings. In scientific contexts, however, their meanings differ. Physicists define velocity as the rate of change of position in a certain direction as a function of time. Velocity is a vector quantity because it describes both a magnitude and a direction of motion. Speed, on the other hand, is defined as the distance traveled by an object as a function of time, regardless of direction. Speed is not a vector quantity because it does not describe an object's direction of motion. In physics, an object that moves in a perfect circle with a constant speed has a variable velocity because its direction of motion around the circle is constantly changing.

Like speed, distance is not a function of the direction of motion and is therefore not a vector quantity. For example, if one walks five meters north or five meters south, the distance traveled is equal to five meters. Displacement, however, is a measurement of the change in position of an object as a function of the direction of motion. Like velocity, displacement is a vector quantity because it includes the direction of motion. A person who walks five meters to the north has a different displacement than a person who walks five meters to the south.

In terms of mathematical descriptions, speed and average velocity can be independently defined. Speed (v), not being a vector, is defined as the distance (d) an object moves divided by the time (t) it takes it to move that distance:

$$v = \frac{d}{t}$$

Average velocity (**v**) is also defined in terms of time, but instead of being a function of the distance traveled, it is a function of displacement (**s**), or change in position:

$$\mathbf{v} = \frac{\mathbf{s}}{t}$$

The velocity and displacement variables are bolded, meaning that they are vector quantities. Sometimes, instead of a bolded variable, a vector quantity is represented by a variable with a right-facing arrow above it.

Both velocity and speed are measured in units of meters per second (m/s). In order to calculate the velocity of an object, its change in position and the amount of time it takes to move must be given. For example, if a car starts moving east and takes 120 seconds (s) to travel 1,000 meters (m) to the next red light, then the average velocity of the car is calculated as follows:

$$v = \frac{s}{t}$$

$$v = \frac{1{,}000 \text{ m}}{120 \text{ s}}$$

$$v = 8.33 \text{ m/s east}$$

While the car was moving down the road, it could have sped up or slowed down, but on average it was moving at a velocity of 8.33 m/s east.

Average velocity is different from instantaneous velocity, which is the velocity of the car at every point in time. This is what a car's speedometer would display if the speedometer measured direction as well as speed. If a person records his or her car's velocity at every second, he or she would have a measurement that is closer to the car's instantaneous velocity. If the person then averages all the values of the instantaneous velocity, the result will be the average velocity. This process is more complicated and requires very precise measurements of the instantaneous velocity at each point in time. It is much easier to calculate average velocity using the displacement and the time it took to move between the two points.

Acceleration

The rate of change of an object's position as a function of time is its velocity. Similarly, the rate of change of an object's velocity as a function of time is a vector quantity known as acceleration. Mathematically, acceleration (a) is a function of the change in velocity (Δv) and the time over which the change took place (t):

$$a = \frac{\Delta v}{t}$$

The change in velocity is simply the initial velocity (v_i) subtracted from the final velocity (v_f):

$$\Delta v = v_f - v_i$$

Acceleration is measured in units of meters per second per second, or meters per second squared (m/s^2). Using the example of the car, imagine that the car is approaching a red light with an initial velocity of 8.33 m/s east. The car must come to a full stop at the red light, so its final velocity will be 0 m/s. It takes the car 1.5 seconds to come to a stop. To calculate the acceleration, first calculate the change in velocity:

$$\Delta v = v_f - vi$$

$$\Delta v = 0 \text{ m/s} - 8.33 \text{ m/s}$$

$$\Delta v = -8.33 \text{ m/s}$$

The change in velocity of the car is negative because the car slowed down, so its final velocity was smaller than its initial velocity. Using the change in velocity, calculate the acceleration:

$$a = \frac{\Delta v}{t}$$

$$a = \frac{-8.33 \text{ m/s}}{1.5 \text{ s}}$$

$$a = -5.55 \text{ m/s}^2$$

Because the acceleration is a vector, the negative sign simply indicates the direction of the vector. The car was previously established to be moving east, so its acceleration when stopping at the red light can be given as 5.55 m/s² west—the opposite of the direction in which it was originally traveling. The word "deceleration" is not used in physics. Objects slow down not because they decelerate but because they accelerate in the opposite direction of their motion. The car stops because it is moving east, yet accelerating west. A ball thrown vertically in the air will slow down on its way up because Earth's gravity is pulling on it, causing it to accelerate downward. Once the ball starts to fall back down, it will pick up speed because it is moving and accelerating in the same direction. As with velocity, this acceleration equation calculates the average acceleration, while instantaneous acceleration is the acceleration at every single point in time. In the example above, the person driving the car might press on the brakes with a varying amount of force, causing the magnitude of the acceleration to be slightly different at each point in time.

Kinematics

Newton's ideas regarding kinematics revolutionized science, especially physics and astronomy. About four centuries after Newton published his *Principia*, humans used these basic laws, and others based on them, to land on the moon. The use of Newton's laws and many other kinematic equations is not just reserved for big leaps in human scientific knowledge.

> **SAMPLE PROBLEM**
>
> A bullet fired from a gun leaves the barrel with a velocity of 375 m/s north. It hits a target 0.5 s later. How far is the bullet from the gun when it hits the target?
>
> **Answer:**
>
> In order to calculate the final position of the bullet, take the equation for average velocity and solve it for the displacement (s) by multiplying both sides by the time elapsed (t):
>
> $$v = \frac{s}{t}$$
>
> $$v \times t = s$$
>
> Insert the given velocity and time into the equation, and multiply:
>
> $$s = (375 \text{ m/s})(0.5 \text{ s})$$
>
> $$s = 187.5 \text{ m north}$$
>
> The bullet is 187.5 m north of the gun.

Every time someone calculates how long it will take to reach a destination based on a desired velocity, that person is putting kinematic equations to use.

Such equations are also used by Global Positioning System (GPS)–enabled devices to estimate distance traveled and time of arrival.

—*Angel G. Fuentes, MS*

BIBLIOGRAPHY

Capecchi, Danilo. *The Problem of the Motion of Bodies: A Historical View of the Development of Classical Mechanics.* Cham: Springer, 2014. Print.

Giambattista, Alan, and Betty McCarthy Richardson. *Physics.* 3rd ed. New York: McGraw, 2015. Print.

Hecht, Eugene. "Origins of Newton's First Law." Physics Teacher 53.2 (2015): 80–83. Web. 17 June 2015.

Jones, Matt. "Linear Motion." *Brightstorm.* Brightstorm, 2015. Web. 12 June 2015.

Nave, Carl R. "Description of Motion in One Dimension." *HyperPhysics.* Georgia State U, n.d. Web. 12 June 2015.

Newton, Isaac. *The Principia: Mathematical Principles of Natural Philosophy.* Originally published 1687. Berkeley: University of California P, 2014. Digital file.

Young, Hugh D., Roger A. Freedman, A. Lewis Ford, and Francis Weston Sears. *Sears & Zemansky's College Physics.* 14th ed. Boston, Mass: Pearson, 2016. Print.

LOAD

FIELDS OF STUDY

Classical Mechanics

SUMMARY

Load in classical physics typically refers to "mechanical load"—the resistance acting against a mechanical system or machine such as a pulley or a motor. This resistance is a force, measured in newtons. It is what must be overcome by the machine in order to move the target object. In most earthbound systems, the target object's weight—the downward force imparted to it by the planet's gravity—is a major component of the load.

PRINCIPAL TERMS

- **energy:** a property of matter and objects that can be transferred and transformed but never created or destroyed, sometimes described as the ability to do work; measured in joules (J).
- **force:** any interaction, such as a push or pull, that changes the motion of an object; measured in newtons (N).
- **mass:** how much matter there is in an object; measured in kilograms or pounds. Mass determines the effects of gravitation and inertia. Unlike weight, which is dependent on gravitation, an object's mass remains constant throughout the universe.

- **weight:** the downward force imparted to an object by gravity acting on its mass; measured according to the International System of Units (SI) in newtons, though it is normally expressed in units of mass for everyday objects on Earth.

Mechanical Load Is a Force

A load, colloquially, refers to some objects or other quantity that needs to be moved, manipulated, or otherwise worked with. In classical physics, load typically refers to mechanical load—that is, the resistance that a machine (such as a pulley or a motor) needs to overcome to do its job. This resistance is a force, an interaction that tends to change the motion of an object. When something or someone attempts to lift an object, for example, gravity creates a resistant force.

Because load is a force, it is best quantified in terms of newtons (N), the standard unit of force. One newton is defined as the force needed to accelerate a mass of one kilogram at a rate of one meter per second over one second of time (kg·m/s²). This is roughly the force necessary to lift a medium-sized apple from the ground to chest height.

Machines, Energy, and Force

Machines, at a basic level, transfer and transform energy. Energy is a fundamental property of matter that can be transferred or transformed but never destroyed. An automobile combustion engine, for example, transforms the chemical energy in gasoline into kinetic energy. Then it transfers this energy from the engine to the wheels. Energy is the capacity for doing work, and work is the application of a force over a distance (that is, moving something). Machines transform the potential for work contained in energy (such as electrical energy from a power plant or chemical energy produced in the human body) into work by creating force.

When calculating mechanical load and the ability of machines to perform work, it is necessary to understand the forces in play. The net force is the sum of all the forces acting on the target object.

Weight versus Mass

Weight is the downward force applied to an object by gravity. As such, weight is often an important factor in the load of a machine, especially if the system is trying to lift something. Even when moving an object horizontally, however, the target object's weight will need to be compensated for by the machine.

On Earth, weight is often used interchangeably with mass, which is the amount of matter in an object and is not affected by gravity. Scales, for instance, measure weight but offer readings in terms of mass units—kilograms or pounds. What scales on Earth are actually measuring is the downward force of gravity, but mass units are much more familiar to people than units of force (newtons). On Earth, it

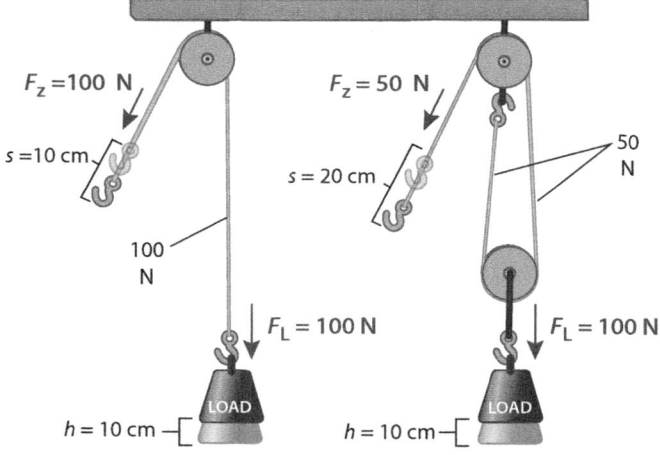

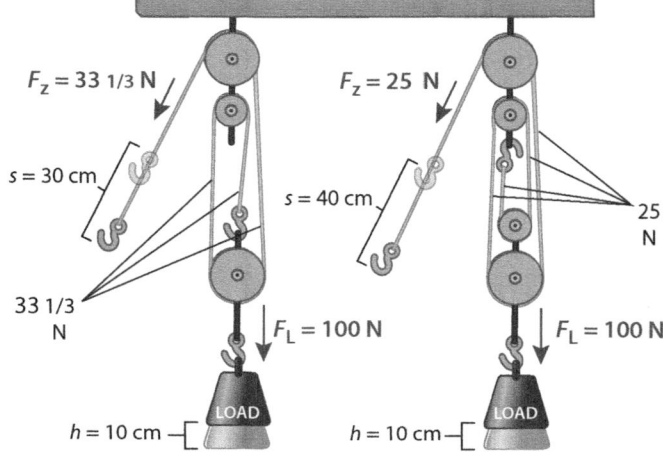

Four pulley systems with an increasing number of pulleys require less force to lift the same load to the same height.

> **SAMPLE PROBLEM**
>
> The maximum load that a given machine can handle can be calculated if one knows the force generated by the machine. Consider a simple single-pulley system attached to a weight. The pulley system does not move and serves only to alter the direction of the force applied. It is arranged vertically, so that a person can stand next to the weight and pull straight downward on one end of the rope running through the pulley. This produces an opposite, upward force on the weight at the other end of the rope. Assume a perfectly frictionless pulley. The person operating the pulley "weighs" seventy-five kilograms. What is the maximum load the person can lift? In other words, how much force can the person generate?
>
> **Answer:**
> No matter how hard the person pulls, he or she will never be able to generate more force than his or her weight. So if the person jumps up and grabs the rope, letting his or her full weight act on it, the force applied to the rope will be equal to the person's weight. Use the formula for weight (W) as a force (F), where m is the mass and g is the acceleration due to gravity:
>
> $$W = F = mg$$
>
> Plug in the known values and solve:
>
> $$F = 75 \text{ kg} \times 9.8 \text{ m/s}^2$$
>
> $$F = 735 \text{ kg·m/s}^2 = 735 \text{ N}$$
>
> More advanced calculations of load and maximum load may be influenced by mechanical advantage (an amplification in force produced by many machines) or sources of resistance other than gravity (such as friction or water resistance). However, the key to determining load will always be in determining what force needs to be overcome in order to achieve the desired effect.

does not matter much, because the force of gravity is essentially a constant.

Once other planets and satellites come into play, however, the differences between weight and mass become apparent. Weight is actually a function of the mass of an object and the gravity acting on it. It changes when the gravity changes. So a person on Earth might step on a scale and see a weight of two hundred pounds (approximately ninety-one kilograms), but on the moon that person would weigh only around thirty-three pounds (fifteen kilograms).

The reality is, the downward force acting on this person is indeed different at the two locations. It is a function of gravity, and the moon's gravity is much smaller than Earth's. However, technically, the units are wrong. Kilograms and pounds are units of mass. Mass quantifies the amount of matter in an object, and mass remains constant throughout the universe regardless of gravitation; a ninety-one-kilogram person on Earth still contains ninety-one kilograms of matter on the moon.

Force, Weight, and Acceleration due to Gravity

Weight is a force. Force (F), in classical physics, is equal to mass (m) multiplied by acceleration (a):

$$F = ma.$$

Therefore, weight can be written as F_g, the force of gravity. Similarly, the acceleration in the above formula can be replaced with the acceleration due to gravity (g). To calculate an object's weight (W) in newtons using its mass, then, simply multiply the mass in kilograms times the acceleration due to gravity (g) acting on the object in meters per second squared:

$$W = F_g = mg.$$

For example, acceleration due to gravity on Earth is 9.8 meters per second squared, so a two-hundred-pound (ninety-one-kilogram) person on Earth would have a weight of

$$W = mg$$

$$W = 91 \text{ kg} \times 9.8 \text{ m/s}^2$$

$$W = 891.8 \text{ kg·m/s}^2.$$

Finally, a kilogram-meter per second squared is a newton, so:

$$W = 891.8 \text{ N}.$$

LOADS IN DAILY LIFE

Simple machines, and thus the loads they move, permeate every aspect of daily life. In industries like shipping, pulleys and inclined planes in the form of ramps often help lift large loads. Understanding the maximum load capacity of a given machine is vital to the safety of both the products being moved and the machine operator.

—Kenrick Vezina, MS

BIBLIOGRAPHY

Coolman, Robert. "What Is Classical Mechanics?" *Live Science.* Purch, 12 Sept. 2014. Web. 30 Apr. 2015.

Edgar, Tricia. "The Single Fixed Pulley System." Education.com. Education.com, 5 Nov. 2013. Web. 17 June 2015.

Henderson, Tom. "Motion in Two Dimensions." *The Physics Classroom:* Physics Classroom, 2012. Digital file.

"Kinematics." *Encyclopaedia Britannica.* Encyclopaedia Britannica, 5 June 2013. Web. 30 Apr. 2015.

Simanek, Donald. "Kinematics." Brief Course in Classical Mechanics. Lock Haven U, Feb. 2005. Web. 28 Apr. 2015.

Tipler, Paul Allen, and Gene Mosca. *Physics for Scientists and Engineers.* 6th ed. Vol. 1-2. New York, NY: W.H. Freeman, 2008. Print.

LOUDNESS AND SOUND INTENSITY

FIELDS OF STUDY

Acoustics; Classical Mechanics; Fluid Mechanics

SUMMARY

The perceived loudness of a sound is directly related to its intensity. Intensity is an objective measure of sound power in the air in a given place. It is expressed in watts per square meter. Higher-intensity sounds seem louder, but the sensory ability of the ear and brain limits the perceived loudness of sounds. Loudness is influenced by the frequency of a sound. Higher-frequency sounds seem louder and higher pitched.

PRINCIPAL TERMS

- **amplitude:** a quantifying wave property measured from a point of rest to a point of maximum displacement; related to sound power and intensity of sound waves.
- **decibel:** a logarithmic unit that describes the power of a given sound in relation to the threshold of human hearing; abbreviated dB.
- **frequency:** how often a complete wave cycle occurs, which is directly proportional to the energy content of a wave and inversely proportional to wavelength; measured in hertz (Hz).
- **just-noticeable difference (JND):** the amount a parameter, such as sound intensity, must be changed so that the difference is noticeable at least half the time.
- **logarithm:** the power to which a fixed numerical base (the default is 10) must be raised to produce a given number.
- **phon scale:** a unit of measure for the perceived loudness of sounds; compares all sounds to a baseline sound of 1,000 hertz (1 kilohertz) frequency.
- **sensitivity:** the ability of an ear or a mechanical device to pick up and interpret sound; highly sensitive devices will have a smaller just-noticeable difference values.
- **sone:** a unit for quantifying perceived loudness; one sone equals forty phons.
- **wavelength:** the distance between crests of a wave; all electromagnetic radiation is transmitted as waves, with longer wavelengths corresponding to lower frequencies and less energy and vice versa.

HEARING A SOUND

What someone perceives as sound is actually the product of a complex process wherein a sound wave traveling through the air (or another medium such as water) interacts with the delicate biological machinery of the ear and is then interpreted by the brain.

The perceived loudness of a sound largely depends on the intensity of the sound. This is directly related to the amplitude of the sound wave. Amplitude is one of the three major properties used to describe all waves, the other two being wavelength and frequency. These properties apply to mechanical waves, which pass through a medium, such as when sound waves pass through air or water, or to electromagnetic waves, such as when light or heat pass through empty space. The amplitude of a sound wave is determined by the degree to which the wave displaces the medium it is traveling through.

Sound forms compression (longitudinal) waves in which the wave's motion is parallel to the direction the energy is being transferred. Amplitude in compression waves is measured as the maximum displacement of the particles of the medium from their normal resting state. Simply put, amplitude reflects how much energy a wave is carrying.

INTENSITY AND LOUDNESS

Sound intensity defines the sound power per unit area at a given point in a medium, typically in watts per square meter (W/m²). In practical acoustics, however, sound pressure level—which is directly related to the sound power—is very often used to describe intensity. For this discussion, assume "intensity" is determined in terms of sound pressure. In either case, intensity is an objective measure independent of an individual's ability to hear.

Loudness, on the other hand, is a perceived quality of the interaction between the sound waves in the air and the human ear as interpreted by the brain. Different people have different sensitivities to sound and can perceive varying differences in intensity or frequency.

Because loudness is a perceived quality of sound, it is more difficult to quantify than an objective quality like intensity or frequency. Just-noticeable difference (JND) is an important element to consider when measuring a perceived quality. JND is the minimum amount a quality or parameter must be changed in order to notice the change at least half of the time.

Several units and measurement scales have been developed to describe loudness while taking

$$I_0 = 10^{-12} \text{ watts}/\text{m}^2$$

Intensity in Decibels

$$I(dB) = 10 \log_{10} \left[\frac{I}{I_0} \right]$$

Common Sounds and Their Loudness	
Sound	Decibels (dB)
Human Hearing Threshold	0
Rainfall	50
Normal Speech	60
Washing Machine	75
Heavy Traffic	85
Hair Dryer	90
Gas Mower	95
Tractor	98
Headphones	102
Chainsaw	107
Rock Concert	110
Leaf Blower	115
Jackhammer	120
Ambulance	125
Jet Plane (from 100 ft.)	133
Gunshot	140
Fireworks	145
12-Gauge Shotgun	165

Sound intensity (I) is a measure of the sound's power per unit of area, usually watts/m². Often the intensity of a sound is measured relative to the intensity of the standard threshold of human hearing (I_0). The ratio of a sound's intensity to the threshold of hearing intensity falls along the logarithmic scale called the decibel scale. Decibels (dB) are the units for measuring the concept of loudness. Common sounds and their decibel value are listed, beginning with the threshold of human hearing, 0 dB. Normal speech is around 60 dB, an ambulance is around 125 dB, and fireworks are around 145 dB.

into account human perception and JND values for intensity.

SAMPLE PROBLEM

With the rule of thumb for loudness in mind, find the change in perceived loudness when an alarm clock's output changes from 4 decibels in intensity to 16 decibels in intensity. Assume the frequency of the alarm does not change.

Answer:

Begin by calculating the change in intensity—final intensity (I_f) relative to its starting intensity (I_s)—or I_f / I_s. Then, express the two decibel values using the equation for decibels given above:

$$16 \text{ dB} = 10 \log (I_f / I_0)$$

$$4 \text{ dB} = 10 \log (I_s / I_0)$$

Use these expressions to compare the starting and final intensities:

$$16 \text{ dB} - 4 \text{ dB} = 10 \log (I_f / I_0) - 10 \log (I_s / I_0)$$

$$12 \text{ dB} = 10 [\log (I_f / I_0) - \log (I_s / I_0)]$$

According to the rules for subtraction using logarithms, this can be further rewritten as:

$$12 \text{ dB} = 10 \log [(I_f / I_0) \div (I_s / I_0)]$$

The two I_0 values cancel each other out, and the decibel unit can be left behind since only the dimensionless ratio is of interest. To solve for I_f / I_s, isolate the log and then take the inverse log of each side:

$$12 = 10 \log(I_f / I_s)$$

$$1.2 = \log(I_f / I_s)$$

$$10^{1.2} = I_f / I_s$$

Use the rule of thumb to convert this change in intensity into a change in loudness can be written mathematically so that change in loudness (L_f / L_s) is expressed in terms of change in intensity (I_f / I_s):

$$\frac{L_f}{L_s} = 2^{\log\left(\frac{I_f}{I_s}\right)}$$

$$\frac{L_f}{L_s} = 2^{\log\left(10^{1.2}\right)}$$

$$\frac{L_f}{L_s} = 2^{1.2} \approx 2.3$$

The alarm will sound about 2.3 times louder at its final intensity than it did at its starting intensity.

Decibels: Relative Sound Power

The decibel (dB) is the most common unit used to describe sound intensity, sound pressure level, and loudness. A decibel, however, has no defined quantity. Instead, it is used to describe the relationship between two values of power. For example, a rocket during takeoff produces about one quadrillion times as much sound power as a human breath. However, in describing the relationship using decibels, which follow a logarithmic scale, there is a 190 dB difference: a rocket at take-off produces about 200 dB of sound and the human breath produces about 10 dB.

The logarithm (log) of a given number is the power to which a fixed base (the default is 10) must be raised to produce the given number. For instance, the log of 100 is 2 because $10^2 = 100$. To calculate sound intensity in decibels [$I(dB)$], using intensity (I) as measured in watts per meter squared, the following equation is used, where I_0 represents the threshold value of sound intensity (in watts per meter squared) for human hearing:

$$I(dB) = 10 \log (I / I_0).$$

The value of I_0 at the standard frequency of 1 kilohertz (1,000 hertz) is 10^{-12} watts per meter squared. Because they describe a ratio, decibels only make sense relative to this base value. Typically, the threshold of human hearing (near silence) is set as zero, with decibel values determined relative to this baseline. Decibels have the added advantage that they roughly correspond to the JND difference in loudness for human hearing.

Some typical sounds and their decibel values include:
- silence: 0 dB
- whisper: 20 dB
- public library: 40 dB
- dishwasher: 80 dB
- thunderclap: 120 dB

Exposure to very powerful sounds can cause a variety of problems. The threshold for annoying noise is usually somewhere between 70 to 80 dB, with discomfort increasing with power. Any sound at 80 dB or higher can cause hearing damage with long-term exposure. At 110 dB, average humans will begin to experience physical pain.

Phons and Sones

The phon and sone were developed in pursuit of a true unit of loudness. The phon was developed experimentally by playing two sounds to listeners. Both sounds always had the same intensity as measured in decibels. However, one sound would have a frequency of 1,000 hertz (Hz) and the other sound could be of any frequency. In this way, a scale was developed that compensated for the human ear perceiving sounds of different frequency as sounds that differ in intensity despite the sounds being of the same intensity. The phon relates all sounds to a baseline curve of sounds with equal perceived loudness but at a frequency of 1,000 Hz.

The sone was also developed using experiments in sound perception. In the sone experiments, listeners were asked to adjust the intensity of a sound (decibels) until they perceived it to double in loudness. The experiment found that a ten-decibel increase in sound pressure level roughly corresponds to a doubling of loudness, and this relationship forms the basis of the sone scale.

Both the phon and sone are relatively specialized units. Converting between phons, sones, and decibels is complex, but the basic relationships between the three are useful:

Loudness in phons is equal to sound intensity in decibels for sounds with 1,000 Hz frequency.

Forty phons equal 1 sone, and every 10 phons thereafter equals a doubling in sone value.

Doubling the sone value equals a doubling in loudness, which equals an increase of 10 phons, which equals an increase in sound intensity of 10 dB.

The "Rule of Thumb" for Loudness

The relationship between the objective measure of sound via intensity and the subjective, perceived measure of sound via loudness can be roughly understood using one simple "rule of thumb." To double the loudness of a sound, increase its intensity tenfold. In other words, ten alarm clocks sound twice as loud as one alarm clock that is making the same sound. This is not a precise rule that applies across all situations, but for everyday calculations it is quite accurate.

Sound Amplitude in Everyday Life

An understanding of sound intensity and perceived loudness enables audio technicians and engineers to design devices that produce sounds ranging from the enjoyable output of an amplified guitar to the annoying-but-effective buzz of an alarm clock. It also allows for the creation of devices that can help protect human ears in otherwise dangerous sound environments.

—*Kenrick Vezina, MS*

BIBLIOGRAPHY

Berg, Richard E. "Sound." *Encyclopaedia Britannica*. Encyclopaedia Britannica, 3 June 2014. Web. 9 June 2015.

Hass, Jeffrey. "An Acoustics Primer: What Is Loudness?" *Introduction to Computer Music*. Indiana U Jacobs School of Music, Center for Electronic and Computer Music, 2013. Web. 9 June 2015.

Russell, Daniel A. "Longitudinal and Transverse Wave Motion." *Acoustics and Vibration Animations*. Pennsylvania State U, 18 Feb. 2015. Web. 27 May 2015.

"Sound Intensity." *Encyclopaedia Britannica*. Encyclopaedia Britannica, 2015. Web. 17 June 2015

Wolfe, Joe. "dB: What Is a Decibel?" *Physclips*. School of Physics, UNSW, n.d. Web. 17 June 2015.

Zielinski, Ellen, Courtney Faber, and Marissa H. Forbes. "Lesson: Waves and Wave Properties." *Teach Engineering*. Regents of the University of Colorado, 2013. Web. 27 May 2015.

MAGNIFICATION

FIELDS OF STUDY

Optics; Spectroscopy

SUMMARY

The properties of mirrors and lenses are used to magnify objects. Magnification creates an enlarged image of an object. Various lenses can cause light rays to converge or diverge. Magnification is a basic principle of optical devices, such as cameras, telescopes, microscopes, and projectors.

PRINCIPAL TERMS

- **angular magnification:** the angle subtended at the eye by a magnified image of an object divided by the angle of the object being viewed by the naked eye without magnification.
- **concave:** having surfaces that curve inward, like a bowl.
- **convex:** having surfaces that curve outward, like a ball.
- **focal length:** the distance from the center of a lens or mirror to the focal point, where transmitted or reflected light rays converge.
- **linear magnification:** the ratio of the apparent height of an object's magnified image to the actual height of the object.
- **resolution:** the smallest identifiable dimension that a lens can differentiate.
- **transverse magnification:** also known as lateral magnification; synonymous with linear magnification.
- **virtual image:** an image that forms at the point where the paths of rays cross when projected backward from a lens.

Lenses and Mirrors

Lenses transmit light and are capable of reproducing an image. Mirrors reflect light and are capable of reproducing or distorting an image. A flat mirror reflects light back to an observer. Observers see a virtual image produced by the reflected light as if they were seeing the reflected object behind the mirror. The virtual image seen in a flat mirror cannot be projected onto another surface because the light rays do not pass through the perceived image. Mirrors can also magnify or reduce an image, depending on the shape of their reflective surface. A concave mirror has a reflective surface that curves inward. A convex mirror has a reflective surface that curves outward. Concave mirrors distort the virtual image to make it appear smaller than the actual object. Convex mirrors distort the virtual image to make it appear larger. In a flat mirror, the dimensions of the virtual image and the dimensions of the object are the same. Magnification is the ratio of the image dimensions to the object dimensions. Therefore, the images reflected by a flat mirror have a magnification of 1. Images reflected by a concave mirror have a negative magnification, and images reflected by a convex mirror have a positive magnification.

If one thinks of a concave or convex mirror as being one section of a circle, the center point of that circle is called the center of curvature. The principal axis of that circle would intersect the mirror at a point known as the vertex. The distance from the vertex to the center of curvature is known as the radius of curvature. The midway point between the vertex and the center of curvature is called the focal point. The distance from the vertex to the focal point is called the focal length. The focal length of a lens describes how strongly it converges or diverges light. The index of refraction is the ratio of the speed of light in a vacuum to its speed in a material. The focal length of a lens is determined by the material's index of refraction, that of air, the radius of curvature for the incoming side of the lens, and the radius of curvature for the outgoing side.

Lenses are also described as convex, concave, or flat. When parallel light rays pass through a lens, they

Magnification

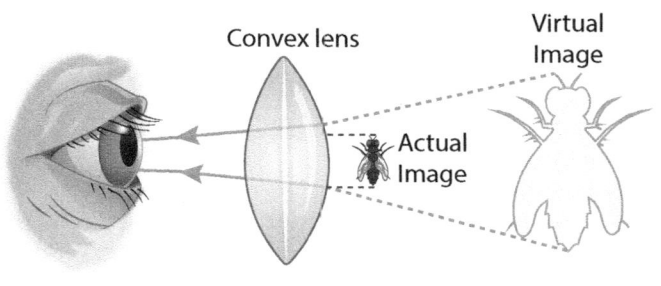

A convex lens causes light passing through it to refract so that it appears that it is coming toward the eye from a different angle. The eye interprets the virtual image as a much larger object than the actual image.

may either converge to a single point beyond the lens or appear to diverge from a single point in front of the lens. The most important function of a lens is magnification. When light passes through a double convex lens, the light rays are refracted by the lens material and change direction according to the material's index of refraction. When the light rays pass through the opposite surface of the lens, they are refracted again and converge at a single focal point on the other side of the lens. Parallel rays traveling through a double concave lens refract from the lens and diverge, never intersecting on the other side of the lens. Thus, they appear to diverge from a single point in front of the lens. For a flat lens, such as the glass in a window, the light rays pass directly through and emerge parallel.

Lenses and Magnification

Lens shape determines how the image the lens produces will be magnified. A simple magnifying glass uses a single convex lens to produce an image that is larger than the object itself. The ratio of the apparent size of the image to the actual size of the object at the focal length of the lens determines the linear magnification (also known as transverse magnification or lateral magnification). An inverted image has a negative value of linear magnification. The angular magnification of a lens or system of lenses denotes the angle between the observer's line of sight and the bottom of the object and the same point in its image.

A concave lens produces a virtual image at the focal length of its curved surface. Light rays passing through a concave lens spread apart, or diverge. Concave and convex lenses can be combined in a complementary fashion so that the image from one lens becomes magnified by other lenses. This is the basic operating principle of binoculars, telescopes, and microscopes, and can achieve large degrees of magnification. Cameras and projectors also depend on this function. In cameras, light enters through a system of lenses and is directed to a light-sensitive surface, such as photographic film or the electronic sensors used in digital cameras. In projectors, light is projected through the image and on through a system of lenses to a surface where it can be seen in a much enlarged format. Typically, the image becomes inverted as the light passes through the focal point.

Reflection, Refraction, and Diffraction

Reflection involves a change in the direction as light rays bounce off the surface of a mirror. Light waves

---

SAMPLE PROBLEM

A coin has a diameter of 1.27 centimeters (0.5 inches). When examined with a simple magnifying glass it appears much larger, and the largest image is seen when the coin and the observer's eye are both at the focal length of the lens, a distance of 25.4 centimeters (10 inches). Using a ruler on the face of the lens, the coin appears to be 5.6 centimeters (2.2 inches) in diameter. What is the magnification of the lens?

Answer:

The actual image is 1.27 centimeters (0.5 inches) in diameter (d), and the virtual image is 5.6 centimeters (2.2 inches). The linear magnification is the ratio of the apparent size of the virtual image to the size of the real object.

$$d_{virtual} / d_{real} = 5.6 \text{ cm} / 1.27 \text{ cm} = 4.4$$

Therefore, the lens has a magnifying power of 4.4, indicating that it can make objects appear 4.4 times larger.

follow the law of reflection: the incident angle, or the angle at which the light approaches the mirror, is equal to the angle of reflection. Refraction and diffraction can affect light as it passes from one medium to another. Diffraction results in a change in the direction of light rays as they pass through an opening or encounter an obstacle. Diffraction can separate light rays according to frequency. One effect of diffraction is the rainbow, in which bands of different colors appear as white light from the sun is separated into different visible wavelengths. Because the focal lengths for different wavelengths of light differ slightly, diffraction can cause a chromatic aberration. Chromatic aberrations can be seen as colored halos around the image in the lens. This effect can make fine details difficult or impossible to see.

The dimension of the smallest detail that can be seen clearly through a magnifying lens determines the resolution of the lens. Typically, the greater the magnifying power of the lens is, the finer its resolution is, as it enables ever smaller details to be seen. Resolution is limited by chromatic aberration and distortions due to refraction and reflection. At each surface of the lens, a certain amount of the light is reflected rather than transmitted. Reflection within the lens material actually produces two images rather than just one image, separated by a distance determined by the thickness of the lens. This can be seen by carefully examining the edge of a reflection in a mirror, where two reflections will actually be seen. The clearest image is obtained at the average focal length of the two surfaces and is termed the circle of least confusion.

—*Richard M. Renneboog, MSc*

BIBLIOGRAPHY

Darrigol, Olivier. *A History of Optics from Greek Antiquity to the Nineteenth Century*. New York: Oxford University Press, 2012. Print.

Ghatak, Ajoy. *Optics*. 5th ed. New Delhi: McGraw, 2012. Print.

Sharma, K. K. *Optics Principles and Applications*. Burlington: Academic, 2006. Print.

Ersoy, Okan K. *Diffraction, Fourier Optics and Imaging*. Hoboken: Wiley, 2007. Print.

Laikin, Milton. *Lens Design*. 4th ed. Boca Raton: CRC, 2007. Print.

Duree, Galen, Jr. *Optics for Dummies*. Hoboken: Wiley, 2011. Print.

MASS AND WEIGHT

FIELDS OF STUDY

Classical Mechanics; Geophysics

SUMMARY

This article discusses the relationship between mass and weight. Mass is an intrinsic property of matter and exists independent of any outside influence. Weight is an extrinsic property of matter and depends on the interaction of mass with the external force of gravitational acceleration.

PRINCIPAL TERMS

- **density:** a measure of the mass of a quantity of matter relative to the volume of space that it occupies.
- **extrinsic value:** any characteristic property of matter that is due to the interaction of the matter with an external factor.
- **gravitational acceleration:** the rate of increase of velocity experienced by matter under the influence of a gravitational field; often referred to as the force of gravity.
- **intrinsic value:** any characteristic property of matter that is due solely to the nature of the matter itself.
- **kilogram:** the base unit of mass in the International System of Units (SI), equal to 1,000 grams (2.2 pounds).
- **matter:** anything that can be characterized as having mass and can be measured by some criterion of measurement.
- **pound:** a standard unit of mass in the imperial and US customary measurement systems, equivalent to 0.45359237 kilograms; also used as a unit of force.

- **slug:** a unit of mass in the foot-pound-second (FPS) system, defined as the amount of mass that will experience an acceleration of one foot per second squared under one pound of force; equivalent to approximately 14.5939 kilograms or 32.1740 pounds.

INTRINSIC AND EXTRINSIC PROPERTIES

Physics is the study of the interaction of matter with its environment. Matter is defined as anything that has mass and can be measured in some way. The base unit of matter is the atom. An atom consists of the three principal subatomic particles, called protons, neutrons, and electrons. All of the matter in the known universe is made up of a limited number of different atoms. The number of protons in each atom determines its identity as a chemical element. While as many as 118 different elements have been identified, only 98 are known to occur naturally, albeit some only in trace amounts and for infinitesimal periods of time. Of these elements, the simplest two, hydrogen and helium, are the most common. Hydrogen and helium make up more than 99 percent of all observable matter in the universe.

How different atoms interact determines the unique properties of any type of matter. Some properties are common to all forms of matter, due to the simple fact of its physical existence. Such properties are called intrinsic properties. Intrinsic properties are constant regardless of any external factors. Mass is one such property; all atoms have mass. The mass of any amount of matter is the sum total of the mass of the atoms that make up that matter. Density is another property common to all matter. Density is determined by both the mass of the matter and the volume of space that it occupies.

PLANET	Earth	Moon	Mercury	Venus	Mars
GRAVITATIONAL ACCELERATION	9.81 m/s^2	1.63 m/s^2	3.70 m/s^2	8.89 m/s^2	3.69 m/s^7
MASS (kg)	WEIGHT (N)				
60	588	97.6	222.2	533.3	221.6
70	686	113.8	259.3	622.2	258.6
80	784	130.1	296.3	711	295.5
90	882	146.4	333.3	799.9	332.5
100	980	162.6	370.4	888.8	369.4
110	1078	178.9	407.4	977.7	406.4
150	1470	244	555.6	1333.2	554.1
200	1960	325.3	740.8	1777.7	738.9

Weight = mass • gravitational acceleration
$W = m \cdot g$

The mass of an object is an intrinsic value; it remains constant regardless of which planet the object is on. Weight of an object will change on each celestial body because it is dependent on the acceleration of gravity on that celestial body. The gravitational acceleration on Earth is 9.81 m/s^2, the moon has a gravitational acceleration of 1.63 m/s^2, on mercury it is 3.70 m/s^2, on Venus it is 8.89 m/s^2, and on Mars it is 3.69 m/s^2. A table shows the weight of an object on different celestial bodies given its mass; weights (in Newtons) are calculated by multiplying the mass by the gravitational acceleration of that celestial body.

Principles of Physics — Mass and weight

> **SAMPLE PROBLEM**
>
> A celestial explorer on Mars collects a measured weight of 50 pounds of Martian rock samples and brings them back to Earth. The gravitational acceleration of Earth is approximately 9.81 m/s^2, but that of Mars is only about 3.69 m/s^2. What is the measured weight of the Martian rocks on Earth in grams? Recall that 1 pound equals 0.45359237 kilograms, or 453.59237 grams.
>
> **Answer:**
>
> The weight of an object is the product of its mass and the gravitational acceleration that it experiences. Since the mass of the rocks is the same on Earth as on Mars, their weight will depend on the gravitational acceleration that they experience. Their weight on Mars is given by the expression
>
> $$W_M = m a_M$$
>
> where W_M is the weight of the rocks on Mars, m is the mass of the rocks, and a_M is the gravitational acceleration on Mars. The same equation can be used to determine the weight of the rocks on Earth, substituting W_E for W_M and a_E for a_M:
>
> $$W_E = m a_E$$
>
> Because the mass, m, is constant, the two equations can each be solved for m and then combined into one:
>
> $$\frac{W_E}{a_E} = m$$
>
> $$\frac{W_M}{a_M} = m$$
>
> $$\frac{W_E}{a_E} = \frac{W_M}{a_M}$$
>
> Solve the equation for W_E:
>
> $$W_E = \frac{W_M a_E}{a_M}$$
>
> $$W_E = \frac{(50 \text{ lb})(9.81 \text{ m/s}^2)}{3.69 \text{ m/s}^2}$$
>
> $$W_E = 132.927 \text{ lb (rounded)}$$
>
> To convert this weight to grams, multiply the weight in pounds by the conversion factor of 453.59237 grams per pound:
>
> $$W_E = (132.927 \text{ lb})(453.59237 \text{ g/lb})$$
>
> $$W_E = 60{,}294.673 \text{ g}$$
>
> The Martian rocks weigh approximately 60,295 grams on Earth.

Some properties of matter are affected by the influence of external forces. These properties are known as extrinsic properties. Weight is one example of an extrinsic property. Matter has weight because its mass is affected by gravitational acceleration, or the force of gravity. Though the mass of any matter is universally constant, its weight varies depending on the amount of gravitational acceleration it experiences. The most common form, or isotope, of hydrogen consists of a single proton, and so its mass is equal to the mass of one proton everywhere in the universe. One mole (approximately 6.022×10^{23}

atoms) of this isotope, which is known as hydrogen-1 or protium, has a mass of exactly 1 gram. On Earth, the weight of one mole of protium is also 1 gram. On the moon or on Jupiter or anywhere else in the universe, one mole of protium will have the exact same mass, but its weight will be different due to differences in gravitational acceleration.

The relationship between mass and weight is defined by the equation $W = mg$, where W is weight, m is mass, and g is the standard acceleration due to gravity. On Earth, the mass of a quantity of matter is generally the same as its weight, because one unit of standard gravity (g) has been defined as equal to the acceleration due to Earth's gravity, which is approximately 9.81 meters per second per second (m/s²). Therefore, to use this equation to calculate weight elsewhere in the universe, the gravitational acceleration in that location would have to be converted to standard gravity units, according to the equation $1\,g = 9.81$ m/s².

UNITS OF MASS AND WEIGHT

In the British or imperial system of measurement, which is the system most commonly used in the United States, the base unit of length is the foot, and the base unit of mass and weight is the pound. In the International System of Units (SI), the modernized form of the metric system, the base unit of length is the meter, and the base unit of mass and weight is the kilogram. The foot-pound-second (FPS) system is a variation on the imperial system that was historically used for scientific applications prior to the adoption of the metric or SI system.

In the FPS system, the pound can serve as a unit of either mass or force. To distinguish one from the other, these units are referred to as pound mass and pound force. The pound mass is the same as the pound used to measure mass in the imperial system and is equal to 0.45359237 kilograms. The pound force is defined as the weight of one pound mass experiencing the acceleration of one unit of standard gravity, following the equation $W = mg$. The unit of mass that corresponds to pound force is the slug. The slug is defined as the amount of matter that will accelerate at one foot per second per second (ft/s²) when propelled by one pound force. As the SI system became more widely accepted, most of these nonstandard units became obsolete.

WEIGHTS AND MASSES IN OTHER PLACES

Gravity is one of the four fundamental forces in nature. The other three fundamental forces are the electromagnetic force, the weak nuclear force, and the strong nuclear force. Gravity is an intrinsic property of matter. Every atom exerts gravitational force. The magnitude of an atom's gravitational force is determined solely by its mass. However, gravity is the weakest of the four forces. A great deal of mass is required before the effects of gravity can be felt. The gravitational force exerted by a planet is the cumulative sum of the gravity of every atom in that planet's structure. The more mass a planet has, the greater its gravity and the greater the weight of any mass in its gravitational field.

Yet the total gravitational force of a planet does not necessarily correspond to the amount of gravity experienced on that planet's surface. Gravity is a function of both mass and distance, where distance is measured from the center of the object exerting the force. Thus, the mass of Saturn is more than ninety-five times that of Earth, but its surface gravity is only slightly greater than on Earth, approximately 11.44 m/s². This is because the distance from Saturn's surface to the center of its mass is so great.

MASS, MOMENTUM, AND ENERGY

One of the dangers of space exploration is impact by micrometeoroids. Without proper shielding, a micrometeoroid the size of a grain of sand can blast a hand-sized hole in the outer wall of a shuttle. This damage is due to the kinetic energy of the micrometeoroid. Kinetic energy is a property that depends on both its mass and its velocity, according to the equation

$$E_k = \frac{1}{2}mv^2$$

where E_k is the kinetic energy, m is the mass, and v is the velocity at which the mass is traveling. Even though the mass of a micrometeoroid is very small, the relative velocity of such particles is typically around 10 kilometers per second, or 36,000 kilometers (22,369 miles) per hour, giving them a very large amount of kinetic energy.

Another intrinsic property that is closely related to kinetic energy is momentum. An object's momentum is also defined by its mass and its velocity, according to the expression

$$p = mv$$

where p is momentum, m is mass, and v is velocity. In mathematical terms, the momentum of an object is said to be the derivative of its kinetic energy with respect to velocity. The law of conservation of energy states that when two or more objects collide, any energy lost by one object must be gained by the other object(s). That transfer of energy will be reflected in their relative speeds before and after the collision.

—Richard M. Renneboog, MSc

BIBLIOGRAPHY

Grissom, Thomas. *The Physicist's World: The Story of Motion and the Limits to Knowledge.* Baltimore: Johns Hopkins University Press, 2011. Print.

Gupta, S. V. *Mass Metrology.* Heidelberg: Springer, 2012. Print.

Kumar, B. N. *Basic Physics for All.* Lanham: University Press of America, 2009. Print.

Myers, Rusty L. *The Basics of Physics.* Westport: Greenwood, 2006. Print.

Ostdiek, Vern J., and Donald J. Bord. *Inquiry into Physics.* 7th ed. Boston: Brooks, 2013. Print.

Shireman, Myrl. *Strengthening Physical Science Skills for Middle & Upper Grades.* Greensboro: Twain, 2008. Print.

MASS SPECTROMETRY

FIELDS OF STUDY

Electromagnetism; Thermodynamics

SUMMARY

Mass spectrometry utilizes the interaction of a charged particle with an applied magnetic field. The path of a charged particle traveling through a magnetic field is precisely described mathematically and is directly proportional to the mass of the charged particle. Mass spectrometry is a commonly used and highly precise method for identifying matter from as little as a single molecule, and can differentiate between isotopes of the same element.

PRINCIPAL TERMS

- **aperture:** an adjustable opening in a barrier through which light or other electromagnetic emission can pass.
- **distinguishability:** the ability of particles with the same energy to be differentiated from one another.
- **electromagnet:** a device that becomes magnetic due to the presence of an electric current.
- **electromagnetic field:** a physical field consisting of a combined electric field (generated by stationary electric charges) and magnetic field (generated by moving electric charges) that affects the behavior of charged objects in its vicinity.
- **electron volt:** the amount of energy carried by a single electron moved across an electric potential difference of one volt; equal to 1.6×10^{-19} joules.
- **magnetic flux:** the measure of the strength of the magnetic field surrounding an active electrical conductor or a magnet, equal to the product of the external magnetic field (B), the area of the affected surface (A), and the cosine of the angle between them ($\cos \theta$).
- **trajectory:** the path of a thrown or falling object, such as a baseball.

IONS AND MAGNETIC FIELDS

In 1897, J.J. Thomson identified electrons as subatomic particles by their behavior in cathode rays. The fact that the rays could be deflected by a magnetic field demonstrated the interaction of charged particles with magnetic fields. The existence of Thomson's cathode rays also showed that electrons can be ejected from a source material in a controlled manner. This necessarily leaves the source material with a positive charge, and essentially all of the mass of the source material. This was also observed by Thomson, in what were termed canal rays, streams of positively charged particles that are much more massive than electrons.

The mathematical relationship that describes the trajectory of such charged particles in a magnetic field was determined through experimental observation. The relationship related the mass of the particle, the charge that it carried, the strength of the magnetic field, and the radius if the circular path that the particle followed at a particular kinetic energy. Mass spectrometry relies exclusively on this relationship for the identification of atomic and molecular species.

THE MASS SPECTROMETER

A mass spectrometer consists of just three principal sectors, all of which function in the gas phase. In the first sector, a source material is subjected to ionizing radiation or other ionization principle that serves to remove an electron from the atoms or molecules of the source material. The ions that are formed in this way are accelerated to a specific energy stated in electron-volts by an applied electric field that also directs them through an aperture and into the second sector of the device.

The second sector is essentially just an empty chamber that is maintained under high vacuum conditions and subjected to a magnetic field of known strength. Because it is an electromagnet field, the magnetic flux can be tuned so that ions of different masses can be made to follow a circular trajectory of exactly the same radius. This is the feature that determines the distinguishability of ions of different mass-to-charge ratio.

The third sector of the device is the detector, which receives the ions that have traveled through the evacuated chamber. The detector essentially just counts the number of ions of a particular mass that it receives in a specific interval of time.

The counts are recorded as a histogram. All mass spectrometers function in exactly this same manner, although the technology used to generate ions from the source material and to detect those ions after passage through the evacuated chamber varies considerably.

THE RADIUS OF CURVATURE

The radius of curvature defines the radius of the circular path of a charged particle, an ion, in a magnetic field. An ion in a magnetic field follows a curved path

---

SAMPLE PROBLEM

In designing a mass spectrometer, calculate the magnetic field strength required to produce a radius of curvature of 25 centimeters for ions having a kinetic energy of 120 electron-volts, and mass-to-charge ratio of 256.

$$R^2 = 2V(m/e) / B^2$$

Therefore

$$B^2 = 2V(m/e) / R^2$$

$$B^2 = 2(120)(256)/(25)^2$$

$$= 98.304 \text{ G}^2$$

$$B = 9.915 \text{ G}.$$

---

having a specific radius that is directly related to the velocity, and therefore the kinetic energy, that the ion possesses when it enters the magnetic field. The radius of curvature is defined by the equation

$$R^2 = 2V(m/e) / B^2$$

where R is measured in centimeters, V is measured in volts, B is measured in Gauss, and m/e is the mass-to-charge ratio of the ion. In practical mass spectrometry, the charge e is the unit charge equivalent to the loss of one electron, and m is therefore just the atomic or molecular weight of the ion. It should be remembered that through this relationship, mass spectrometry also differentiates between isotopes and not just elements.

MASS SPECTROMETRY IS EVERYWHERE

The principles that define the behavior of ions in the magnetic field of a mass spectrometer are fundamental universal principles that apply to every charged particle in every magnetic field in every part of the universe. Charged particles that enter Earth's magnetic field and produce the Northern and Southern aurorae follow the same laws that govern the movement of ions in cathode ray tubes (CRTs) such as television and radar screens. Similarly, medical and other types of diagnostic imaging that utilize ion sources such as alpha and beta radiation depend strictly on the same principles to control and direct the movement of those ions.

—Richard M. Renneboog M.Sc.

BIBLIOGRAPHY

De Hoffmann, Edmond, and Vincent Stroobant. *Mass Spectrometry Principles and Applications.* 3rd ed. Somerset: Wiley, 2013. Print.

Gross, Jürgen H., and Peter Roepstorff. *Mass Spectrometry A Textbook.* 2nd ed. Berlin: Springer Berlin, 2014. Print.

Matthiesen, Rune, and Humana Press *Mass Spectrometry Data Analysis in Proteomics.* New York: Humana, 2013. Print.

Lebedev, Albert. *Comprehensive Environmental Mass Spectrometry.* Hertfordshire: ILM Publications, 2012. Print.

Watson, Jack T., and Orrin David, Sparkman. *Introduction to Mass Spectrometry : Instrumentation, Applications and Strategies for Data Interpretation.* 4th ed. Chichester [u.a.]: Wiley, 2011. Print.

Vékey, Károly, András Telekes, and Akos Vertes. *Medical Applications of Mass Spectrometry.* Amsterdam; Boston: Elsevier, 2008. Print.

MECHANICAL OR ELECTRICAL LOAD AND WORK

FIELDS OF STUDY

Classical Mechanics; Electronics

SUMMARY

Mechanical and electrical systems are designed to perform work functions either to augment human effort or to perform work that humans are not physically capable of performing. Both electrical and mechanical systems perform work by acting on a load that opposes the work being done. Both depend on different aspects of material properties such as compressive strength and electrical resistance. Advanced applications include biomechanics and robotics.

PRINCIPAL TERMS

- **compressive strength:** the ability of a material to resist deformation or structural failure when experiencing a force of compression (squeezing).
- **demand:** the load on an electrical supply system over time.
- **dummy load:** a device applied to an electrical circuit or other system in order to provide a corresponding load without performing an output function.
- **impedance:** the opposition to electrical current flow produced by a voltage.
- **mechanical advantage:** the ratio of the output work done by a system or machine to the input work required for a function to be carried out.
- **tensile strength:** the ability of a material to resist structural failure when experiencing a force of tension (pulling).
- **work:** a force successfully moving an object, or the successful transfer of energy. The International System of Units unit of work is the joule.

Work, Energy, and Load

Mechanical and electrical systems are designed to perform work. Mechanical systems typically provide a large mechanical advantage compared to human effort alone. Simple machines such as levers and wheels are the central components of most machines. The work delivered as the output of levers and wheels can be many orders of magnitude greater than the work input. An elevator is a good example of mechanical advantage. The operator presses a button and machine actions then carry out the work of raising or lowering large weights a certain distance. The machinery that carries out the function of an elevator is both electrical and mechanical. The cables that lift the elevator car must have enough tensile strength that they will not break under the weight of the elevator. The structural components that support the elevator must have enough compressive strength that they will not fail under the crushing weight of the elevator. Tension (pulling) and compression (squeezing) are examples of an axial load, since they function in a direction parallel to the central axis of a component.

A load that is applied across the central axis is called a transverse load. A load applies a twisting movement to a component about its axis is called a torsional load. The electrical systems that supply voltage and current to the electric motors and other devices required for the operation of the elevator must be able to meet the demand placed on them when the machinery is in operation. All such systems are designed to perform physical work.

ELECTRICAL AND MECHANICAL WORK

Work is defined in physics as the operation of a force over a distance. In the elevator example, the lifting force of the machinery displaces the elevator by a certain distance, and the output work can be easily calculated using the force, the displacement of the object, and the cosine of the angle between the direction of the force and the direction of the displacement.

Electricity is defined as the movement of electrons through a conductor. The force that moves the electrons through the system is called the electromotive force (emf), or the applied voltage. A 12-volt battery, for example, supplies an emf of 12 volts over the length of an electrical system. The output performance of batteries and other electrical systems can be tested by the use of a dummy load. A dummy load applies a normal load to the system to test its potential output without actually performing any of the work functions of the system. A battery provides a constant voltage or emf to produce a constant flow of electrons, or a direct current (DC), in only one direction. Components of the system provide load by resisting the flow of electrons through them.

Most electrical systems rely on alternating current (AC). In AC, the electrons travel in one direction for a short time and then travel in the opposite direction for a longer period of time. AC is produced by an emf that varies in a sinusoidal manner, typically at 50 to 60 hertz (1 hertz = 1 cycle per second). AC moves through a maximum positive value and a maximum negative value. Therefore, AC has a value of zero twice in each cycle. The resistance provided by components in an AC system also varies in a cyclic manner as a result. This resistance is called the impedance. Current, voltage, and resistance values determine the electrical power of the devices, the rate at which the device performs work. The power consumed by any device is defined as the product of its resistance and the square of the current passing through the device.

Electrical work is converted into mechanical work by a motor. The electrical power required to operate the motor depends on the mechanical resistance that it experiences in the conversion of electrical work to mechanical work. In the elevator example above, the electric motor that supplies the mechanical power to raise the car against the force of gravity has to do more electrical work when there are more people in the elevator. Similarly, the cables and other mechanical components have to resist a greater force when there is more weight in the elevator car. The elevator car is thus a load acting in opposition to both the electrical and mechanical work performed in the system.

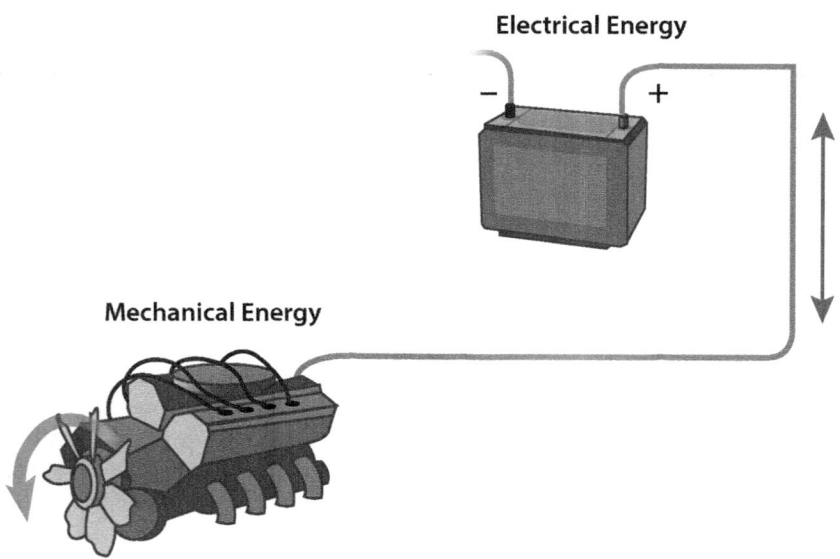

A mechanical system, which includes a turbine and motor/generator, connected to a battery. In an electromechanical system electrical energy may convert into mechanical energy, such as a battery powering a motor that turns a rotor. Or mechanical energy may convert into electrical energy, such as a turbine turning a rotor in a generator that produces electrical energy to store in a battery.

SAMPLE PROBLEM

An electric stove operates at 1,500 watts, an air conditioner operates at 1,000 watts, and four light bulbs operate at 60 watts each. Calculate the individual resistances of these devices and the load on the electrical system of the home in which they operate. (Standard operating voltage for stoves is 220 volts and 110 volts for the other devices.)

Answer:
The load on the electrical system is the sum of the individual loads. In this case,

$$1,500 + 1,000 + (4 \times 60) = 2,740 \text{ watts}$$

The resistances are found using the two formulas $E = IR$ and $P = I^2R$, and rearranging them to solve for R as

$$R = E/I$$

$$I^2 = P/R$$

Set the equations equal to one another:

$$(E/R)^2 = P/R$$

$$E^2/R^2 = P/R$$

$$E^2 = PR$$

$$R = E^2/P$$

Therefore,

- the resistance of the stove is $(220)^2/(1,500) = 32.27\ \Omega$ (Ω or omega is the standard symbol for ohms)
- the resistance of the air conditioner is $(110)^2/1,000 = 12.1\ \Omega$
- the resistance of each light bulb is $(110)^2/60 = 201.67\ \Omega$

Note that a calculation of the combined resistance of these devices is complicated because it depends on whether they are connected in series or in parallel with each other and their impedance rather than their resistance.

Voltage, Current, and Power

Electrical systems function when an applied voltage drives a current through a load resistance. The three are related by Ohm's law. Georg Simon Ohm (1787–1854) discovered their relationship in 1827. Ohm's law states that the voltage (E) applied to a system is equal to the product of the resistance (R) and the current (I) flowing in the system, or

$$E = IR.$$

The power (P) produced by the system is the product of the resistance and the squared value of the current, or

$$P = I^2 R.$$

These simple formulas are fundamental to the design of electrical systems and have their mechanical counterparts in classical mechanics. Voltage corresponds to force in mechanical systems. Resistance corresponds to friction and other factors that act against the movement of a mechanical component. Current may correspond to the number of mechanical components that function in the system between the work input and a specific work output. By this analogy, the force required to achieve a specific mechanical result would be the product of the number of components and the resistive factors of each component.

Applied Electromechanics

Perhaps the most fascinating application of electrical and mechanical loading is biomechanical engineering, including robotics. Industrial robots are relatively simple mechanical systems that function using electrical and hydraulic power systems. They are designed to carry out a programmed set of precise movements using specified spatial coordinates. Therefore, they cannot vary from that pattern unless reprogrammed. Biomechanics, however, is based on the principle that muscles function as motors and bones function as structural elements. The challenge that this poses is to construct artificial devices or prosthetics that permit functioning with the same range and type of motion as the limbs or organs that they replace. This requires an unequaled amount of design calculation to specify electrical and mechanical components with structural and performance capabilities comparable to those of the actual living system in which they function.

—Richard M. Renneboog, MSc

BIBLIOGRAPHY

Chabay, Ruth W., and Bruce A. Sherwood. *Matter and Interactions*. 4th ed. Hoboken: Wiley, 2015. Print.

"Estimating Appliance and Home Electronic Energy Use." Energy.gov. US Dept. of Energy, 10 May 2015. Web. 12 Aug. 2015.

Gibbons, Patrick C ,. *Physics*. 2nd ed. Hauppauge, N.Y.: Barron's, 2008. Print. Barrons EZ 101 Study Keys.

Gross, Charles A. *Electric Machines*. Boca Raton: CRC, 2007. Print.

Pfeiffer, Friedrich. *Mechanical System Dynamics*. Corrected 2nd ed. Berlin: Springer, 2010. Print. Lecture Notes in Applied and Computational Mechanics.

Robbins, Allan H., and Wilhelm C. Miller. *Circuit Analysis, Theory and Practice*. 5th ed. Clifton Park: Delmar, 2013. Print.

Vukosavic, Slobodan N. *Electrical Machines*. New York: Springer, 2013. Print.

MÖSSBAUER EFFECT

FIELDS OF STUDY

Atomic Physics

SUMMARY

The Mössbauer effect is the recoilless emission of gamma rays from the nucleus of an atom. Mössbauer observed that atoms that did not emit gamma rays in the gas phase did so when they were immobilized in a solid form. The Mössbauer effect is an expression of the law of conservation of momentum, and the energy resonances observed in Mössbauer spectroscopy provide information about quantum state transitions in the nucleus of specific atoms. Most other analytic methodologies provide information about the quantum states of electrons but not of the nucleus.

PRINCIPAL TERMS

- **electron:** a negatively charged subatomic particle that is often bound to the positive charge of the nucleus but can also exist in a free state in an atom.
- **electron volt:** the amount of energy carried by a single electron moved across an electric potential difference of one volt; equal to 1.6×10^{-19} joules.
- **energy:** a property of matter and objects that can be transferred and transformed but never created or destroyed, sometimes described as the ability to do work; measured in joules (J).
- **gamma radiation:** electromagnetic radiation with a wavelength shorter than 1×10^{-11} meters; gamma (γ) rays typically emitted during gamma decay, a subtype of radioactive decay.
- **gamma ray:** a high-energy photon that is usually emitted by an unstable nucleus via radioactive decay; it may also be produced by various other means, including particle-antiparticle annihilation and the interaction of atmospheric particles with cosmic rays.
- **quantum mechanics:** the branch of physics that deals with matter interactions on a subatomic scale, based on the concepts that energy is quantized, not continuous, and that elementary particles exhibit wavelike behavior.
- **quantum state:** the condition of a physical system as defined by its associated quantum attributes.
- **resonance:** the response of an elastic body to a force acting on the body at its natural frequency.
- **ultraviolet (UV) radiation:** electromagnetic radiation with more energy than visible light but less than x-rays, in the wavelength range of 4×10^{-7} to 1×10^{-7} meters.

THE MÖSSBAUER EFFECT

The Mössbauer effect depends on the emission of energy as a gamma ray by the nucleus of an atom transitioning from one quantum state to another. In any system that is described by quantum mechanics, transitions from one quantum state to another are associated directly with the emission or absorption of electromagnetic energy. That energy has a specific wavelength and frequency, depending on the quantum states involved in the transition. When a nucleus is irradiated with a range of gamma ray frequencies, the nucleus will absorb the appropriate energy to become promoted to a higher energy quantum state. The specific wavelengths of gamma radiation appropriate to the transitions are emitted from the nucleus as it reverts to the original quantum state, and can be detected by the proper instrumentation.

This process of absorption and re-emission is called resonance, and is the common principle of several different spectroscopic techniques such as nuclear magnetic resonance (NMR) and electron spin or paramagnetic resonance (ESP or EPR), and ultraviolet photoelectron spectroscopy (UV-PES) that uses ultraviolet (UV) radiation. The Mössbauer effect was first observed in 1958 by Rudolf Mössbauer, who was investigating the lack of gamma ray resonance in gases. What Mössbauer found was that gamma ray resonance was observed when the material had been immobilized as a solid.

Mössbauer Resonance

The Law of Conservation of Momentum is a fundamental physical law that applies on the quantum scale as well as on the Newtonian or macroscopic scale. The principle is generally expressed: "For every action there is an equal and opposite reaction." Accordingly, emission of energy by a nucleus is accompanied by the recoil of the nucleus due to the force exerted by the emission. On a large scale, this is equivalent to the recoil of a cannon as it ejects a cannonball. The force of the ejection is exerted equally against the cannonball and the cannon, driving them in opposite directions with the same momentum but with different velocities. Similarly, when gamma ray photons are emitted from a nucleus, the force of their emission also acts against the mass of the nucleus.

In a free or unbound state, this energy is absorbed by the nuclear motion and so is not observed externally as the net energy difference of a resonance frequency. Conservation of momentum is an adjunct property of the Law of Conservation of Energy, which requires that the total energy in a system remains constant. Thus, the total momentum in a system must also remain constant. When the difference in mass becomes so great that the larger mass remains

---

SAMPLE PROBLEM

Calculate the energy in electron volts for photons having a wavelength of 3000A° (3000 X 10^{-8} cm), given that the energy of a single photon is defined by the relationship $e = h\nu$, where h = Planck's constant, 6.624 X 10^{-27} erg.sec, one electron volt = 1.602 X 10^{-12} erg, and the speed of light is 2.99 X 10^{10} cm.sec^{-1}.

Answer:

The frequency ν is the reciprocal value of the wavelength λ.

Therefore,

$$e = h\nu = hc/\lambda$$

$$= (6.624 \times 10^{-27} \text{ erg.sec})(2.99 \times 10^{10} \text{ cm.sec}^{-1})/(3000 \times 10^{-8} \text{ cm})$$

$$= 0.0066 \times 10^{-9} \text{ erg}$$

$$= 6.6 \times 10^{-12} \text{ erg}$$

$$1 \text{ eV} = 1.602 \times 10^{-12} \text{ erg.}$$

Therefore, 6.6 X 10^{-12} erg = 4.12 Ev.

---

immobile, and therefore has no momentum, the additional energy and momentum must transfer to the system component that has motion.

In the Mössbauer effect, those components are the gamma ray photons being emitted. By immobilizing the nucleus under study as a solid or in a solid matrix, the mass difference becomes so great that all of the momentum of the recoil is transferred to the photons and is expressed as the energy of those photons. This net energy difference is detected as a resonance energy, measured in units of electron volts.

Conservation of Momentum

The momentum of any particle in motion is determined as the product of its mass and its velocity. Conservation of momentum for two recoiling entities requires that they both have the same momentum if both are in motion. Thus magnitude of the momentum of the recoiling nucleus (the cannon) must be the same as the magnitude of the momentum of the gamma ray photons (the cannonballs). This can be expressed mathematically as

$$|P_R| = |P_\gamma|$$

and the corresponding energy relationship is given by

$$E_\gamma^2 = 2Mc^2 E_R.$$

where E_γ and E_R are the energy of the gamma ray photons and of the recoiling mass, M is the mass of the recoiling body, and c is the speed of light (the velocity of gamma ray photons).

Mössbauer Spectroscopy

The Mössbauer effect can be utilized analytically in either the absorption or emission mode, both of which provide unique information about nuclear quantum transitions that is not available by any other methods. The methodology of Mössbauer spectroscopy, however, requires that the material being used as the source of the gamma rays must be of the same isotopic type as the material being examined. This is a restriction put in place by the very nature of quantum mechanics as the basis of quantum transitions. Thus, for example, examination of ^{57}Fe atoms in a certain material requires use of radioactive ^{57}Co as the source of gamma ray radiation. Since effectively all elements have radioactive isotopes, either naturally or artificially, the applicability of Mössbauer spectroscopy extends to a great many areas of science and technology as an analytical technique. The detailed information about nuclear quantum states and transitions from Mössbauer spectroscopy is obtained from analysis of the different types of signal splitting that are observed. Each type relates to different quantum aspects and states of the particular nucleus, and include magnetic splitting, isomer (or chemical) shift, and electric quadrupole splitting. The signal splittings that are observed provide information about the structure and behavior of the nucleus in quantum state transitions, and thus about the relationship between protons and neutrons in the nucleus. Other forms of spectroscopy, other than NMR, generate information about the electrons in the atoms and molecules under study, but not about the nucleus.

—*Richard M. Renneboog M.Sc.*

Bibliography

Dickson, Dominic P E, and Frank J, Berry. *Mössbauer Spectroscopy*. Cambridge; New York: Cambridge UP, 1986. Print.

Gütlich, Philipp, Rainer Link, and Alfred Trautwein. *Mössbauer Spectroscopy and Transition Metal Chemistry Fundamentals and Applications*. Heidelberg; Dordrecht; London: Springer-Verlag, 2011. Print.

Maddock, A. G. *Mössbauer Spectroscopy : Principles and Applications*. Chichester: Horwood, 1997. Print.

Yoshida, Yutaka, and Guido Langouche. *Mössbauer Spectroscopy Tutorial Book*. Berlin, Heidelberg: Springer Berlin Heidelberg : Imprint: Springer, 2013. Print.

NET FORCE

FIELDS OF STUDY

Classical Mechanics

SUMMARY

This article describes net force and how it is used to quantify the total force acting on an object and the motion that results. Individual forces, such as frictional, tension, normal, applied, or gravitational forces, may be used to calculate net force on an object. Newton's laws of motion are used to explain the relationship between force, mass, and motion—the foundation for the study of classical mechanics.

PRINCIPAL TERMS

- **acceleration:** the rate of change of an object's velocity.
- **friction:** the force created by the resistance to relative motion between solid surfaces.
- **gravity:** the force that describes the attraction between one body and another.
- **normal force:** the force exerted on an object perpendicular to the surface of contact.
- **right-hand rule:** the rule that shows the orientation of vector quantities normal to a surface by using the shape of the right hand.
- **tension:** the force directed along the length of a wire, string, or cable pulled at opposite ends.
- **vector:** a quantity that has both magnitude and direction.
- **velocity:** the rate of change of an object's displacement.

Newton's Laws of Motion

Force can be thought of simply as a push or a pull. An object may have several forces acting on it at a given time. In order to quantify the net force, or total force, one must consider the vector sum of all the forces acting on an object. Forces are vector quantities because they possess both magnitude and direction.

Newton's three laws of motion explain the relationship between force, mass, and motion. The first law states that if the net force on a resting object is zero, that object will remain at rest, and if the net force on a moving object is zero, that object's velocity will remain constant—that is, it will continue to move with the same speed in the same direction. (A net force of zero means that either there are no forces acting on an object or that the forces acting on the object cancel each other out.) This does not mean that if, for example, a person pushes a box across the floor and then stops, the box will continue to move. Even after the person stops pushing, there will still be a force acting on the box: friction. The box will stop moving because the friction between it and the floor is great enough to halt its movement. Pushing the same box across a frictionless surface would result in the box continuing to move at a constant velocity forever, or until it encountered another force.

Newton's second law states that the net force is equal to the mass of an object multiplied by its acceleration. Acceleration, like force and velocity, is a vector quantity. Mass is the measure of an object's quantity of matter. It is a scalar quantity, not a vector, because it has magnitude but no direction. Mass can also be thought of as an object's resistance to being accelerated. When there is no net force acting on an object, its acceleration is zero. This means that the object's velocity, whether zero or nonzero, will remain constant, because acceleration is change in velocity. When a net force does act on an object, it will cause the object to accelerate. The direction of the acceleration will be the same as the direction of the net force.

Newton's third law states that when one object exerts a force on another object, the second object will exert an answering force on the first object that is equal in magnitude but opposite in direction. For example, if a shoe is leaning against a box, the shoe is exerting a horizontal force on the box, while at the same time the box is exerting a horizontal force on the shoe. These two forces are equal in magnitude, but they act in opposite directions.

Identifying Different Forces

In order to quantify the net force, one must consider all of the various forces acting on an object. There are many different types of force. In the International System of Units (SI), these forces are measured in newtons (N). One newton is equal to the mass of the object in kilograms (kg) multiplied by its acceleration due to force in meters per second per second, or meters per second squared (m/s^2).

The force of gravity is the pull exerted by one physical body on another physical body. Generally, when one speaks of gravitational force on a body, one is referring to the force that pulls the body toward Earth. If an object is in free fall and the effects of air resistance are neglected, then gravitational force is the only force acting on that object.

Newton's second law can be expressed in the form

$$F = ma$$

where F is force, m is mass, and a is acceleration. This equation can be used to calculate gravitational force by replacing a with acceleration due to gravity (g), which on Earth is approximately 9.81 m/s^2. The magnitude of the gravitational force experienced by a body is equal to the mass of the body multiplied by g. This quantity is also what is meant by the term "weight." In physics, weight is defined as the force exerted on a body due to gravity. That body does not have to be in free fall for this relationship to work. An object at rest on a tabletop still experiences the same magnitude of gravitational force.

When an object is placed on a table, its weight is countered by what is called a normal force, which pushes up against the object. This force is always perpendicular to the surface on which the object rests. The greater the weight of the object, the greater the normal force exerted by the table to maintain its shape and resist being compressed. For the object on the table, the magnitude of the normal force is equal to the object's weight. However, in other situations, the normal force can be greater or less than the object's weight.

Friction is the resistance to motion that occurs as an object slides over a surface. It is caused by small irregularities that add to the roughness of the surface. Frictional force is parallel to the surface and acts in the opposite direction of the motion. In some situations, a surface will be considered frictionless, so that frictional forces can be neglected in calculations.

Tension is the pulling force exerted by a rope, cable, or similar cord that is pulled taut by an object at each end. To exert a tension force on an object, the

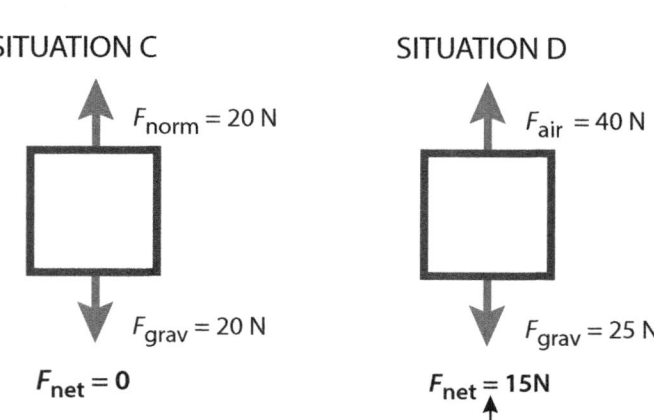

The net force (F_{net}) of a system is a summation of all the forces acting on the object(s) in the system. In some situations, the net force will be zero. In other situations, the net force will have magnitude and direction.

rope must be in contact with the object. Tension force is directed away from the object and along the rope.

FREE-BODY DIAGRAMS AND DETERMINING NET FORCE

To find the net force from two or more forces acting on an object, one needs to add the individual vector forces. Because forces can act on an object in different directions, each of the three physical dimensions—height, width, and depth—must be considered.

Free-body diagrams are sketches that show all of the different forces acting on an object. The object may be drawn as a box or as a point particle. The number of forces drawn will depend on the specific situation. Each force is represented by an arrow pointing in the direction in which the force is acting, with a label indicating its force type. A three-dimensional coordinate system can then be used to resolve each force into components so that Newton's second law can be applied to each component separately.

The most commonly used system is the Cartesian coordinate system, which uses numbers in the form (x, y, z) to represent width, height, and depth. The right-hand rule is a helpful way to remember how these three dimensions, or axes, are positioned relative to one another. The first or index finger represents the x axis, the second or middle finger represents the y axis, and the thumb represents the z axis. (Alternatively, the thumb can represent the x axis, the index finger can represent the y axis, and the middle finger can represent the z axis.) When these fingers are spread apart, their position resembles the orientation of the three axes. The right-hand rule can be used as a basis for finding vectors.

Consider a box sitting at rest on a table. There are two forces acting on this box: the normal force from the table surface and the gravitational force from Earth. The normal force is perpendicular to the surface of the table and drawn up through the top of the object. The gravitational force is directed toward Earth and is drawn from the object directly down. The normal force is equal to the object's weight, because the object is at rest and no other forces are acting on it. By Newton's second law, because the forces are balanced and there is no acceleration, the net force is zero.

Now, consider that same box with a force applied to it from the right. In addition to the normal and gravitational forces, one must take into account the magnitude of the force used to push the box and the frictional force opposing this motion. In the free-body diagram, the force pushing the box is drawn pointing from right to left and is positive, and the frictional force is drawn pointing from left to right and is negative. If the force used to push the box is greater than the frictional force, then there is a net force, and the box experiences an acceleration.

APPLICATIONS OF NET FORCE

Newton's laws of motion are an inescapable part of everyday life. Throwing a ball, driving a car, and simply walking are all activities that obey Newton's laws. Understanding these laws and how net force relates to motion forms the foundation for the study of classical mechanics and dynamics and allows one to quantify and predict outcomes of certain events. For example, English astronomer Edmond Halley (1656–1742) used Newton's laws to correctly predict the periodicity of a comet, which was later named Halley's comet in his honor.

—*Casey M. Schwarz, PhD*

BIBLIOGRAPHY

Chabay, Ruth W., and Bruce A. Sherwood. *Matter and Interactions*. 4th ed. Hoboken: Wiley, 2015. Print.

Cutnell, John D., Kenneth W, Johnson, Shane Stadler, and David Young. *Physics*. 10th ed. Hoboken, NJ: Wiley, 2015. Print.

Halliday, David, Robert Resnick, and Jearl Walker. *Fundamentals of Physics*. 10th ed. Hoboken, NJ: Wiley, 2014. Print.

Henderson, Tom. "Determining the Net Force." *The Physics Classroom*. Physics Classroom, n.d. Web. 12 Apr. 2015.

Knight, Randall Dewey. *Physics for Scientists and Engineers: A Strategic Approach*. San Francisco: Pearson/Addison Wesley, 2004. Print.

Tipler, Paul Allen, and Gene Mosca. *Physics for Scientists and Engineers*. 6th ed. Vol. 1-2. New York, NY: W.H. Freeman, 2008. Print.

Walker, James S. *Physics*. 5th ed. San Francisco: Pearson/Addison Wesley, 2016. Print

NEWTON-METER

FIELDS OF STUDY

Classical Mechanics

SUMMARY

This article describes the newton-meter, an International System of Units (SI) unit used to measure torque or energy. It also discusses the Newton meter, a device that measures applied force. The measurement of torque and energy is essential when discussing work, rotational dynamics, and kinetic energy, among other important concepts in the field of classical mechanics.

PRINCIPAL TERMS

- **dyne:** a unit of force in the centimeter-gram-second (CGS) unit system, equal to 10–5 newtons, or 10 micronewtons.
- **energy:** the capacity of a system to do work.
- **joule:** the derived SI unit of work, energy, or heat, equal to the energy expended by applying one newton of force over a distance of one meter.
- **moment:** the combination of a physical quantity, such as force, and the distance of that quantity from a given reference point.
- **Newton meter:** a device used to measure force; also called a "force meter" or a "force gauge."
- **torque:** a moment that measures the ability of a force to rotate an object about an axis; also called "moment of force."
- **work:** the displacement of an object by the application of force.
- **work-energy theorem:** the principle that the work performed on an object is equal to the change in that object's kinetic energy.

TORQUE

In the metric system, or International System of Units (SI), the newton-meter (N·m) is a unit of torque, also called moment of force. One newton-meter is equal to the torque generated when one newton of force is applied perpendicularly to a one-meter moment arm. A "moment arm" is the distance between the line of force (the point at which force is applied) and the axis of rotation (the point about which the object will rotate).

The newton-meter can be converted to a number of other units. These include the joule (J), which is equal to 1 N·m, and the dyne-centimeter (dyn·cm), which is equal to 10^{-7} N·m. The joule is also an SI unit. The dyne-centimeter belongs to the centimeter-gram-second (CGS) system of units, which is another variation on the metric system. The dyne is the CGS unit of force.

The newton-meter can also be used to measure work or energy. In this case, the "meter" part of "newton-meter" refers to the displacement of an object due to the application of force. However, the official SI unit of energy is the joule, defined as the amount of energy used when one newton of force is applied over a distance of one meter.

Quantitatively, the newton-meter and the joule are equivalent. That is why newton-meters can be used to measure energy as well. Both are derived SI units, meaning that they can be expressed in terms of the seven base SI units (meter, kilogram, second, ampere, kelvin, candela, and mole). Thus, both one newton-meter and one joule are equal to the product of one kilogram and one square meter, divided by one second squared:

FROM	TO			
	Newton-Meter	Foot-Pound	Joule	Dyne-cm
Newton-Meter	1	0.738	1	1×10^7
Foot-Pound	3.356	1	1.356	1.356×10^7
Joule	1	1.738	1	1×10^7
Dyne-cm	1.00×10^{-7}	7.375×10^{-8}	1.00×10^{-7}	1

Energy Conversion table for units of torque, including newton-meters, foot-pounds, joules, and dyne-centimeters.

$$N \cdot m = J = (kg \times m^2) / s^2$$

The work-energy theorem relates work to kinetic energy and, by extension, to speed. This theorem states that the total work done on an object—that is, the net force applied— is equal to the change in that object's kinetic energy. Generally, when the total work is positive, the object's speed increases, and when it is negative, the object's speed decreases.

Force

Two classic examples of the use of torque are children on a see-saw and a mechanic using a wrench to turn a lug nut. In both examples, less force is required when it is applied farther from the pivot point or axis of rotation.

The Newton meter, or force meter, is a device that measures force. One type of Newton meter is a spring balance. It works by measuring how much a spring stretches when a force is applied.

—*Casey M. Schwarz, PhD*

Bibliography

Cutnell, John D., Kenneth W., Johnson, Shane Stadler, and David Young. *Physics*. 10th ed. Hoboken, NJ: Wiley, 2015. Print.

Halliday, David, Robert Resnick, and Jearl Walker. *Fundamentals of Physics*. 10th ed. Hoboken, NJ: Wiley, 2014. Print.

Knight, Randall Dewey, Brian Jones, and Stuart Field. *College Physics: A Strategic Approach*. San Francisco: Pearson/Addison Wesley, 2007. Print.

Serway, Raymond A., and Chris Vuille. College Physics. 9th ed. Boston: Brooks, 2012. Print.

Tipler, Paul Allen, and Gene Mosca. *Physics for Scientists and Engineers*. 6th ed. Vol. 1-2. New York, NY: W.H. Freeman, 2008. Print.

Walker, James S. *Physics*. 5th ed. San Francisco: Pearson/Addison Wesley, 2016. Print

"What Are Forces?" BBC Bitesize. BBC, n.d. Web. 3 Apr. 2015.

NORMAL FORCE

FIELDS OF STUDY

Classical Mechanics

SUMMARY

This article defines normal force and describes how it relates to the weight of an object, frictional forces, and an inclined plane. Normal force is a contact force related to Newton's third law of motion. The normal force measurement is used to determine the net force acting on an object.

PRINCIPAL TERMS

- **acceleration:** the rate of change of an object's velocity over time.
- **frictional resistance:** the force created by the resistance to relative motion between solid surfaces; it is normally proportional to the roughness of the surfaces as well as the force squeezing the surfaces together.
- **gravity:** the attractive force of one body on another.
- **net force:** the vector sum of every force acting on an object.
- **perpendicular:** set at 90 degrees to a line or surface, forming a right angle.
- **vector:** a quantity that has both a magnitude and a direction.
- **velocity:** the rate of change of an object's displacement over time.
- **weight:** the force due to gravity acting upon an object.

The Normal Force due to Gravity

Normal force is related to Newton's third law of motion, which states that for every action there is an equal and opposite reaction. An object at rest on a table has an acceleration of zero, so its net force is also zero. The force of the table pushing up on the object and opposing the downward force of gravity is referred to as the "normal force." The normal force is always perpendicular to the surface on which an object is resting. The weight (W) of an object is equal

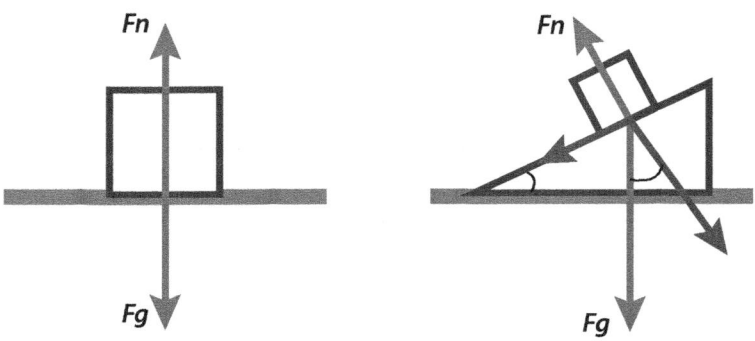

The normal force (*Fn* or *N*) acts on an object away from the point of contact at an angle perpendicular to the contact surface. It is not necessarily equal and opposite to the gravitational force (F_g). On an inclined plane, a portion of the object's gravitational force continues to pull the object downward.

to the object's mass (*m*) in kilograms (kg) multiplied by gravity (*g*) in units of meters per second squared (m/s²):

$$W = mg$$

When an object is placed on a horizontal surface with no other forces involved, the normal force (*N* or *Fn*) is directly equal to the object's weight:

$$N = mg$$

The weight and therefore the normal force are vector quantities measured in newtons (N). The greater the weight of an object placed on a table, the greater the normal force the table must exert to oppose being compressed by the object.

The normal force is not always directly equal to an object's weight; it can be greater or less. For example, if someone is pulling a crate by a handle at angle above the horizontal, the force exerted to pull the crate has an upward component that counteracts part of the crate's weight. In this case, the normal force would be less than the weight of the crate.

Normal Force on an Incline

For an object on an incline, the normal force is still perpendicular to the surface but is no longer vertical. When drawing a free-body diagram (a vector diagram that depicts all forces for a given situation) for object on an inclined surface, the *x*- and *y*-axes are typically oriented so that the *x*-axis is parallel and the *y*-axis is perpendicular to the surface. The normal force points in the positive *y* direction, perpendicular to the surface. The weight points downward at the same angle with respect to the negative *y*-axis in which the surface is inclined. The *x* and *y* components of weight are now:

$$W_x = W\sin\theta = mg\sin\theta$$

$$W_y = -W\cos\theta = -mg\cos\theta$$

Frictional Forces

Frictional resistance refers to forces that are parallel to a surface and oppose the direction of the object's motion. Kinetic friction occurs when surfaces slide against each other at a certain velocity. The force of kinetic friction (f_k) is equal to the normal force (*N*) multiplied by the coefficient of kinetic friction (μ_k):

$$f_k = \mu_k N$$

Static friction prevents two surfaces from moving against each other. The force due to static friction (f_s) can have values between 0 and $f_{s,max}$, where $f_{s,max}$ is a maximum limit to the force that can be delivered by static friction. If an applied force exceeds $f_{s,max}$, an object may start to slide and then kinetic friction takes over. Like kinetic friction, static friction is proportional to the normal force. The force of static friction (f_s) is equal to the normal force (*N*) multiplied by the coefficient of static friction (μ_s):

$$f_{s,max} = \mu_s N$$

The coefficient for kinetic and static friction is a positive number typically between 0 and 1 and varies depending on the material.

Normal Forces and Apparent Weight

Depending on the specific situation or other forces involved, the normal force can be less than or greater than the object's weight. While riding on an elevator, one experiences apparent weight. If the elevator is moving with an upward acceleration, one feels heavier since the normal force is greater than one's weight. If the elevator is moving with a downward acceleration,

SAMPLE PROBLEM

A 50-kg box is placed on a plank of wood. When the wood is tilted to an angle of 25 degrees, the box begins to slide. Find the coefficient of static friction between the wood and the box and the magnitude of the force of static friction on the box.

Answer:
First, draw a free-body diagram where with the coordinate system aligned with the incline of the wood. Choose the positive x direction to point down the incline and the positive y direction to be perpendicular to the wood.

There are three forces that act on the box, the normal force (N), the force from static friction (f_s), and the box's weight (W). These forces are resolved into the following x and y components:

Since there is only a normal force in the y direction, the normal force in the x direction is set to zero.

$$N_x = 0$$

$$N_y = N$$

When an object is on the verge of slipping, the static friction is at its maximum value ($f_s = f_{s,max} = \mu_s N$). Since there is only a force due to static friction acting in the x direction, the force due to static friction in the y direction is set to zero.

$$f_{s,x} = -f_{s,max} = -\mu_s N$$

$$f_{s,y} = 0$$

The weight of the box has components in both the x and y directions.

$$W_x = mg\sin\theta$$

$$W_y = -mg\cos\theta$$

Since the box is at rest, the acceleration of the box is zero in both the x and y directions, and therefore the net force in both the x and y directions equals zero.

$$F_y = ma_y = 0$$

The net force in the y direction is a sum of all the forces acting on the box in the y direction and is equal to the y components of the normal force (N_y), static friction ($f_{s,y}$), and the weight (W_y):

$$F_y = N_y + f_{s,y} + W_y$$

Substitute in the values for static friction and weight and zero for F_y. Then solve for the normal force (N):

$$0 = N + 0 - mg\cos\theta$$

$$N = mg\cos\theta$$

(Sample problem continued on next page)

(SAMPLE PROBLEM CONTINUED)

The net force in the x direction also equals zero:

$$F_x = ma_x = 0$$

The net force in the x direction is a sum of all the forces acting on the box in the x direction and is equal to the x components of the normal force (N_x), static friction ($f_{s,x}$), and the weight (W_x):

$$F_x = N_x + f_{s,x} + W_x$$

Substitute in the values for static friction and weight and zero for F_x and N_x:

$$0 = 0 - \mu_s N + mg\sin\theta$$

Now, $mg\cos\theta$ can be substituted for N, giving

$$0 = 0 - \mu_s mg\cos\theta + mg\sin\theta$$

Then solve for the coefficient of static friction (μ_s):

$$\mu_s mg\cos\theta = mg\sin\theta$$

$$\mu_s = \frac{mg\sin\theta}{mg\cos\theta} = \tan\theta$$

$$\tan 25° = 0.47$$

Note that this result does not depend on the mass of the object. To find the force of static friction acting ($f_{s,max}$) on the box, plug in the coefficient (μ_s), the mass (m), the gravitational force (g), and the angle ($\cos\theta$). Then solve:

$$f_{s,max} = \mu_s N$$

$$f_{s,max} = \mu_s mg\cos\theta$$

$$f_{s,max} = (0.47)(50\text{ kg})(9.8\tfrac{m}{s^2})(\cos 25)$$

$$f_{s,max} = 209 \tfrac{kg \cdot m}{s^2} = 209 \text{ N}$$

one feels lighter since the normal force is less than one's weight. While training astronauts, the National Aeronautic and Space Administration uses this effect to simulate an experience of weightlessness.

—*Casey M. Schwarz, PhD*

BIBLIOGRAPHY

Cutnell, John D., Kenneth W, Johnson, Shane Stadler, and David Young. *Physics*. 10th ed. Hoboken, NJ: Wiley, 2015. Print.

Halliday, David, Robert Resnick, and Jearl Walker. *Fundamentals of Physics*. 10th ed. Hoboken, NJ: Wiley, 2014. Print.

Ohanian, Hans C., and John T. Markert. *Physics for Engineers and Scientists*. 3rd extended ed. New York: Norton, 2007. Print.

Tipler, Paul Allen, and Gene Mosca. *Physics for Scientists and Engineers*. 6th ed. Vol. 1-2. New York, NY: W.H. Freeman, 2008. Print.

"Types of Forces" Physics Classroom. Physics Classroom, n.d. Web. 20 June 2015.

Walker, James S. *Physics*. 5th ed. San Francisco: Pearson/Addison Wesley, 2016. Print

NUCLEAR FISSION, FUSION, AND MASS DEFECT

FIELDS OF STUDY

Nuclear Physics; Particle Physics; Quantum Physics

SUMMARY

Fission, fusion and mass defect are properties of the atomic nucleus. Fission refers to the splitting apart of a nucleus into smaller particles and nuclei with a corresponding release of energy. Fusion refers to the combining of nuclei into a larger nucleus. Mass defect is the difference in mass between an intact nucleus and the total mass of its component particles. All three are essential concepts in the control and manipulation of nuclear reactions.

PRINCIPAL TERMS

- **alpha particle:** a particle consisting of two protons and two neutrons, identical to a helium nucleus, which is emitted from an unstable nucleus via radioactive decay.
- **alpha radiation:** alpha (α) particles typically emitted during alpha decay, a subtype of radioactive decay.
- **background radiation:** the total amount of ionizing radiation to which Earth is constantly exposed from both natural and artificial sources.
- **beta particle:** an electron that is emitted from an unstable nucleus via radioactive decay.
- **half-life:** the average time it takes for half of the unstable nuclei in a radioactive element to undergo radioactive decay, transforming into a lighter element and giving off radiation.
- **isotopes:** variants of a chemical element with differing numbers of neutrons; they are often unstable and radioactive.
- **neutrons:** subatomic particles that, with protons, make up the mass of an atom's nucleus; they have functionally the same weight as protons but no electric charge.
- **x-ray:** electromagnetic radiation with a wavelength between 1×10^{-10} and 1×10^{-8} meters.

NUCLEAR SCIENCE

The modern theory of atomic structure was developed in part to explain observed properties such as background radiation and the decay of radioactive materials according to the half-life of different elements. The concept of atoms having a nucleus composed of a specific number of protons and neutrons is essential to describing the mechanisms of those phenomena. The background radiation is understood to be the result of naturally-occurring nuclear fission reactions of various isotopes existing within the material of Earth, as well as from non-terrestrial sources. The half-life of a radioactive isotope is an expression of the rate at which that isotope undergoes nuclear fission to eventually become a non-radioactive isotope, generally of a different element. While nuclear fission effectively describes these processes, physical and chemical observation of Sol (Earth's Sun) and other stars does not provide evidence of elements undergoing nuclear fission. The vast majority of the mass of stars consists of simple hydrogen, and given that only nuclei larger than simple hydrogen can undergo fission, the inevitable conclusion is that hydrogen in stars is undergoing the opposite process of fusion to form the nuclei of heavier atoms. The central goal of nuclear science is to fully understand these processes in order to be able to control and utilize the characteristics of their functioning and to harness the tremendous amounts of energy that they can provide.

FISSION, FUSION AND MASS DEFECT

Nuclear fission occurs naturally in isotopes of most elements. Each atom of an element has exactly the same number of protons in each nucleus. However, atoms of the same element can have different numbers of neutrons. The stability of isotopes is determined by the ratio of protons to neutrons, and isotopes that have either too many or too few neutrons relative to the optimum number that is in the atoms of stable isotopes can spontaneously split apart, or undergo fission. Unstable isotopes are radioactive, and emit different types of nuclear radiation. Alpha radiation occurs when the nucleus ejects an alpha particle. Other processes emit beta radiation by emitting beta particles (electrons, from the conversion of a neutron into a proton and an electron), and high-energy X-rays. Each radioactive decay process releases energy, as heat, light or higher energy electromagnetic waves such as x-rays and gamma radiation. In a nuclear chain reaction, neutrons released by one decay process trigger the release of many more from other nuclei, and the amount of energy that is thus released becomes very large, very quickly. Nuclear fusion reactions release a great deal more energy than do fission reactions, but they also require the input of a great deal of energy to make them take place. In a nuclear fusion reaction, nuclei must be brought together with such high energy that the strong forces of repulsion between the nuclei are overcome, allowing the two nuclei to combine into one and release the nuclear binding energy. This is characterized by a quantity known as the mass defect, which represents the difference in the mass of the whole nucleus and the total mass that the individual protons and neutrons that make up that nucleus would have as separate entities. The Einstein equation, $e = mc^2$, describes the equivalent relationship between the mass defect and the nuclear binding energy.

USING THE HALF-LIFE

The half-life of radioactive decay from nuclear fission follows a very precise mathematical relationship.

SAMPLE PROBLEM

The $t_{1/2}$ of C-14 is 5,730 years. An artifact made of wood is found to have just 55% of the amount of C-14 relative to a living sample. Calculate the age of the artifact.

Answer:

Relating the two quantities using the rate equation yields the formula

$$(-0.693)t = t_{1/2}\, ln(A/A_o)$$

and $A/A_o = 0.55$.

Therefore,

$$t = (5730\ yr)(-0.5978)/(-0.693)$$

$$= 4943\ years$$

The calculation of time based on the half-life of the isotope carbon-14 is typical and relates the amount of material present after a certain amount of time has passed to the amount of the material that was originally present. Carbon-14 is produced in the atmosphere at a constant rate through interactions of carbon dioxide with cosmic ray particles entering Earth's atmosphere. Accordingly, the C-14 that is produced is incorporated into living matter at a constant rate. Measurement of the amount of C-14 present in an artifact, and knowing the rate at which C-14 undergoes nuclear fission, it is possible to calculate the number of half-lives that have passed since the artifact ceased to incorporate new C-14. The relationship is described by the equation

$$A_t = A_o e^{-kt}$$

At the half-life, A_t is exactly one-half of A_o, and the logarithmic calculation is greatly simplified.

APPLICATIONS OF NUCLEAR SCIENCE

Nuclear science provides many useful materials and applications that range from the study of biochemical mechanisms and medical treatments using radioactive isotopes, x-ray imaging and the generation of electricity through nuclear fission. Nuclear fusion, apart from being the source of all energy on Earth, offers the possibility of unlimited energy if it can be controlled. Elsewhere, fusion reactions in high energy physics experiments provide an ever deeper understanding of the basic structure and principles of the physical universe.

—*Richard M. Renneboog M.Sc.*

BIBLIOGRAPHY

Choppin, Gregory R., Jan-Olov Liljenzin, Jan Rydberg, and Christian Ekberg. *Radiochemistry and Nuclear Chemistry*. 4th ed. Amsterdam: Academic, 2013. Print.

Clery, Daniel. *A Piece of the Sun: The Quest for Fusion Energy*. New York: Overlook, 2014. Print.

Krappe, Hans J., and Krzysztof Pomorski. *Theory of Nuclear Fission: A Textbook*. Berlin [u.a.]: Springer, 2012. Print.

Povh, Bogdan, Klaus Rith, Christoph Scholz, and Frank Zetsche. *Particles and Nuclei: An Introduction to the Physical Concepts.* 7th ed. Berlin; Heidelberg [u.a.]: Springer, 2015. Print.

Evans, Robert L. *Fueling Our Future: An Introduction to Sustainable Energy.* Cambridge: Cambridge University Press, 2008. Print.

Singer, Neal. *Wonders of Nuclear Fusion: Creating an Ultimate Energy Source.* Albquerque: University of New Mexico, 2011. Print.

Shultis, J. Kenneth, and Richard E. Faw. *Fundamentals of Nuclear Science and Engineering.* 2nd ed. Boca Raton: CRC, 2008. Print.

OPEN SYSTEMS

FIELDS OF STUDY

Classical Mechanics; Thermodynamics

SUMMARY

A system is a portion of the universe singled out for study. Anything from a single atom to the entire universe may be viewed as a system. A system may be open, closed, or isolated. An open system allows both matter and energy to pass in and out.

PRINCIPAL TERMS

- **closed system:** a system that may exchange energy, but not matter, with its surroundings.
- **conservation of energy:** the principle that energy cannot be created or destroyed, only transformed.
- **dynamic equilibrium:** the state in which reversible reactions occur at equal rates in opposite directions, balancing each other and resulting in no net change.
- **entropy:** a measure of a system's level of disorder, which in physics refers to the potential amount of states the molecules of the system may assume.
- **radiant energy:** energy consisting of electromagnetic radiation; an important mechanism for the transfer of energy in or out of some systems.
- **system boundary:** a physical or conceptual delineator between a system and the outside environment.
- **system efficiency:** the proportion of a system's input it converts to the intended output.

What Is a System?

A system in physics is a portion of the universe that has been chosen for study. A system consists of components viewed as a whole, and systems may be identified at any scale. An atom is a system, as is the entire universe. Systems are defined so that they single out the group of interactions a scientist is interested in studying.

A system may be open, closed, or isolated depending on how it exchanges energy. Open systems, which allow the transfer of both energy and matter in and out, are most common. Most natural systems and created objects are open systems. A closed system allows energy, but not matter, in or out. Earth is often viewed as a closed system, receiving radiant energy from the sun but neither losing nor gaining any matter (other than negligible amounts from meteorites and spacecraft). Isolated systems do not allow matter or energy in or out. The universe is thought to be an isolated system. Perspectives of the types of systems also vary between fields of study.

Open System Characteristics

An open system is characterized by the ability to transfer energy and matter through the system boundary. The human body is a classic biological open system because of the intake of outside matter (food and water) and its subsequent exit. A mechanical system such as a car, which takes in fuel and gives off exhaust, is also open. In both examples, matter and energy are output back into the environment, obeying the principle of conservation of energy.

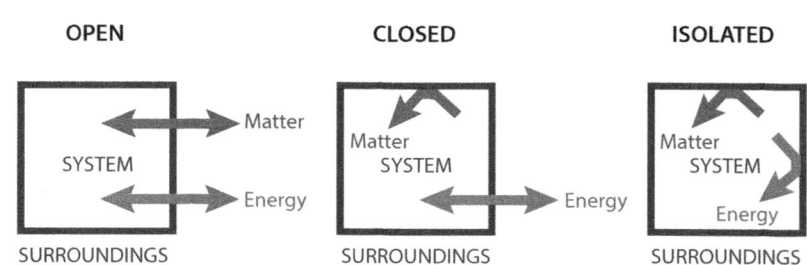

There are three classifications of systems based on whether energy and/or matter is exchanged between the system and it's surroundings. Open systems exchange energy and matter with their surroundings. Closed systems exchange only energy, while isolated systems do not exchange energy or matter with their surroundings.

Typically, when the scope of study is narrow, a system will be open. If the system is broadened to include elements of the environment, it may become closed. Systems may also experience dynamic equilibrium while remaining open.

All all closed systems not at thermodynamic equilibrium tend towards states of increased entropy (randomness) as time goes on. This causes all systems to wear down over time, diminishing system efficiency.

—*Kenrick Vezina, MS*

BIBLIOGRAPHY

Henderson, Tom. "Isolated Systems." *Physics Classroom*. Physics Classroom, 2015. Web. 26 Aug. 2015.

Lee, Esther, Jiaxu Wang, and Jonathan Wang. "Dynamic Equilibrium." UC Davis ChemWiki. Regents of the University of California, 10 June 2014. Web. 26 Aug. 2015.

Miller, James Grier, and Jessie L. Miller. "The Earth as a System." Primer Project. Internatl. Soc. for the Systems Sciences, 10 Nov. 1997. Web. 25 Aug. 2015.

Nave, R. "Conservation Laws." HyperPhysics. Georgia State U, 2012. Web. 26 Aug. 2015.

Sanctuary, Bryan. "Open, Closed, and Isolated Systems in Physical Chemistry." Foundations of Quantum Mechanics and Physical Chemistry. MCH Multimedia, 11 July 2011. Web. 26 Aug. 2015.

"What Is Entropy?" Energy Concepts Primer. New Mexico Solar Energy Assoc., 2011. Web. 26 Aug. 2015.

P

PARTICLE ACCELERATORS

FIELDS OF STUDY

Particle Physics; Electromagnetism; Quantum Physics

SUMMARY

Particle accelerators began as cathode rays, as streams of electrons are accelerated as they pass through an electric field. Accelerators are constructed either in linear or circular configurations. Electromagnetic fields are used to accelerate charged particles to energies measured in TeV. Low energy accelerators are commonly used in devices such as mass spectrometers and in many other applications. High-energy accelerators are used to examine the fundamental relationships between matter and energy by particle beam collisions. Accelerators have applications in many fields.

PRINCIPAL TERMS

- **alpha particle:** a particle consisting of two protons and two neutrons, identical to a helium nucleus, which is emitted from an unstable nucleus via radioactive decay.
- **bremsstrahlung:** radiation generated when a particle is slowed via a collision with another particle; from the German word for "braking radiation."
- **electron volt:** the amount of energy carried by a single electron moved across an electric potential difference of one volt; equal to 1.6×10^{-19} joules.
- **laminar flow:** the even and stable flow of a fluid; opposite of turbulent flow.
- **quantum chromodynamics:** a quantum field theory that describes the interactions of quarks and gluons, subatomic particles that are responsible for the strong interaction.
- **quantum field theory:** a theory that explains interactions between subatomic particles as the result of a field extending between them.
-
- **quark:** an elementary fermion that combines with other quarks to form a baryon, such as a proton or a neutron, or with an antiquark to form a particle called a meson.
- **synchrotron radiation:** radiation emitted by the acceleration of a charged particle traveling in a magnetic field at near light speed.

PARTICLE INTERACTIONS

Particle accelerators began as relatively simple cathode ray tubes, in which cathode rays were produced within an evacuated tube and directed by passing them as a stream through an electric field. In 1897 Thomson identified cathode rays, and their counterpart canal rays, as streams of charged particles rather than a form of continuous electromagnetic radiation. He termed the cathode ray particles electrons and the canal ray particles protons, and arbitrarily assigned them as having negative and positive charge, respectively. This also marked the beginning of the modern theory of atomic structure based on experimental evidence.

In the period from 1917 to 1919, Ernest Rutherford experimentally verified the natural process of nuclear disintegration resulting from the collision of alpha particles with the nuclei of other atoms. An alpha particle is the nucleus of a helium atom, consisting of two protons and two neutrons and therefore carrying two positive charges. In the particular experiment that was used, a radioactive material that was known to emit alpha particles was placed in a container filled with pure nitrogen gas. Analysis of the gas after a period of time showed that both oxygen and hydrogen had been produced. Since the oxygen atom has a higher atomic number and atomic weight than the nitrogen atom, the only way that the oxygen atoms could have been produced is by collision of the alpha particles with the nuclei of nitrogen atoms, adding part of their mass to the nitrogen nuclei as the additional proton and neutron in the oxygen atom and ejecting a proton to form a hydrogen atom, as

$$\alpha + {}^{14}_{7}N \rightarrow p + {}^{16}_{8}O,$$

where p represents a proton, which is the nucleus of the hydrogen atom. The important feature of this observation is that it took place in a closed system, eliminating any other possible source for the oxygen and hydrogen that was formed. This proved that nuclear collisions take place that can alter the structure of the nuclei and change their identity as elements. Rutherford's experiment marked the beginning of research into the nature of nuclear reactions. In his experiment, Rutherford used a material source that naturally emits alpha particles, but there are other ways to produce alpha particles.

An alpha particle is just a helium atom that has been stripped of its electrons, leaving the bare nucleus having two positive charges, as He^{2+}. Alpha particles can therefore be readily produced by subjecting helium gas to ionizing radiation and using an electric field to accelerate the He^{2+} ions through an aperture with energies measured in units of the electron-volt. This is in fact a common technique of generating ions of all kinds, most commonly in mass spectrometry. In a particle accelerator, the ions are delivered into a chamber under high vacuum, where controlled electromagnetic fields are used to accelerate the particles to speeds very close to the speed of light. At such speeds, the particles are effectively single beams of energy, formed in accord with the theory of relativity. Given the desired kinetic energy, the beams can be directed to cross paths and where the beams coincide with the right geometry, collisions between take place converting the energy back into particles of the corresponding masses and quantum properties.

At lower energies, collisions reform the nuclei, as in Rutherford's experiment, which has been the method used to identify atoms of new elements. At high energies, the collisions can have sufficient energy to completely shatter the nuclei into its elementary constituents. In the construction and operation of such devices, it is absolutely essential that the movement of the particle beams through the system proceeds with laminar flow, without turbulence or aberration. Accordingly, particle accelerators are the most precise tools or instruments ever made by humans. The Large Hadron Collider (LHC) facility at CERN, in Switzerland, is also the largest machine ever constructed. The high precision and controllability of the devices is necessary because the production of high-energy collisions requires two beams that are moving through the system in opposite directions. Thus a high-energy accelerator may actually consist of two or more separate accelerators within a common structure.

The information to be obtained from particle collisions comes from the bremsstrahlung, from synchrotron radiation and from other traces recorded in the detectors that are built into the accelerator structures where the particle beams are made to cross. It is the interpretation of this data that has provided the experimental support for quantum field theory and quantum chromodynamics.

Types of Accelerators

Particle accelerators function on two basic principles. These are linear acceleration of charged particles by electric or electromagnetic field or circular cyclotron acceleration of charged particles by magnetic fields. Electric field acceleration of charged particles to high energies was achieved by Cockcroft and Walton using a DC voltage-multiplier system that accelerated protons to 800kV (8 X 10^5 volts). In 1932 they used this system carry out the first disintegrations of atomic nuclei by bombarding lithium atoms with protons to produce alpha particles, as

$$p + {}^{7}_{3}Li \rightarrow 2\ He^{2+}.$$

Their DC voltage-multiplier system was constructed using capacitors and rectifier tubes. A rectifier tube is the vacuum tube equivalent of the semiconductor-based diode, and restricts the flow of electricity to a single direction in an electrical circuit. At each successive stage of the device, the effective voltage adds together such that the output voltage is equal to the input voltage multiplied by the number of stages in the device.

Particle acceleration is achieved by introducing charged particles into the high-potential electric field so that they are ejected into a collision chamber or magnetic field chamber under high vacuum. A second method of producing electric field acceleration is the Van de Graaff generator. This device uses a rapidly moving continuous belt to carry static charge from a source point at its lower end to a hollow sphere at its upper end. The build-up of static charge between the hollow sphere and the grounded lower end of the system can generate as much as 15 million volts of potential difference for the acceleration of charged particles in an attached system. This type of accelerator is not used for high energy experiments, however, as the potentials needed for such applications have far exceeded the capabilities of the devices. The magnitude of the electrical potentials that they generate are also great enough to cause physical damage to the materials from which the devices are constructed, rendering them imprecise and unstable. They are, however, commonly used for low energy ion acceleration in mass spectrometers and similar devices due to the high sensitivity with which they can be controlled reliably.

More recently, though, the linear accelerator concept was greatly advanced using radar microwave technology, and in 1966 the 3.2 kilometer (2 mile) long Stanford Linear Accelerator became active. In the 1930s, a new form of accelerator called a cyclotron was demonstrated, using a time-varying magnetic field to accelerate charged particles at a constant radius. The betatron, first demonstrated in 1940, uses magnetic fields at RF frequencies about an evacuated chamber to accelerate electrons (beta particles) to energies of as much as 300 MeV.

Cyclotrons that were subsequently developed from the betatron concept were rather limited in their capabilities at energies above 25MeV, due to the increase in relativistic mass of the particles being accelerated. A type of cyclotron called a microtron was developed and operates on the principle of phase stability by synchronizing the period of the rotation of the particles in the accelerator with the frequency of the RF modulation of the magnetic field. The particles are accelerated by the electric field that is induced by the time-varying magnetic field, and each rotation is to a larger radius. The synchroton accelerator remains the most advanced and the most powerful type of accelerator in the present time, and is now capable of producing particle energies measured in TeV (1 TeV = 10^{12} eV). The power of such accelerators is apparently limited only by the cost of building them. These large systems utilize a storage ring structure to generate particle beams that are then introduced into the main ring for their experimental purposes. The experiments that they are used to carry out are typically high-energy collisions between particles, as beams rotating through the system in opposite directions are brought into focus at a common point.

MAGNETIC FOCUSING

Focusing of the beams within the accelerator is essential if collisions between the particles and the material of the chamber walls is to be avoided. Although the actual mass of the particles is small, they have very high kinetic energy and collisions with container material can cause significant damage that has the potential to bring about the failure of the device, requiring time-consuming and expensive repairs. More importantly, however, collisions between accelerated particles and the container wall material result in the introduction of undesired nuclei and other particles into the accelerated particle beam.

The beam of particles being accelerated represents a material of extremely high purity and isotopic composition, which enables and facilitates the predictability and interpretation of particle collision

---

SAMPLE PROBLEM

Calculate the separation d required of two thin lenses having focal lengths of 24 cm and 50 cm to produce an overall focal length F of 36 cm.

Answer:

The appropriate equation to use is this:

$$1/F = 1/f_1 + 1/f_2 - d/f_1 f_2.$$

Rearrange to obtain an expression for d as follows:

$$d/F = 1/f_1 + 1/f_2 - 1/F$$

$$d/f_1 f_2 = 1/f_1 + 1/f_2 - 1/F$$

$$d = f_1 f_2/f_1 + f_1 f_2/f_2 - f_1 f_2/F.$$

Substitute in the appropriate values to obtain

$$d = 24 \text{ cm} + 50 \text{ cm} - (24 \text{ cm} \times 50 \text{ cm})/36 \text{ cm}$$

$$= 40.667 \text{ cm}.$$

results. For example, if a beam of gold nuclei is being accelerated, collisions with container wall material would introduce iron, chromium, nickel, carbon and various other undesired nuclei into the accelerated stream. There they would yield collisions with the target nuclei that do not represent collisions with gold nuclei, raising the signal to noise ratio in the detectors and making identification and interpretation of the data more difficult.

Beams of charged particles have very predictable behavior as they move through a magnetic field. By adjusting the nature of the magnetic field so that particles experience a uniform gradient field rather than a constant magnetic field, it is possible to focus the particles in much the same way that lenses focus beams of light. An alternating series of magnetic gradients are used to focus the particle beams in cyclotrons to produce and maintain a collimated stream of particles. The analogy from the field of optics is given by the focal length F of a pair of individual lenses, each with its own focal length f_n, separated by a distance d, as

$$1/F = 1/f_1 + 1/f_2 - d/f_1 f_2.$$

When the focal lengths of two lenses are equal, but in opposite directions, the relation reduces to

$$F = f_2/d.$$

The magnetic field gradients used to focus a stream of particles in a cyclotron have essentially the same effect, although the mathematical description is very different, involving integral calculus on magnetic field vectors.

Potential of Particle Accelerators

The applications of particle accelerators are numerous, and reach into many areas as new methodologies are developed. Accelerators are used to generate X-rays and neutron beams that are used for the development of advanced materials and materials testing, for the exploration of hydrogen storage technology and sequestering of carbon from CO_2 emissions. The development of solar energy harvesting methods and materials is another area of application. Through nuclear transmutation, particle accelerators offer a means of safely nuclear waste disposal, or alternatively the production of effective, short-lived nuclear fuels. The production of ionizing radiation from particle accelerators can be used to detoxify and purify waste water as well as drinking water. One of the most important applications of accelerator technology is in the production of radioisotopes used to prepare labeled medicines, medical imaging, and most recently in the development of particle beam therapies in the treatment of diseases such as cancer. Electron beams from accelerators are widely used industrially for a great many purposes, and new applications are constantly under development.

—*Richard M. Renneboog M.Sc.*

Bibliography

Wilson, E.J.N. *An Introduction to Particle Accelerators*. Oxford: Oxford UP, 2006. Print.

Wiedemann, Helmut. *Particle Accelerator Physics*. 4th ed. Cham: Springer International, 2015. Print.

Conte, Mario, and William W. MacKay. *An Introduction to the Physics of Particle Accelerators*. 2nd ed. Hackensack, NJ: World Scientific, 2013. Print.

Amaldi, Ugo, and Adele La Rana. *Particle Accelerators: From Big Bang Physics to Hadron Therapy*. Cham [Switzerland]: Springer, 2015. Print.

Raghavan, Jayakumar. *Particle Accelerators, Colliders, and the Story of High Energy Physics: Charming the Cosmic Snake*. Berlin: Springer Berlin, 2011. Print.

DeLaney, Thomas F., and Hanne M. Kooy. *Proton and Charged Particle Radiotherapy*. Philadelphia: Wolters Kluwer Health/Lippincott Williams & Wilkins, 2008. Print.

PARTICLE DETECTORS

FIELDS OF STUDY

Particle Physics; Electronics

SUMMARY

Particle detectors have many different applications beyond high energy physics research. They range from simple exposure badges that monitor radiation and hazardous materials exposure to electrostatic dust removal from household air and industrial exhaust stacks. Detectors are designed in many different configurations according to their application. There are two basic geometries for detectors of particle collisions. Many different devices rely on particle detection for their operational value.

PRINCIPAL TERMS

- **angular momentum:** the rotational momentum of an object around an axis, defined as the product of its moment of inertia and its angular velocity.
- **conservation of energy:** a fundamental law of physics that states that the amount of energy in a system remains constant over time. Although the energy can be transformed or transferred, it cannot be created or destroyed.
- **conservation of momentum:** in physics, the principle that the total momentum in a closed system is always constant.
- **positron:** a subatomic particle that has the same mass as an electron as well as an equal-but-opposite electric charge, that is, a positive charge.
- **quantum field theory:** a theory that explains interactions between subatomic particles as the result of a field extending between them.
- **quark:** an elementary fermion that combines with other quarks to form a baryon, such as a proton or a neutron, or with an antiquark to form a particle called a meson.
- **spin:** an intrinsic form of angular momentum carried by elementary particles, composite particles (hadrons), and atomic nuclei.
- **wave-particle duality:** the idea that a particle can behave as either a particle or a wave, depending on how and when it is being observed.

ACCELERATORS AND DETECTORS

It could be argued that the first actual particle detector was the apparatus used in Young's "double slit" experiment, which served to demonstrate the wave-particle duality of photons of light (though it did not resolve the debate at the time). Photons of light, however, are not accelerated particles in the conventional sense. Accelerators began as the cathode ray tube, in which Thomson first identified the "cathode rays" and "canal rays" as streams of discrete particles bearing opposite charges in 1897. His designation of the charge on the "electrons" of the cathode rays, and on the "protons" of the canal rays, as negative and positive, respectively, was entirely arbitrary. However, the important feature was that this would not have been possible had Thomson not had the means to detect what was happening in his apparatus.

Any particle accelerator system has absolutely no value without a means of detecting, identifying and quantifying the particles that are accelerated by the device. By coating the inside of the tube with phosphorus compounds that would glow and emit light when struck by the charged particles, the cathode ray tube became a viable scientific instrument in its own right, and could be used to examine, however crudely, the energy of emitted particles. Some fifty years later, the cathode ray tube was in the majority of American homes as the central component of the television. It also had numerous other uses as radar and sonar screens, point-of-sale terminals in stores, and most recently as the part of the personal computer that humans looked at until digital technology replaced the CRT screen with LCD and LED displays.

In scientific experiments, the cathode ray tube remains an important particle detector because of its ability to provide a visual display of particle interactions in real time through the effect that the particles have on the screen phosphors. Not all accelerators produce particles that can be detected by CRT screens, and this has resulted in the design and construction of very specific and sensitive particle detection systems for specific experiments.

PARTICLE PHYSICS

Thomson's observations mark the beginning of particle physics as a distinct field of study and application, and a foundation point for the development of the "standard model" of the universe and quantum mechanics. The theory of quantum mechanics, and quantum field theory, is based on the premise that energy is quantized, and that the law of conservation of energy and the law of conservation of momentum are universally applicable. In addition, systems that are described by a property termed spin are also quantized, and their quantum state is defined by angular momentum quantum numbers. Given these basic premises, particle physicists have derived several relationships that demonstrate the existence of different subatomic particles such as the positron, and the most fundamental particle called the quark. Quarks are deemed to come in six flavors termed up, down, top, bottom, strange and color. In the standard model, particles are divided into several groups whose existence was posited by quantum theory. The unique properties of particles such as the neutron and the neutrino, as well as the antiparticles that are produced in fermion pair production, demanded that special detectors be designed and constructed in order to gain evidence for their existence. Such proofs serve both to support the predictions of quantum mechanical theory and quantum field theory, and to provide information with which the theory can be revised as needed.

DETECTOR SYSTEMS
Detectors come in a variety of shapes and configurations that depend on their mode of action as much as on the type of particle they are to detect. Detectors that function on a broad scale include such simple devices as exposure badges used to monitor exposure to airborne particles and radioactive contamination, Geiger counters, and scintillation counters. Somewhat more precise detectors are used in instruments such as mass spectrometers to detect atomic and molecular ions that are formed in the device during analytical procedures. In the highly precise devices, known as particle accelerators, the particle detector systems must have the precision and sensitivity to detect individual subatomic or elementary particles. These detectors fall into two geometric classes: fixed target geometry or collider geometry.

Detectors having fixed target geometry are much like those used in mass spectrometers; the particles produced by bombarding a fixed target are directed through a magnetic field to a muon filter where they are identified by observation. The muon filter is collinear with the target so that, in effect, the particles that are produced are driven forward from the target and into the muon filter.

Detectors having collider geometry have the detection filter as a cylindrical structure surrounding the point of impact in the collision. In this geometry, particles that are produced during collision are driven into trajectories transverse to the direction of the colliding beams of particles. The magnetic enclosure of the collider geometry detectors can have either of the solenoid structure, in which the magnetic field is collinear with the beam, or the toroid structure, in which the magnetic field is orthogonal to the beams. The ATLAS detector at the CERN facility, in Switzerland, is of the toroid type, and is approximately 8 meters (25 feet) in diameter.

The high precision with which such devices are constructed is essential to ensure that the analysis of particle collisions can proceed with accurate data regarding the energy and momentum of the particles. The trajectories that are calculated from the data must agree with the mathematics of the theoretical principles involved if the particles are to be properly identified. The momentum of the particles is demonstrated in characteristic trajectories that are determined precisely by the magnetic field strength and the specific charge on the particles.

In high energy collisions that produce quarks and certain other elementary particles, partial charges may be involved rather than the whole charges that exist on electrons, positrons, protons, antiprotons and similar particles. Accordingly, imprecise control of the system would lead to erroneous interpretation of the results of an experiment. Particles can only be detected if they transfer energy to the matter of the detector, and specifically to the electrons at the surface of the matter of the detector. Electrons that gain energy in this way are ejected from the detector matter, and it is the energy and momentum of the ejected electrons that is actually measured. The momentum and energy of the particles that impacted are inferred from the properties of the ejected electrons. The calculations involved in the analysis of collision data are extremely complex, and are typically written in bra-ket notation. Overall, the structure of particle detectors represents an entire engineering science in its own right, relying heavily on both conventional and solid state electronics and materials.

APPLICATIONS

The opportunity to work with detectors of the type used at CERN and other facilities such as the Stanford Linear Accelerator (SLAC) is a rare experience reserved for established researchers. That does not mean that particle detectors are without more commonplace applications. In fact, the devices are quite common, typically functioning on a much smaller scale than the detectors used at research facilities. Many analytical devices employ the same principles of particle detection as are used in the massive detectors of large-scale devices such as CERN's Large Hadron Collider.

The applications of particle detectors are generally specific to the purpose for which they are used. The particle detection system in mass spectrometers, for example, is used just for the detection of relatively low energy particles, typically of between 50 and 100 electron-volts, since that is a typical energy used to generate the molecular ions to be detected. That also makes them ideal for detecting particles produced in atmospheric interactions with lightning, background radiation, cosmic particles that impinge on Earth's atmosphere and many other applications.

Forensic analysis and other laboratories rely a great deal on mass spectrometers and other devices that measure quantities and compositions by the production of ions. Flame ionization is another important method of identification of specific components in a material. The character of charged particle detectors also has broad scope in more mundane applications such as the electrostatic removal of dust and other materials from household air and industrial exhaust stacks. In short, particle detection is a very broad field, with many different applications well beyond those of high energy physics.

—*Richard M. Renneboog M.Sc.*

BIBLIOGRAPHY

Grupen, Claus, Boris A. Shwartz, and Helmuth Spieler. *Particle Detectors*. Cambridge: Cambridge UP, 2011. Print.

Enss, Christian. *Cryogenic Particle Detection*. 2nd ed. Berlin: Springer, 2005. Print.

Blum, Walter, Werner Riegler, and Luigi Rolandi. *Particle Detection with Drift Chambers*. Berlin: Springer, 2010. Print.

Grupen, Claus, and Irène Buvat. *Handbook of Particle Detection and Imaging*. Berlin: Springer, 2012. Print.

Kirkland, Kyle. *Physical Sciences : Notable Research and Discoveries*. New York: Facts on File, 2010. Print. Frontiers of Science.

PHASE

FIELDS OF STUDY

Acoustics; Electromagnetism; Classical Mechanics

SUMMARY

This article defines and describes phase as it relates to waves. Waves can interfere with each other to produce waves with increased or decreased amplitudes. This interference is useful for many applications, including increasing or reducing sound intensities.

PRINCIPAL TERMS

- **antiphase:** a 180-degree or π-radian phase difference.
- **constructive interference:** when two or more waves of the same phase combine to form a larger amplitude.
- **destructive interference:** when two or more waves of different phases combine to form a smaller amplitude.
- **instantaneous phase:** the time-variant angle of a sinusoidal function.
- **phase offset:** also called phase difference; the time interval or phase angle that results when one wave is ahead of or behind another.
- **pi:** the ratio of the circumference of a circle to its diameter, symbolically represented as π. Its numerical value is approximately 3.14159.
- **sinusoidal function:** a curve that is like a sine wave but experiences a shift in amplitude or phase.

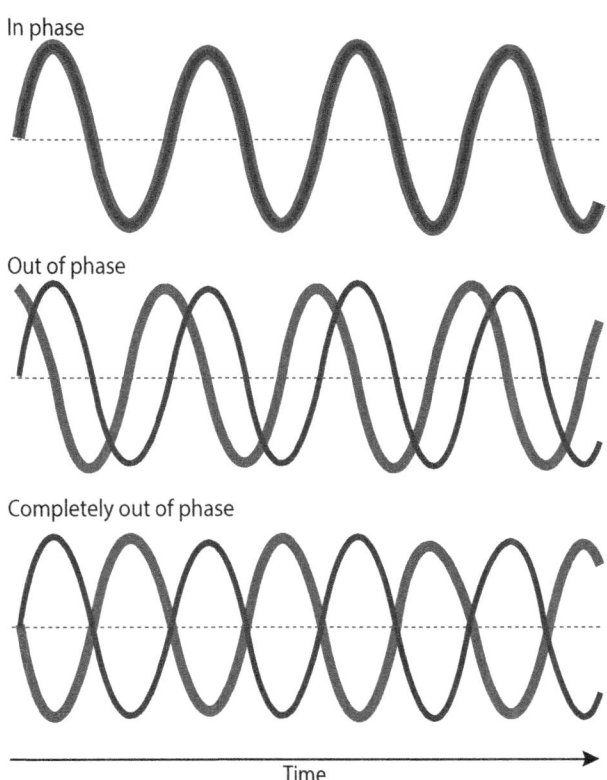

The phase of a wave is its cycle pattern. Two waves can either be in phase, sharing the same pattern, or out of phase, having different patterns. Three sets of waves are shown. One set is in phase, so that both waves reach a crest at the same time and reach a trough at the same time. One set is a little out of phase, so that the two waves crest at different times. The last set is completely out of phase, so that one wave reaches a crest as the other reaches a trough.

The Phase of a Wave

The phase of a wave denotes the current location of a wave relative to a reference point. Amplitude (A), frequency (f), and phase (φ) of a wave can be related by the sinusoidal function

$$x(t) = A \cos(2\pi f t + \phi)$$

or, if shifted,

$$y(t) = A \sin(2\pi f t + \phi)$$

where π is pi and t is time. Phase can also be used to refer to the time-variant angle ($2\pi f t + \varphi$), or instantaneous phase.

A phase difference, or phase offset, occurs when waves of the same frequency travel ahead of or behind each other. Two waves with same frequency are considered in phase with each other if they have no phase difference and out of phase if they have a phase difference. Phase offset can be expressed in degrees or radians, with 360 degrees (2π radians) indicating one wavelength.

Interference

Superposition occurs when two or more waves combine to form a wave displacement equal to the sum of the individual waves' displacements. Consider two waves traveling with the same amplitude, frequency, and wavelength in the same direction. Constructive interference occurs when the two waves are in phase with each other and produces a resulting wave with an amplitude equal to twice that of the individual waves. Destructive interference occurs when the two waves are antiphase and the waves cancel out. This total displacement of zero results because the positive displacement of one wave momentarily adds to the negative displacement of the second. When dropping stones in a pond, for example, the ripples moving outward from the stones interfere constructively in some places to produce a wave that has greater amplitude and, in other places, interfere destructively to produce an area that appears undisturbed. Destructive interference is commonly used to reduce the intensity of noise in places such as an airplane cabin.

—*Casey M. Schwarz, PhD*

Bibliography

Halliday, David, Robert Resnick, and Jearl Walker. *Fundamentals of Physics.* 10th ed. Hoboken, NJ: Wiley, 2014. Print.

Hass, Jeffrey. "What Is Phase?" *Acoustics Primer.* Center for Electronic and Computer Music, Indiana U, Bloomington, 2003. Web. 20 May. 2015.

Herman, Russell L. *A Course in Mathematical Methods for Physicists.* Boca Raton: CRC, 2014. Print.

"Interference of Waves" *Physics Classroom.* Physics Classroom, 1996–2015. Web. 20 May. 2015.

National Geographic Society. *The Science of Everything: How Things Work in Our World.* Washington: Natl. Geographic Soc., 2013. Print.

Walker, James S. *Physics.* 5th ed. San Francisco: Pearson/Addison Wesley, 2016. Print

PHONON

FIELDS OF STUDY

Quantum Physics; Classic Physics; Harmonics

SUMMARY

Phonons are the quantum equivalent in solid state vibrational states to photons in electromagnetic energy states. Phonons are density waves in solid matter, and are responsible for the transmission of sound as well as the heat capacity in the material. The minimum wavelength of phonons in a material is determined by the equilibrium distance between atoms, but the frequency is determined by the speed of sound in the particular material. Phonons are designated as acoustic or optical, and their interaction with electrons is a mechanism for superconductivity. Nanotechnology and quantum computers are expected to rely heavily on phonons.

PRINCIPAL TERMS

- **acoustics:** the study of sound; also, the qualities of a space that affect how sound is heard within that space.
- **amplitude:** a quantifying wave property measured from a point of rest to a point of maximum displacement; related to sound power and intensity of sound waves.
- **frequency:** the number of complete wavelengths that occur within one unit of time, typically expressed as hertz (Hz; cycles per second).
- **mass-spring system:** a system consisting of an elastic object connected to an object with mass.
- **quantum state:** the condition of a physical system as defined by its associated quantum attributes.
- **thermodynamics:** the study of the relationships between heat, energy, and work in a system.
- **wave propagation:** the manner in which a wave travels.

Quantum Vibrations

Phonons are a property of condensed, or solid, matter. The phonon model describes the vibrations of solids at the quantum level. Within a solid material such as a crystal the component atoms are locked in place by the electric force between ions, dipole-dipole interactions, van der Waals forces and the covalent bonds between atoms. Accordingly, vibrational motions within the array are limited. The behavior of the system can be analogized by a mass-spring system, in which the component atoms are the masses and the forces acting between them are the springs. Within such a system, the vibrational movement of one atom in a direction results in the movement of the adjacent atoms in concert. The resulting wave propagation is a phonon described mathematically either by classic vector properties or by quantum mechanical principles. The amplitude of the phonon wave is the distance by which the components are displaced from their equilibrium position. The frequency of the phonons is an important physical consideration, as longer wavelength phonons are responsible for sound conduction in the material. Shorter wavelength phonons are associated with the atomic-scale vibrations recognized as heat and the property of heat capacity in thermodynamics.

ACOUSTIC AND OPTICAL PHONONS

The vibrational energy levels of the lattice structure of solid matter require that a specific amount of energy be supplied in order to raise the vibration of the lattice to the next energy level, or quantum state. When the atoms move in phase with each other, the phonons are termed acoustic, because the waves are density compression waves, like those of sound waves in air in the science of acoustics. Such waves have only a longitudinal direction component. When the motions are not in phase with each other, and have transverse direction components as well as the longitudinal component, they have interaction with electromagnetic waves such as infrared light. They are accordingly termed optical phonons. The interaction

> **SAMPLE PROBLEM**
>
> The C—C bond distance in diamond is 154.4 picometers, and in graphite is 141.5 picometers. Calculate the minimum phonon wavelength for both materials.
>
> **Answer:**
>
> The minimum phonon wavelength is twice the equilibrium separation of atoms in the crystal lattice. Therefore, the minimum phonon wavelength for diamond is
>
> $$2 \times 154.4 \text{ pm} = 308.8 \text{ pm, or } 3.08 \times 10^{-10} \text{ m}$$
>
> and for graphite is
>
> $$2 \times 141.5 \text{ pm} = 283 \text{ pm, or } 2.83 \times 10^{-10} \text{ m}.$$

of phonons with electrons is also believed to be the mechanism whereby a material has the properties of a superconductor. The vector description of phonons generates sets of phonon wave vectors with various wave numbers that determine Brillouin zones. The set of phonon wave vectors having the smallest magnitude of the wave number define a region termed the first Brillouin zone. The minimum possible phonon wavelength is twice the equilibrium separation between atoms. The frequency of the phonon is determined by the velocity of sound in the specific material.

Basic Material Science

The phonon model is of increasing importance as nanotechnology develops and becomes more mainstream. At the nanometer scale of nanotechnology, interaction between electrons and phonons has a significant influence on the overall functioning of nanoscale devices. On a macroscopic scale, superconductivity is a difficult condition to achieve, but in the phonon model, it is the result of the interaction between electrons and phonons. It is conceivable that nanodevices may incorporate structures that are functionally superconducting at that scale. Phonons also have the ability to transport information between two points, an important aspect in the development of quantum computers. Production of solar energy can also be enhanced by the development of solar cells that harness the photoelectric effect with phonon-electron interactions.

—*Richard M. Renneboog M.Sc.*

Bibliography

Wolfe, James P. *Imaging Phonons: Acoustic Wave Propagation in Solids*. Cambridge: Cambridge University Press, 2005. Print.

Lou, Liang-fu. *Introduction to Phonons and Electrons*. Singapore; River Edge, N.J.: World Scientific, 2003. Print.

Hamaguchi, Chihiro. *Basic Semiconductor Physics*. 2nd ed. Berlin: Springer Berlin, 2014. Print.

Kitamura, Toyoyuki. *Liquid Glass Transition: A Unified Theory from the Two Band Model*. Chennai: Elsevier, 2013. Print.

Ridley, B.K. *Quantum Processes in Semiconductors*. 5th Rev ed. Oxford: Oxford University Press, 2013. Print.

PHOTON

FIELDS OF STUDY

Particle Physics; Quantum Physics

SUMMARY

Photons are massless elementary particles that are the smallest possible units of electromagnetic radiation. Einstein proposed that light traveled in concentrated bundles called photons to explain the photoelectric effect. Electrons in atoms and molecules make transitions between quantum energy levels by absorbing or emitting precise amounts of energy as photons.

PRINCIPAL TERMS

- **absorption coefficient:** a value characteristic of a particular medium that represents the amount of light or sound it absorbs from a wave passing through it.

- **electromagnetic spectrum:** the full range of electromagnetic radiation, sorted into segments with similar properties by wavelength; x-rays occupy one of these segments.
- **hertz:** the SI unit of frequency; one hertz (Hz) is equal to one cycle (complete orbit) per second.
- **photoelectric effect:** a phenomenon that describes the emission of electrons from matter (typically metal) upon exposure to electromagnetic radiation.
- **quantum mechanics:** the branch of physics that deals with matter interactions on a subatomic scale, based on the concepts that energy is quantized, not continuous, and that elementary particles exhibit wavelike behavior.
- **quantum state:** the condition of a physical system as defined by its associated quantum attributes.
- **visible light:** electromagnetic radiation that human eyes can see, with a wavelength between 4×10^{-7} to 7×10^{-7} meters.
- **wave-particle duality:** the idea that a particle can behave as either a particle or a wave, depending on how and when it is being observed.

SAMPLE PROBLEM

A sample of sodium metal is prepared under vacuum to prevent surface oxidation and tested for the photoelectric effect. When irradiated with light at 7×10^{14} Hz (in the yellow-green part of the visible spectrum), the stopping potential V_O is measured to be 1.0 volts. Calculate the kinetic energy of the photoelectrons.

Answer:

The kinetic energy of the photoelectrons is given by

$$E_K = eV_O,$$

where $e = 1.602192 \times 10^{-19}$ coulomb. Substitute the values of e and V_O to obtain

$$E_K = (1.602192 \times 10^{-19} \text{ coul})(1.0 \text{V})$$

$$= 1.602192 \times 10^{-19} \text{ electron-volts (eV)}$$

PLANCK'S QUANTUM HYPOTHESIS

In 1900 Max Planck postulated that the energy states of an oscillator must be integral multiples of the frequency v of the electromagnetic radiation that it emits and a constant (h) that is now known as Planck's constant. In accord with this hypothesis, the smallest energy difference that is allowed for such an oscillator is given by the equation

$$E = hv.$$

This also defines the photon, or quantum, of energy corresponding to the particular frequency v, although Planck did not use that term. Bohr was able to utilize Planck's quantum principle to account for the formation of the Balmer series of absorption lines that appear in the emission and transmission spectra of hydrogen atoms. These appear as distinct black lines in the spectrum of visible light that passes through the gas. The realization was that the lines were separated by multiples of Planck's constant and could be attributed to the transfer of electrons between energy levels within the atoms. This is the basis of quantum mechanics as it describes the energy states of electrons in atoms.

Electrons in atoms and molecules can transfer from one energy level to another only by the absorption or emission of photons of the appropriate energy. The energy required for such a transmission may be from any part of the electromagnetic spectrum, but is very specific to the particular transition. A transition that requires a photon with energy corresponding to a wavelength of 6.4628×10^{14} hertz will not be brought about by a photon with a different frequency.

WAVES AND/OR PARTICLES

The concept of the photon evolved directly from the wave-particle duality of light, demonstrated in part by Young's "double slit experiment" and in part by the ability to describe aspects of the behavior of light using the mathematics that describe particle behavior. The photon unites these as components of quantum mechanics by the interaction of electromagnetic radiation with electrons in atoms. The specific wavelengths at which atoms and molecules absorb energy to effect the transition of electrons from one quantum state to another are commonly used as an analytical technique for the identification and quantification of materials. The absorption coefficient of a substance is particularly useful in determining the concentration or amount of material present in a solution. A great many chemical processes occur by photochemical reaction processes in which molecules absorb photons of the appropriate energy to promote electrons to a higher energy state that facilitates a particular type of reaction.

THE PHOTOELECTRIC EFFECT

The photon is perhaps most important in a physical process in the photoelectric effect, in which absorption of a photon of the appropriate energy causes the release of an electron. These electrons produce an electric current in an appropriate circuit. To explain this effect, Einstein proposed the energy in a beam of light traveled through space in concentrated bundles that he termed "photons." The energy of the electrons that are emitted in this way can be measured by applying an opposing voltage to the circuit. At a voltage called the stopping potential (V_O), the current produced by the photoelectrons is balanced and no current flows. The kinetic energy of the electrons that are emitted is determined by multiplying the stopping potential by the charge of an electron, as follows:

$$E_K = eV_O.$$

In practice, the photoelectric effect does not occur with a material below a specific frequency of the incident light, and it is independent of the intensity, or brightness, of the light. Milliken studied the photoelectric effect extensively, and in 1923 was awarded the Nobel Prize in Physics for his work.

PHOTONS AND ELECTRONS

Modern technology is structured on the movement of electrons in electronic devices and electromagnetic radiation for communication. The influx of photons and charged particles from the Sun and other extraterrestrial sources influence many aspects of technology and communications. In the atmosphere, absorption of photons by airborne molecules produces numerous free radical processes, including the formation of the ozone that serves to protect the surface of Earth from fatal exposure to ultraviolet radiation. Currently, research is progressing toward the development of devices that process digital information as photons rather than electrons.

—*Richard M. Renneboog M.Sc.*

BIBLIOGRAPHY

Creath, Katherine, Chandrasekhar Roychoudhuri, and Al F. Kracklauer. *The Nature of Light: What Is a Photon?* Boca Raton: CRC, 2008. Print.
Keller, Ole. *Light - The Physics of the Photon*. Boca Raton: CRC, 2014. Print.
Jauch, Josef M. *Theory of Photons and Electrons*. Berlin [S.l.]: Springer-Verlag, 2012. Print.
Lewerenz, Hans-Joachim. *Photons in Natural and Life Sciences: An Interdisciplinary Approach*. Berlin: Springer, 2012. Print.

PITCH

FIELDS OF STUDY

Acoustics

SUMMARY

This article describes pitch as it relates to sound, musical tones, and frequency. Frequency and pitch are related but are not the same. Frequency is an absolute value, while pitch is more subjective and may depend on other tone qualities.

PRINCIPAL TERMS

- **definite pitch:** a sound in which the pitch is easily detected.
- **frequency:** the number of times a repeated event happens over time.
- **harmony:** the combination of two or more different musical notes played simultaneously that create a pleasant sound.
- **indefinite pitch:** a sound in which the pitch is not easily detected.
- **loudness:** a quality of sound related to amplitude.

- **scale:** in music, a group of notes arranged by pitch or frequency.
- **timbre:** the quality of a tone or sound that is unique from its loudness and pitch and allows one to distinguish between different sound sources.
- **tone:** a vocal or musical sound with a specific pitch and loudness.

PITCH, LOUDNESS, AND QUALITY

Pitch is the quality of sound related to its perceived highness or lowness. Pitch is associated with frequency but is not the same. Low pitches are associated with low frequencies, and high pitches with high frequencies. However, frequency refers to an absolute value, while pitch is subjective, depending on sound perception. The frequency of a sound wave is determined and quantified by its number of vibrations in a period of time and is measured in hertz (Hz).

Pitch relationships are perceived according to fixed ratios. Musical tones can be ascribed to relative positions on a musical scale based on frequency. The interval between the first and last notes on a scale is referred to as an "octave." When two frequencies have a 2:1 ratio, they sound as though they are an octave apart. In other words, when a frequency is halved or doubled, one hears a different octave. When listening to harmony, one usually hears a single pitch related to the lowest frequency, although some listeners can distinguish separate pitches.

Pitch also depends on other sound attributes such as loudness and timbre. If the loudness of a low pitch increases, it will be perceived as becoming lower. If the loudness of a high pitch increases, it will be perceived as becoming higher. Timbre, or tone color, allows a person to distinguish one musical sound from another of identical pitch and loudness.

DEFINITE AND INDEFINITE PITCH

A listener can easily identify a particular pitch when a sound or note is of definite pitch. A definite pitch has a sound that is steady with a frequency that is measurable. Indefinite pitches are nearly impossible for a listener to discern as a certain pitch. Percussion instruments such as snare drums are indefinite pitch

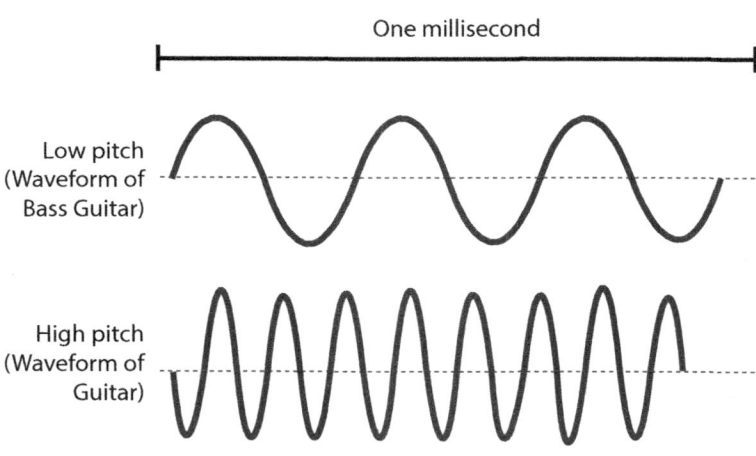

A sound wave's frequency determines its pitch. Two waves of different pitch across the same time span will go through a different number of cycles. Low-pitch sounds go through fewer cycles compared to high-pitch sounds, which have higher frequencies and go through more cycles.

instruments because they produce sounds lacking notes with clear pitches.

PITCH AND FREQUENCY

Pitch and frequency describe the same phenomenon but are not interchangeable. Pitch depends on the listener's interpretation of a sound's frequency. A real-world example relating to both pitch and frequency is band-pass filters. Band-pass filters are used in wireless transmitters and receivers and allow certain frequency ranges to pass while rejecting others and maintaining quality.

—*Casey M. Schwarz, PhD*

BIBLIOGRAPHY

Ballora, Mark. "Pitch vs. Frequency." Electronic Musician. NewBay Media, 1 Sept. 2006. Web. 3 June 2015.

Halliday, David, Robert Resnick, and Jearl Walker. *Fundamentals of Physics.* 10th ed. Hoboken, NJ: Wiley, 2014. Print.

Nave, Carl R. "Pitch." HyperPhysics. Dept. of Physics and Astronomy, Georgia State U, 2012. Web. 3 June 2015.

"Pitch and Frequency." Physics Classroom. Physics Classroom, 1996–2015. Web. 3 June 2015.

Spilsbury, Louise, and Richard Spilsbury. *Why Can't I Hear That? Pitch and Frequency.* Chicago: Heinemann, 2014. Print.

Walker, James S. *Physics.* 5th ed. San Francisco: Pearson/Addison Wesley, 2016. Print

PLANCK'S CONSTANT

FIELDS OF STUDY

Quantum Physics; Thermodynamics; Nuclear Physics

SUMMARY

Planck's constant is deemed to be a fundamental and universal constant on a par with the speed of light. Planck proposed the constant as a mathematical trick to solve certain problems that did not have a classical solution. Extension of his postulate by other physicists, notably De Broglie and Einstein, brought Planck's constant to be the foundation relationship of quantum theory.

PRINCIPAL TERMS

- **blackbody radiation:** radiation emitted by a body solely as a result of its temperature, regardless of its composition or any previously absorbed radiation.
- **line spectrum:** the lines of color that represent the characteristic frequencies at which atoms emit electromagnetic radiation.
- **photoelectric effect:** a phenomenon that describes the emission of electrons from matter (typically metal) upon exposure to electromagnetic radiation.
- **photon:** a massless elementary particle that is the smallest possible unit, or quantum, of light and other electromagnetic radiation.
- **Planck's law:** a mathematical description of the amount of radiation emitted at different frequencies by a blackbody at a given temperature.
- **Stefan-Boltzmann law:** a mathematical description of the total radiant energy emitted by a blackbody, relating it to the temperature of the blackbody raised to the fourth power.
- **ultraviolet (UV) radiation:** electromagnetic radiation with more energy than visible light but less than x-rays, in the wavelength range of 4×10^{-7} to 1×10^{-7} meters.
- **ultraviolet catastrophe:** the erroneous prediction, based on the laws of classical physics, that a blackbody would emit an infinite amount of energy at short wavelengths, starting around the ultraviolet region.

BLACK BODY RADIATION

Before the quantum theory was proposed physicists sought explanations for observed properties that did not yield to analysis by classical methods. One such feature was the line spectrum of different gases. This was the appearance of black lines in the spectrum of light passed through the gas. The lines appeared as regular patterns, but no theoretical model of atomic structure could account for them. Another physical enigma was the so-called blackbody radiation, which was dependent only on the temperature of the body.

To solve the mathematics of the relationship Planck postulated that the smallest unit of light, which he called a quantum, was equal to the frequency of the light multiplied by some constant. It was a mathematical trick that Planck employed strictly to be able to solve the mathematics of the line spectrum and of blackbody radiation in a way that would avoid the ultraviolet catastrophe that plagued the other attempted solutions of the day. Without the necessary and correct interpretation of the physics, the inevitable conclusion of the Stefan-Boltzman law, which related the emitted energy to the temperature raised to the fourth power.

Logically, as the wavelength of light emitted by a hot body decreased (as the temperature increased), by the time the body emitted ultraviolet light it would be radiating an infinite amount of energy, which is impossible, hence the term ultraviolet catastrophe. The constant that he postulated came to be known as Planck's constant, having the value of 6.6262×10^{-34} joule.sec and his mathematical description of the energy emitted in black body radiation is known as Planck's law.

> **SAMPLE PROBLEM**
>
> Calculate the energy of a photon of light having a wavelength $\lambda = 540$ nanometers (5.40×10^{-7} m).
>
> **Answer:**
> The wavelength of light is related to its frequency by the speed of light, $c = 2.99 \times 10^8$ m.sec^{-1}, as follows:
>
> $$c = f\lambda.$$
>
> Therefore,
>
> $$f = c / \lambda$$
>
> $$= (2.99 \times 10^8 \text{m.sec-1}) / (5.40 \times 10^{-7}\text{m})$$
>
> $$= 5.54 \times 10^{14} \text{ sec}^{-1}$$
>
> and
>
> $$E = hf$$
>
> $$= (6.6262 \times 10^{-34} \text{ joule.sec})(5.54 \times 10^{14} \text{ sec}^{-1})$$
>
> $$= 3.67 \times 10^{-19} \text{ joule}.$$

Photoelectric Effect

Another puzzling phenomenon of Planck's time was the photoelectric effect, in which a metal irradiated with ultraviolet light was observed to emit electrons. An explanation of this phenomenon was proposed by Einstein in 1905 based on the postulate that the absorption of a photon of light of the correct quantum of energy cold account for the ejection of the electron from the metal atom. Ultraviolet radiation of a particular frequency could be equated with the energy of the electron through Planck's constant, as

$$E = hf.$$

The concept was further developed by De Broglie, who subsequently postulated that the quantum of energy for any wavelength was proportionated as the product of Planck's constant and the corresponding frequency. This is the fundamental principle of all aspects of quantum mechanics.

Fundamental Physics

Planck's constant is considered to be a universal constant on a par with the speed of light in quantum mechanics. It is an integral component of the mathematical basis that has become the standard model of particle physics, and appears in essentially all of the necessary calculations. As such it is used in all of the physical sciences at a very basic level.

—*Richard M. Renneboog M.Sc.*

Bibliography

Cheng, Ta-Pei. *Einstein's Physics : Atoms, Quanta, and Relativity Derived, Explained, and Appraised.* Oxford: Oxford University Press, 2014. Print.

Kafatos, Minas C., and Robert Nadeau. *The Conscious Universe: Part and Whole in Modern Physical Theory.* New York: Springer, 1990. Print.

Uzan, Jean-Philippe, and Bénédicte Leclercq. *The Natural Laws of the Universe: Understanding Fundamental Constants.* Berlin; Heidelberg; New York: Springer: Praxis, 2008. Print.

Susskind, Leonard, and Art Friedman. *Quantum Mechanics: The Theoretical Minimum.* New York: Basic, 2014. Print.

Al-Khalili, Jim. *Quantum: A Guide for the Perplexed.* London: Phoenix, 2012. Print.

Atkins, Peter, and Julio de Paula. *Atkins' Physical Chemistry.* 10th ed. Oxford: Oxford *University Press,* 2014. Print.

POTENTIAL ENERGY

FIELDS OF STUDY

Classical Mechanics; Electromagnetism; Nuclear Physics

SUMMARY

There are many kinds of potential energy present in an object at once. These forms of potential energy can come from varied sources. They all affect the object's behavior in different ways, depending on their properties. Some of these potential energies include electric, gravitational, and elastic energy. This article defines these and provides examples on each and their mathematical descriptions.

PRINCIPAL TERMS

- **displacement:** difference in the position of an object from its initial position to its final position, regardless of the path it takes.
- **electric potential energy:** energy that is present in particles due to their charge and closeness to other charges.
- **force field:** the effect of a field force and a function of the relative position of the object from the force field source; the object and the source do not need to be in physical contact for this effect to occur.
- **gravitational force:** the pull that objects exert on each other due to their masses and the separation between their centers of mass.
- **Hooke's law:** the pull or push on objects attached to springs due to the relative stiffness of the spring and the amount of stretch or compression of the spring.
- **joule:** a derived unit of energy from base units of the International System of Units (SI), or metric system, equal to a kilogram times one meter per second squared.
- **kinetic energy:** energy due to any kind of motion, be it rotation, vibration, or translation.
- **law of conservation of energy:** the total amount of energy in a closed system is always the same; therefore, energy can neither be created nor destroyed.

Storing Energy

Potential energy is defined as the energy that an object possesses due to its physical position. Using this energy to do work converts it into another type of energy, such as kinetic energy. Scottish physicist William John Macquorn Rankine (1820–72) proposed the term "potential energy" in 1853, based on the research of English physician and physicist Thomas Young (1773–1829). Building on previous work from other physicists and mathematicians, Young had mathematically defined energy as the product of an object's mass (m) and the square of its velocity (v). Though partially correct, this definition was later expanded and corrected by French Italian mathematician Joseph-Louis Lagrange (1736–1813), who included a factor of one-half before the mass and velocity. The energy Young and Lagrange described was in fact kinetic energy, but no such distinction was made at the time. As a result, Rankine used the word "potential" to distinguish it from what he called "actual, or sensible energy." The term "kinetic" was later applied by William Thomson, Lord Kelvin (1824–1907).

All forms of energy, whether kinetic, potential, or other, are measured in joules (J). Named after English physicist James Prescott Joule (1818–89), the joule is part of the International System of Units (SI). It is a derived unit, comprising one or more of the seven SI base units. One joule is equal to the energy required to apply a force of one newton (N)—or, because the newton is also an SI derived unit, one kilogram-meter per second squared ($kg \cdot m/s^2$)—over a distance of one meter (m). Thus, in SI base units, 1 J = 1 $kg \cdot m/s^2$.

Among the most important aspects of potential energy is its ability to move objects from one point to another. Physicists define the difference between an object's starting and end points, regardless of the path taken, as the object's displacement. If a plane flies from New York to London, whether it travels east across the Atlantic Ocean or west across North America, the Pacific Ocean, Asia, and Europe, in both cases the plane will have the same displacement, even though the distance traveled was different. When potential energy is used to displace an object by any amount, it is said that work was done on that object.

THE POTENTIAL OF FORCES

As the definition of energy was expanded over the years, different forms of potential energy were identified for different forces. A form of potential energy exists for every type of conservative force. A conservative force is one that conserves mechanical energy by turning potential energy into kinetic energy, or vice versa, rather than some other form of energy. Examples of conservative forces include gravitational force and the restoring force of a spring. Friction is an example of a nonconservative force, because it converts mechanical energy into heat.

When an object is held above Earth's surface, the gravitational force is acting on it in order to bring it back down. Gravitational force (F_g) is the pull that objects exert on each other due to their masses. It is defined by the equation

$$F_g = \frac{Gm_1m_2}{r^2}$$

where G is the universal gravitational constant, m_1 and m_2 are the masses of the objects, and r is the distance between their centers of mass. The center of mass is the weighted average position of the center of the particles that make up an object. The value of the universal gravitational constant is 6.67384×10^{-11} m^3/kg·s^2.

Like other types of force, the gravitational force is a distance or field force. This means that the objects do not need to be in contact for the force to take effect. Field forces are important because they produce force fields around the objects that create them. In this case, the field produced by the gravitational force

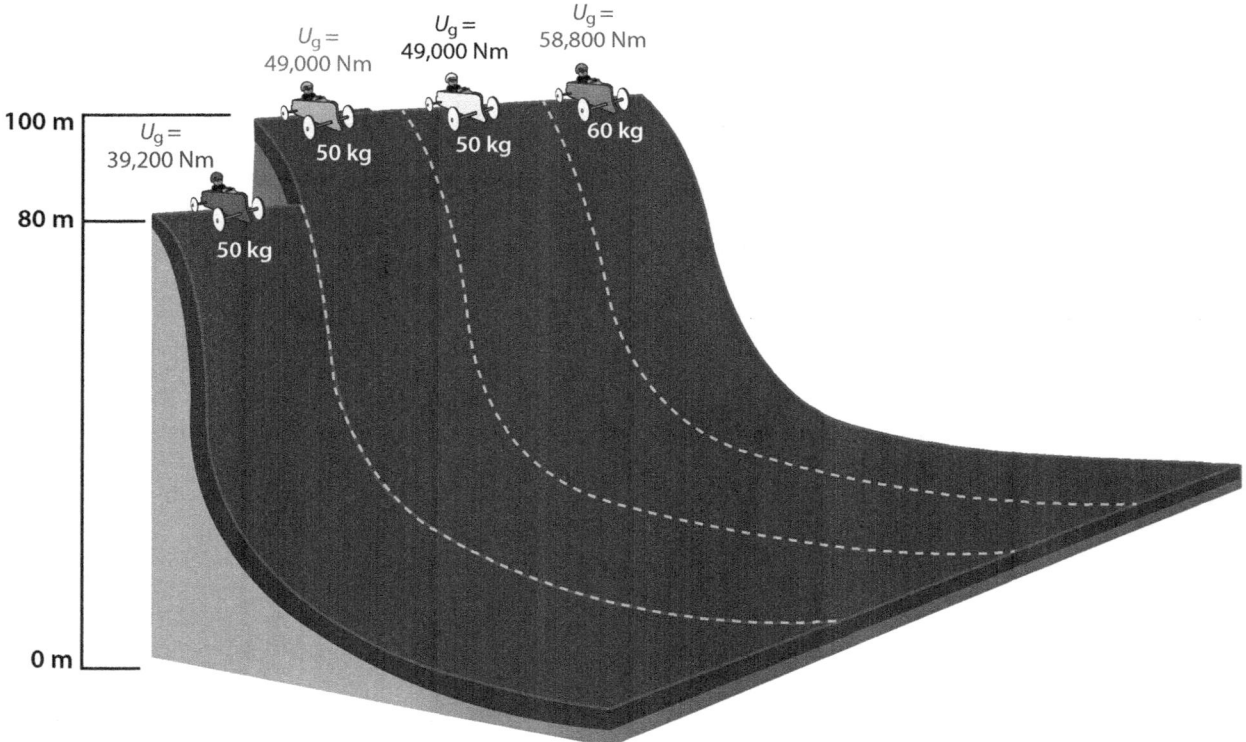

$$U_g = mgh$$

Gravitational potential energy is one type of potential energy. An object's gravitational potential energy is calculated by multiplying the object's mass (m) by its height (h) and the acceleration due to gravity ($g = 9.8$ m/s^2). Four soapbox derby cars are at the top of their assigned slopes. The red car is on a shorter slope, so it has a smaller potential energy than the other cars. The green car has a greater mass, so it has a larger potential energy than the others. Both the blue and the white cars have the same mass and are starting at the same height, so they have the same potential energy.

is known as gravity, or the gravitational field. On Earth, it has a value of about 9.8 meters per second squared.

The spring force, or elastic force, is called a restoring force because when a spring attached to a mass is either extended or compressed, the spring force acts to restore the mass to its original or equilibrium position. The mass must be in contact with the spring for this force to act on it, so it is not a field force and hence cannot be described by a force field. In 1660, English physicist Robert Hooke (1635–1703) noticed that the stiffer the spring, the greater the restoring force. He quantified spring stiffness as the spring constant (k), which is measured in newtons per meter and is unique to each spring. In 1678, he described the linear relationship between the spring force and the distance of the mass on the spring from equilibrium (x) and defined the spring force (F_s) as the product of k and x. This definition, now known as Hooke's law, is represented as

$$F_s = -kx$$

The negative sign in this equation represents the restoring nature of the force. It is acting in the opposite direction of the displacement.

Gravitational and Elastic Potential Energies

One of the main types of potential energy is gravitational potential energy. An object in a gravitational field acquires gravitational potential energy by changing position. The gravitational potential energy (U_g) of an object within another object's gravitational field is defined by

$$U_g = -\frac{Gm_1m_2}{r}$$

Again, G is the gravitational constant, m_1 and m_2 are the objects' masses, and r is the distance between their centers of mass. This general equation works in every situation. However, when an object is within Earth's gravitational field near the planet's surface, this equation can be simplified, because the force of Earth's gravity (g) is already known:

$$U_g = mgh$$

SAMPLE PROBLEM

A parallel-plate capacitor has a capacitance (C) of 6×10^{-6} F. The plates are separated by a distance (d) of 3 mm and produce an electric field (E) with a magnitude of 9 N/C. What is the electric potential energy (U_e) stored by the capacitor?

Answer:

To calculate the energy stored in a capacitor, its capacitance and the electric potential difference between its plates must be known. Given the electric field and the distance, the electric potential difference can be obtained. First, convert the distance from millimeters to meters by dividing by 1,000:

$$\frac{3\,\text{mm}}{1{,}000} = 0.003\,\text{m}$$

Use this value and the electric field to calculate the potential difference:

$$\Delta V = Ed$$

$$\Delta V = (9\,\text{N/C})(0.003\,\text{m})$$

$$\Delta V = 0.027\,\text{N·m/C} = 0.027\,\text{V}$$

Then use this value and the capacitance to calculate the electric potential energy. Recall that F = C/V and V = J/C.

$$U_e = \frac{1}{2}C\Delta V^2$$

$$U_e = \frac{1}{2}(6\times10^{-6}\,\text{F})(0.027\,\text{V})^2$$

$$U_e = \frac{1}{2}(6\times10^{-6}\,\tfrac{C}{V})(0.027\,\text{V})(0.027\,\text{V})$$

$$U_e = 2.187\times10^{-9}\,\text{C}\times\text{V}$$

$$U_e = 2.187\times10^{-9}\,\text{J}$$

There is 2.187×10^{-9} joule of electric potential energy stored in the capacitor.

Here, m is the mass of the object being affected and h is its height above Earth's surface.

The potential energy associated with the spring or elastic force is called elastic potential energy (U_s). Like the spring force, it is also a function of the spring constant and the distance of the mass from its equilibrium position:

$$U_s = \frac{1}{2}kx^2$$

ELECTRIC POTENTIAL ENERGY

Another form of potential energy is electric potential energy. Electric potential energy derives from the electrostatic force, which is a conservative force. The electrostatic force is what causes two opposite electric charges to attract one another and two like electric charges to repel. It was first described mathematically by French physicist Charles-Augustin de Coulomb (1736–1806) in 1785. Coulomb defined the electrostatic force (F_e) between two electrically charged particles as a function of the charge of each particle (q_1 and q_2), the distance between them (r), and the electric force constant (k_e):

$$F_e = \frac{k_e |q_1||q_2|}{r^2}$$

This equation is known as Coulomb's law. The electric force constant, also called Coulomb's constant, is about 8.988×10^9 newton-meters squared per coulomb squared (N·m²/C²). The coulomb is the SI derived unit of electric charge, equal to one angstrom-second (A·s). The angstrom, an SI base unit, is used to measure electric current.

Like the gravitational force, the electrostatic force is a field force. This force produces a field called the electric field. Just as an object's gravitational field is dependent on its mass, the magnitude of an electric field (E) is dependent on the charge that generates it. The value of E is defined as the force that would be applied to a charged test particle within the field. It is measured in newtons per coulomb (N/C). The equation for determining E at a given point is derived from Coulomb's law, where one of the two charged particles is treated as the particle generating the field (q) and the other is considered to be the test particle:

$$E = \frac{k_e |q|}{r^2}$$

The value of r remains the same. Here, it represents the distance of the test particle from that generating the field.

Because the electrostatic force is a conservative force, a charge within an electric field can acquire electric potential energy. Notably, both the gravitational and electrostatic forces are equal to a constant multiplied by the basic property of the objects that are responsible for the force, the product of which is then divided by the square of the distance between them. This can be applied to the equation for electric potential energy as well. Take the general equation for gravitational potential energy, remove the negative sign (which is simply a convention indicating that gravitational force is attractive), then replace the masses m_1 and m_2 with the charges q_1 and q_2 and the universal gravitational constant (G) with the electric force constant (k_e):

$$U_e = \frac{k_e q_1 q_2}{r}$$

The available amount of electric potential energy affects everyone. The electricity that powers buildings and appliances is stored in miniature devices and circuit elements. Some of these small elements are called capacitors. A capacitor can be most simply defined as two parallel plates separated by some distance, each having an equal but opposite electrical charge. Between the plates is some type of insulator, such as air, a vacuum, or a poor conductor. An electric field is produced between the plates. The electric potential (V) of a charge at a given point within the field is equal to the electric potential energy (U_e) of the charge divided by the charge itself (q):

$$V = \frac{U_e}{q}$$

Electric potential is different from electric potential energy. It is the available amount of electric potential energy that a charged particle can obtain, equal to electric potential energy per unit charge. It is measured in volts (V), an SI derived unit equal to one joule per coulomb (J/C).

However, electric potential is not meaningful on its own. In order to obtain a useful quantity, one must measure the change in electric potential between two points—or, in a capacitor, between the two plates. This quantity is called potential difference, better known as voltage, and is also measured in volts. Using the magnitude of the electric field (E) and the distance between the plates (d), the potential difference (ΔV) between the plates can be calculated:

$$\Delta V = Ed$$

The ability of a capacitor to store charge is proportional to the potential difference between the plates. This ability, called capacitance, is measured in farads (F). One farad is equal to one coulomb of electric charge per volt (C/V).

Capacitors store electric potential energy in a manner analogous to how springs store elastic potential energy. If a mass attached to a spring is pulled so that the spring is extended, the mass is attracted to its equilibrium position but cannot return, resulting in potential energy being stored in the mass. Similarly, because the plates of a capacitor have opposite charges, they are attracted toward each other, but because there is no connection between the plates, they cannot. As with gravitational force, the equation for the electrical potential energy of a capacitor resembles that for the elastic potential energy of a mass on a spring. Simply replace the spring constant with the capacitance (C) and the distance from equilibrium with its potential difference (ΔV) to obtain the electric potential energy (U_e):

$$U_e = \frac{1}{2} C \Delta V^2$$

Electric energy, like all other forms of energy, must obey the law of conservation of energy. No matter what processes take place within a system, the sum of the energy that system has at the start must be equal to the sum of its energy at the end. If one form of energy is expended, it must become a different form of energy. As an electron moves toward a proton, it loses electric potential energy but gains the same amount of kinetic energy, causing it to speed up.

Energy and the Future

Understanding potential energy is a problem that everyone faces. Electricity is everywhere, whether produced by natural or artificial means. Lightning is an example of how electrical energy is stored and released. Computers, cell phones, and televisions all have capacitors in them that store charges and electric potential energy. An understanding of potential energy is necessary for many other areas of science and technology as well. Aircraft and spacecraft must be able to overcome the gravitational influence of Earth. To maintain a constant flow of electricity to houses, electric companies must calculate how much electric energy to release during different times of the day. Solar energy systems must store energy obtained from solar cells during the day in batteries so it can be used at night. In these ways and many others, potential energy plays a major role in people's daily lives.

—*Angel G. Fuentes, MS*

Bibliography

Giambattista, Alan, and Betty McCarthy Richardson. *Physics*. 3rd ed. New York: McGraw, 2015. Print.

Henderson, Tom. "Electric Field and the Movement of Charge." *Physics Classroom*. Physics Classroom, 1996–2015. Web. 7 May 2015.

Khan, Sal. "Conservation of Energy." Khan Academy. Khan Acad., n.d. Web. 7 May 2015.

Khan, Sal. "Electric Potential Energy." Khan Academy. Khan Acad., n.d. Web. 7 May 2015.

Nave, Carl R. "Energy Stored on a Capacitor." HyperPhysics. Georgia State U, 2012. Web. 7 May 2015.

Rankine, William John Macquorn. "On the General Law of the Transformation of Energy." Miscellaneous Scientific Papers: By W. J. Macquorn Rankine, CE, LLD, FRS. Ed. W. J. Millar. 1881. Lexington: Elibron, 2005. 203–8. Print.

Simanek, Donald E. "Conservation Laws." A Brief Course in Classical Mechanics. Lock Haven U, Jan. 2011. Web. 7 May 2015.

Young, Hugh D., Roger A. Freedman, A. Lewis Ford, and Francis Weston Sears. *Sears & Zemansky's College Physics*. 14th ed. Boston, Mass: Pearson, 2016. Print.

POWER

FIELDS OF STUDY

Classical Mechanics

SUMMARY

Power is the rate of work performed or energy consumed. Work, in turn, is the displacement of an object by the application of force. When speaking of electric power, "work" refers to the movement of a charged particle across an electric field. The International System of Units (SI) unit for power is the watt, which is equal to one joule of work done or energy transferred per second. Nonstandard units such as horsepower are sometimes used instead.

PRINCIPAL TERMS

- **current:** the rate at which an electric charge, usually in the form of electrons, moves through a wire or other conductive material.
- **force:** a change in the motion of an object caused by its interaction with another object.
- **horsepower:** an alternative unit of power, in theory based on the amount of power an average horse can produce; commonly used to describe engines.
- **mechanical advantage:** the amplification of force using a tool, device, or machine, such as a lever or pulley.
- **simple machine:** a simple mechanical device that redirects or amplifies force.
- **voltage:** the difference in electric potential between two points, measured in volts; electric current flows naturally from the higher-voltage point to the lower-voltage point.
- **watt:** the International System of Units (SI) unit of power, equal to one joule of work or energy expended per second.
- **work:** the displacement of an object as a result of the application of force.

WHAT IS POWER?

In physics, power is the rate at which work is performed or energy is consumed over time. The International System of Units (SI) unit for measuring power is the watt (W), which is equal to one joule of energy consumed per second (J/s). Watts are a familiar unit, found on many electric devices to denote their power consumption over time. However, they have a much broader use within physics.

In some instances, particularly when dealing with engines, horsepower may be used instead of watts. Horsepower is a non-SI unit that is based on the ability of an average horse to do work over time. It is poorly standardized, with different types of horsepower equating to different measurements in watts. For example, one metric horsepower is equal to approximately 735.5 watts, one mechanical horsepower is approximately 745.7 watts, and one electric horsepower is exactly 746 watts. The most extreme difference is boiler horsepower, one of which is equal to 9,809.5 watts.

POWER, WORK, AND ENERGY

Power is simply a measure of work done over time. Work is a way of quantifying the successful application of force or transfer of energy. The relationship between power (P), work (W), and time (t) can be expressed as follows:

$$P = \frac{W}{t}$$

To fully understand the factors that go into calculating power, it helps to break down this basic formula into the smallest possible component pieces, beginning with work.

Work is the product of the force (F) acting on an object and the displacement (s) of the object from its original position as a result of that force. (Displacement is the shortest distance between an object's initial position and its final position, regardless of the path taken between the two.) Assuming that the force in question is the source of all displacement, work can be calculated as follows:

$$W = Fs.$$

A force, meanwhile, is any interaction that changes the motion of an object—in simple terms, a push or a pull. Force is quantified in terms of the mass (m)

of the object experiencing the force and its resulting acceleration (a):

$$F = ma.$$

Acceleration is a measure of the change in velocity (Δv) of an object over a period of time (t):

$$a = \frac{\Delta v}{t}$$

These formulas for force and work can be plugged back into the original formula for power, resulting in an equation that defines power in terms of mass, velocity, displacement, and time:

$$P = \frac{W}{t}$$

$$P = \frac{Fs}{t}$$

$$P = \frac{mas}{t}$$

$$P = \frac{m(\frac{\Delta v}{t})s}{t}$$

$$P = \frac{m(\Delta v)s}{t^2}$$

Note that this equation assumes that the force that ultimately generates the power is entirely responsible for the object's displacement. Real-world calculations involving work, and therefore power, must account for any difference between the direction of the force and the overall direction of movement.

Machine Power

Machines work by taking an applied force and then transforming and transmitting it. The blades of a windmill capture the force of wind blowing across a plain. As the blades turn, a series of connected devices turns the wind force into the grinding force of a millstone.

In addition to redirecting and transforming a force, a machine may also amplify it. The degree to which a machine amplifies applied force is called its mechanical advantage. Devices such as levers, pulleys, and wedges are called simple machines because they are the simplest possible devices that can generate a mechanical advantage. More complicated machines often consist of a series of interconnected simple machines.

It is important to note that the amount of work, and thus power, input into a machine must be the same as its output. Energy can be neither created nor destroyed, and work is simply the transfer of energy. If a machine amplifies a force, the displacement that results from that force must decrease in order for work to remain constant.

No machine is perfect, so real-world machines often have a unique power coefficient to denote what percent of the total available power the machine can actually use. If a machine is said to operate at 80 percent efficiency, it can only use 80 percent of the power available to it. Such a machine would have a power coefficient of 0.8.

Electric Power

Electric power is fundamentally similar to mechanical power, but the terminology is different. Electric power is generated by a current, which is the movement of charged particles (usually electrons) across

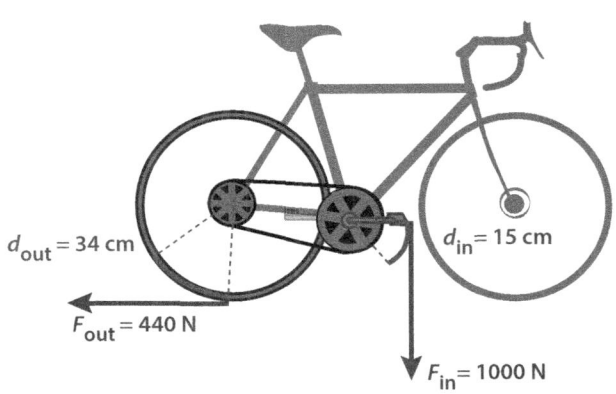

$$P = (Fd)/t$$
Power = (Force • distance)/time

Power$_{out}$ = (440 N • 0.34 m)/1s   Power$_{in}$ = (1,000N • 1.5 m)/1 s

Power$_{out}$ = 149.6 Nm/s   Power$_{in}$ = 150 Nm/s

Power is directly proportional to the force applied to an object and the displacement of that object over a set period of time. In a system such as a bicycle, the power put into pedaling is transferred to the wheels, causing them to rotate. In a perfect system, power is conserved; in reality, minimal power may be lost to frictional forces.

SAMPLE PROBLEM

Consider a hydroelectric dam with a water-powered turbine that operates at 60-percent efficiency. The blades of the turbine have a surface area of 5 meters squared (m²). The flow of water through the dam is carefully regulated so that it maintains a constant velocity of 2 meters per second (m/s). Assuming that the water has the average density of fresh water—approximately 1,000 kilograms per cubic meter (kg/m³)—calculate how much energy the dam generates in a single day.

Answer:

The equation for calculating the total power generated by a fluid flow is

$$P_{total} = \frac{1}{2}\rho A v^3$$

Plug in the values given for fluid density (ρ), blade surface area (A), and fluid velocity (v), then solve, paying attention to the units:

$$P_{total} = \frac{1}{2}(1,000 \text{ kg/m}^3)(5 \text{ m}^2)(2 \text{ m/s})^3$$

$$P_{total} = \frac{1}{2}(1,000 \times 5 \times 8)\frac{\text{kg} \cdot \text{m}^2 \cdot \text{m}^3}{\text{m}^3 \cdot \text{s}^3}$$

$$P_{total} = 20,000 \frac{\text{kg} \cdot \text{m}^2}{\text{s}^3}$$

The units of kg·m²/s³ may seem troublesome, but one watt is actually equivalent to one kg·m²/s³. Thus, in a perfect world, the total power available to the turbine would be 20,000 watts. To adjust for real-world imperfection, this value must be multiplied by the power coefficient (C_p). If the dam operates at 60-percent efficiency, then its power coefficient is 0.6.

$$P_{usable} = C_p P_{total}$$

$$P_{usable} = (0.6)(20,000 \text{ W}) = 12,000 \text{ W}$$

The dam generates 12,000 watts of power. Because one watt is equal to one joule per second, this means that 12,000 joules of energy are being produced each second. To calculate how much energy is produced in one day, simply multiply this amount by the number of seconds in a day:

$$(24 \text{ h})(60 \text{ min/h})(60 \text{ s/min}) = 86,400 \text{ s}$$

$$E_{daily} = (12,000 \text{ J/s})(86,400 \text{ s})$$

$$E_{daily} = 1,036,800,000 \text{ J}$$

The dam generates 1,036,800,000 joules, or 1.0368 gigajoules (GJ), of energy per day.

---

a conductive material. Currents flow between points with differing electric potentials.

"Electric potential" refers to the electric potential energy that a single charge would have at a given point in an electric field. It is equal to the work that would need to be done to carry the charge to that point when traveling against the field. On its own, the electric potential of a single point is meaningless. It only becomes useful when there is a second point in the electric field for the charge to travel to (or from). The difference between the electric potentials of these two points is called voltage. It is a measure of the energy that will be released if a charge is allowed to travel from the point of higher electric potential to the point of lower electric potential. The SI unit of electric potential is the volt, so the electric potential difference is also measured in volts.

One way of generating electric power is with a hydroelectric dam. Moving water carries kinetic energy, just like any other moving substance, and the wide, flat blades of a water-powered turbine provide a large surface area across which this energy can be captured. In theory, the total power that a turbine could extract from the kinetic energy of moving water—or from any other fluid, such as wind—can be calculated with the formula

$$P_{total} = \frac{1}{2}\rho A v^3$$

where ρ is the density of the water (or other fluid), A is the area of the turbine blade (or other energy-capturing device), and v is the velocity at which the water is flowing. However, like any other machine in the real world, a turbine cannot be 100-percent efficient. To account for the inefficiency of a particular power generator, a power coefficient (C_p) must be added to the above formula:

$$P_{usable} = \frac{1}{2}C_p \rho A v^3$$

Again, the power coefficient is a decimal representation of the percent of the total power generated that a machine is able to extract and convert.

Power All Around

Although the average individual is unlikely to need to calculate the energy output of a hydroelectric dam, understanding the principles of how power is calculated is tremendously useful. Electric power in particular is a vital part of everyday life, with electric bills accounting for a major household expense. Consumers should pay close attention to the wattage and power ratings of various appliances. The principles of power, work, force, and energy also underpin all machines, from the simple wheel-and-axle arrangement of a doorknob to the more complicated components found in combustion engines or washing machines.

—*Kenrick Vezina, MS*

Bibliography

"Estimating Appliance and Home Electronic Energy Use." Energy.gov. US Dept. of Energy, 10 May 2015. Web. 23 July 2015.

Henderson, Tom. "Vectors: Motion and Forces in Two Dimensions." *The Physics Classroom*. Physics Classroom, n.d. Web. 23 July 2015.

International Committee for Weights and Measures. The International System of Units (SI). Updated 8th ed. Sevres: BIPM, 2014. Bureau International des Poids et Mesures. Web. 23 July 2015.

"Kinematics." Encyclopaedia Britannica. Encyclopaedia Britannica, 6 June 2013. Web. 23 July 2015.

Siegel, Ethan. "What Does Torque in a Car Do?" Starts with a Bang! ScienceBlogs, 21 Apr. 2009. Web. 23 July 2015.

Simanek, Donald E. "Kinematics." A Brief Course in Classical Mechanics. Lock Haven U, Feb. 2005. Web. 23 July 2015.

PRISMS

FIELDS OF STUDY
Optics; Electromagnetism

SUMMARY

Prisms are transparent objects that refract (bend) light that passes through them. Some prisms also reflect light, while others—particularly the common triangular prisms—disperse visible light into its component wavelengths, producing a rainbow-like effect. Prisms work under the principles of electromagnetism. When passing from one medium to another, the speed and direction of travel of electromagnetic waves are changed.

PRINCIPAL TERMS

- **dispersion:** the process by which light is split into its spectrum of wavelengths as it passes through a medium.
- **electromagnetic spectrum:** the range of wavelengths that make up electromagnetic radiation. These include visible light, infrared, and ultraviolet light.
- **optics:** the branch of physics devoted to the study of light and vision.
- **polarization:** the process by which the motion of electromagnetic waves, which is perpendicular to the direction of energy transfer, is brought into alignment in the same plane.
- **refraction:** the change in a wave's path, speed, or wavelength when the wave passes between two media.
- **visible light:** a form of electromagnetic radiation that represents the limited range of the electromagnetic spectrum that the human eye can see.
- **wavelength:** the full length of a complete wave cycle, measured as the distance between adjacent crests or troughs of a wave.

How Prisms Work

Light undergoes refraction, or a change in its direction, speed, or wavelength, as it passes through a prism, which is a transparent object. In common parlance, "prism" often refers specifically to a small, triangular object made of glass or plastic that refracts visible light in such a way that it splits into a rainbow-like series of colors. Sir Isaac Newton (1642–1727) was one of the first to understand this phenomenon through the use of a prism in the seventeenth century. In optics, a prism may be made of any transparent material, come in a variety of shapes, and have a variety of effects including reflection or magnification.

The angle of refraction, or bending, of light varies by wavelength in many media. Short-wavelength visible light, such as blue light, may be refracted at a greater angle than long-wavelength visible light, such as red. As a result, light passing through a prism ends up sorted by wavelength. This produces a rainbow-like effect wherein the colors that comprise visible light are separated by their differing wavelengths. This process is referred to as dispersion.

Visible Light and the Electromagnetic Spectrum

What humans see as visible light is in fact a form of electromagnetic (EM) radiation. EM radiation is energy that travels as waves. Examples include infrared radiation, ultraviolet radiation, and radio waves. The electromagnetic (EM) spectrum sorts all forms of EM radiation into categories by its wavelength. Within visible light, the distinct colors recognizable to the human eye each occupy their own subrange of wavelengths.

Not all prisms are dispersive. Some are used to redirect light without dispersion, bouncing light that enters in a desired direction—this is often the case with prisms used in binoculars to ensure the image reaches the user's eye right-side-up. Prisms can also split beams of light into multiple beams.

Prisms and Polarization

EM waves are a type of transverse wave: the wave moves perpendicular to the direction of energy transfer. For example, ocean waves move up and down while transferring energy horizontally across the surface of the ocean. EM waves are a bit different in that they do not typically move in the uniform manner of ocean waves. Instead, EM waves may take any number of orientations.

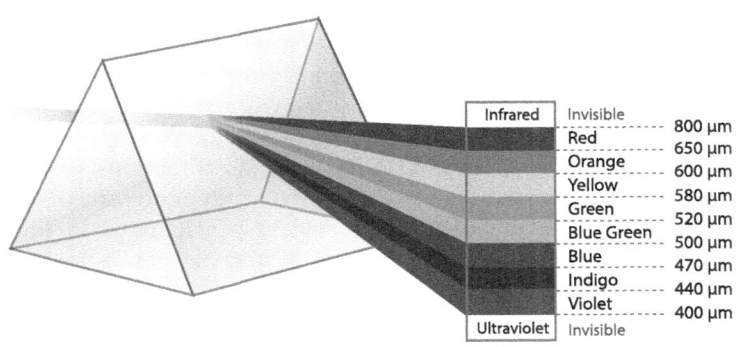

Visible light passing through a prism is refracted. White light is a combination of all of the visible light spectrum; however, each wavelength will refract at a different angle. The light leaving a prism is spread out as a rainbow so that light with larger wavelengths like red and orange light are bent less than light with smaller wavelengths like indigo and violet light.

Prisms can be used in the polarization of light waves. The Nicol prism, for example, uses double refraction to split a beam of light into two paths. This allows one of the light waves, which has become polarized, to be restricted and filtered to a single plane. Similarly, polarized sunglasses filter out partially polarized light to eliminate glare.

PRISMS IN EVERYDAY LIFE

Prisms can be found in a variety of everyday devices, from the laser in a computer's CD drive, to the interior of a pair of binoculars, to a backyard telescope.

Prisms are used in conjunction with lenses to capture, focus, and redirect light. This makes it possible to see things that an unaided human eye would be incapable of detecting.

—*Kenrick Vezina, MS*

BIBLIOGRAPHY

"Anatomy of an Electromagnetic Wave." Mission: Science. NASA, n.d. Web. 13 July 2015.

Bass, Michael, ed. Handbook of Optics. 3rd ed. 5 vols. New York: McGraw, 2010. Print.

"Electromagnetic Radiation." Encyclopaedia Britannica. Encyclopaedia Britannica, 26 Nov. 2014. Web. 13 July 2015.

"Introduction to Optical Prisms." Edmund Optics Worldwide. Edmund Optics, n.d. Web. 13 July 2015.

Morgan, Erinn. "Polarized Sunglasses." All about Vision. Access Media Group, May 2014. Web. 14 July 2015.

Parry-Hill, Matthew, and Michael W. Davidson. "Birefringent Polarizing Prisms." Microscopy Resource Center. Olympus America, n.d. Web. 13 July 2015.

Yacoubian, Araz. Optics Essentials: An Interdisciplinary Guide. CRC, 2015. Print.

PROJECTILES

FIELDS OF STUDY

Classical Mechanics

SUMMARY

Physical objects moving by inertia within a gravitational field are called projectiles. The study of their motion is called ballistics. The mathematics that describes ballistic motion applies to objects as small as a subatomic particle and as large as an intercontinental missile. The physical size of the projectile makes no difference. This enables the use of ballistics in many applications of physical science.

PRINCIPAL TERMS

- **acceleration:** the rate at which the velocity of an object increases over time.
- **ballistic trajectory:** the motion described by a projectile traveling by inertia in a gravitational field.
- **deceleration:** the rate at which the velocity of an object decreases over time.
- **inertia:** the resistance of an object to changes in its velocity.
- **initial velocity:** the velocity of an object at the start of some interval of time.
- **parabolic:** refers to a shape (a parabola) that can be described by an equation of the form $y = ax^2 + b$.
- **range:** the horizontal distance that a projectile can travel before striking the ground.

- **terminal velocity:** the velocity at which the acceleration due to gravity of an object falling freely through a fluid medium, such as air or water, is exactly balanced by the deceleration of the object due to resistance from that medium.
- **vertex:** the uppermost point in a ballistic trajectory.

WHAT IS A PROJECTILE?

A projectile is any object that is shot through the air by an initial force. After this initial force, the projectile's motion is affected only by the force of gravity. Examples of projectiles are numerous, including sand and stones thrown up by the spinning wheels of a car, a baseball struck by a bat, a bullet fired from a gun, and a satellite launched into orbit. Even very small particles such as ions (electrically charged atoms) traveling through a vacuum chamber are considered projectiles.

After a projectile has been given an initial velocity by some outside force, it maintains its horizontal velocity due to inertia. After that, it only changes its vertical position and speed because of gravity. When the initial force stops acting on the object, such as when a football leaves the hand of the person throwing it, the object becomes a projectile. The subsequent path it follows is called a ballistic trajectory. That trajectory is determined by the angle and velocity of the object when it first becomes a projectile. It is also determined by the force of gravity acting on the object.

The ballistic trajectory of a projectile is parabolic in shape. Its motion has only two directions, vertical and horizontal, which operate independently of one another. Generally, a projectile will move in both these directions, although it may have no horizontal movement at all. For example, a ball that is thrown straight up in the air and comes down in exactly the same spot has moved vertically but not horizontally.

BALLISTICS

A normal parabola is a symmetrical curve. Its shape is described by the mathematical formula $y = ax^2 + bx + c$, where y represents the vertical displacement, x represents the horizontal displacement, and a represents the change in vertical position relative to horizontal position. The coefficients b and c simply change the position of the curve. The simplest form of this equation is $y = ax^2$, where b and c both equal 0.

A ballistic trajectory differs somewhat from a normal parabola. Technically, a parabola is a curve whose ends go on forever. A ballistic trajectory, however, has distinct start and end points, separated by a horizontal range. The highest point of a ballistic trajectory is called the vertex. This is the point at which the projectile's vertical velocity is zero, as it transitions from traveling upward to falling back down. Ideally, if the start point and the end point are at the same elevation, the vertex will be exactly halfway between the two.

The vertical velocity and the horizontal velocity of a projectile can be calculated separately, as neither affects the other. In an idealized model, as the projectile travels, its horizontal velocity will remain constant, while its vertical velocity will steadily decrease as it travels upward. This deceleration is due to the force of gravity acting on the projectile. When the vertical velocity reaches zero, the projectile can go no higher. It then begins to fall back to the ground, accelerating at the same rate that it decelerated previously. This rate is the acceleration due to gravity, which on Earth

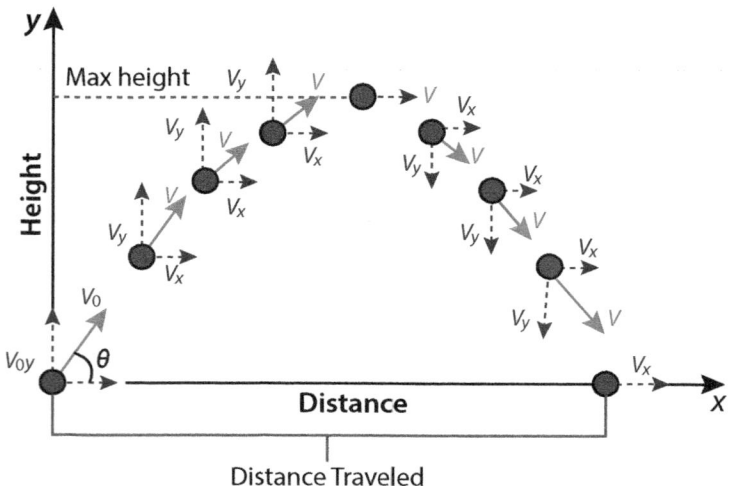

The velocity of a projective can be separated into its horizontal (vx) and vertical (vy) components, or combined as the projectiles total velocity (v). A cannon ball fired from a cannon at an upward angle (θ) will reach a maximum height at which point the vertical velocity will become zero and the total velocity will be equal to the horizontal velocity. Changes in magnitude of the horizontal and vertical components result in a change in the total velocity's magnitude and direction.

SAMPLE PROBLEM

A person standing at the top of a 30-meter-high cliff throws a rock straight out at a zero-degree angle. The rock is released from a point 2 meters above the edge of the cliff and travels with an initial horizontal velocity of 25 meters per second. Ignoring the effects of air resistance, how far will the rock travel horizontally before striking the ground 32 meters below? Recall that the acceleration due to gravity is 9.81 meters per second per second.

Answer:

The simplest way to solve this problem is to calculate the amount of time it would take for the rock to fall straight down from a height of 32 meters, then calculate how far the rock will travel horizontally during that time at a velocity of 25 meters per second. If air resistance is ignored, it can be assumed that the horizontal velocity will remain constant throughout the flight time.

The initial vertical velocity of the rock is 0 m/s. Its final vertical velocity of the rock is not known, but it is known that the rock travels a distance of 32 m in free fall, accelerating at a rate of 9.81 m/s². The final velocity of an object in free fall can be calculated using the following equations, known as equations of motion:

$$v_f^2 = v_0^2 + 2as$$

$$v_f = v_0 + at$$

where v_0 is the initial velocity, v_f is the final velocity, a is acceleration, s is displacement (in this case, vertical displacement), and t is time. The variable v_0 can be eliminated, because the initial velocity is 0. Therefore,

$$v_f^2 = 2as$$

$$v_f = at$$

Solve the first equation for v_f, combine both equations into one, and solve for t:

$$v_f = \sqrt{2as}$$
$$at = \sqrt{2as}$$
$$t = \frac{\sqrt{2as}}{a}$$
$$t = \frac{\sqrt{2(9.81 \text{ m/s}^2)(32 \text{ m})}}{9.81 \text{ m/s}^2}$$
$$t = 2.554 \text{ s}$$

Multiply t (2.554 s) by the horizontal velocity (25 m/s) to determine the horizontal distance traveled:

$$2.554 \times 25 = 63.85$$

The rock traveled a horizontal distance of about 63.85 meters.

is approximately 9.81 meters per second per second (m/s^2).

However, this model of ballistic trajectory does not take into account the medium that the projectile travels through. The medium creates resistance as objects pass through it, and the greater the density of the medium, the greater the resistance. A bullet from a high-powered rifle, for example, can travel several hundred meters through the air without losing much horizontal velocity. However, the same bullet fired into water will lose almost all of its horizontal velocity after just two or three meters.

A parachutist jumping from an airplane experiences the same thing as he or she falls through the air. When the parachutist exits the airplane, the initial horizontal velocity will be the same as that of the airplane, and the initial vertical velocity will be zero. Air resistance will cause the parachutist's horizontal velocity to decrease until it is essentially zero and the parachutist is falling straight down. Meanwhile, the parachutist's vertical velocity will increase at a rate of 9.81 m/s^2 until the force of the air resistance is equal to the force of gravitational acceleration, at which point he or she can fall no faster. The speed at which this occurs is called terminal velocity and is typically reached long before the parachutist will deploy his or her parachute.

Practical Ballistics

The phenomenon of medium resistance means that in reality, the ballistic trajectory of a projectile will not have a perfectly parabolic shape. Resistance will cause the projectile's horizontal velocity to decrease slightly over time, reducing its range. Also, the vertex will not be exactly halfway between the start point and the end point. It will be closer to the end point, because the decrease in horizontal velocity means that the projectile will travel a shorter horizontal distance after passing its vertex. This is a very important consideration for pilots, parachutists, target shooters, baseball players, basketball players, and anyone else who relies on ballistic motion to achieve the goal of hitting a target with a projectile.

Another important consideration is the flight time of a projectile following a ballistic trajectory. Because the horizontal velocity is not affected by the force of gravity, the projectile will continue to travel horizontally until the force of gravity has brought it into contact with the ground. This means that if a ball is thrown straight ahead in such a way that its initial vertical velocity is zero, it will travel through the air for the same amount of time as if it had been simply dropped from the same height. This is because the ball is accelerating toward the ground at the same rate in both cases. This length of time is the flight time, and it is determined solely by the force of gravity acting on the projectile. The greater the horizontal velocity of the projectile, the farther it will travel during the flight time. A slow projectile and a fast projectile will both strike the ground at the same time, however, as long as they are both thrown from the same point at a zero-degree angle.

Equations of Motion

Like all of physics, the description of ballistic trajectories relies heavily on mathematics. It is a fairly simple exercise to watch a projectile travel through the air and relate its motion to mathematics. This can be as simple as watching a stream of water come from a garden hose or tossing a ball into the air. The equations that describe the movements of objects through a constant gravitational field are called "kinematic equations" or "equations of motion." They can be remembered using the acronym VAST, which stands for velocity (v), acceleration (a), displacement (s), and time (t). As demonstrated in the sample problem, these equations can be rearranged and combined in such a way that if just two of these quantities are known, they can be used to calculate the other two.

—*Richard M. Renneboog, MSc*

Bibliography

Cross, Rod. *Physics of Baseball and Softball*. New York: Springer, 2011. Print.

Denny, Mark. *Their Arrows Will Darken the Sun: The Evolution and Science of Ballistics*. Baltimore: Johns Hopkins University Press, 2011. Print.

Grissom, Thomas. *The Physicist's World: The Story of Motion and the Limits of Knowledge*. Baltimore: Johns Hopkins University Press, 2011. Print.

Hamilton, Sue L. *Forensic Ballistics: Styles of Projectiles*. Edina: ABDO, 2008. Print.

Kinard, Jeff. *Artillery: An Illustrated History of Its Impact*. Santa Barbara: ABC-CLIO, 2007. Print.

Stahler, Wendy, Dustin Clingman, and Kaveh Kahrizi. *Beginning Math and Physics for Game Programmers*. Indianapolis: New Riders, 2004. Print.

Taylor, John R. *Classical Mechanics*. Sausalito: U Science, 2005. Print.

White, Colin. *Projectile Dynamics in Sport: Principles and Applications*. New York: Routledge, 2011. Print.

PSYCHOPHYSICS

FIELDS OF STUDY

Acoustics; biophysics; optics/light physics

SUMMARY

Also called "sense physiology." Scientists who work in this field study the quantitative relationship between events that are psychological and events that are physical. They examine sensations produced by the mind and the physical stimuli that produce these sensations. For some of the senses, scientists can accurately measure, on a physical scale, the magnitude of a stimulus. Then they can measure the minimum stimulus that provokes a sensation.

PRINCIPAL TERMS

- **acoustics:** the study of sound; also, the qualities of a space that affect how sound is heard within that space.
- **frequency:** the number of complete wavelengths that occur within one unit of time, typically expressed as hertz (Hz; cycles per second).
- **just noticeable difference (JND):** the amount a parameter, such as sound intensity, must be changed so that the difference is noticeable at least half the time.
- **loudness:** the intensity of sound waves, which depends on the wave's amplitude; measurements of loudness or volume are expressed in decibels.
- **perception of sound:** the ability to perceive mechanical waves in one's environment (e.g., air, water) as sounds, limited by the physiology of the ear and the brain's ability to interpret information from sound waves.
- **sensitivity:** the ability of an ear or a mechanical device to pick up and interpret sound; highly sensitive devices will have smaller just-noticeable difference values.
- **tone:** a vocal or musical sound with a specific pitch and loudness.

The Origins of Psychophysics

Gustav Theodor Fechner, a German scientist and philosopher, coined the term "psychophysics" to describe his experiments and investigations into this realm of experimental psychology which explores the relationship between physical stimuli and the effects they produce on the mind. He studied as a physicist in early life but in later life, he became interested in metaphysics and looked for a way to describe the relationship between the physical and spiritual world. His work built on the work of Ernst Heinrich Weber, even though they discovered these principles separately.

"Weber's law" states that if two weights are different by an amount that is just noticeable when separated by a given increment, for the difference to remain noticeable, that increment must be proportionally increased. So, to perceive a difference between a background (x) and that background and a stimulus ($x + dx$), the size of the difference must be proportional to the background ($dx = kx$, where k is a constant).

Fechner created a formula that described that the magnitude of a stimulus must be increased geometrically for the magnitude of sensation to increase arithmetically. Fechner applied Weber's law to the measurement of a sensation in response to a stimulus. In Fechner's equation, the relationship between a stimulus (x) and the perceived sensation ($s(x)$) is $s(x) = \log(x)$. His work made it possible to measure sensation in relationship to a measured stimulus and started a body of work called "scientific quantitative psychology."

Sensation versus Perception

The mind receives millions of pieces of input each day. The brain is only capable of making sense of a fraction of these inputs and elicits reactions to those pieces of information that it deems most important. For example, there are vast wavelengths on the electromagnetic spectrum, but the human eye only "sees" a small portion of those, wavelengths between about 400 and 750 millimicrons. The human ear only hears frequencies in about the 20 to 20,000 cycles-per-second range. Psychophysicists attempt to map out the relationship between sensations and perceptions to provide a basic understanding of how the senses function and to quantitatively determine how much of a stimulus the brain can detect through sensory systems such as vision, hearing, taste, smell, and pain.

To determine how the brain processes this information, scientists use tests involving detection (properties that a stimulus must possess for the brain to be aware of it), identification (how the brain knows what the stimulus is), discrimination (how different stimuli need to be for the brain to distinguish between them), and scaling (how the brain judges the magnitude of the stimulus and whether it is the same as or different from another stimulus). Sometimes scientists use reaction time to determine how different two stimuli are from each other. Large differences lead to immediate reaction times, while small differences often take some time to determine.

Sensation means that the sensory areas in the cerebral cortex have received a nerve impulse. Perception is what the brain understands from that impulse. The difference between these two give rise to some interesting contradictions, such as optical illusions. Scientists who study sensation and perception use two types of thresholds to describe what is happening:
- Absolute threshold: The minimum intensity of a stimulus that can be detected
- Difference threshold: The minimum difference in intensity that can be detected between two stimuli

A hearing test is an example of absolute threshold in the acoustics sense. This type of test tests for loudness as well as frequency and is a measure of the sensitivity of the person's ear. The person whose perception of sound is being tested presses a button if he or she hears a tone, releases it when the tone is no longer heard, then presses the button again when the tone is again heard. The intensity of the tone at which one hears it is that person's absolute threshold for that particular tone. Weber defined absolute threshold in this type of situation as the intensity at which 50 percent of people could detect the tone. Anything below this absolute threshold is called "subliminal" or below threshold.

Difference threshold measures the point at which one can tell that a changed stimulus is either more or less intense than the standard stimulus. The lower difference threshold is determined at the point when 25% of people perceive the stimulus to be more intense. The upper difference threshold is determined at the point when 75% of people perceive the stimulus to be more intense. These two differences are then averaged to determine the difference threshold. Weber's law states that the difference threshold is a constant proportion of the initial stimulus intensity.

Fechner called the difference threshold the "just noticeable difference." He stated that all just noticeable differences seem like the same amount of change in stimulus intensity, and that the perceived intensity of a stimulus changes in proportion to the logarithm of the physical stimulus intensity, not in direct proportion. At the higher end of the intensity scale, we are almost insensitive to changes in intensity, while at the lower end, we are highly sensitive to changes. For example, when a car's headlights are on in the daytime, they are barely noticeable. However, when the headlights are turned on after nightfall, they are very noticeable. This relationship, known as Fechner's Law, formed the basis of the decibel scale, a measure of loudness.

Just noticeable difference varies from sense to sense. For example, the increment of change noticeable to most people is about 1 percent for brightness, about 2 percent for loudness, and about 20 percent for saltiness.

PRACTICAL USES OF PSYCHOPHYSICS
Psychophysicists use these methods to study sensations both in an experimental psychology sense and in a practical sense. For example, these studies are used in comparisons and evaluations of products that involve the senses such as cleaning products and personal healthcare products (perfumes, shampoos, soaps, etc.). This type of testing may also be used in psychological and personnel testing for job applicants. As another example, subjects may be asked to judge the magnitude of color differences against a standard color difference for industrial purposes. Or, subjects may be asked to match colors to help scientists understand the differences in color perception between individuals. Another method involves subjects evaluating images to see how different types of displays or imaging devices (such as televisions or

computer monitors) affect image quality. Other practical applications of psychophysics involve vision and hearing testing, as discussed above.

—*Marianne Moss Madsen, MS*

BIBLIOGRAPHY

Baird, John C. *Sensation and Judgment: Complementarity Theory of Psychophysics (Scientific Psychology Series)*. Psychology Press, 2014. Electronic.

Cunningham, Douglas W., and Christian Wallraven. *Experimental Design: From User Studies to Psychophysics*. A K Peters/CRC Press, 2011. Print.

Jauhiainen, Tapani, Veikko Hakkinen, and Dieter Schaffrath. *From Psychophysics and Psychophysiology to Phenomonenology of Perception: The Ontological and Epistemological Approach of Yrjo Reenpaa*. Amazon Digital Services, 2015. Electronic.

Kingdom, Frederick A.A., and Nicolass Prins. *Psychophysics: Second Edition: A Practical Introduction*. Academic Press, 2016. Print.

Lu, Zhong-Lin, and Barbara Dosher. *Visual Psychophysics: From Laboratory to Theory*. The MIT Press, 2013. Print.

Mather, George. *Foundations of Sensation and Perception, 3rd Edition*. Psychology Press, 2016.

Roederer, Juan. *The Physics and Psychophysics of Music: An Introduction*. Springer, 4th ed., 2008. Print.

Zwislocki, Josef J. *Sensory Neuroscience: Four Laws of Psychophysics*. Springer, 2009. Electronic.

Q

QUANTUM CHROMODYNAMICS

FIELDS OF STUDY

Quantum Field Theory; Particle Physics; Quantum Physics

SUMMARY

Quantum chromodynamics (QCD) is the quantum field theory underlying the strong interaction, one of the four fundamental forces in the standard model of particle physics. The strong interaction holds quarks together to form hadrons, and it also holds together protons and neutrons in atomic nuclei. This article will explain the principles of QCD and its context within particle physics.

PRINCIPAL TERMS

- **color force:** the force of the strong interaction that operates on the quark level.
- **fermion:** one of two main classes of particles, characterized by adherence to Fermi-Dirac statistics; includes quarks, leptons, and any particles that contain an odd number of quarks or leptons, such as protons and neutrons.
- **gluon:** the elementary particle that carries the strong interaction between quarks.
- **hadron:** a composite particle consisting of either three quarks or one quark and one antiquark, held together by the color force.
- **quark:** an elementary fermion that combines with other quarks to form a baryon, such as a proton or a neutron, or with an antiquark to form a particle called a meson.
- **quantum field theory:** a theory that explains interactions between subatomic particles as the result of a field extending between them.
- **standard model:** the generally accepted theory of particle physics that deals with the strong, weak, and electromagnetic interactions.
- **strong interaction:** the fundamental process of particle interaction that binds quarks into hadrons and hadrons into nuclei.

FUNDAMENTAL FORCES

All interactions between matter are governed by forces that act on specific properties universal to all matter. There are four fundamental interactions, or forces, in the standard model of physics: the strong interaction, the weak interaction, gravity, and electromagnetism. The strong interaction is what binds quarks together to form hadrons and holds together the hadrons—specifically, protons and neutrons—in atomic nuclei. It is the most powerful of the fundamental forces, able to overcome the electromagnetic repulsion of the positively charged protons.

Each of the four fundamental interactions is believed to operate according to a quantum field theory. A quantum field theory is a theory in which particles are treated not as individual entities but as excitations in an underlying physical field. Quantum chromodynamics (QCD) is the quantum field theory of the strong interaction. When the strong interaction operates on the level of quarks, it is also known as the color force. The force that holds protons and neutrons together is a residual effect of the color force and is often called the "residual strong force" or the "strong nuclear force."

"Color" in QCD has nothing to do with visual color. It simply refers to one of the main properties of quarks, along with spin and electric charge. The existence of this property, known as "color charge," was suggested in 1964 to explain how quarks with otherwise identical properties could coexist in the same hadron without violating the Pauli exclusion principle. (This principle states that no two fermions, of which quarks are one type, that occupy the same particle can have completely identical properties.) The color analogy was chosen because color charge has three possible values, just as the human eye can detect three primary colors of light. These colors—red, green, and blue—are also the names of the three color charges. An antiquark, or antimatter particle of a quark, has the anti-color charge of the quark to which it corresponds: anti-red, anti-green, or anti-blue.

The function of color charge in QCD is analogous to that of electric charge in quantum electrodynamics (QED), the quantum field theory of electromagnetism. In order to be affected by the strong interaction, a particle must have a color charge, just as a particle must have an electric charge to be affected by electromagnetism. The strong interaction is carried by massless particles known as gluons, so called because they "glue" quarks together. Gluons have color charge but no electric charge or mass, meaning that they are also affected by the very interaction that they carry. This is in contrast to photons, their QED counterparts, which carry electromagnetism but have no electric charge themselves.

THE STANDARD MODEL

In order to understand the context of QCD, it is necessary to have a basic understanding of the standard model of particle physics.

Particles can be characterized in a number of different ways. An elementary particle is one that cannot be broken down further into component particles. The two classes of elementary particles are fermions, which consist of quarks and leptons, and bosons, which consist of gauge bosons and scalar bosons. The difference between the two classes is the rules that govern their behavior. Fermions obey a set of rules called Fermi-Dirac statistics, while bosons follow Bose-Einstein statistics. Gauge bosons are the particles that carry the fundamental interactions. They travel between the particles affected by those interactions, transmitting the energy that makes up their respective fields (color field, electromagnetic field, etc.). The gluon and the photon are both gauge bosons.

All other particles are composite particles, formed from some combination of elementary particles (or antiparticles). Composite particles can also be classified as either fermions or bosons. If a particle contains an odd number of fermions, it too is a fermion; if it contains an even number of fermions, it is a boson. For example, a hadron is any composite particle that consists of quarks held together by the color force. There are two types of hadrons: mesons and baryons. Mesons contain one quark and one antiquark, so they are bosons. Baryons contain three quarks, so they are fermions. Protons and neutrons are both types of baryons. The type of meson or baryon is determined by the types, or "flavors," of the quarks they contain.

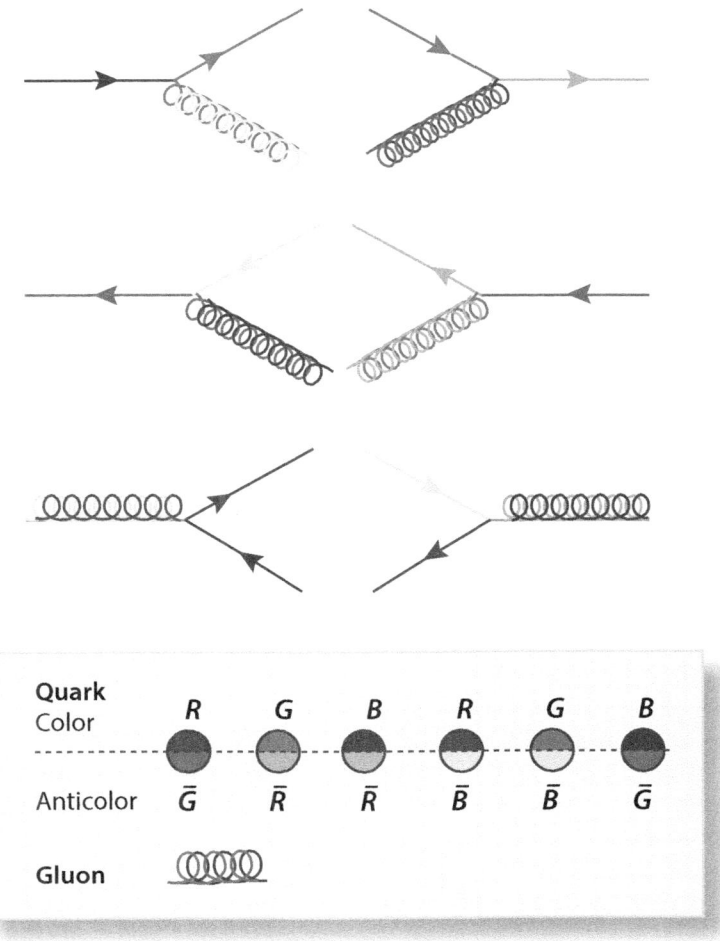

The theory of the strong interaction between quarks and gluons is called quantum chromodynamics (QCD). Quarks and gluons carry something called the color charge, which is analogous to the electric charge in quantum electrodynamics. Quark colors are red, green, and blue. Quark anti-colors are anti-red (cyan), anti-green (magenta), and anti-blue (yellow).

Color Charge and the Residual Strong Force

Quarks and gluons are the only types of particles that have color charge. Even though hadrons are made of quarks, all hadrons must be "colorless"—that is, the colors of their constituent quarks must cancel one another out. In a meson, this means that the quark and the antiquark must be a color and its anti-color, such as red and anti-red. A baryon, on the other hand, must contain one quark of each color. The three colors combined neutralize each other, much like the combination of all visible colors of light appears white to the human eye.

While quarks have a single color charge and antiquarks have a single anti-color charge, gluons carry both a color charge and an anti-color charge. A gluon's color charge does not necessarily have to correspond to its anti-color charge. Quarks within the same hadron are constantly exchanging gluons between them, which causes the quarks to change color. This is always done in such a way that the overall colorlessness of the hadron is maintained. For example, if a red quark emits a red/anti-blue gluon, its color will change to blue, while the blue quark will absorb the gluon and become red.

Although hadrons have zero color charge, the protons and neutrons in a nucleus are still held together by the strong interaction. This is where the residual strong force comes in. Some of the color force between the quarks in a hadron "leaks" out and affects the quarks in nearby hadrons. This residual force is much weaker than the color force and has a much shorter range. However, within the confines of an atomic nucleus, it is still stronger than the electromagnetic repulsion that would otherwise drive the protons apart.

Color Confinement

One particularly puzzling aspect of QCD is the apparent fact that particles with a color charge cannot be observed in isolation. There is no known way to separate a hadron into its constituent quarks. A proton and an electron can be pulled apart because the electromagnetic force between them decreases exponentially with distance. In contrast, the extremely strong color force between quarks remains constant regardless of distance. Any attempt to pull a quark from a meson, for example, will "stretch" the gluon field between the quark and the antiquark more and more, increasing the amount of energy it contains. Eventually the field will "snap" and convert the excess energy into a new quark and antiquark. These particles will immediately combine with the two that have been separated, producing two new mesons to replace the original.

This phenomenon is known as "color confinement" because quarks must always be confined within hadrons. While not fully understood, it is believed to be the result of gluons having color charge as well as carrying the color force. Color confinement is the reason why no free quarks have ever been found.

—*Gina Hagler, MBA*

Bibliography

Cao, Tian Yu. *From Current Algebra to Quantum Chromodynamics: A Case for Structural Realism.* New York: Cambridge University Press, 2010. Print.

Finston, David R., and Patrick J. Morandi. *Summary Algebra: Structure and Application.* Cham: Birkhäuser, 2014. Print.

Gattringer, Christof, and Christian B. Lang. *Quantum Chromodynamics on the Lattice: An Introductory Presentation.* Heidelberg: Springer, 2010. Print.

Robinson, Matthew. *Symmetry and the Standard Model: Mathematics and Particle Physics.* New York: Springer, 2011. Print.

Schwartz, Matthew D. *Quantum Field Theory and the Standard Model.* New York: Cambridge University Press, 2014. Print.

Srednicki, Mark. *Quantum Field Theory.* New York: Cambridge U, 2007. Print.

Wilczek, Frank. "QCD Made Simple." *Physics Today* Aug. 2000: 22–28. Academic Search Complete. Web. 31 July 2015.

QUANTUM FIELD THEORY

FIELDS OF STUDY

Quantum Field Theory; Quantum Physics

SUMMARY

Quantum field theory has its origins in quantum mechanics. It is a mathematical model designed to explain the behavior of subatomic particles in terms of physical fields. Examples of quantum field theories include quantum chromodynamics, quantum electrodynamics, and conformal field theory. Valid descriptions of the behavior of electrons and other matter at the atomic scale in electronic devices is an important application of quantum field theory.

PRINCIPAL TERMS

- **conformal field theory:** a quantum field theory that is independent of the scale of the system and supports only massless excitations.
- **degrees of freedom:** the number of physical parameters required to specify the position and configuration of a particle or other body.
- **quantum chromodynamics:** a quantum field theory that describes the interactions of quarks and gluons, subatomic particles that are responsible for the strong interaction.
- **quantum electrodynamics:** a quantum field theory that describes the interactions of photons and electrically charged particles, which are responsible for electromagnetism.
- **quantum mechanics:** the branch of physics that deals with matter interactions on a subatomic scale, based on the concepts that energy is quantized, not continuous, and that elementary particles exhibit wavelike behavior.
- **renormalization:** a mathematical procedure by which processes occurring on very different scales can be aligned.
- **standard model:** a generally accepted unified framework of particle physics that explains electromagnetism, the weak interaction, and the strong interaction as products of interactions between different types of elementary particles.
- **superposition:** in quantum mechanics, the concept that a particle exists in all possible states at the same time until either its position or its energy is known.

QUANTIZATION OF PHYSICAL FIELDS

One significant aspect of quantum mechanics is the tendency of subatomic particles to behave as waves under some conditions and as particles under others. This concept is called wave-particle duality. Quantum field theory arises from the idea that such behavior can be described as the product of continuous physical fields rather than as interactions between individual entities.

A physical field is a region of space-time in which each point has a specific value for some physical quantity, such as electric potential or gravitational force. The voltage difference between two points in space, such as the terminals of a battery, is an example of an electric field, because electric potential exists between the two points. Similarly, the region between two magnetic poles is an example of a magnetic field, because magnetic potential exists between the poles. A quantum field theory explains particle interactions by treating particles not as discrete units but as energy excitations, or "field quanta," in fields such as these.

There are four fundamental forces that underlie all of physics: electromagnetism, the strong interaction, the weak interaction, and gravity. The standard model of particle physics is a unifying framework that brings together the quantum field theories of the first three of those forces. The theory of electromagnetism is quantum electrodynamics (QED), which describes the behavior of charged particles in an electromagnetic field. The theory of the strong interaction is quantum chromodynamics (QCD); it describes the interactions of elementary particles called quarks and gluons, which come together to form hadrons. (Hadrons are relatively massive composite particles, such as protons and neutrons.) The theory of the weak interaction is the electroweak theory, which unites it with electromagnetism.

As yet, there is no complete quantum field theory describing gravity. A model has been proposed, but it relies on hypothetical particles called gravitons, the existence of which has yet to be confirmed. A viable quantum field theory of gravity would be the first step toward a unified "theory of everything" that would incorporate gravity into the standard model. The lack of such a theory is one of the major unsolved problems in physics.

One subset of quantum field theory is conformal field theory (CFT). A quantum field theory is considered to be a CFT if it is scale invariant—that is, if the interactions it describes are independent of the scale of the system. In other words, if a CFT is either increased or decreased in size, the physics of the system it describes will remain the same. A CFT can only support massless excitations (i.e., particles), because having mass would make the system behave differently at different scales.

MATHEMATICS OF QUANTUM FIELDS

In general, a quantum mechanical system will contain a finite number of particles, each with a finite number of degrees of freedom. Degrees of freedom are the number of parameters required to precisely identify and locate an individual particle. For example, a point particle in three-dimensional space has three degrees of freedom, because three coordinates (x, y, z) are necessary to describe its location. If the point particle were instead a rigid body, it would have six degrees of freedom, because another three parameters would be needed to describe its orientation in space.

However, because fields are continuous, they have an effectively infinite number of degrees of freedom. Moreover, quantum fields can support any number of particles. One way to avoid dealing with these infinities is renormalization. This involves "integrating out," or removing from consideration, any parameters that do not apply to the length and energy scales being studied. Renormalization allows for the creation of an effective field theory, which is an approximation of a field theory that holds true for the lengths and energies of interest.

Quantum field theories are mathematics-intensive models that have had success in predicting the behaviors of subatomic particles. However, as models, they do not constitute proof of the behaviors they predict. The validity of these models must constantly be tested against experimentation and observation.

The observation of effects on the quantum scale requires very high-energy collision experiments, such as those carried out at the European Organization for Nuclear Research (CERN) facilities in Switzerland and at the Stanford Linear Accelerator Center (SLAC) facility in Connecticut. Calculation of wave properties in quantum mechanics begins with a statement of the general wave equation relating position and velocity. For a one-dimensional application only, this equation is relatively simple. However, it becomes increasingly complex as more degrees of freedom are involved.

One problem in quantum mechanics is that exact solutions involving more than one particle are effectively impossible to calculate. The principle of superposition can be used to simplify the situation by invoking a certain uniformity. Because the equation is linear, any combination of linear solutions is also a linear solution of the appropriate wave equation. The entire range of linear solutions to a particular wave equation represents the field corresponding to that wave equation.

FUNDAMENTAL PHYSICS

For the vast majority of people, quantum field theory is of no consequence. Because quantum theory deals with matter at the most fundamental level, classical Newtonian physics is sufficient on a macroscopic scale. However, ongoing developments in quantum theory have increasing significance as technology becomes functional at ever-smaller scales.

As the transistor circuits imprinted on silicon computer chips approach the lower size limits dictated by atomic size, quantum field effects become more important. Understanding these effects will be vital to ensuring that digital electronics of this type will be functional. New materials such as graphene raise the possibility of more efficient functioning on an even smaller scale. Development of devices using such materials will depend on the ability of theoretical models to predict the quantum behavior of particles within them.

Quantum field theory can also address questions on a much larger scale. It may help scientists understand the processes occurring within stars and the interactions of cosmic particles with Earth's atmosphere. Accordingly, seemingly esoteric theoretical models do in fact have relevance in the real world.

—*Richard M. Renneboog MSc*

BIBLIOGRAPHY

Huang, Kerson. *Quantum Field Theory from Operators to Path Integrals.* 2nd ed. Weinheim: Wiley, 2010. Print.

Itzykson, Claude, and Jean-Bernard Zuber. *Quantum Field Theory.* 1980. Mineola: Dover, 2005. Print.

Lahiri, Amitabha, and Palash B. Pal. *A First Book of Quantum Field Theory.* 2nd ed. Boca Raton: CRC, 2005. Print.

Lancaster, Tom, and Stephen J. Blundell. *Quantum Field Theory for the Gifted Amateur.* Oxford: Oxford University Press, 2014. Print.

Teller, Paul. *An Interpretive Introduction to Quantum Field Theory.* Princeton: Princeton University Press, 1995. Print.

Zee, A. *Quantum Field Theory in a Nutshell.* 2nd ed. Princeton: Princeton University Press, 2010. Print.

QUANTUM MECHANICS

FIELDS OF STUDY

Quantum Physics

SUMMARY

Quantum physics deals with phenomena at the atomic level and smaller. In the past, physicists assumed that the forces and reactions that occurred at the observable level would hold true at the atomic and subatomic levels. However, in the early twentieth century, that notion was disproved, and the field of quantum physics was born.

PRINCIPAL TERMS

- **blackbody radiation:** radiation emitted by a body solely as a result of its temperature, regardless of its composition or any previously absorbed radiation.
- **line spectrum:** the lines of color that represent the characteristic frequencies at which atoms emit electromagnetic radiation.
- **Schrödinger equation:** a wave equation that describes the quantum state of a particle or system and can be used to predict its most likely future behavior.
- **uncertainty principle:** the idea, proposed by Werner Heisenberg, that one can determine with high precision either the position of a particle at a given time or its momentum, but not both.
- **wave-particle duality:** the idea that a particle can behave as either a particle or a wave, depending on how and when it is being observed.

BEYOND CLASSICAL MECHANICS

As the start of the twentieth century, scientists believed they were on the cusp of a theory that would unify all known forces and provide an understanding of the universe as a whole. With Isaac Newton's (1642–1727) three laws of motion accounting for the behavior of objects at rest and in motion and James Clerk Maxwell's (1831–79) work with electromagnetism having unified the fields of electricity and magnetism, physicists believed that only a few pieces of the puzzle remained. However, the field of physics was about to change dramatically.

In the nineteenth century, physicists had discovered that the line spectra of different atoms show radiation being emitted at certain frequencies that are characteristic of each element. However, they did not know why this is so. In 1913, Danish physicist Niels Bohr (1885–1962) proposed a new atomic model that offered a solution. He theorized that an atom's electrons could only travel around its nucleus in certain discrete orbitals. They could only move from one orbital to another by absorbing or emitting specific amounts of energy. This energy is the source of the atom's characteristic spectrum lines.

Fundamentally, quantum mechanics is based on the idea that at the smallest possible level, certain physical properties are quantized—that is, made up of individual, discrete units, rather than an unbroken continuum. (A quantum is the smallest possible unit of a physical property.) The idea was first proposed in general terms by Austrian physicist Ludwig Boltzmann (1844–1906) in 1877. It was later picked up by German physicist Max Planck (1858–1947), who, though suspicious of Boltzmann's idea and its implications, nevertheless applied it to his study of blackbody radiation.

THE QUANTIZATION OF ENERGY

Planck's law of blackbody radiation, proposed in 1900, defines the energy emitted by an oscillating particle in a blackbody as

$$E = nh\nu$$

where n is a positive integer, h is a constant value known as Planck's constant, and ν is the frequency of the particle. In other words, E can only ever equal multiples of $h\nu$. Planck believed that this quantization of energy was merely a mathematical convenience and did not reflect reality. However, German-born physicist Albert Einstein (1879–1955) disagreed. In 1905, he published a paper on the photoelectric effect—the emission of precise streams of electrons by certain metals when struck by certain wavelengths of light—in which he suggested that quantization is an inherent property of energy. Einstein argued that energy is in fact composed of individual "energy quanta," each with the energy $h\nu$. These quanta were later called photons.

Elements emit photons at very specific wavelengths. Hydrogen, for example, emits visible light at four distinct wavelengths. (It also emits at several other wavelengths outside the visible spectrum.) These emissions occur in discrete bursts, as opposed to continuous rays. Scientists observed that if they bombarded hydrogen atoms with photons of energies that corresponded to the emission wavelengths, the photons would be absorbed and later released. The energies of the photons were equal to the energy differences between the electron orbitals, which are specific to each element. Photons that are not of these energies simply bounce off.

This phenomenon accounts for the different colors in "neon" signs. In a neon sign, electricity is run through a tube of gas, exciting the electrons and causing them to jump to higher energy levels. The electrons then emit the excess energy in the form of photons that correspond to specific wavelengths of visible light. Neon itself produces a reddish-orange light. Other gases, such as xenon, argon, and helium, are used to produce other colors.

WAVES OR PARTICLES?

In the 1920s, scientists were vexed by problems explaining the motion of electrons inside atoms. They did not seem to obey the laws of classical physics. Bohr's atomic model solved some of these problems, but it did not explain why electrons behave as they do. In 1924, French physicist Louis de Broglie (1892–1987) proposed a revolutionary idea that would lead to the development of modern quantum mechanics.

Scientists had long debated whether light was a particle or a wave. Einstein's 1905 paper on the photoelectric effect demonstrated that it is, in fact, both. Light can behave as either a wave or a particle, depending on the circumstances under which it is observed. This concept came to be known as wave-particle duality. In his doctoral thesis, de Broglie suggested that electrons can also behave as both waves and particles. This was confirmed by a 1927 experiment that showed electrons undergoing diffraction, a phenomenon characteristic of waves. It was later found that all matter exhibits wave-particle duality, not just electrons.

The behavior of any wave can be described by a wave equation. Austrian physicist Erwin Schrödinger (1887–1961) decided to find the wave equation of electrons. He published his result, now known as the Schrödinger equation, in 1926. The Schrödinger equation applies not just to electrons but to all isolated systems. Solving it results in a wave function,

The double-slit experimental design is used to show the wavelike behavior of particles. A screen with two slots is placed in front of a recording screen. Particles are shot at the slotted screen, pass through the slots, and hit the recording screen. Results from this experiment show that the particles move in a wavelike fashion. The path of individual particles cannot be accurately predicted, but the frequency of hits across the length of the recording screen indicate where particles are more likely to hit.

which is a mathematical function that describes the probability of finding a given particle in the system in a certain position at a certain time.

Another problem in the study of electrons was that the act of observing an electron involved exposing it to light. This caused the electron to interact with a photon, which could then alter the very qualities being observed. German theoretical physicist Werner Heisenberg (1901–76) discovered this problem while trying to observe an electron with a microscope. The short wavelengths of light required to accurately measure the electron's position exposed it to high-energy photons, which disturbed its momentum. The opposite was also true: any wavelength long enough to observe an electron's momentum without disturbing it would be too long to use for determining its position. This led Heisenberg to publish his uncertainty principle in 1927. The uncertainty principle essentially states that there are hard limits on what can be simultaneously known about any given quantum system. Heisenberg attributed this to the observer effect, wherein the act of observing a system causes that system to change in some way. However, it was later determined that while the observer effect often plays a role in the uncertainty principle, it is not its sole cause. Rather, uncertainty is an inherent property of all wavelike systems. And because all matter exhibits wave-particle duality, this could be extended to apply to all of reality, on the macro as well as the micro and quantum scales. Philosophically, this would require giving up the idea of a deterministic universe. Instead, reality would be considered to be probabilistic, describable only in terms of probability rather than as a certainty.

On the face of it, this idea sounds absurd. Indeed, the classic thought experiment of Schrödinger's cat, in which a cat in a box is considered to be simultaneously alive and dead until the box is opened, was intended to demonstrate this absurdity. The so-called Copenhagen interpretation of quantum mechanics, based on the work of Bohr and Heisenberg, left open the possibility of uncertainty on a macro scale but did not explicitly endorse it. Other physicists, including Einstein and Schrödinger, firmly opposed the idea. Since then, various other interpretations have also supported the probabilistic view.

QUANTUM MECHANICS AND THE UNIVERSE
Quantum mechanics remains an actively researched field. It has revealed that reality, at the smallest scales, occurs in discrete units, with mysterious gaps between the infinite set of possible positions. This contrasts to Newtonian physics; in quantum physics, even if everything is known about a system, at best, only a set of outcomes can be predicted. Until that outcome is decided, all of the possibilities exist in superposition.

Quantum physics is nonlocal, meaning that things can affect each other at long range with seemingly no contact. Particles can, and indeed must, jump between places without passing through the intervening space. Reality is a wave function, a set of possibilities and probabilities down to the deepest level. However, beyond the massive implications for ontology and the cutting edge of particle physics, it has pervasive practical uses.

For example, understanding quantum mechanics allowed for the invention of the transistor, needed to build miniaturized electronics, including computers. Quantum mechanics is also used by plants. They use quantum teleportation to transport electrons and allow the electrons to try out all the possible paths to choose the most efficient one. Quantum mechanics can be used in chemistry to better account for how reactions occur. Improvements in electronics have resulted from quantum mechanics. Examples include flash memories in USB drives. Even a simple light switch, where layers of oxidation would otherwise prevent transmission of current between the contacts, is subject to the reality of quantum mechanics.

—Gina Hagler, MBA

BIBLIOGRAPHY
Al-Khalili, Jim. *Quantum: A Guide for the Perplexed.* London: Phoenix, 2012. Print.

Brumfiel, Geoff. "Common Interpretation of Heisenberg's Uncertainty Principle Is Proved False." *Scientific American.* Scientific Amer., 11 Sept. 2012. Web. 10 June 2015.

Fayer, Michael D. *Absolutely Small: How Quantum Theory Explains Our Everyday World.* New York: AMACOM, 2010. Print.

Ford, Kenneth William. *101 Quantum Questions: What You Need to Know about the World You Can't See.* Cambridge: Harvard University Press, 2011. Print.

Halpern, Paul. *Einstein's Dice and Schrödinger's Cat: How Two Great Minds Battled Quantum Randomness to Create a Unified Theory of Physics.* New York: Basic, 2015. Print.

Susskind, Leonard, and George Hrabovsky. *The Theoretical Minimum: What You Need to Know to Start Doing Physics.* New York: Basic, 2013. Print.

QUANTUM STATISTICS

FIELDS OF STUDY

Particle Physics; Quantum Physics

SUMMARY

Quantum statistics describes the probable distribution of identical particles within a system with reference to their quantized energy states. Particles that have identical energy characteristics cannot be differentiated in an absolute manner. This makes the probability that they occupy a certain energy level an important aspect of their distribution functions.

PRINCIPAL TERMS

- **Boltzmann factor:** the ratio of the probability of finding a particle or system at a certain energy level to the probability of finding it at another energy level; proportional to the probability of the system being in a particular quantum state.
- **boson:** an elementary particle that carries a specific type of force, or a composite particle containing an even number of fermions, having an integer spin.
- **bra-ket notation:** a system of mathematical notation developed by physicist Paul Dirac to manipulate very large vector equations more easily.
- **distinguishability:** the ability of particles with the same energy to be differentiated from one another.
- **distribution function:** a mathematical function that describes the probability that a certain variable, such as the energy state of a particle, will have a given value or range of values.
- **fermion:** an elementary particle, either a quark or a lepton, that is a fundamental unit of matter, or a composite particle containing an odd number of other fermions, and that has a half-integer spin.
- **phase space:** a space containing all possible states of the particles in a given system, wherein each state is represented by a single point.
- **quantum state:** the particular condition of a physical system as defined by its various quantum attributes.

PARTICLES AND PROBABILITY

In a system containing a large number of particles, the average overall behavior of the system may be known. However, it is not possible to know for certain the exact state of each particle. In order to understand the behavior of the system on a macroscopic scale, the behavior of the various particles can be described in terms of probability. Each possible microscopic state, or microstate, of the system is represented as a single point in a multidimensional Summary space known as phase space. A distribution function can then be used to calculate which possible states correspond to the observed macroscopic state, or macrostate, and thus are most likely to be true. These probable states make up what is known as a "statistical ensemble."

The disconnect between microstates and macrostates can be illustrated using the example of an inflated balloon. At a constant temperature, the total energy of all the air molecules in the balloon can be known, as can the average energy of each molecule. However, because the air a person exhales, and thus the air within the balloon, consists of different molecules with different masses, the individual molecules will in fact have different energies that correspond to their respective momenta and positions. Collisions between molecules can impart greater momentum, and therefore greater energy, to one molecule while decreasing the momentum and energy of the other. Thus, while the total energy of the system remains constant, it is the sum of the different energies possessed by the air molecules within the balloon. The actual distribution of energies among the different molecules would be difficult, if not impossible, to

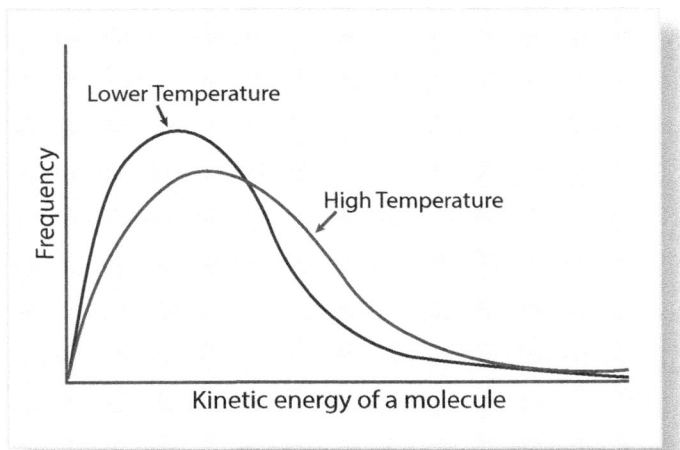

The Maxwell-Boltzman distribution shows the probability that particles will have a particular kinetic energy. The distribution is dependent on temperature. Higher temperatures cause the distribution to have a larger variance (wider range), and particles will be more likely to have higher kinetic energy; lower temperatures cause the distribution to be less varied, and particles will be more likely to have lower kinetic energy.

determine. Despite this, the most probable distributions can be calculated given what is known about the system.

This use of probability theory to determine probable microstates of macroscopic systems (or to extrapolate macroscopic behavior from microstates) is called "statistical mechanics." The rules that govern particle behavior in a system are known as "statistics." In a classical mechanical system, where quantum effects are negligible, the system can be described using Maxwell-Boltzmann statistics, named after Scottish physicist James Clerk Maxwell (1831–79) and Austrian physicist Ludwig Boltzmann (1844–1906).

Below a certain level of complexity, however, quantum effects must be taken into account. This is particularly true with elementary and other subatomic particles. According to the uncertainty principle, proposed by German physicist Werner Heisenberg (1901– 76), one can know either the position or the momentum of a quantum particle, but not both at once. Thus, the behavior of particles in the quantum realm is not readily describable by classical mechanics. Instead, such particles are best described by a wave function, which can be used to determine the probability that a particle will be in a given position (or, if its position is known, have a given momentum) at a particular time. Quantum particles cannot be described using Maxwell-Boltzmann statistics. Instead, they must be described using one of two systems of quantum statistics: Bose-Einstein statistics or Fermi-Dirac statistics.

Distribution Functions

Particles in a system may be distributed among various possible quantum states, and the corresponding energy values of those states, in many different ways. Each potential distribution forms a microstate. While these microstates contain a massive amount of information, the most important is how many particles are in a given quantum state (and therefore at a given energy level). In this respect, "quantum state" and "energy level" mean the same thing.

In classical mechanics, all particles have distinguishability. Even particles of the same type with identical intrinsic properties, such as mass and spin, can be distinguished based on their trajectory or other aspects of their behavior. As a result, exchanging any two particles in a classical system causes changes to the system. In such a system, the distribution of identical (but distinguishable) particles among different energy states is described by the Maxwell-Boltzmann distribution function

$$f(E) = \frac{1}{A e^{E/kT}}$$

where $f(E)$ is the probability that a given particle has the energy E, A is a normalization constant, e is Euler's number, k is the Boltzmann constant (roughly 1.38 joules per kelvin), and T is the absolute temperature of the system. This function can also be written as follows:

$$f(E) = A e^{-E/kT}.$$

Here, the element "$e^{-E/kT}$" is the Boltzmann factor of the system. It is proportional to the probability of finding the system in a given microstate of energy E.

In quantum mechanics, particles do not have distinguishability. Particles are classified as either bosons, which have whole-integer spins, or fermions, which have half-integer spins. The primary difference between them, aside from their spins, is that no two identical fermions can occupy the same quantum state at the same time. This principle, known as the "Pauli exclusion principle," does not apply to bosons.

In a quantum mechanical system, bosons cannot be distinguished from other bosons, and fermions cannot be distinguished from other fermions. Thus, the Maxwell-Boltzmann distribution does not apply.

Bosons are so named because they follow Bose-Einstein statistics, named for Bengali physicist Satyendra Nath Bose (1894–1974) and German-born physicist Albert Einstein (1879–1955). The energy distribution of bosons is described by the Bose-Einstein distribution function:

$$f(E) = \frac{1}{Ae^{E/kT} - 1}$$

While very similar to the first form of the Maxwell-Boltzmann distribution function, this function subtracts 1 from the denominator to account for particle indistinguishability. As bosons are not subject to the Pauli exclusion principle, this function allows an unlimited number of bosons to occupy the same energy level.

Like bosons, fermions got their name from the statistics they follow. Fermi-Dirac statistics was named for Italian physicist Enrico Fermi (1901–54) and English physicist Paul Dirac (1902–84). In some respects, fermions may be thought of as the opposite of bosons. This is reflected in the Fermi-Dirac energy distribution function:

$$f(E) = \frac{1}{Ae^{E/kT} + 1}$$

This function is almost identical to the Bose-Einstein function, with the only difference being that the 1 is added, not subtracted. Because fermions are subject to the Pauli exclusion principle, this function only permits one fermion to occupy each energy level.

Probability and Distinguishability

In classical systems, identical particles are distinguishable by their position and momentum. This normally requires a frame of reference or a coordinate system in which the positions and movements of particles can be described as vectors. A vector is a quantity that describes both the magnitude of a property and its direction.

It is possible to perform mathematical operations on such classical systems independently of the origin point of any coordinate system. The equations for such operations state the relationship between a set of operators and the vector characteristics on which they operate. This enables vector operations within a system to be described in a relative sense rather than an absolute sense. In other words, the effect of an operation on a vector property can be determined regardless of where that property is located within the overall system. Though such mathematical relations are highly complex, they can be greatly simplified using bra-ket notation, introduced by Dirac.

Bra-ket notation is also commonly applied to quantum systems, although a number of restrictions arise. In matrix operations, for example, only certain values are allowed. Such values, called "eigenvalues," produce an answer that is simply a multiple of the original vector. This relationship is what determines the opposite designations of bosons and fermions in mathematical expressions and limits elementary fermions to spin values of +1/2 and −1/2. (Composite fermions can also have spin values of +3/2 or −3/2.) It is also what prevents two identical fermions from occupying the same energy level.

It is important to understand how the number of particles in different energy levels relates to observable physical properties, such as absorption and emission spectra. Such spectra reflect the transitions of electrons (a type of lepton, or elementary fermion) from one allowed energy level to another. Knowing the probability that electrons of a certain energy occupy certain allowed energy levels enables one to predict whether a specific energy transition will be observed in a system at a given temperature or energy level. Similar considerations are important in the study of subatomic particle interactions during high-energy nuclear collision experiments.

—*Richard M. Renneboog, MSc*

Bibliography

Bub, Jeffrey. *Interpreting the Quantum World.* 1997. New York: Cambridge University Press, 1999. Print.

Hansen, Klavs. *Statistical Physics of Nanoparticles in the Gas Phase.* Dordrecht: Springer, 2013. Print.

Longair, Malcolm. Quantum *Concepts in Physics: An Alternative Approach to the Understanding of Quantum Mechanics.* New York: Cambridge University Press, 2013. Print.

Mandl, Franz. *Quantum Mechanics.* 1992. New York: Wiley, 2013. Print.

Pathria, R. K., and Paul D. Beale. *Statistical Mechanics.* 3rd ed. Burlington: Butterworth, 2011. Print.

Swanson, D. G. *Quantum Mechanics: Foundations and Applications.* Boca Raton: CRC, 2007. Print.

QUARKS

FIELDS OF STUDY

Quantum Electrodynamics; Quantum Field Theory; Quantum Mechanics; Relativity; String theory; Superstring theory

SUMMARY

A quark is a fundamental particle. It has an antiparticle called an antiquark. Quarks join together to form hadrons, such as protons and neutrons. Quarks are not independent; they are always in combination with other quarks. Because they are never alone, quarks have never actually been seen, but their behavior has been observed. There are six types, or "flavors," of quarks: Up, down, strange, charm, bottom, and top.

PRINCIPAL TERMS

- **color charge:** a property of quarks that distinguishes quarks and gluons from each other.
- **color force:** the force of the strong interaction that operates on the quark level.
- **elementary particle:** one of the fundamental constituents of matter.
- **fermion:** one of two main classes of particles, characterized by adherence to Fermi-Dirac statistics; includes quarks, leptons, and any particles that contain an odd number of quarks or leptons, such as protons and neutrons.
- **hadron:** a subatomic particle that is made of either three quarks or one quark and one antiquark and held together by the strong force.
- **quantum chromodynamics:** a quantum field theory that describes the interactions of quarks and gluons, subatomic particles that are responsible for the strong interaction.
- **spin:** an intrinsic form of angular momentum carried by elementary particles, composite particles (hadrons), and atomic nuclei.
- **strong force:** the force that holds quarks together.
- **strong interaction:** the fundamental process of particle interaction that binds quarks into hadrons and hadrons into nuclei.

DISCOVERY OF QUARKS

In 1964, Murray Gell-Mann and George Zweig postulated that the hundreds of particles known at that time, that were thought to be the foundation of the universe, could all be made from combinations of only three elementary particles, fundamentals that would form the basis of everything and were so small that they couldn't be broken up into anything smaller. They chose the word "quark" based on a nonsense word from James Joyce's novel, *Finnegan's Wake* ("Three quarks for Muster Mark!"). To make the math regarding the electrical charges that they were observing work correctly, they assigned fractional electrical charges, which was a new concept at the time, to the quarks. For quite some time, quarks were thought to be just an imaginary way to explain something that had no other explanation. Eventually, experiments proved that quarks do exist, and that there are at least six types of quarks. The last to be discovered was the top quark in 1995.

Quarks were first observed at the Stanford Linear Accelerator National Laboratory in 1968. There, scientists used particle accelerators to smash subatomic particles into small bits by accelerating them to incredible speeds, then smashing the particles into a target material. These high-speed crashes used so much energy that they made the protons and neutrons inside the atoms break apart. Scientists could then observe particle "tracks" that scattered in just the way they had predicted that quarks would behave. Scientists believe that one can't see a quark in isolation because of the color force holding them together so tightly. If one uses enough energy to separate quarks from each other, they form quark/antiquark pairs before they are far enough apart to be seen as a separate entity.

Types and Properties of Quarks

The study of quarks is called quantum chromodynamics. Quarks are only one of the types of the particles which make up matter. They belong to a class of particles called fermions. There are six quarks, which are known by their "flavor," but since they are always paired, scientists usually talk about them in three groups: Up/down, charm/strange, and top/bottom. Most matter is made of protons and neutrons, which are made from quarks. For each quark, there is also an anti-quark.

Hadrons are composite particles made of quarks. The two types of hadrons are as follows:
- Baryons: a hadron made of three quarks, such as protons (made of two up quarks and one down quark) and neutrons (made up of one up quark and two down quarks);
- Mesons: a hadron made of one quark and one antiquark, such as a pion (made of an up quark and a down antiquark, a very unstable combination).

In 2013, scientists claimed to have found a new type of particle: One made of four quarks. Now there are even claims of a five-quark particle, which, through further testing, appear to be false. None of these particles fit the theory of the standard model of physics, which is the framework that physicists use to describe all the currently known elementary particles, at least in theory.

Quarks have mass, but because they are never observed alone, it is impossible to isolate them and measure the mass of each flavor of quark in a direct manner. The mass of quarks is implied from the scattering experiments described above.

Quarks have a fractional electric charge, which is unusual. For example, an up quark has a charge of +2/3 and a down quark has a charge of -1/3. Therefore, a proton, which is made of two up quarks (+2/3) + (+2/3) and a down quark (-1/3) has a positive electrical charge of +1. Quarks also have spin. Their spin is ½, which means they are fermions, as anything with a spin of 1 is a boson.

Quarks have color charge, though one should not confuse this naming convention with the traditional understanding of color. This is just a way to explain how quarks can behave according to a well-known principle stating that no two identical objects can occupy the same place (the Pauli Exclusion Principle). Quarks can be red, blue, or green, and antiquarks can be anti-red, anti-blue, or anti-green. Quarks making up the same hadron must have different colors; for example, all three quarks in a baryon are different colors, and a meson contains a colored quark and its corresponding anti-colored quark. This means that a baryon, with all three colors, is color neutral, just the way that if red, green, and blue light are emitted together, the light is white. Mesons are color neutral because they contain one color and its corresponding anticolor. According to the standard model, quarks can only combine in certain ways based on their charge and color.

The strong force holds quarks together with the strong interaction. The electrical charges of the quarks work along with gluons (the particles that carry the strong force) to keep the hadrons such as protons and neutrons, and, therefore, the entire universe, together. Gluons also have color charge, but composite particles made of quarks and gluons are color neutral. When two quarks get close enough, they exchange gluons, which binds them together with a very strong color force field. Surprisingly, as quarks get further apart, the bond becomes stronger and stronger, and as they move closer together, the bond becomes weaker. Quarks are constantly changing their color charges as they continuously exchange gluons with other quarks.

Why Are Quarks Important?

Quarks are one of the basic particles that make up our universe. By understanding them and the forces that hold them together and the other properties that they exhibit, we can have a better understanding of our universe as a whole. Scientist do not know if there is anything even smaller than a quark, but the search continues.

—*Marianne Moss Madsen, MS*

Bibliography

Bahr, Benjamin, Boris Lemmer, and Rina Piccolo. *Quirky Quarks: A Cartoon Guide to the Fascinating Realm of Physics.* Springer, 2016.

Blaha, Stephen. *The Periodic Table of the 192 Quarks and Leptons in The Theory of Everything: The U(4) Layer Group, Physics is Logic.* Pingree-Hill Publishing, 2016.

Blake, Calvin. *World of Quarks*. Amazon Digital Services, LLC, 2015. Electronic.

Close, Frank. *The New Cosmic Onion: Quarks and the Nature of the Universe*. CRC Press, Revised Edition. 2006.

Economou, Eleftherios N. *From Quarks to the Universe: A Short Physics Course*. Springer, 2015.

Heilbron, J.L. *Physics: A Short History from Quintessence to Quarks*. Oxford University Press. 2016.

Kisak, Paul F. (ed.). *The Quark: A Fundamental Constituent of Matter.* CreateSpace Independent Publishing Platform, 2016.

Riggs, Shelton. *The Nuclear Force: The Force Which Binds Quark Trios Into Protons and Neutrons*. Amazon Digital Services, LLC, 2015. Electronic.

RADIANS AND DEGREES

FIELDS OF STUDY

Classical Mechanics; Electronics; Harmonics

SUMMARY

From the development of astronomy by ancient civilizations to the internal combustion engine, rotations about a center have been an integral part of the development of scientific theories and advances in engineering. Though its origins are not known exactly, the degree as a unit of measurement was originally developed by Babylonians and Persians. The concept of the radian was developed in the early eighteenth century by British mathematician Roger Cotes. This article explains these basic units of angular separation.

PRINCIPAL TERMS

- **angle:** the separation along a curved path between two rays originating at the same point.
- **arclength:** the length of an arc defined by an angle and two radii.
- **circumference:** the distance around a circle defined by a radius.
- **cosine:** in a right triangle, the ratio of the length of the side adjacent to an acute angle to the length of the hypotenuse.
- **pi:** the ratio of the circumference of a circle to its diameter.
- **radius:** distance from the center of a circle to any point along the circle.
- **sine:** in a right triangle, the ratio of the length of the side opposite an acute angle to the length of the hypotenuse.
- **tangent:** in a right triangle, the ratio of the length of the side opposite an acute angle to the length of the side adjacent to the same angle.

Origins of Degrees and Radians

For centuries, scientists and mathematicians used the unit of degrees (°) to measure the size of an angle. The origin of the degree is not exactly known. One theory is that it originated in ancient astronomy. Previous civilizations started to notice that the positions of the stars were the same every 360 days—very close to a year. To represent a year, Babylonian and Persian mathematicians divided the circumference of a circle into 360 spaces or units. This unit became known in ancient Greece as the *moira*, which was translated into Arabic as *daraja*, meaning a step on a scale or ladder. In Latin, this became *de gradus*, and in English, *degree*. With the development of

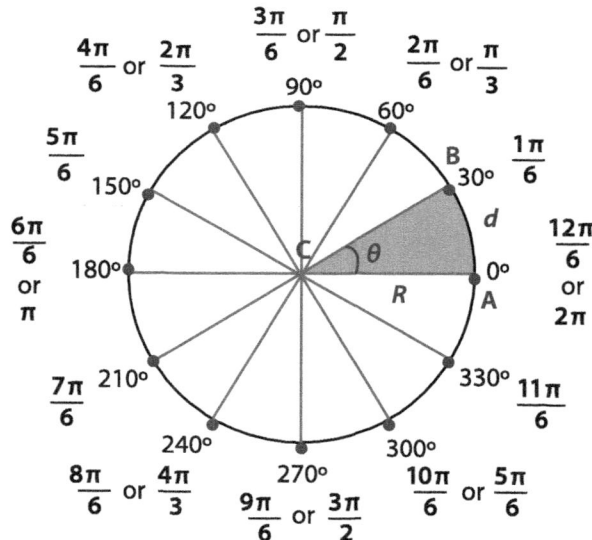

$$d(A,B) = R\theta = R(\pi/6)$$

A circle divided into twelve sections with angle measurements in degrees and radians for each section. Each section is 30 degrees or 1/6th of the value for pi. Although both of these units are used to measure angles, they cannot both be used in formulas for other calculations. The length of an arc (d) can be calculated by multiplying the radius by the angle in radians, but not in degrees.

> **SAMPLE PROBLEM**
>
> Consider two angles, one 125° and the other $5\pi/7$ rad. What are the values of the angles in radians and degrees, respectively, given that π has an approximate value of 3.14?
>
> **Answer:**
> To calculate the first angle in radians, multiply 125° by $\pi/180$:
>
> $$\theta_r = \theta_d \frac{\pi}{180}$$
>
> $$\theta_r = 125° \frac{\pi}{180}$$
>
> $$\theta_r = \frac{(125°)(3.14)}{180}$$
>
> $$\theta_r = 2.18 \text{ rad}$$
>
> The first angle measures approximately 2.18 radians.
>
> To calculate the second angle in degrees, multiply $5\pi/7$ rad by $180/\pi$:
>
> $$\theta_d = \theta_r \frac{180}{\pi}$$
>
> $$\theta_d = \left(\frac{5\pi}{7} \text{ rad}\right)\left(\frac{180}{\pi}\right)$$
>
> $$\theta_d = \frac{5 \times 180}{7}$$
>
> $$\theta_d = 128.57°$$
>
> The second angle measures approximately 128.57 degrees.

trigonometry, the mathematics of angles, mathematicians and scientists found they needed a way to represent angular measurements as numeric values without arbitrarily determined units such as degrees.

In 1714, British mathematician Roger Cotes (1682–1716) published a paper titled "Logometria." In it, he developed the idea of an angular unit derived from the ratio of an arclength along the circumference of a circle to the radius of the circle. He defined this unit, now known as the radian, as the angle corresponding to an arclength that is equal to the radius of the circle. This measurement of angular separation is approximately equal to 57.3 degrees. The unit of radians uses no arbitrary definitions as to what it means. It can therefore easily be used in mathematical applications ranging from trigonometry to calculus. In trigonometry, the basic functions of the sine, cosine, and tangent of an angle are defined as the ratios between the lengths of two sides of a right triangle. Since the radian is a ratio of two lengths, it has no units and can be used in trigonometry. In calculus, the need for a unit-less quantity was clear when trying to calculate simple harmonic motions of molecules and other simple oscillators.

CALCULATING RADIANS

The circumference of a circle is equal to $2\pi r$, where r is the radius of the circle and π is pi, a mathematical constant approximately equal to 3.14. Because a radian corresponds to the central angle of an arclength equal to one radius, there are 2π radians in a circle. A full circle is 360 degrees around. For a circle with a radius equal to one, therefore, one radian is equal to $360/2\pi$ degrees, or $180/\pi$ degrees. Conversely, one degree is equal to $\pi/180$ radians. Multiply the measurement of an angle in degrees (θ_d) by $\pi/180$ to convert the angle to radians (θ_r):

$$\theta_r = \theta_d \frac{\pi}{180}$$

For example, if an angle measures 120°, that same angle in radians (rad) is

$$\theta_r = \theta_d \frac{\pi}{180}$$

$$\theta_r = 120° \frac{\pi}{180}$$

$$\theta_r = \frac{2\pi}{3} \text{ rad}$$

To convert radians to degrees, simply multiply by $180/\pi$:

$$\theta_d = \theta_r \frac{180}{\pi}$$

For example, if an angle measures $\pi/2$ rad, that angle in degrees is

$$\theta_d = \theta_r \frac{180}{\pi}$$

$$\theta_d = \left(\frac{\pi}{2} \text{ rad}\right)\left(\frac{180}{\pi}\right) = \frac{180}{2}$$

$$\theta_d = 90°$$

Applications of Radians and Degrees

The radian and degree both remain useful units of measurement. The degree is still used in navigation and astronomy. It is also used by the general public to explain rotations and revolutions. On the other hand, the radian holds a prominent role in scientific and mathematic developments. It is the International System of Units (SI) derived unit of angular measurement. Without the use of radians, scientists would not be able to properly calibrate electric generators and the angular frequencies at which they spin.

—*Angel G. Fuentes, MS*

Bibliography

Giambattista, Alan, and Betty McCarthy Richardson. *Physics*. 3rd ed. New York: McGraw, 2015. Print.

Hemphill, Boyd E., and John C. Polking. "Degree/Radian Circle." Math.Rice.edu. Polking, 4 May 1998. Web. 21 July 2015.

"Introduction to Radians." Khan Academy. Khan Acad., 2015. Web. 21 July 2015.

Joyce, David E. "Dave's Short Trig Course: Measurement of Angles." Clark University. Dept. of Mathematics and Computer Science, Clark U, 2013. Web. 21 July 2015.

"Measuring Angles in Degrees." Khan Academy. Khan Acad., 2015. Web. 21 July 2015.

Weisstein, Eric W. "Degree." Wolfram MathWorld. Wolfram Research, 1999–2015. Web. 21 July 2015.

Young, Hugh D., Roger A. Freedman, A. Lewis Ford, and Francis Weston Sears. *Sears & Zemansky's College Physics*. 14th ed. Boston, Mass: Pearson, 2016. Print.

RADIATION

FIELDS OF STUDY

Electromagnetism; Atomic Physics; Nuclear Physics

SUMMARY

Radiation is energy moving as electromagnetic waves or subatomic particles. Most radiation falls within the electromagnetic spectrum. Electromagnetic radiation is divided into categories based on wavelength. Some common activities associated with radiation include receiving music translated into radio waves or applying sunscreen to prevent skin damage from ultraviolet radiation.

PRINCIPAL TERMS

- **alpha radiation:** alpha (α) particles typically emitted during alpha decay, a subtype of radioactive decay.
- **beta radiation:** beta (β) rays emitted during beta decay, a subtype of radioactive decay.
- **gamma radiation:** electromagnetic radiation with a wavelength shorter than 1×10^{-11} meters; gamma (γ) rays typically emitted during gamma decay, a subtype of radioactive decay.
- **inverse-square law:** radiation emanating from a single point has an intensity inversely proportional to the square of the distance the radiation has traveled from its source.
- **radio waves:** electromagnetic radiation with a wavelength between 1×10^{-3} and 1×10^{5} meters; able to travel long distances without being broken up by atmospheric interference.
- **thermal radiation:** electromagnetic radiation generated by charged particles in matter being moved around by heat; typically associated with infrared radiation and a frequency of 7×10^{-7} meters to 1×10^{-3} meters.
- **ultraviolet (UV) radiation:** electromagnetic radiation with more energy than visible light but less than x-rays, in the wavelength range of 4×10^{-7} to 1×10^{-7} meters.
- **visible light:** electromagnetic radiation that human eyes can see, with a wavelength between 4×10^{-7} to 7×10^{-7} meters.
- **wavelength:** the distance between crests or troughs of a wave.
- **x-ray:** electromagnetic radiation with a wavelength between 1×10^{-10} and 1×10^{-8} meters.

The Electromagnetic Spectrum

The energy transmitted through space from the sun to Earth, from a radio station to a radio, and from a television to an eye all fall somewhere on the electromagnetic (EM) spectrum. The EM lays out the various forms of radiation according to wavelength.

All electromagnetic energy is transmitted as a wave. Each wave has a crest (high point) and a trough (low point). Wavelength measures the distance between two crests or two troughs. This is the distance the wave moves in one complete cycle. The frequency of these cycles—waves per second—is measured in hertz. One hertz is equivalent to one cycle per second.

The energy carried by a wave is directly proportional to its frequency. Higher frequency means more energy; lower frequency, less energy. Furthermore, energy and frequency are both inversely related to wavelength. Longer wavelengths have less energy and lower frequencies. EM waves emitted from a single point (for example, a radio tower) lose their intensity according to an inverse-square law. The farther they travel from their point of origin, the weaker they get.

Maxwell, Rutherford, and Villard

The history of electromagnetic radiation in modern physics begins with Scottish physicist James Clerk Maxwell. In 1865, he published a theory to explain unifying electricity, magnetism, and light as a single phenomenon (known as EMR). He demonstrated that the interaction between electrical fields and magnetic fields can transmit energy through space as a wave. In the 1880s, German physicist Heinrich Hertz used Maxwell's theory to prove the existence of radio waves.

In the late 1880s, physicists Ernest Rutherford and Paul Villard used ingenious experiments to detect and describe radioactive decay and the radiation it produces. Their work laid the groundwork for modern nuclear physics. Rutherford tested the penetrating power of the radiation from uranium. He detected the presence of alpha and beta particles. These particles transmit alpha and beta radiation. Meanwhile, Villard used similar experiments with radium and discovered an extremely penetrating form

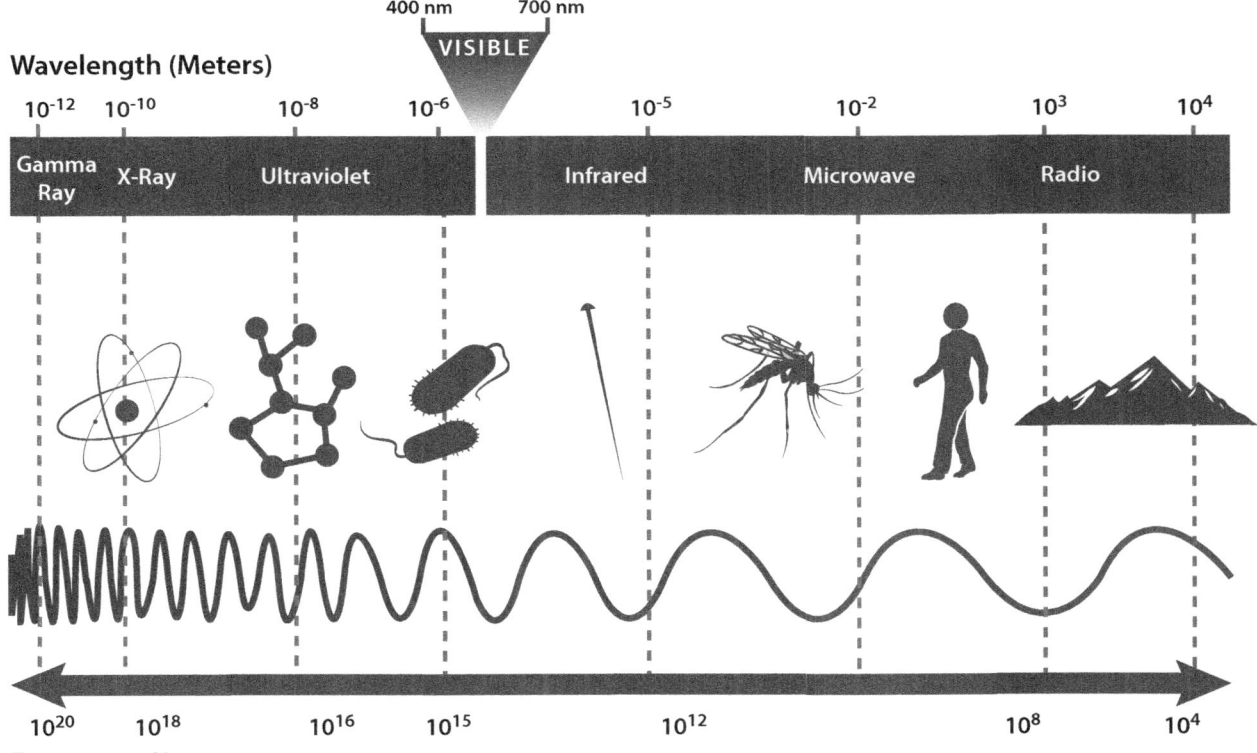

The electromagnetic spectrum is the range of all possible electromagnetic radiation wavelengths. Visible radiation in the form of visible light falls between about 4.0×10^{-7} m to 7.0×10^{-7} m. Radio waves can be as large as mountains. Gamma radiation has a short wavelength, smaller than atoms.

of EMR. Rutherford dubbed these gamma rays (also known as gamma radiation).

TYPES OF RADIATION

Each type of electromagnetic radiation has unique properties that set it apart from others. These are described in ascending order of wavelength.

Gamma radiation has a wavelength of less than 1.0×10^{-11} meters. It comes from gamma decay. In this type of radioactive decay a photon (high-energy particle) is ejected from the nucleus of an atom. On Earth, this is emitted from radioactive isotopes of elements like radium. It may also occur as the result of the tremendous energy of a lightning strike.

X-rays have wavelengths between 1×10^{-10} and 1×10^{-8} meters. The wavelengths of x-rays are just small enough that they can pass through human flesh with little interference. They cannot pass through more dense materials such as bone. This is why x-rays are safe for medical use.

Ultraviolet radiation (UV rays or UV light) has a wavelength of 4×10^{-7} to 1×10^{-7} meters. The most familiar source of UV rays is the sun. UV rays are what cause skin tans or burns, as well as skin cancer.

Visible light (simply "light" in everyday speech) is the narrow band of EMR that human eyes can detect, with wavelengths between 4×10^{-7} to 7×10^{-7} meters. This range is further subdivided into all the colors of the rainbow, which correspond to specific wavelengths. For example, blue has the shortest wavelengths of visible light. Red has the longest. The "visible" designation is somewhat arbitrary, as several animals are able to see parts of the EMR spectrum humans cannot. For example, bees can see into the ultraviolet range.

Infrared has a wavelength of 7×10^{-7} meters to 1×10^{-3} meters and is closely associated with radiating heat (thermal radiation) that is invisible to the naked eye. This is why people can feel heat coming off their skin on a cold day. Some objects emit thermal radiation as visible light as well (objects so hot they glow).

Radio waves (which include microwaves) have wavelengths between 1×10–3 and 1×10^5 meters. These versatile waves are able to cover large areas of the earth at once and so have wide use as a method of communication. Radio waves can transmit information as variations in the waves. Radios, televisions, and other devices pick these up and turn them back into usable information. Any EMR with a longer wavelength than radio (greater than 1×10^5 meters) is referred to by the generic term "longwave" radiation.

RADIATION AND RADIOACTIVITY

The term "radioactivity" was coined by Marie and Pierre Curie as a result of their intensive study of radioactive materials such as the ore called pitchblende. "Radioactivity" refers to a specific type of radiation, ionizing radiation, which occurs in the EMR at wavelengths shorter than those of visible light. An object or substance is radioactive only if it emits ionizing radiation. This is the dangerous radiation associated with nuclear bombs, uranium, and overexposure to the sun. In short, ionizing radiation has so much energy that it can remove electrons from atoms, including those of living things. Changing the atomic makeup of a living thing may have catastrophic results. These results can include an increased risk of cancer as a result of ionizing radiation damaging DNA and inducing mutations. Fortunately, ionizing radiation at low doses is essentially harmless. Life on Earth is subjected to some varying level of functionally harmless background radiation, including ionizing radiation, at all times.

RADIATION'S MANY USES

EMR has become increasingly central to communications technology. EMR is used to send information to and from satellites as microwave radio waves, to broadcast television channels (including high definition) to satellite dishes and antennas in homes, to disinfect objects using UV radiation, and even to produce art by using paints that only reflect certain colors of light from the visible spectrum. Whether microwaving a meal, viewing a famous painting, or using x-rays to detect a broken bone, EMR makes possible the transmission of energy and information without physical contact.

—*Kenrick Vezina, MS*

BIBLIOGRAPHY

"Anatomy of an Electromagnetic Wave." *Mission: Science*. Natl. Aeronautics and Space Administration, 2010. Web. 7 May 2015.

Davidson, Michael W. "The Rutherford Experiment." *Molecular Expressions: Electricity and Magnetism*. Florida State U, 28 Feb. 2014. Web. 7 May 2015.

Errede, Steven. "A Brief History of the Development of Classical Electrodynamics." *UIUC Physics 245*. Loomis Laboratory of Physics, University of Illinois at Urbana-Champaign. 2007. Web. 7 May 2015. Lecture.

Radosevich, James A. *UV Radiation: Properties, Effects, and Applications*. New York: Nova, 2014. Print.

Spector, Dina. "How Do Microwaves Cook Food?" *Business Insider*. Business Insider, 10 June 2014. Web. 7 May 2015.

Yip, Sidney. *Nuclear Radiation Interactions*. Hackensack: World Scientific, 2014. Print.

RESONANCE: NUCLEAR MAGNETIC RESONANCE AND ELECTRON SPIN RESONANCE

FIELDS OF STUDY

Quantum Physics; Atomic Physics; Nuclear Physics

SUMMARY

Resonance in regard to nuclear and electron spin resonance (NMR and ESR) refers to the absorption of energy by nuclei and by electrons while in a magnetic field. In both cases the component of the spin magnetic moment in the direction of the applied field is quantized and changing this component requires the absorption or emission of a photon of the necessary energy. Both methodologies provide structural information about molecules with unpaired electron or nuclear spin, and both are important analytical techniques in chemical, forensic and medical analysis and diagnostics.

PRINCIPAL TERMS

- **angular momentum:** the rotational momentum of an object around an axis, defined as the product of its moment of inertia and its angular velocity.
- **electromagnetic field:** a physical field consisting of a combined electric field (generated by stationary electric charges) and magnetic field (generated by moving electric charges) that affects the behavior of charged objects in its vicinity.
- **magnetic flux:** the measure of the strength of the magnetic field surrounding an active electrical conductor or a magnet, equal to the product of the external magnetic field (B), the area of the affected surface (A), and the cosine of the angle between them ($cos\theta$).
- **quantum mechanics:** the branch of physics that deals with matter interactions on a subatomic scale, based on the concepts that energy is quantized, not continuous, and that elementary particles exhibit wavelike behavior.
- **quantum state:** the condition of a physical system as defined by its associated quantum attributes.
- **resonance:** the response of an elastic body to a force acting on the body at its natural frequency.
- **spin:** an intrinsic form of angular momentum carried by elementary particles, composite particles (hadrons), and atomic nuclei.

ELECTRONS AND PROTONS IN ATOMS AND MOLECULES

The modern theory of atomic structure, which is supported by a great deal of experimental observation spanning more than a century, is based on the theory of quantum mechanics. The theory is able to describe the structure of atoms in a way that accounts well for, or predicts, the behavior of atoms and molecules. Each atom is ascribed a particular quantum state that is defined by various quantum numbers. The basic principle of the quantum state is that the components of the atom can possess only specific amounts, or quanta, of energy that correspond to a specific energy level. Any transitions to or from a different energy state require the absorption or emission of the specific quantum of energy that corresponds to the energy difference between those states. No other energies are allowed.

The quantum is essentially the smallest particle of energy that corresponds to a specific wavelength of electromagnetic radiation. One of the specific quantum numbers associated with a particular quantum state of an atom is termed spin. Another is the angular momentum of the particle. In a particular quantum state, an electron is assigned either a positive or a negative spin quantum number. The

Pauli exclusion principle restricts the number of electrons that can occupy the same region of space about the atomic nucleus, or orbital, to a maximum of two. Furthermore, the two electrons must have opposite spin quantum numbers. A similar condition applies to the protons in the nucleus of the atom in regard to their magnetic quantum number, although there is no analogous restriction to the Pauli exclusion principle.

A nucleus with a magnetic moment will interact with a magnetic environment. The ability of the quantum state of electrons and nuclei to be changed by an external force is the basic principle of the processes known as electron spin resonance (ESR) and nuclear magnetic resonance (NMR). Electron spin resonance is also known as electron paramagnetic resonance, or EPR. In both ESR and NMR, resonance occurs when the atoms in a molecule absorb energy corresponding to specific wavelengths when irradiated with a range of electromagnetic wavelengths while in a constant electromagnetic field. The pattern of absorption peaks is characteristic to the molecular structure and electronic configuration of the molecule, and analysis of the patterns provides essential information about the molecule

NMR typically is used to determine the three-dimensional structure of molecules and can be used to monitor the progress of reactions if the materials involved have distinct peaks that can be unequivocally assigned to a specific compound. ESR, on the other hand, does not provide structural information about a molecule. Rather, the methodology provides detailed information about the disposition of electrons within the molecule, and produces signals only with compounds that have an unpaired electron.

Structural and Electronic Analysis

NMR is an extremely useful analytical technique. The method relies on the resonance of the magnetic moment of a nucleus with the particular radiofrequency of electromagnetic radiation. In the technique, a sample of a test material is prepared and set to spin within a field having a constant magnetic flux. The sample is then irradiated with the designed electromagnetic frequency. The magnetically susceptible atoms within the molecule will absorb specific energies from the range of frequencies used. The base frequency of the instrument depends on the particular nucleus being used.

The most common NMR spectrometers are used to examine the hydrogen atoms within a molecule, and is commonly referred to as proton NMR. However, many other nuclei are active in NMR. Some others that are commonly used to examine organic (carbon-based) compounds are ^{13}C and ^{31}P. Inorganic compounds may also be analyzed using NMR based on specific metal atoms.

Each nuclear type has its own base frequency range, and all absorption assignments are made relative to an internal standard compound. For example, absorptions due to protons do not appear in the range of ^{31}P NMR absorptions, and vice versa. The pattern of peaks in an NMR spectrum of any kind is determined by the interaction of the particular susceptible nucleus with its neighboring susceptible nuclei. Each has its own local magnetic environment that is affected by the interaction with its neighbors when an absorption occurs. In a ^{1}H-NMR spectrum, for example, the pattern of absorption peaks follows an $n + 1$ rule, in which n represents the number of adjacent hydrogen atoms with the same magnetic environment. A lone methyl group ($-CH_3$) has no neighboring protons, and so appears as a single sharp peak. An ethyl group ($-CH_2CH_3$), however, has a more complicated pattern. The methyl group has

---

SAMPLE PROBLEM

The NMR spectrum of a certain compound obtained at a resolution of 60 MHz showed a doublet of peaks at δ 1.01 and a quartet at δ 1.76. The integration ratio of the two sets of peaks was 3:2, indicating an ethyl group. However, the spectrum of the same compound obtained from a spectrometer of higher resolution revealed that the sets of peaks were each actually composed of peaks exhibiting fine splitting. What does this indicate?

Answer:

Adjacent protons groups interact with each other and split their corresponding NMR signals equally. In this case, the integration ratio indicates an ethyl group. However, because one set of peaks exhibits mutual splitting and the other set does not, there must be two separate ethyl groups in structural environments that are very similar but not exactly identical.

two neighboring protons, and therefore appears as a triplet while the methylene protons have the three neighboring protons of the methyl group and so appear as a quartet of peaks.

The integration function of an NMR spectrometer measures the area under each absorption peak as a measure of the proportional number of protons represented by the peaks. Integration of the peak pattern of the ethyl group, for example, would exhibit a ratio of two to three, for the two methylene protons and the three methyl protons. Note that this same proportionality holds no matter how many ethyl groups may be present in a molecule, and other considerations are needed to differentiate between them.

The internal standard used for 1H-NMR and other spectra is a compound called tetramethylsilane, or TMS. The peaks that are recorded in the absorption spectrum are assigned by their location downfield relative to the sharp, distinct TMS peak. The relative locations of the peaks are given in δ (delta) values, with each δ being the equivalent of 1 part per million of the base frequency. Generally, the richer the magnetic environment is in electrons, the farther downfield the particular peaks are removed from the TMS peak. The protons attached to a benzene ring, for example, absorb at frequencies approximately 7 ppm downfield from TMS, while those of a methyl group are often slightly less than just 1 ppm downfield.

The separation of the peaks in a group of peaks is also an important factor in analyzing an NMR spectrum. The distance between adjacent peaks in a group, termed the splitting of the group, can be used to identify adjacent groups of atoms. In the ethyl group example above, the splitting of the peaks in the methyl triplet is precisely the same as the splitting of the peaks in the quartet. This is because as adjacent groups they influence each other equally. There is a similar effect in an ESR spectrum, and an ESR spectrum typically shows a more-or-less symmetrical oscillating wave pattern centered on a particular energy that is absorbed.

Fine Detail
An NMR spectrum can have a great deal of fine detail information, depending on the resolution of the spectrometer. The higher the base frequency of the spectrometer, the finer is the detail resolution that can be observed. A spectrum obtained from a 30MHz NMR spectrometer, for example, is very crude compared to the spectrum of the same sample that can be had from the same machine running at 100 MHz, and positively primitive compared to one obtained from a machine operating at 500 MHz.

The finer the resolution that is achieved, the more precise is the information that can be obtained from an NMR spectrum. Single peaks in a lower power spectrum often exhibit very fine splitting in the spectrum from a higher powered machine, and all of the additional information has meaning in determining the structural relationships of the atoms within the particular molecule and the manner in which the structure of the molecule fluctuates over time.

The capabilities of NMR spectroscopy have been expanded to produce two and three dimensional spectra. Similarly, the peaks in an ESR spectrum are produced by the absorption of energy by electrons as their quantum spin property aligns with the applied magnetic field in either a positive or negative manner. The width of the peaks depends on the rather random interaction of the particular electron with the local magnetic microenvironments produced by adjacent electrons. Here as well significant structural information, especially in regard to the quantum states of the electrons, can be obtained from the "hyperfine splitting" that is observed.

Applications of NMR and ESR
The technologies have taken on very important roles. NMR, and to a lesser extent ESR, are workhorse techniques of analytical chemistry and forensic analysis. ESR is often utilized by historians and archaeologists as a specialized means of determining the age of artifacts. NMR has become the premier tool of medicine as Magnetic Resonance Imaging (MRI), capable of monitoring and observing chemical processes in the human body and visualizing them in three dimensional images. This particular technique is termed functional MRI, or fMRI.

—*Richard M. Renneboog M.Sc.*

Bibliography
Skrabal, Peter M. *Spectroscopy. An Interdisciplinary Integral Description of Spectroscopy From UV to NMR.* Zürich, SW: Hochschulverlag AG, 2012. Print.

Pochapsky, Thomas C., and Susan Sondej Pochapsky. *NMR for Physical and Biological Scientists*. New York: Taylor & Francis, 2007. Print.

Vlaardingerbroek, Marinus T., and Jacques A. Den Boer. *Magnetic Resonance Imaging: Theory and Practice*. 3rd ed. Berlin [u.a.]: Springer, 2004. Print.

Weil, John A., and James R. Bolton. *Electron Paramagnetic Resonance: Elementary Theory and Practical Applications*. 2nd ed. Hoboken, N.J.: Wiley-Interscience, 2007. Print.

REVOLUTIONS PER MINUTE (RPM)

FIELDS OF STUDY

Classical Mechanics

SUMMARY

Revolutions per minute (RPM) is a term for quantifying rotational movement around a fixed internal point. Although not recognized by the International System of Units (SI), it is a useful measure of rotational speed in certain contexts.

PRINCIPAL TERMS

- **angular velocity:** the speed and direction of movement of a rotating object.
- **hertz:** the SI unit of frequency; one hertz (Hz) is equal to one cycle (complete orbit) per second.
- **International System of Units (SI):** a standardized set of units and measures used by scientists worldwide; the metric system.
- **frequency:** the amount a cyclical event occurs during a set time unit.
- **radians:** (rad) the SI unit of measure for angles, based on relationship between the radius and circumference of a circle.
- **revolution:** describes circular motion wherein an object circles around an external axis (e.g., the moon orbiting Earth); contrast to rotation, wherein the axis is internal (e.g., the moon spinning about its axis).

QUANTIFYING ROTATIONAL MOTION

RPM (revolutions per minute) is a measurement of how many times in one minute an object completes a full rotation around a fixed axis. It is not recognized as an official unit by the International System of Units (SI) due to semantic issues with the word revolution and because it does not follow the SI base unit of time (seconds). Official SI measurements of frequency use the unit hertz (Hz). One hertz is equal to one complete cycle per second, so 1 hertz is equal to 60 RPM. RPM is commonly used in describing rotation in machines such as engines, while other applications use hertz.

ROTATIONAL MOVEMENT VERSUS LINEAR MOVEMENT

In rotational movement, distance traveled is calculated as a proportion of the circle's circumference. This can be visualized as a slice of pizza. The length of the outside crust is distance traveled, the two straight sides are each equal to the radius of the circle, and the angle formed by the two sides is the angle of rotation. In SI, angles are measured in radians (rad), which are based on the relationship of a circle's radius to its circumference. By definition, there are 2π radians in a circle. The radian per second (rad/s) is the SI unit of angular frequency, as well as angular velocity. 2π rad/s is equal to 1 hertz or 60 RPM.

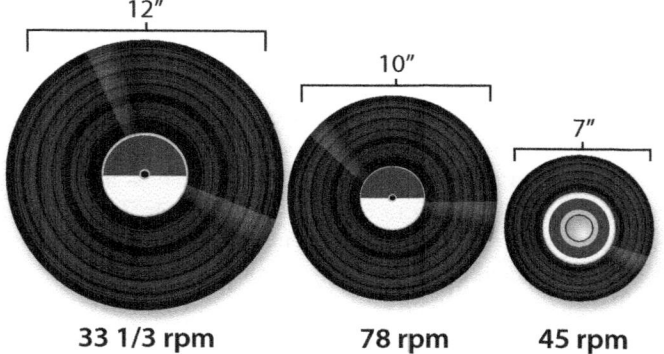

Different sized records were designed to have a needle pass through the grooves at a particular speed so that the sound plays correctly. When a record is spun at a faster RPM than it was written for, the frequency of the sound waves produced are increased. When a record meant to be turned at 45 RPM is turned at 33 1/3 RPM the sound production is slowed, the frequency is lower, and the human ear interprets this as a deeper tone.

Applications

RPM is not a standard unit, but it is still widely used and widely useful. It is most familiar as a measure of automobile engine operating speed. The phenomenon it quantifies, rotational movement, is even more important. It is vital that engineers understand rotational motion in order to build working machines.

—*Kenrick Vezina, MS*

Bibliography

"Angular and Linear Velocity." AlgebraLAB. Mainland High School, 2015. Web. 7 May 2015.

"Frequency and Period." SparkNotes. SparkNotes, 2011. Web. 7 May 2015.

Hall, Nancy. "Angular Displacement, Velocity, Acceleration." NASA. NASA, 5 May 2015. Web. 7 May 2015.

Henderson, Tom. "Motion in Two Dimensions." *The Physics Classroom*: Physics Classroom, 2012. Digital file.

Kahn, Sal. "Kinematics." Encyclopaedia Britannica. Encyclopaedia Britannica, 5 Jun. 2013. Web. 7 May 2015.

Nave, Carl R. "Basic Rotational Quantities." HyperPhysics. C. R. Nave, 2012. Web. 7 May 2015.

"SI Brochure: The International System of Units (SI), 8th Edition." Bureau International des Poids et Mesures. BIPM, 2014. Web. 7 May 2015.

Simanek, Donald. "Kinematics." A Brief Course in Classical Mechanics. Lock Haven U, Feb. 2005. Web. 7 May. 2015.

S

SIMPLE, LINEAR, AND COMPLEX SYSTEMS

FIELDS OF STUDY

Classical Mechanics

SUMMARY

Almost every aspect of modern life is built on the principle of systems. Complex machines used every day, such as bicycles, cars, and even corkscrews, are systems that consist of and depend upon interacting simple machines.

PRINCIPAL TERMS

- **efficiency:** how effective a machine is at transforming or transferring energy, quantified as the ratio of the actual performance of the machine to an idealized, theoretical version of it.
- **joule:** the International System of Units (SI) unit of work and energy, equivalent to the work done by a force of one newton applied over a distance of one meter.
- **mechanical advantage:** the amplification of force provided by many machines, measured as the ratio of the output force to the input force.
- **net force:** the sum of all forces acting on an object.
- **newton:** the International System of Units (SI) unit of force; one newton is equal to the force required to accelerate a one-kilogram mass at one meter per second per second.
- **power:** the rate of work done (energy transfer) over time.
- **system elements:** the individual components that work together and make up an overall system used to complete a task.
- **work:** the force moving an object, or the successful transfer of energy.

SIMPLE MACHINES

A simple machine is used as the easiest way to multiply a force to complete a task. These and other kinds of machines are examples of a system, which is a process that has a defined input and at least one output. The six classic simple machines are the lever, the wheel and axle, the pulley, the inclined plane, the wedge, and the screw. A lever, for example, is simply a rod or plank balanced over a fulcrum. A wheel functions like a circular lever as it revolves about its center. This full circular motion enables a wheel to transfer energy, power, or mass from one position to another.

All simple machines provide mechanical advantage by amplifying the input force to create a greater output force. A lever positioned so that the application of 1 newton of force as input enables the production of 3 newtons of force as output has a mechanical advantage of 3. The trade-off for mechanical advantage is efficiency, the measurement of the machine's effectiveness at transferring energy. It can be thought of as the ratio of the output to the input. The greater the mechanical advantage of a machine and the more elements involved, the lower its efficiency. For example, if the input force of 3 newtons moves through a distance of 3 meters (9.84 feet) to move an output force of 1 newton through a distance of 6 meters (19.68 feet), the machine has an efficiency of 0.67, or 67 percent. The input work would be 9 joules, and the output work would be 6 joules. No machine can be 100 percent efficient. There are always forces that counteract the force being applied, particularly as friction. The actual force that can then be transferred by a machine to produce an output is the net force. This is the sum of the applied force and the opposing forces.

Simple machines can be combined to produce more complex systems that are defined by internal and external boundaries. In a linear system, one device affects one output. In a nonlinear (complex) system, a single component can affect other system elements and produce more than one output. A bicycle is a complex machine/system constructed from several simple machines working as a single unit. The pedals are levers, providing torque to the main gears.

Simple System

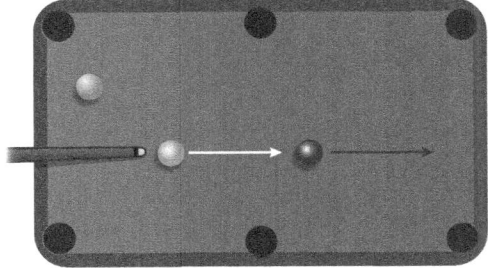

Linear System

Complex System

As more components are added to a system, the system becomes more complex. A cue hitting the cue ball is a simple system. If a cue hits a cue ball sending it in a straight line to collide with another billiard ball, it is a linear system. When more billiard balls are hit they begin to interact with each other in a more dynamic manner, which increases the complexity.

These are connected by a chain system to another set of gears that transfer the torque to the rear wheel. The derailleurs that shift the drive chain from one gear to another are based on the inclined plane for their function. The handlebars also act as levers, allowing the orientation of the front wheel to be changed. This lets the bicycle change direction as it moves. Tension is normally applied to the brake and shift cables by pulling on their attached levers.

FEEDBACK AND CONTROL SYSTEMS

A machine is of no use without an effective means of controlling its actions and functioning. Control systems built with modern digital electronics allow for precise control of the movement and actions of machines. This is accomplished by incorporating feedback mechanisms. A feedback mechanism measures an output stream, such as voltage or pressure, and uses that information to adjust the input stream in a way that maintains the proper desired stream. Input from an external source is generally the

---

SAMPLE PROBLEM

Calculate the overall efficiency of a linear system made up of six components having individual efficiencies of 98 percent, 87 percent, 86 percent, 75 percent, 50 percent, and 66 percent. Calculate the efficiency of a nonlinear (complex) system consisting of the same six components arranged such that the 98 percent efficient component drives an output from the remaining components in three streams made up of the 87 percent and 86 percent, 75 percent and 66 percent, and 50 percent components.

Answer:

The linear system produces an output that reflects the sequential product of the individual efficiencies of the components in the system, as

$$0.98 \times 0.87 \times 0.86 \times 0.75 \times 0.50 \times 0.66 = 0.1815, \text{ or}$$

$$18.15 \text{ percent}$$

The nonlinear system must be treated as three independent systems, since it produces three independent outputs. There will be one stream that has an efficiency of

$$0.98 \times 0.87 \times 0.86 = 0.7332, \text{ or } 73.32 \text{ percent}$$

a second stream having an efficiency of

$$0.98 \times 0.75 \times 0.66 = 0.4851, \text{ or } 48.51 \text{ percent}$$

and a third stream with an efficiency of

$$0.98 \times 0.50 = 0.49, \text{ or } 49 \text{ percent}$$

primary input to a system. A common example is the power delivered by the engine of a car with an automatic transmission. Fluid pressure in the transmission of the car is used to control the power delivered by the engine, increasing it when going uphill and decreasing it when rolling on flat road surfaces or going downhill.

Systems in Real Life

Linear and complex systems are the fundamental structures of modern technology. Every machine that is used is a combination of simple machine structures. Examination of almost every aspect of modern life will reveal an underlying linear or complex system structure involved.

—*Richard M. Renneboog, MSc*

Bibliography

Chabay, Ruth W., and Bruce A. Sherwood. *Matter and Interactions*. 4th ed. Hoboken: Wiley, 2015. Print.

Hicks, Tyler G. *Handbook of Mechanical Engineering Calculations*. 2nd ed. New York: McGraw, 2006. Print.

Parasiliti, Francesco, and Paolo Bertoldi, eds. *Energy Efficiency in Motor Driven Systems*. New York: Springer, 2003. Print.

Seeler, Karl A. *System Dynamics: An Introduction for Mechanical Engineers*. New York: Springer, 2014. Print.

Thumann, Albert, and Harry Franz. *Efficient Electrical Systems Design Handbook*. Lilburn: Fairmont, 2009. Print.

Vukosavic, Slobodan N. *Electrical Machines*. New York: Springer, 2013. Print.

SIMPLE MACHINES AND MECHANICAL ADVANTAGE

FIELDS OF STUDY

Classical Mechanics

SUMMARY

Simple machines are mechanical devices that make tasks—typically, moving an object— easier by redirecting or amplifying an applied force or by increasing the object's velocity or displacement. The classical simple machines include the lever, wheel and axle, pulley, inclined plane, wedge, and screw. Gear trains are also sometimes classified as simple machines, and they can be thought of as having the properties of a lever (movement about a fixed fulcrum) and a wheel and axle (transfer of rotational movement).

PRINCIPAL TERMS

- **actual mechanical advantage:** the ratio comparing the output force of a machine to the input force, taking into account friction and other factors that limit the efficiency of real-world machines. A mechanical advantage of more than one indicates an amplification of force.
- **efficiency:** the measure of how effective a machine is at transforming or transferring energy, quantified as the ratio of the actual performance of the machine to an idealized, theoretical version of it. A perfect machine would have an efficiency value of one.
- **ideal mechanical advantage:** the ratio comparing the output force of a machine to the input force, ignoring friction and other factors that limit the efficiency of real-world machines. A mechanical advantage of more than one indicates an amplification of force.
- **input:** the force (or energy) that is "put in" to a machine; for example, the horizontal force of wind provides the input for a windmill.
- **net force:** the sum of all of the forces acting on an object; note that forces with equal magnitude but opposite directions negate each other. An object moves in the direction of the net force acting on it.
- **output:** the force (or energy) produced by a machine. The machine transforms the input into the output; for instance, a windmill transforms the force of wind (input) into the circular motion of a millstone for grinding (output).
- **power:** the rate of work (energy transfer) over time; the International System of Units unit of power is the watt (W), which equals one joule per second (J/s).

- **work:** a force moving an object, or the successful transfer of energy. The International System of Units unit of work is the joule.

What Is a Simple Machine?

Simple machines are devices that make tasks—typically, moving some target object—easier by redirecting or amplifying some input force into a new output force. A lever is a very simple and effective machine, consisting of nothing more than a rigid plane and a fulcrum on which the lever is balanced. The ancient Greek mathematician Archimedes (ca. 287–212 BCE) came up with the concept of the simple machine. He is famous for having said that, with a place to stand and a lever, he could move the entire world.

Generally, simple machines are distinguished from other, more complex machines by virtue of being the simplest possible ways of generating mechanical advantage—that is, multiplying a force. The six classical simple machines are the lever, wheel and axle, pulley, inclined plane, wedge, and screw. The gear is also a simple machine, often described as a type of special lever combined with a wheel and axel.

In many instances, more complex machines—sometimes called "compound machines"—can be thought of as being an assembly of several simple machines. For instance, a hand-crank can opener works using wedges (the cutting edge), wheels and axles (the crank), and levers (the grips).

Inclined plane—a flat, rigid surface raised at an angle, in other words, a ramp. The ramp that extends behind some moving trucks is a classic example of an inclined plane. A ramp directs and amplifies a forward push into the lifting or lowering of heavy objects, but it reduces distance. It helps move objects vertically by making it easier to push or pull an object across its surface (not acting directly against gravity), but an object must be moved farther as a result.

Lever—a tool that redirects and amplifies (or reduces) an input force. It consists of a stiff plank, bar, or rod—anything straight and rigid—balanced over a fulcrum. The fulcrum is usually wedge shaped or round, but anything that the lever can be pushed or pulled against can work. An impromptu lever can easily be made by laying a sturdy stick across a round rock. Even an action as simple as using a screwdriver to pry open a paint can lid is actually a form of lever use. The rim of the can serves as a fulcrum for the steel rod of the screwdriver. Three types of levers exist and depend upon the location of the fulcrum with respect to the load and the effort. In a first-class lever, the fulcrum is between the load and the effort, as in a see-saw. The second-class lever places the load between the fulcrum and the effort. Lastly, in a third-class lever, the effort is between the load and the fulcrum. Pressing down on a lever with a fulcrum near the target force will result in an amplified upward force.

Pulley—uses one or more wheels (often grooved to keep the rope from slipping) and rope to redirect or amplify an input force. The simplest version consists of little more than a rope draped over a surface that it can slide easily over, such as a smooth metal hook, and serves only to redirect force. For instance, a rope thrown over a rafter in a barn will allow a person to pull down on one end of the rope, aided by gravity, to lift a load upward at the other. This is an impromptu example of a simple pulley system, with the rafter acting as an anchored pulley. More complex arrangements with moving pulleys produce a mechanical advantage, amplifying an input force. In exchange, the person applying the force has to move a greater amount of rope. A two-pulley system, with one anchored and one able to move freely, is the simplest possible way of arranging ropes over wheels to produce a mechanical advantage. The more movable pulleys there are in a system, the greater the advantage.

Screw—an inclined plane with a spiraling groove wrapped around a cylinder, typically with a pointed edge. Screws amplify force and direct a circular force into a linear force along the line of the central column. For example, turning a screw into wood with a screwdriver turns the circular motion of a wrist into linear motion into the wood. Screws are generally used to secure objects tightly together, as in common household construction projects. Other uses for screws include clamping or crushing objects (as in a vice), excavating holes (a drill), and even moving air (a fan). Screws work by using the inclined plane to redirect and increase a turning force (torque) applied to the cylinder into a vertical force parallel to the cylinder.

Wedge—a triangular object with one thick, flat end and two sloped sides that come to a point at the other end made from two inclined planes fused together along their bottoms to create a sharp point. Wedges amplify and redirect an input force applied to their

275

thick end into an amplified force pushing out to either side of the sharp end, perpendicular to the input force. Thus, wedges are often used to cut or pry things apart. Axe heads, for instance, amplify the downward force of a swing and redirect it to either side as the sharp end of the head is driven into wood, forcing the wood apart and chopping it in two. Most cutting surfaces work the same way, but on different scales. Other examples of wedges also railroad spikes, and chisels. Wedges may also be used to secure objects in place, as with railroad spikes or doorstops.

Wheel and axle— a broad disk mounted on a stiff rod so that when one rotates, the other does too. Typically, the rod is attached to the center of the disk. When the wheel spins, the force is amplified and transmitted to the axle. Spinning the wheel transfers the motion to the axle. If the wheel is bigger than the axle, then the force is reduced but the axle moves faster. If the wheel is smaller than the axle, then the force is amplified but the axle spins more slowly. Spinning the axle rather than the wheel reduces the force but moves the wheel a greater distance. Note that gears can be considered an extension of the same basic ideas as the wheel and axle.

Gears—a wheel with teeth around the outer edge. These teeth mesh with the teeth of another gear, or with a chain that meshes with the teeth of another gear. Two or more gears connected in this way form a gear train. This is a simple machine that takes the rotational force applied to the input gear and transfers, redirects, and amplifies or reduces it. This depends on the relative sizes of the gears and the direction of the force transfer. Gears are commonplace in all manner of machinery, particularly those involving a rotational motor. The main types of gears are bevel, spur, rack and pinion, and worm.

POWER AND WORK REMAIN CONSTANT

A force is said to do work if it moves an object. An object will move if the net force on it—the sum of all forces acting upon it—results in a positive force in any direction. If the net force acting on an object is positive in any direction, the object will move in that direction. Therefore, a person standing absolutely still on the surface of the earth is experiencing a net force of zero. The force of gravity is performing no work on him or her. If a person is falling straight down, the net force is positive in the direction of gravity and the gravitational force of the earth is doing work.

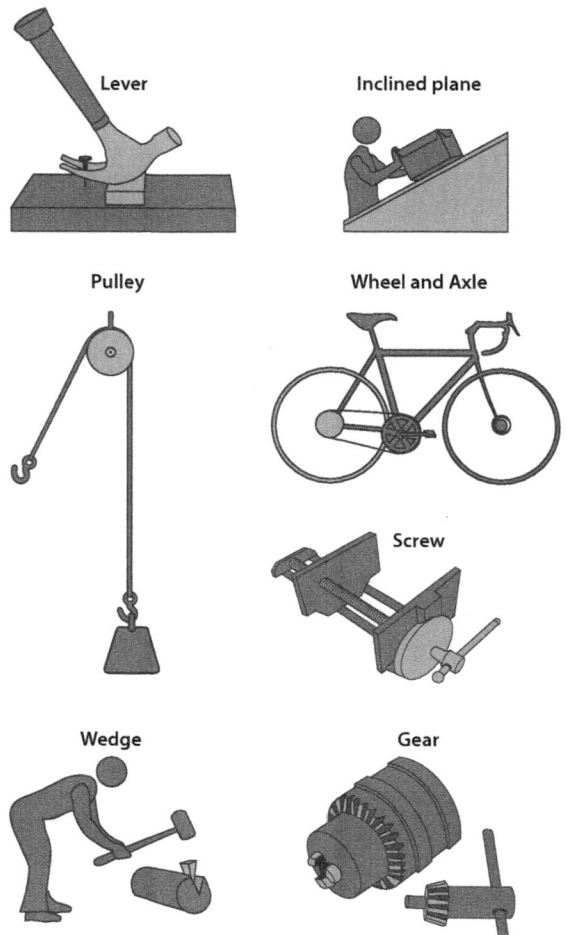

Simple machines are used many ways in our lives. The tail of a hammer is a lever; a ramp is an inclined plane; pulleys are often used to lift heavy objects; a bicycle has many simple machines including a wheel and axle; a bench vise works because of a screw; to split wood, you might use a mallet and wedge; and a power drill chuck and chuck key are a gear system.

A heavy box sitting on a ramp experiences a net force of zero. However, if somebody applies a horizontal force to the box, sliding it upward and away, the net force will be positive in the diagonal direction of movement, because an object moving up or down a ramp always experiences a net force parallel to the incline.

A heavy bale of hay resting on the floor of a barn experiences a net force of zero. If somebody attaches a pulley system and starts lifting the bale upward, however, the net force is positive in the direction of movement.

An axe (wedge) lying flat on a log experiences a net force of zero. If someone swings the axe straight down, the net force is positive in the direction of gravity, and the gravitational force of the earth is doing work on the axe.

In the International System of Units (SI), work and energy are both measured in joules (J). One joule is equal to the work performed (or energy transferred) when a force of one newton (N) moves something a distance of one meter. Since work is what happens when energy is transferred, the law of conservation of energy applies.

The work input into a machine must equal the work output because the energy on either end of the machine must also remain constant. Power, measured in watts (W), is simply the rate of work over time, and it must also remain equal on either side of a machine. One watt is equivalent to one joule per second (J/s).

In physics, work (W) is equal to the product of the strength of the force (F) applied, the displacement of the object from its original position (s), and the cosine of the angle between the directions of force and displacement (θ)

$$w = F \cdot s \cdot \cos\theta.$$

This formula is useful for understanding the force/distance tradeoff inherent to the way machines work. For the same fundamental reasons that energy can only be transformed, not created or destroyed, the total work performed at either end of a simple machine must remain constant. (In other words, "work" and "energy transfer" are essentially the same thing.) To keep the work value constant, a simple machine that amplifies force via mechanical advantage must also reduce the displacement (total distance) caused by that force. Force and displacement are inversely related—increasing one by a certain factor will decrease the other by the same factor.

The formula for work, above, helps one understand the force-distance trade-off inherent in simple machines. For the same reasons that energy can only be transformed, not created or destroyed, the total work done at either end of a simple machine must remain constant. To keep the work value constant, a simple machine that amplifies force via mechanical advantage must reduce the displacement caused by that force. (The angle between the force and displacement determines whether that force caused the displacement.)

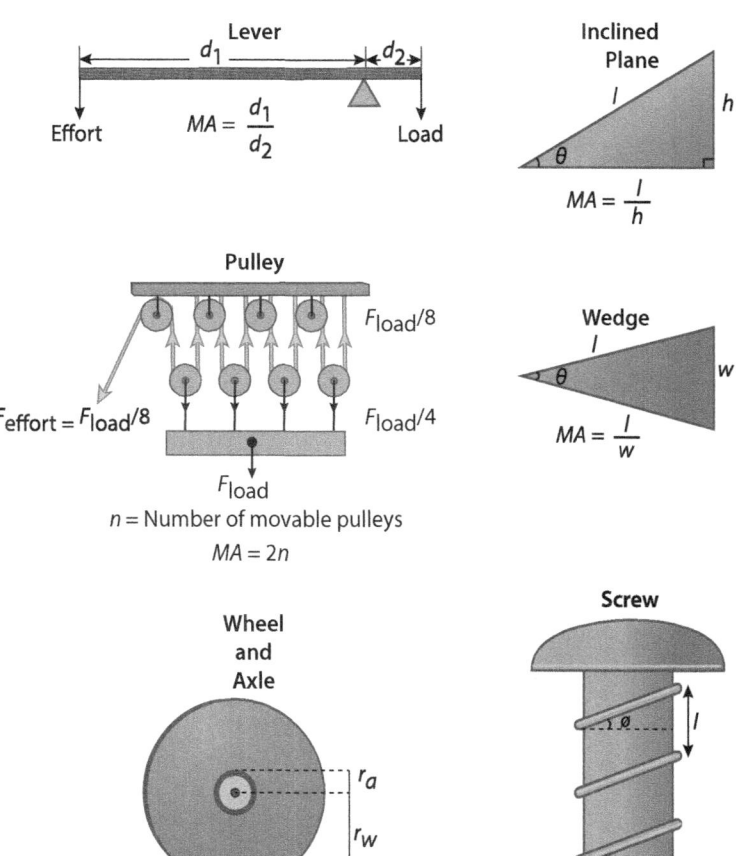

The relationship between the two forces acting on a simple machine, effort and load, is called the mechanical advantage (MA). Theoretical mechanical advantage for each simple machine is calculated differently. Levers are calculated by the ratio of effort arm length (d_1) to load arm length (d_2). A wheel and axle is calculated by the ratio of wheel radius (r_w) to axle radius (r_a). Mechanical advantage of inclined planes is calculated by the ratio of length (L) to height (h). Pulleys are calculated by multiplying the number of moveable pulleys by two. A wedge is calculated similar to the inclined plane as a ratio of length (L) to width (w). The mechanical advantage of a screw is calculated by a ratio of the screw circumference (πd) to thread lead (L). And a gear system is calculated by the ratio of the driven gear's number of teeth to the driver gear's number of teeth.

Pushing a box up a ramp, for example, will not move the box as far as the same push would along flat ground, but the push will carry greater force. A pulley system might enable a farmer to lift a heavy hay bale, but the bale will be displaced a shorter distance per unit of force applied. If a smaller driver gear is meshed with a larger gear, the smaller gear turns at a faster rate and in the opposite direction of the larger gear but with less force.

IMPERFECT MACHINES
A perfect machine would be a machine unimpeded by friction or design flaws, capable of transmitting a force perfectly. In the real world, there is no perfect machine. Even the simplest machines fail to transmit forces perfectly; some energy is always lost to friction. Thus, a distinction is made between ideal mechanical advantage (which assumes a perfect machine) and actual mechanical advantage (to take into account friction and other forces).

The difference between a theoretically perfect machine and its real-world counterpart is measured in terms of its efficiency. Efficiency is the ratio of the actual, measured performance of a machine to its theoretically perfect performance. In other words, efficiency is the ratio of the actual mechanical advantage to the ideal mechanical advantage. A perfect machine would have an efficiency value of 1.

Wedges lose efficiency due to friction between the sloping sides and the surfaces they move against. Gears lose efficiency from the friction between their meshed teeth. In the case of ramps, the smoother the ramp, the easier an object would move on it, as less energy would be wasted overcoming the friction between them. Unfortunately, less friction would also mean that the load could slide down the ramp more easily, negating the work.

CALCULATING MECHANICAL ADVANTAGE IN SIMPLE MACHINES
When a machine amplifies the force put into it, it is said to provide a mechanical advantage. Perhaps the simplest way of generating mechanical advantage is with a lever—a simple machine consisting of a stiff plane balanced on a fulcrum. A playground see-saw is a classic lever. Archimedes (ca. 287–212 BCE), an ancient Greek mathematician and engineer, proved the law of the lever. This law states that the mechanical advantage of a lever is dependent on the relative position of the fulcrum. If it is closer to the output end (where the load is), then the lever will produce a mechanical advantage. If closer to the input, it will instead reduce the input force. The mechanical advantage offered by a lever is directly related to how close the fulcrum is to the output end.

In the real world, there is no such thing as a perfect machine. Even the simplest machines do not transmit forces perfectly; some is lost to friction, or resistance, between moving surfaces. The actual mechanical advantage of a machine is measured against the ideal mechanical advantage, or the theoretically perfect performance, of the same machine. The degree to which a machine achieves its ideal mechanical advantage is its efficiency.

SIMPLE MACHINES ARE EVERYWHERE
Simple machines permeate every aspect of everyday life. One of the first uses of the screw was to transfer water. According to legend, the Archimedes' screw was developed by famed Greek inventor Archimedes (ca. 287–212 BCE) on a visit to Egypt as a method of lifting water into irrigation ditches. Also called a "screw pump," the device is a large screw with a broad thread fitted tightly inside a pipe. As the screw is turned, the threads pull water upward. Similar devices are still used to move water, grain, and other substances.

The "pulling" action of a screw can also be seen in the propellers of ships and propeller planes. These spin blades act like the threads on a screw. Indeed, Leonardo da Vinci (1452–1519) designed a helicopter-like device that used a large, broad-threaded screw to be spun by hand using a lever and to "pull" the device upward into the air. (Unfortunately, it did not work.)

Almost every cutting tool, from a knife to a sword to a pair of scissors, uses a wedge to push apart the surface it is cutting. Wedges can also be used to pin things in place by being driven into a material; instead of splitting the material, the amplified force allows the wedge to be driven securely into place. Old iron railroad spikes are simple bars with a wedged end that were used to hold rail tracks in place.

Gears are extremely common, albeit often hidden from view. Any motor-powered device almost certainly uses a gear train to transfer and amplify the force generated by the motor. Electric screwdrivers do so, for example, as do combustion engines.

Every staircase is like an unevenly built ramp, designed to let people ascend without tiring by moving diagonally instead of straight upward against gravity. Ramps have been used since antiquity to help move large loads and are common in construction and shipping.

Pulleys are common, especially in industries such as shipping, where large loads often need to be lifted. Some examples of pulleys are less obvious. Consider the hook-and-loop strap on a winter glove's cuff. The strap goes through a slot and bends back over itself; pulling on the end of the strap makes the plastic slot act as a tiny movable pulley

Due to the simple construction of the wheel and axle and its easy amplification of force, it has found countless applications throughout modern society. A mechanical winch used to lift a bucket from a well is a wheel-and-axle system, while a doorknob acts as a wheel and axle to unlatch a door. Numerous motor-powered devices feature wheels and axles connected to gear-and-pulley systems to transmit and redirect the force of the motor to its desired ends.

More importantly, the principles of power, work, force and mechanical advantage, along with the basic structures of classical simple machines, form the basis for a deeper understanding of much of the more complicated machinery one may need to interact with, such as automobile engines.

—Kenrick Vezina, MS

BIBLIOGRAPHY

Coolman, Robert. "What Is Classical Mechanics?" *Live Science*. Purch, 12 Sept. 2014. Web. 10 July 2015.

Delson, Nathan. "Gear Ratios and Mechanical Advantage." *Mechanical and Aerospace Engineering*. University of California, San Diego, 2004. Web. 1 July 2015.

"The Elements of Machines." *Inventor's Toolbox*. Museum of Science, 1997. Web. 10 July 2015.

Henderson, Tom. "Motion in Two Dimensions." *The Physics Classroom*: Physics Classroom, 2012. Digital file.

Kahn, Sal. "Mechanical Advantage." *Khan Academy*. Khan Academy, n.d. Web. 10 July 2015.

Simanek, Donald. "Kinematics." A Brief Course in Classical Mechanics. Lock Haven U, Feb. 2005. Web. 28 Apr. 2015.

"Simple Machine." *Encyclopaedia Britannica*. Encyclopaedia Britannica, 26 Aug. 2014. Web. 1 July 2015.

SOLENOID

FIELDS OF STUDY

Electromagnetism

SUMMARY

A solenoid is an electromagnetic device made up of a coiled conducting wire. An applied electrical current produces a strong magnetic field in a solenoid, according to Ampère's law. The fine electronic control of solenoids and electromagnets makes them suitable for a very broad range of applications.

PRINCIPAL TERMS

- **Ampère's law:** the rule stating that the strength of a magnetic field about a current-carrying conductor is directly proportional to the magnitude of the current.
- **electromagnet:** a device that becomes magnetic due to the presence of an electric current.
- **ferromagnetic:** describing material that can be made permanently magnetic by the presence of a magnetic field, such as iron and other ferrous (ironlike) metals.
- **inductance:** the ability to generate a magnetic field when electrical current is flowing.
- **permeability:** the ability with which a magnetized material supports a magnetic field.
- **right-hand rule:** if a wire is grasped in the right hand with the thumb pointing in the direction of current flow, the fingers will point in the direction of the magnetic field around the wire.
- **turns:** the number of times that a conductor is wrapped around to form a helical coil.

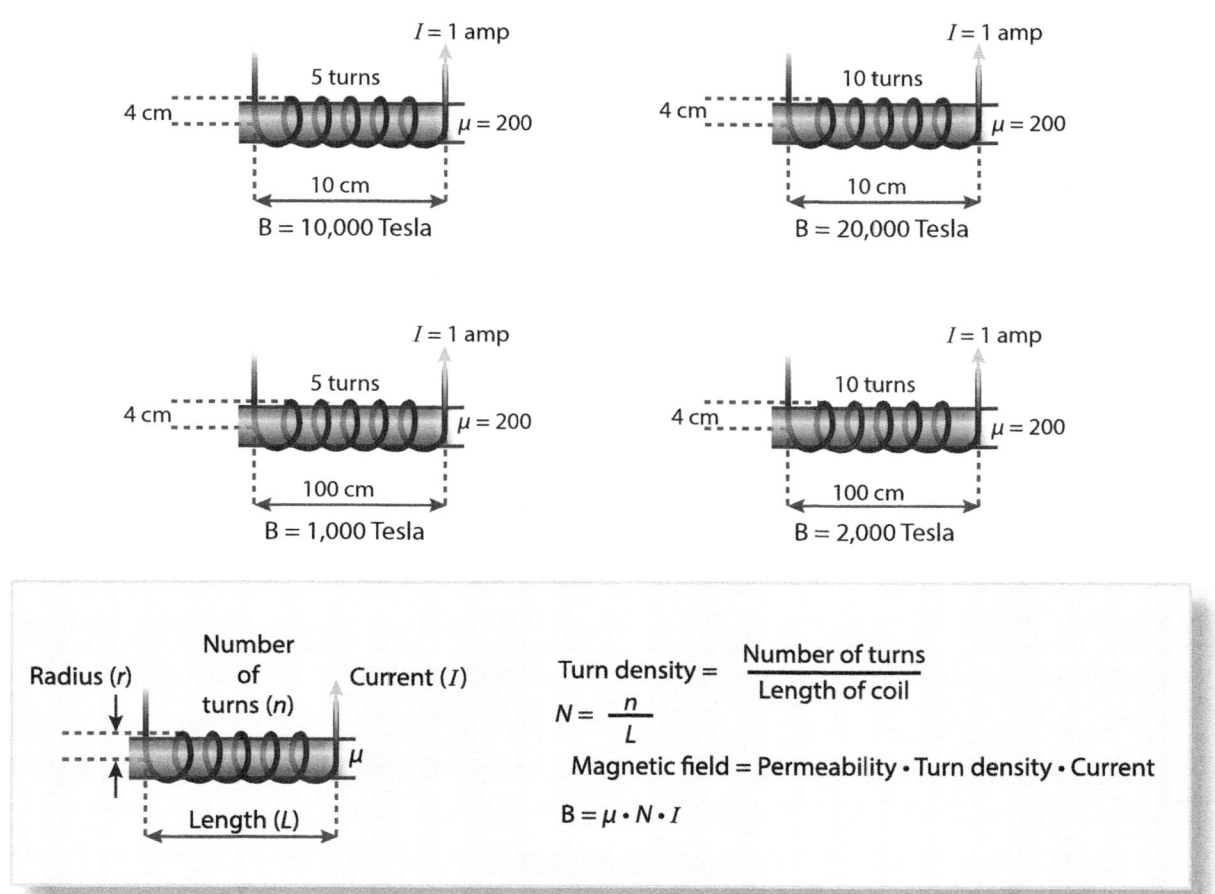

The magnetic field of a solenoid is directly dependant on the product of the turn density (N), permeability (μ), and current running through the solenoid (I). Because turn density is a ratio of the number of turns (n) per length of coil (L), the magnetic field of a solenoid will increase with an increase in the number of turns and will decrease as the length of the coil increases.

Electricity and Magnetism

A solenoid combines the relationship between electricity and magnetism in a way that performs useful functions. In its simplest form, a solenoid consists just of a wire looped into the shape of a helical coil, like a spring. Electrical current flowing through the wire generates a magnetic field, which is what makes the solenoid so useful. Magnets made of a ferromagnetic material such as iron, nickel, or cobalt maintain their magnetism as permanent magnets. A solenoid, on the other hand, is an electromagnet. This means that it is magnetic due to an electric current and only as long as that current operates.

André-Marie Ampère (1775–1836) identified that an electrical current flowing in a conductor causes a magnetic field around the conductor. The right-hand rule can always be used to determine the direction of this magnetic field. According to Ampère's law, the strength of the magnetic field is directly proportional to the magnitude of the current. This is echoed in Faraday's law of magnetic induction, which states that a magnet moving past a conductor induces a voltage and current in the conductor and that the magnitude of these depends on changes in the magnetic field strength as it passes the conductor. The induced electrical current produces a magnetic field around the wire. Lenz's law states that this induced magnetic field opposes the original magnetic field. When the conductor, usually a wire, is wrapped into a helical coil, the magnetic fields about the individual turns of wire add together to form a single large magnetic field. This is a solenoid.

As electrical current flows through the wire of a solenoid, it generates a magnetic field according to Ampère's law. The magnetic field is produced at each point along the length of the coiled wire, just as it would be if the wire were straight. The helical shape causes the magnetic field around the adjacent coils of wire to align, forming a tube of concentrated magnetic flux lines within the coil. The flux lines surrounding the outside of the solenoid are also aligned in the opposite direction, but are widely dispersed. The magnetic field surrounding the outside of an active solenoid is thus much weaker than the internal magnetic field, to the extent that it is negligible. A solenoid's internal magnetic field has much the same characteristics as a permanent bar magnet, but only when electric current is flowing through the wire. Often the wire is coiled around a metal core to increase the strength of the magnetic field produced.

Electromagnets and Inductors

By itself, the coil of a solenoid produces a magnetic field that creates an induced voltage opposite the direction of current flow in the circuit. This is the basic principle of an electronic component called an inductor. These are readily identifiable on any circuit board as they typically are constructed by wrapping a length of fine wire around a small cardboard tube. The purpose of an inductor is to provide a buffer against fluctuations in the voltage applied to the circuit so that the voltage remains constant. When an increase in voltage occurs, it increases the current flowing through the circuit, and this causes the inductor to produce a greater voltage in the opposite sense. Similarly, when applied voltage decreases, so does the counter voltage produced by the inductor. The inductance of an inductor depends on the number of turns of wire, the area of the inside of the coil, the length of the coil, and the permeability of the core material. Inductance is measured in henries (H), a standard unit named for Joseph Henry (1797–1878), according to the formula

$$L = \frac{\mu N^2 A}{\ell}$$

where L is the inductance in henries, l is the length of the solenoid in meters, N is the number of turns of coiled wire, A is the cross-sectional area of the solenoid coil in square meters ($A = \pi r^2$), and μ is the permeability of the core material.

Inductors and inductance coils typically are constructed around a material that is not magnetizable, such as air, plastic, or cardboard. When the solenoid coil is wrapped instead about a core material that is magnetizable, such as an iron rod, an electromagnet is formed. While the material within the core may be permanently magnetized, its magnetic field strength is greatly enhanced by that of the solenoid while current is flowing.

The magnetic flux in the core of an electromagnet depends greatly on the magnetic permeability of the core material. Electricity and magnetism are related at the subatomic level through unpaired electrons in the atoms of the material. Ferromagnetic metals acquire greatly enhanced magnetic properties when acted upon by a solenoid and conduct magnetism much as they would conduct electricity.

The flow of electricity through a conductor is subject to electrical resistance. If the material's resistance is completely eliminated, the material becomes a superconductor. In that state, an induced electric current will continue to flow indefinitely after the magnetic induction field is removed. Electrons moving through a superconducting medium move in pairs like a laser beam rather than individually moving through numerous collisions down the length of the conductor. Since there is no energy lost to collisions, there is no resistance to the movement of the electrons and no heat loss due to friction. Superconducting magnets can be based on the geometry of solenoids.

Solenoids in Action

The simplicity of a solenoid lends itself to a broad array of applications. Solenoids allow the creation of basic electromagnets and inductors. These devices can be controlled electronically with very high precision and are found in many places. One example is the starter motor of an internal combustion engine in a car or truck, which uses a solenoid to take a small current from the ignition switch and relay a stronger current from the battery to start the engine. Scrap yards also often make use of large electromagnetic hoists to move ferrous metals about efficiently. The current is turned on to activate the solenoid's magnetic field, allowing materials to be attached to the magnet. When they have been moved to the desired spot, the current is switched off and the materials are released.

Another common example of a solenoid is a simple electromagnetic lock or switch system. This can be produced using a solenoid and a rod that can be drawn into the core of the solenoid when current is flowing. Such a system is encountered whenever someone has to be remotely let in at an apartment building. A signal from a control panel typically closes a circuit that shunts electrical current into a solenoid. This causes it to generate a magnetic field and draw the lock pin into its open position. When the signal is released, the solenoid ceases to be active and allows a spring to pull the lock pin back into position. Basically all electromagnetic switch systems work this way. The security of such systems typically depends upon the manner in which the control circuit must be accessed, ranging from a simple push button to complex ID protocols.

Solenoids and their magnetic properties are also used in a number of very powerful analytical techniques. Among the most important is nuclear magnetic resonance (NMR) spectrometry. Chemists have used this method to analyze the structure of molecules since its development in the 1950s. The technique involves placing a homogeneous sample of a compound in solution into a strong magnetic field and detecting the energy levels that it absorbs when irradiated with a variable electromagnetic field. The energy and patterns of the absorption depend entirely on the three-dimensional structure of the molecule being examined, and provide very precise information about that structure.

The methodology of nuclear magnetic resonance was expanded with the development of superconducting magnets and magnetic resonance imaging (MRI) for medical diagnosis. The images are obtained by immersing the patient inside a strong magnetic field and plotting the patterns of energy absorption through irradiation by a variable electromagnetic field, just as in NMR spectrometry. The technique has been used routinely to examine living persons, ancient Egyptian mummies, and large zoo animals. Control of the stability and uniformity of the magnetic field is critical to these applications. Superconducting magnets made possible by solenoids help achieve this ability.

—*Richard M. Renneboog, MSc*

BIBLIOGRAPHY

Bird, John. *Electrical and Electronic Principles and Technology.* 5th ed. London [u.a.]: Routledge, 2014. Print.

Chow, Tai L. *Introduction to Electromagnetic Theory. A Modern Perspective.* Boston: Jones, 2006. Print.

Fitzpatrick, Richard. *Maxwell's Equations and the Principles of Electromagnetism.* Hingham: Infinity Science, 2008. Print.

Fleisch, Daniel A. *A Student's Guide to Maxwell's Equations.* Cambridge: Cambridge University Press, 2008. Print.

Gross, Charles A. *Electric Machines.* Boca Raton: CRC, 2007. Print.

Kelly, P. F. *Electricity and Magnetism.* Boca Raton: CRC, 2015. Print.

Rexford, Kenneth, and Peter R. Giuliani. *Electrical Control for Machines.* 6th ed. New York: Thomson Learning, 2004. Print.

Robbins, Allan H. and Wilhelm C. Miller. *Circuit Analysis: Theory and Practice.* 5th ed. Clifton Park: Delmar, 2013. Print.

SOUND AMPLITUDE

FIELDS OF STUDY

Acoustics; Classical Mechanics; Fluid Mechanics

SUMMARY

The amplitude of sound waves determines their intensity; a large amplitude indicates an intense sound. Amplitude is measured by the displacement of the medium (such as air) through which sound waves travel. The sensory ability of the ear and brain limits the perceived loudness of sounds. Loudness and intensity are measured using decibels (dB), which relate sound intensity to human hearing on a logarithmic scale.

PRINCIPAL TERMS

- **decibel:** abbreviated dB; a unit of measure that quantifies sound intensity in relation to human

hearing. It follows a logarithmic scale starting at zero for near silence.
- **displacement:** the distance particles in a medium are moved from their equilibrium by a wave; used to determine amplitude.
- **frequency:** how often a complete wave cycle occurs, directly proportional to the energy content of a wave and inversely proportional to wavelength; measured in hertz (Hz).
- **intensity:** in acoustics, sound power per unit area; measured in decibels or watts per square meter.
- **loudness:** the perceived intensity of sound; the objective intensity of sound in individual variations in hearing ability.
- **noise:** in acoustics, any sound whether wanted or not; may also describe background sound that obscures a desired signal or sound.
- **perception of sound:** the ability to perceive mechanical waves in one's environment (e.g., air, water) as sounds, limited by the physiology of the ear and the brain's ability to interpret information from sound waves.
- **pulse:** a lone disturbance passing through a medium from one place to another, similar to a wave but not cyclical and repeating.
- **wavelength:** the distance between crests of a wave; all electromagnetic radiation is transmitted as waves, with longer wavelengths corresponding to lower frequencies and less energy and vice versa.

THE THREE WAVE PROPERTIES

Amplitude is one of the three major properties used to describe all waves, along with wavelength and frequency. These properties apply whether dealing with mechanical waves, such as sound waves, passing through a medium such as air or water, or electromagnetic waves, such as light and heat, passing through empty space. In mechanical (sound) waves, amplitude is determined by how much a wave displaces the medium it is traveling through.

Sound forms compression (longitudinal) waves, and the motion of the wave is parallel to the direction that energy is being transferred. Rather than up-and-down or side-to-side movement, compression waves travel as repeated, cyclical pulses of compression and rarefaction (expansion) in their medium. Amplitude in compression waves is measured as the maximum displacement of the particles of the medium from their normal resting state. Sound traveling through air is measured by air pressure at various points along the wave.

AMPLITUDE AND IRREGULAR PULSES

A sound wave is cyclical and has a regular pattern of displacement. If something other than a regular pattern of displacement produces sound, such as with the vibration of a guitar string, the individual bursts of sound are considered pulses. A pulse can be thought of as a lone burst of disturbance. Calculating the amplitude of sound being generated as a series of irregular pulses offers unique challenges. In these situations, sound amplitude may vary from pulse to pulse, and attempts to measure amplitude hinge on determining what to base the measurement on. Choices include the average value of amplitude over time, the instantaneous amplitude, or the peak amplitude.

AMPLITUDE, LOUDNESS, AND INTENSITY

In fundamental terms, amplitude reflects how much energy a wave is carrying. Large amounts of energy allow the wave to displace its medium farther. Wavelength and frequency, which are inversely related, also reflect a wave's ability to transfer energy over time. Assuming equal amplitude, a long-wavelength, low-frequency wave will deliver less energy over time than a short-wavelength, high-frequency wave.

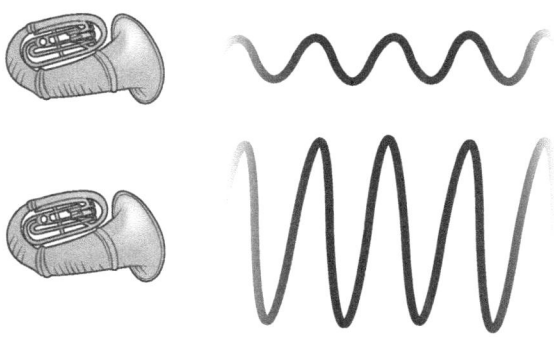

On a waveform, the amplitude is measured from the origin to the wave's crest, or half the wave height. In gen-eral, amplitude is a measure of the intensity of a wave. Amplitude of a sound wave is a measure of the volume of that sound. More airflow through a horn generates more energy, which produces a sound at a higher volume.

The amplitude and energy content of a sound wave is interpreted as loudness by the human ear. Intensity describes the sound power per unit area at a given point, regardless of a person's ability to hear. Humans perceive the amplitude of sound as loudness only when our ears and brains can pick up and successfully interpret the sound waves. Even if a person is unable to hear certain sounds, sounds of sufficient intensity can be felt as vibrations by the rest of the body, as when standing near a sound system at a concert.

Frequency, in the form of wavelengths, forms the other half of the human perception of sound. Pitch, whether a sound is "high" or "low," corresponds to frequency. High-frequency sounds are associated with high pitches, and low-frequency sounds, low pitches.

QUANTIFYING LOUDNESS
Amplitude for all waves is measured using units of distance, such as meters, but the intensity and loudness of sound are quantified using the decibel (dB). Decibels measure intensity and loudness on a logarithmic scale. Its baseline value of zero is set at what a typical human would perceive as near-total silence. A sampling of typical sounds and their decibel values include:
- silence: 0 dB
- whisper: 20 dB
- public library: 40 dB
- dishwasher: 80 dB
- thunder: 120 dB

Exposure to intense sounds can cause a variety of problems. The threshold for annoying noise is usually somewhere between 70 to 80 dB, with discomfort increasing along with intensity. Any sound at 80 dB or higher can cause hearing damage with long-term exposure. At 110 dB, average humans will begin to experience physical pain from the sound.

PHYSICAL EFFECTS OF HIGH-INTENSITY SOUND
Sound waves of a high enough intensity can carry sufficient energy to influence their environment in other ways than producing sound. Sounds of sufficient intensity, for instance, can damage the sensitive organs of the ear. High-intensity ultrasound (very high frequency sound) can be a benefit too. It is often used therapeutically to penetrate organ tissues, and narrow, high-intensity pulses of sound are capable of breaking up kidney stones with minimal damage to surrounding tissue. On the other hand, sufficiently high intensity sound can cause unwanted tissue damage—up to and including the total rupture of an eardrum membrane at intensities of 150 dB and above. Standing within twenty-five meters of a jet taking off or near firecrackers or shot-gun blast will produce sound at 150 dB or above.

SOUND AMPLITUDE IN EVERYDAY LIFE
Coupled with frequency, amplitude enables musicians and engineers to accurately describe everything from the twang of a guitar to the thump of a bass drum. For physicists, amplitude provides valuable information about the energy content of a sound wave. Understanding the relationship between sound amplitude, intensity, loudness, and potential hearing damage enables engineers to design devices that produce sounds that are safe or devices that can help protect human ears in otherwise dangerous sound environments. Measurements of amplitude and frequency enable biologists to describe the songs of birds and note how they change over time due to evolution. The amplitude of sound waves is an important quantification of what humans commonly perceive as loudness of sounds.

—*Kenrick Vezina, MS*

BIBLIOGRAPHY
Goldsmith, Mike. *Sound: A Very Short Introduction.* New York: Oxford University Press, 2015. Print.

Berg, Richard E. "Sound: Physics." *Encyclopaedia Britannica.* Encyclopaedia Britannica, 3 June 2014. Web. 9 June 2015.

Hass, Jeffrey. "What Is Amplitude?" Indiana Univ. Jeffrey Hass, 2003. Web. 9 June 2015.

Kennedy, J. E., G. R. Ter Haar, and D. Cranston. "High Intensity Focused Ultrasound: Surgery of the Future?" *British Jour. of Radiology* 76.909 (2014): 590–99. Print.

Reisberg, Daniel, ed. Auditory Imagery. New York: Psychology, 2014. Print.

Russell, Daniel A. "Longitudinal and Transverse Wave Motion." *Acoustics and Vibration Animations.* Pennsylvania State U., 18 Feb. 2015. Web. 27 May 2015.

SPEED

FIELDS OF STUDY

Acoustics; Harmonics; Classical Mechanics

SUMMARY

Speed is the distance traveled divided by the amount of time it takes to travel that distance. Waves are a disturbance that travels through a medium and transports energy. The speed of a wave is the distance traveled by a given point on the wave in a given amount of time.

PRINCIPAL TERMS

- **frequency:** the number of oscillations, wavelengths, or cycles per unit of time.
- **sound barrier:** the effects created when an object travels faster than the local speed of sound.
- **speed of sound:** a sound wave's traveled distance per unit of time.
- **velocity:** the rate of change of position in a specified direction of motion.
- **wave propagation:** the manner in which a wave travels.
- **wavelength:** the distance between adjacent crests or troughs in the wave.

Speed and Velocity

The speed of an object is the rate at which it travels a certain distance. Velocity is speed with a direction of motion associated with it. Speed and velocity both have the dimensions of a length divided by a time. The International System of Units (SI) unit of speed is the meter per second.

Speed falls under a variety of physics disciplines including classical mechanics, astronomy, electromagnetism, and acoustics and can be related to many physics fundamentals, such as acceleration, force, and mechanical power. Newton's laws of motion, harmonic motion, and vibration all involve speed.

Italian scientist Galileo Galilei (1564–1642) was one of the first to investigate speed. Galileo measured

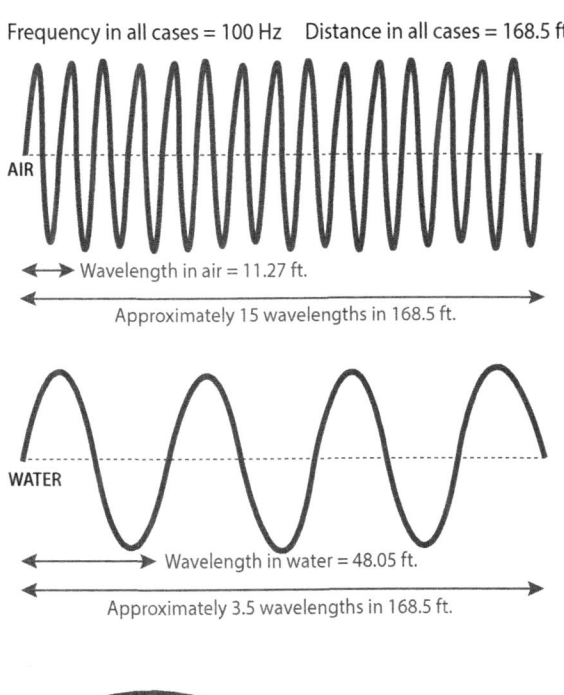

Speed is the distance traveled over the time it takes to travel; for a sound wave, it is a measure of the distance of one wave cycle divided by the time it takes for one wave cycle. Wave speed can be calculated by multiplying the wavelength by the frequency of a wave. If the frequency of a sound is kept constant at 100 Hz in air, water, and steel, the wave's speed will increase as the wavelength is lengthened.

speed by observing the change in distance an object covered and the time it took to do so.

WAVE PROPAGATION

A wave can be described as a disturbance that travels from one place to another carrying energy. Wave propagation is the manner in which waves travel. Wave propagation through a medium is slower than through a vacuum because the atoms of the medium absorb some of the energy of the wave as it passes through.

There are three main types of waves: mechanical, electromagnetic, and matter waves. Matter waves, or De Broglie waves, are propagated in particles and electrons. Mechanical waves include water waves, seismic waves, and sound waves. They are governed by Newton's laws and can only propagate through a material medium. Electromagnetic waves include light waves and radio waves. They can propagate through a vacuum or a medium.

WAVE SPEED

A wave can be described by its speed, wavelength, and frequency. For a mechanical wave, the wave speed is determined by the properties of the material through which it travels. Wave speed is the wavelength multiplied by wave frequency. For example, if a sound wave travels through different materials but its frequency remains constant, then the wave's speed increases as the wavelength increases.

SPEED OF SOUND

Mechanical waves are used every day in music, Doppler radar, and medical imaging, among other applications. Understanding wave properties, including speed, wavelength, and frequency, is necessary for predicting waves' effects and uses. For example, the speed of sound, in dry air under normal atmospheric pressure and temperature, is 343 meters per second (1,127 feet per second). If an object such as an airplane travels faster than the speed of sound, it breaks the sound barrier. This event can be accompanied by a shock wave that produces a burst of sound, or sonic boom.

—*Casey M. Schwarz, PhD*

BIBLIOGRAPHY

Cutnell, John D., Kenneth W. Johnson, Shane Stadler, and David Young. *Physics*. 10th ed. Hoboken, NJ: Wiley, 2015. Print.

Halliday, David, Robert Resnick, and Jearl Walker. *Fundamentals of Physics*. 10th ed. Hoboken, NJ: Wiley, 2014. Print.

Knight, Randall Dewey, Brian Jones, and Stuart Field. *College Physics: A Strategic Approach*. San Francisco: Pearson/Addison Wesley, 2007. Print.

Ohanian, Hans C., and John T. Markert. *Physics for Engineers and Scientists*. 3rd extended ed. New York: Norton, 2007. Print.

Walker, James S. *Physics*. 5th ed. San Francisco: Pearson/Addison Wesley, 2016. Print

Young, Hugh D., and Roger A. Freedman. *University Physics*. 13th ed. Harlow: Pearson Education Limited, 2014. Print.

SPRINGS

FIELDS OF STUDY

Classical Mechanics

SUMMARY

This article describes spring systems and their relation to Hooke's law, the conservation of energy, and harmonic oscillators. Hooke's law appears in all fields on physics and engineering because it can be applied to most solid objects in addition to springs. Harmonic oscillators are important systems in classical mechanics; they are found in nature and used for devices.

PRINCIPAL TERMS

- **compression:** the pushing forces applied to an object in order to diminish its size or volume.
- **elongation:** the lengthening of an object under stress.
- **harmonic oscillations:** the motion resulting from a system that when displaced from its equilibrium point exerts a restoring force that is proportional to the displacement.
- **Hooke's law:** the principle that states that the force necessary to compress or stretch a spring by a certain displacement is proportional to that displacement.

- **kinetic energy:** the energy associated with motion.
- **potential energy:** stored energy due to position or configuration.
- **spring constant:** a mathematical value that defines the stiffness of a spring.
- **tension:** the pulling force exerted by the ends of a rope, wire, rod, or similar object.

Hooke's Law

Elasticity is the ability of an object or material to return to its normal shape after experiencing elongation (lengthening) or compression (shortening) due to tension. An object is considered more elastic if it can be returned to its original shape more precisely. One example of an elastic object is the spring. When a spring is stretched or compressed, it exerts a restoring force that returns it to its beginning length. Hooke's law states that this restoring force (F) is proportional to the amount (x) by which it is stretched or compressed:

$$F = kx.$$

The spring constant (k) is a constant of proportionality that refers to the stiffness of the spring. The greater the value of k, the stiffer the spring. Force (F) has units of newtons (N), displacement (x) has units of meters (m), and therefore the spring constant (k) has units of newtons per meter (N/m). It is important to note that this equation gives magnitude only. To be more exact, one may establish the origin of the x-axis to be at the equilibrium length of a spring. If one stretches the spring in the positive x direction ($x > 0$), the spring exerts a force in the negative x direction with a magnitude of kx:

$$Fx = -kx.$$

Likewise, if one compresses the spring in the negative x direction ($x < 0$), the spring exerts a force in the positive x direction with a magnitude of kx. The restoring force is opposite to the direction of the displacement. Hooke's law only works for small stretches and compressions, however. If a spring is stretched excessively far, then it will reach a point where it will become permanently deformed and will not return to its original shape.

Hooke's law can apply to forces other than those associated with springs. The force that holds atoms together can be modeled using Hooke's law. These forces are responsible for vibrations and oscillations, normal force, and wave motion.

Energy of a Spring

Hooke's law is also an example of the first law of thermodynamics. A spring conserves energy when it is compressed or stretched. The potential energy (U) of a spring when it has been displaced from its equilibrium by an amount x is:

$$U = \tfrac{1}{2}kx^2.$$

The potential energy of a spring at its equilibrium point is zero and always positively increases as it is displaced from its equilibrium point. This displacement can be due to either stretching or compression. The potential energy of the spring can be converted

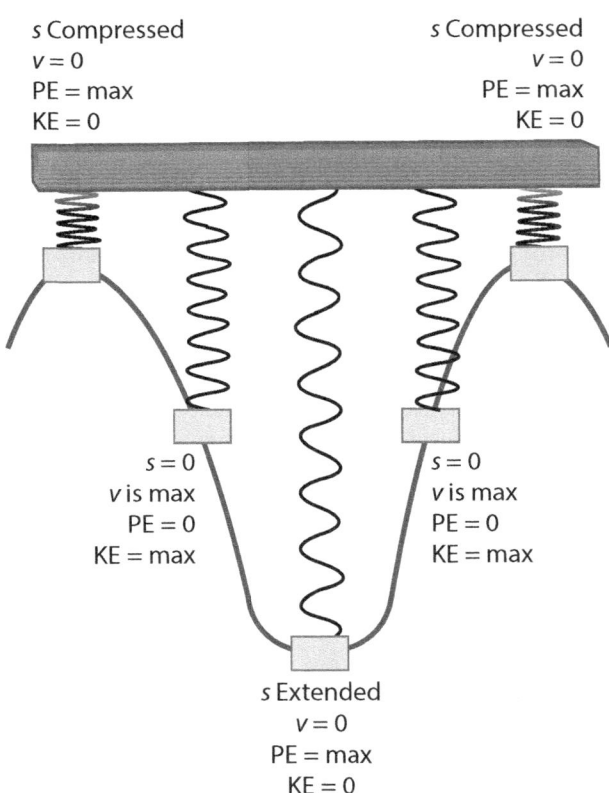

A spring will hold potential energy when it is compressed or stretched. As a spring passes through its neutral position, it reaches its maximum kinetic energy and its maximum velocity.

> **SAMPLE PROBLEM**
>
> A spring is compressed 3 centimeters by an applied force of 75 newtons. Find the spring constant.
>
> **Answer:**
> Use Hooke's equation to solve for the spring constant (k). Plug in the values for force (F) and displacement (x), remembering to convert displacement from centimeters (cm) to meters (m):
>
> $$F = kx$$
>
> $$k = \frac{F}{x}$$
>
> $$k = \frac{75 \text{ N}}{0.030 \text{ m}}$$
>
> $$k = 2{,}500 \text{ N/m}$$
>
> The spring constant is 2,500 N/m. By measuring the displacement of the spring from its original length by an applied force, one can determine the stiffness of the spring.

to kinetic energy (K), both measured in joules (J). Kinetic energy depends on an object's mass (m) and velocity (v):

$$K = \tfrac{1}{2}mv^2.$$

Consider a spring with one end attached to a support and the other end attached to a mass. Now one compresses the spring by a certain distance, then releases it on a frictionless surface. The total energy (E) of this system is the sum of the potential and kinetic energy:

$$E = U + K.$$

The mass oscillates back and forth, and the energy changes from potential to kinetic energy. The maximum potential energy occurs when the spring is at maximum compression and extension at the endpoints of the oscillation. At the endpoints, the kinetic energy is zero. The maximum kinetic energy happens when the mass travels through the equilibrium position. One can find the maximum velocity (v_{max}) of the mass by setting the maximum potential energy (U_{max}) equal to the maximum kinetic energy (K_{max}) and solving for v_{max}:

$$U_{max} = K_{max}$$

$$\tfrac{1}{2}kx_{max}^2 = \tfrac{1}{2}mv_{max}^2$$

$$v_{max} = x_{max}\sqrt{\frac{k}{m}}$$

HARMONIC OSCILLATOR

A spring with a mass attached to one end demonstrates simple harmonic motion and is a classic example of a harmonic oscillator. Consider a mass on a spring that has been stretched or compressed, then let go. The mass will then exhibit harmonic oscillations, moving back and forth about its equilibrium position. The period (T) of a mass on a spring is:

$$T = 2\pi\sqrt{\frac{m}{k}}$$

The unit for the period is the second (s). The period is independent of the amplitude and gravitational acceleration and is the same for horizontal and vertical spring systems.

SPRINGS IN EVERYDAY USE

There are many different types of springs that come in various sizes and shapes that provide different functions. Large springs used in railroad cars are heavy and stiff and used to smooth the ride of the car. Small delicate spiral springs are used in mechanical watches. The fact that these small springs obey Hooke's law with a frequency determined by the mass and the spring stiffness allows for accurate mechanical watches and clocks to be manufactured.

—*Casey M. Schwarz, PhD*

BIBLIOGRAPHY

Cutnell, John D., Kenneth W. Johnson, Shane Stadler, and David Young. *Physics*. 10th ed. Hoboken, NJ: Wiley, 2015. Print.

Halliday, David, Robert Resnick, and Jearl Walker. *Fundamentals of Physics*. 10th ed. Hoboken, NJ: Wiley, 2014. Print.

Knight, Randall Dewey, Brian Jones, and Stuart Field. *College Physics: A Strategic Approach*. San Francisco: Pearson/Addison Wesley, 2007. Print.

"Motion of a Mass on a Spring." *Physics Classroom*. Physics Classroom, n.d. Web. 4 July 2015.

Walker, James S. *Physics*. 5th ed. San Francisco: Pearson/Addison Wesley, 2016. Print

Williams, Matt, "What is Hooke's Law?" *Universe Today*. Universe Today, 13 Feb. 2015. Web. 6 July 2015.

STRING THEORY

FIELDS OF STUDY

Particle Physics; Quantum Physics

SUMMARY

String theory and its variants, including M-theory and superstring theory, are mathematical models that attempt to describe the universe's basic structure and account for all the matter, energy, and forces in it. The mathematics of the theory require a universe consisting of eleven dimensions to account for observed particle properties.

PRINCIPAL TERMS

- **bosons:** also known as gauge bosons and mesons; elementary particles that mediate a specific type of force between hadrons and leptons, and whose spin can only have integer values.
- **bosonic string:** in string theory, the structure of bosons as one-dimensional strings rather than dimensionless points.
- **fermions:** elementary particles whose spin can have only half-integer values.
- **graviton:** the gauge boson that mediates the gravitational force.
- **M-theory:** a physical model of the structure of the universe that encompasses all consistent versions of superstring theory.
- **perturbation theory:** the application of known solutions to simple quantum mechanical systems (such as that of the hydrogen atom) to more complex systems that cannot be resolved easily.
- **standard model:** the model of atomic structure in which atoms are composed of smaller particles called leptons, hadrons, and bosons.
- **supersymmetry:** the theory that all bosons have partner fermion particles.

DIMENSIONS AND STRINGS

Humans long perceived the universe to exist in three physical dimensions: length, width, and height. However, as observation and investigation probed deeper into the underlying structure and composition of the universe, that view became insufficient. Early Greek philosophers such as Plato (ca. 428–348 BCE) and Aristotle (384–322 BCE) contemplated what composes the essential "matter" of the universe. The observation that dividing a material into ever smaller pieces did not change the essential nature of each piece led to the idea that atoms are the essential form of matter. Each kind of matter was thought to have its own character and therefore its own kind of atom. Further considerations such as "elements" (for example, earth, air, fire, and water) were added to describe the properties observed when materials changed form. Philosophers also wondered from what atoms were formed. Generally, it came to be thought that the universe was composed of an unknowable number of points having neither length nor width nor height.

In the twentieth century CE, Albert Einstein (1879–1955) postulated the equivalence of matter (m) and energy (E), in relation to the speed of light (c). The equation $E = mc^2$ led to the modern theory of atomic structure and, eventually, the standard model. In the standard model, atoms are made of particles called leptons (or light particles), hadrons (heavy particles) formed from particles called quarks, and bosons and fermions. Supersymmetry theorizes that each boson has a fermion that is its natural partner, and vice versa. Electrons are classed as leptons. Protons and neutrons are classed as hadrons. The standard model specifies that leptons and hadrons interact through the exchange of gauge bosons. It also states that their various combinations construct all known matter in the universe. Each gauge boson mediates a specific kind of force within the standard model. Photons mediate electromagnetic force. Therefore, they affect all charged particles. Gluons mediate the strong nuclear force that affects quarks. Weakons mediate the weak nuclear force. Gravitons mediate the force of gravity. Therefore, they are the basis for quantum gravity.

Essentially, a model in which the universe is composed of dimensionless points posits that the universe is composed of nothing—a concept that breaks down. Paradoxically, the mathematics derived from this concept is able to predict many things that can be experimentally observed. But the limitations of the concept cannot be reconciled with the fact that the angular momentum of elementary particles is in direct proportion to the square of their corresponding energies. At least one dimension seems to be required. This is the basis of string theory. In string theory, bosons take the form of one-dimensional bosonic strings rather than dimensionless points. Bosonic strings allow for a mathematical description of the observed angular momentum and energy relationships not adequately described by points. Ironically, the one-dimensional bosonic strings can only be described in the context of a universe with ten theoretical dimensions, of which just three can be physically detected.

Historical Development of String Theory

In 1897, J. J. Thompson (1856–1940) identified electrons as independent charged particles within atoms. This discovery marked the origins of the standard model of atomic structure. In 1911, Ernest Rutherford (1871–1937) discovered the proton. Calculations related to atomic mass indicated that a third, neutral subatomic particle, the neutron, should be present in the atomic nucleus. Because the neutron lacked an electrical charge, it was impossible to observe directly. The existence of the neutron was not demonstrated until 1932, by James Chadwick (1891–1974). Even then, the evidence for the existence of the neutron was indirect. With these three particles, theoretical physicists refined a theory of atomic structure that could be reconciled with observed behaviors and energies of electrons within atoms.

Key to the development of string theory was the need to reconcile the interaction of atoms with light and to explain the dual nature of light itself. That atoms absorb and emit light of specific wavelength and energy suggested that light behaves as though composed of particles. In particular, the photoelectron effect was a troublesome problem. In the photoelectron effect, light striking a metal surface could cause the metal to emit electrons through space. Seemingly, this phenomenon could be true only if light was an energy particle that could be absorbed by the atoms of the metal. However, the "double-slit experiment" of Thomas Young (1773–1829) had demonstrated that light also behaves as waves. This dichotomy became known as "wave-particle duality." In 1905, Einstein reconciled this paradox with the equivalence of mass and energy in his famous equation. From this development, quantum mechanics was formulated. Quantum mechanics describes the behavior of electrons in atoms on the basis of discrete units of energy ("quanta"). An electron can absorb or emit just a single quantum of energy as a photon of light in changing its position within the atom.

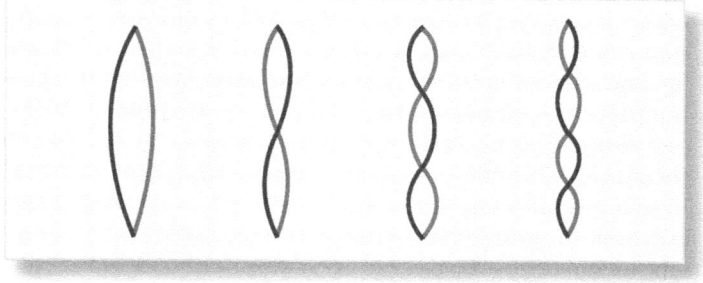

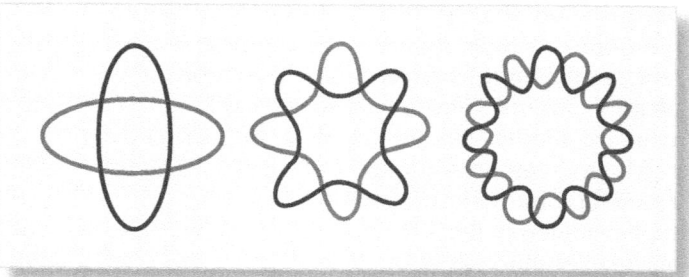

This explanation of string theory includes diagrams of different loops described by the theory.

String Theory in the Twentieth Century

Wolfgang Pauli (1900–58) further refined theory regarding the electronic structure of atoms. Pauli's exclusion principle states that only electrons that have opposite "spin" quantum numbers can occupy the same energy level in an atom or molecule; these spin numbers refer to the electron's intrinsic angular momentum. Werner Heisenberg (1901–76) proposed his uncertainty principle that states that an observer can know precisely an electron's position or energy, but not both. The basic premise of that principle is that measurement of one property inevitably changes the other. These few principles essentially make up the modern atomic theory. They do not involve nuclear matter directly, although the structure of the atomic nucleus determines the configuration of the surrounding electrons.

Both nuclear reactions and nuclear theory developed rapidly in the first half of the twentieth century. These culminated with the atomic bomb during World War II and the postwar development of nuclear energy for electrical power generation. This rapid development posed more questions about the structure of the atomic nucleus than it answered. Therefore, many nuclear scientists began studying the makeup of subatomic particles in order to understand the forces involved and to establish a "theory of everything" (or the "grand unified theory"). The mathematical models used to generate and to study the structural components of subatomic particles were still based on quantum mechanics. However, as more, finer results were obtained from high-energy collision experiments, mathematical descriptions of atomic structure using quantum mechanics were impossible both to solve accurately and to reconcile with observation. To bypass this problem, perturbation theory was developed. This theory uses known quantum mechanical solutions for the behavior of particles in simple systems to devise approximate solutions to more complex problems. For example, it uses the hydrogen atom as the basis for developing accurate approximations for the behavior of particles in complex systems that cannot be calculated directly. Throughout the 1960s and 1970s, theories were expanded both to encompass newly identified subatomic particles and to include gravity as a fundamental property.

In the 1980s, string theory was developed to account for the observed relationships between angular momentum and energy in particles. String theory bases the structure of the universe on "strings" that have only one dimension of length, rather than on points that have no dimensions at all. The vibrational frequencies of actual strings can be used as an analogy for the bosonic strings of string theory. A string of any particular length has a specific set of frequencies at which it will vibrate throughout its length. The "particle in a box" approach of quantum mechanics imposes boundary conditions on how particles follow a sinusoidal path within the confines of the box. Similarly, string theory imposes boundary conditions for vibrational frequencies by establishing the length of the vibrating string, just as the bridge and frets of a guitar determine the vibrational frequencies of a guitar string.

Structures of Other Dimensions and the Theory of Everything

String theory accounts for several of the observed characteristics of particle physics that do not fit within the quantum mechanical model. The mathematics of string theory require that a universe have ten dimensions, rather than the four dimensions of quantum mechanics (length, width, depth, and time). Superstring theory, which incorporates quantum gravity into its structure, requires a total of eleven different dimensions. M-theory is another theory to describe the structure of the universe. This theory unifies multiple string theories. M-Theory uses two-dimensional structures called "branes" (an abbreviation of "membrane") as the basic structural component. Within eleven dimensions, structures can have any dimensional feature ranging from the zero-dimensional point of quantum theory (the so-called 0-brane) to the one-dimensional strings of string theory (1-brane) and a basic eleven-dimensional structure (11-brane). Such structures can have various shapes. However, as strings, they must be either open or closed much like either a single piece of thread or a piece of thread with its ends joined to form a loop.

—*Richard M. Renneboog, MSc*

BIBLIOGRAPHY

Becker, Katrin, Melanie Becker, and John Schwarz. *String Theory and M-Theory*. A Modern Introduction. New York: Cambridge University Press, 2007. Print.

Blumenhagen, Ralph, Dieter Lüst, and Stefan Theisen. *Basic Concepts of String Theory*. New York: Springer, 2013. Print.

Calle, Carlos I. *Superstrings and Other Things: A Guide to Physics*. 2nd ed. Boca Raton: CRC, 2010. Print.

Cappelli, Andrea, et al. *The Birth of String Theory*. New York: Cambridge University Press, 2012. Print.

Gubser, Steven S. *The Little Book of String Theory*. Princeton: Princeton University Press, 2010. Print.

Schwarz, Patricia M., and John H. Schwarz. *Special Relativity: From Einstein to Strings*. New York: Cambridge University Press, 2004. Print.

Vayenas, Constantinos, and Stamatios N.-A. Souentie. *Gravity, Special Relativity, and the Strong Force: A Bohr-Einstein-DeBroglie Model for the Formation of Hadrons*. New York: Springer, 2012. Print.

STRONG FORCE

FIELDS OF STUDY

Quantum Electrodynamics; Quantum Field Theory; Quantum Mechanics; Relativity; String Theory; Superstring Theory

SUMMARY

The strong force is one of the four fundamental forces (electromagnetic, gravitational, strong, and weak) that work throughout the universe. It is effective only over a very short range and only at the level of subatomic particles, but it is the strongest of these forces. It holds quarks and/or antiquarks, and thus protons and neutrons, close to each other, forming the basis of atoms and, by extension, everything in the universe.

PRINCIPAL TERMS

- **color force:** the force of the strong interaction that operates on the quark level.
- **hadron:** a subatomic particle that is made of either three quarks or one quark and one antiquark and held together by the strong force.
- **quantum chromodynamics:** a quantum field theory that describes the interactions of quarks and gluons, subatomic particles that are responsible for the strong interaction.
- **quantum field theory:** a theory that explains interactions between subatomic particles as the result of a field extending between them.
- **quark:** an elementary fermion that combines with other quarks to form a baryon, such as a proton or neutron, or with an antiquark to form a particle called a meson.
- **strong interaction:** the fundamental process of particle interaction that binds quarks into hadrons and hadrons into nuclei.

THE STRONG FORCE AND QUARKS

Before scientists understood quantum physics, they still understood that atoms were made of smaller particles. Back in the 1800s, electrons were noted as a part of an atom. Because electrons were negatively charged and removing an electron from an atom left it with a positive charge, scientists believed that atoms were held together with electromagnetic forces. Then experiments showed that atoms had a positively charged nucleus that contained both positively charged protons and neutrally charged neutrons and were surrounded by negatively charged electrons. Clearly, electrons and nuclei were held together electromagnetically, but what was holding the protons so closely together when, logically, their charges should make them repulse each other?

Scientists first thought that neutrons were functioning as some kind of glue. But, since neutrons have no charge, they theorized that there must be another force strong enough to overcome electromagnetic force. In the 1930s, Hideki Yukawa theorized that a new type of quantum particle existed that provided this strong interaction. Then in the 1960s, when Murray Gell-Mann and George Zweig began to promote their theory of fundamental particles called quarks, the strong force began to come into focus. In the Gell-Mann/Zweig model, quarks have electrical charges of 1/3 or 2/3 as well as possessing a color force. This theory sparked a new field of study they called quantum chromodynamics, part of the quantum field theory which explains the behavior of subatomic particles. The strong force is about 100 times stronger than electromagnetism. It acts directly only on quarks and gluons, the basic exchange particle that acts as a mediator force between quarks (and called gluons because they act like glue).

"Color" in this sense doesn't refer to an actual color that one can see, but is a way to help one understand the way the strong force works. The strong force has three types of charges: Red, blue, and green, and three opposite charges: Anti-red, anti-blue, and anti-green. To make a strong neutral particle, such as a hadron, the "colors" must combine together to form a neutral, such as a particle made of a red, a green, and a blue quark or a particle made of a red quark and an anti-red quark. Because quarks are always exchanging gluons, they have color, but don't have a specific color at any one time. This extremely complex dance of color between quarks means that one never sees a quark on its own—the strong force keeps quarks together in groups of two or three that are always color neutral. This is called "color confinement." Color force only directly affects particles inside a hadron.

Even after decades of study, scientists are still learning things about the strong force, and any calculations involving them must be solved by supercomputers.

What's Important about the Strong Force?

The strong force holds quarks, one of the basic particles that make up our universe, together. By understanding this force, we can have a better understanding of our universe as a whole. The breaking of the strong force bond is what releases energy when heat is generated in a nuclear power plant or when a nuclear weapon is detonated.

—*Marianne Moss Madsen, MS*

Bibliography

Greene, Brian. *The Elegant Universe Superstrings, Hidden Dimensions, and the Quest for the Ultimate Theory*. New York: Vintage, 2010. Print.

Lang, Thomas G. *Our Fluid Universe: A Unified Theory of Physics that Physically Describes Photons, Matter, Gravity, Electromagnetic Forces, the Strong and Weak Forces, Quantum Theory, and Much More*. CreateSpace Independent Publishing Platform. 2014. Electronic.

Murray, Raymond L., and Keith E. Holbert. *Nuclear Energy an Introduction to the Concepts, Systems, and Applications of Nuclear Processes*. 7th ed. Amsterdam: Elsevier, 2015. Print.

Randall, Lisa. *Higgs Discovery: The Power of Empty Space*. Ecco. 2013.

Riggs, Shelton. *The Nuclear Force: The Force Which Binds Quark Trios Into Protons and Neutrons*. Amazon Digital Services, LLC, 2015. Electronic.

Susskind, Leonard, and Art Friedman. *Quantum Mechanics: The Theoretical Minimum*. New York: Basic, 2014. Print.

Tyson, Neil DeGrasse, and Donald Goldsmith. *Origins: Fourteen Billion Years of Cosmic Evolution*. New York: W.W. Norton, 2014. Print.

Vavenas, Costas G., and Souentie, Stamatios N.-A. *Gravity, Special Relativity, and the Strong Force: A Bohr-Einstein-de Broglie Model for the Formation of Hadrons*. Springer, 2012.

Vayenas, Constantinos G., and Stamatios N. A. Souentie. *Gravity, Special Relativity, and the Strong Force: A Bohr-Einstein-de Broglie Model for the Formation of Hadrons*. New York: Springer, 2012. Print.

SUPERCONDUCTOR

FIELDS OF STUDY

Electronics; Atomic Physics; Quantum Physics

SUMMARY

Superconductors have been known since 1911, but it is only recently that the property has been found to exist at temperatures up to 158°C (316°F). Superconducting materials have no electrical resistance, and an electrical current initiated in a superconducting circuit will flow indefinitely. Application of a critical magnetic field at a temperature below a superconductor's critical temperature will cause the material to exhibit normal conductivity rather than superconductivity. Superconductor technology has great potential for rapid transportation, energy transmission, and electromagnetic field generation.

PRINCIPAL TERMS

- **conductor:** a material that has a low resistance to electric charges, allowing them to move through it easily.
- **continuity:** a clear path for electricity from point A to point B.
- **current:** the rate at which an electric charge, usually in the form of electrons, moves through a wire or other conductive material.
- **eigensystem:** the set of all eigenvectors of a matrix paired with their respective eigenvalues.
- **electron:** a negatively charged subatomic particle that is often bound to the positive charge of the nucleus but can also exist in a free state in an atom.
- **quantum state:** the condition of a physical system as defined by its associated quantum attributes.
- **standard temperature and pressure (STP):** standard reference conditions when dealing with gases, defined by the International Union of Pure and Applied Chemistry as a temperature of 273.15 kelvins (0 degrees Celsius or 32 degrees Fahrenheit) and pressure of 101.3 kilopascals (1 atmosphere); used in chemistry and physics to establish a standardized set of conditions for experimentation.
- **wave function:** a function that describes the quantum state of a system and represents the probability of finding the system in a given state at a given time.

CONDUCTORS AND NONCONDUCTORS

The term *conductor* refers to the ability of a material to facilitate the transport of some property from one point to another. The medium of transport is matter in one form or another; matter can conduct heat, sound, and electrical current, primarily. An electrical current exists when electrons flow through a conductor from a point of higher electrical potential to a point of lower electrical potential. All conductors have a characteristic resistance to the flow of electrons through them. This typically results in the generation of heat, and more than one fire has been started by electrical conductors that have become overheated as the result of carrying more current than is safe.

Nonconductors, as the name suggests, do not conduct electrical current well, if they conduct it at all, and are generally referred to as insulators. Electrical components called resistors lie somewhere in between these two designations, because they are designed to carry an electrical current while providing a specific resistance to the flow of electrons.

A superconductor is a material that will conduct electrical current with absolutely no resistance to the flow of electrons through it. Metals are the most widely known conductors, having relatively low resistance to current flow. Recently, research has also identified several polymeric plastic materials that can carry an electrical current at least as well as metals.

Both metal and plastic conductors have the same weakness—namely that their ability to conduct electrical current is temperature-dependent. In general applications, typical conductors function best in conditions that are near standard temperature and pressure (STP). As the temperature of the material increases, however, its ability to conduct electrical current decreases. Reducing the temperature instead of raising it has its own kinds of problems, as the physical structure of the material changes to become more crystalline and brittle.

An ideal superconductor would exhibit none of these effects. Both metal and polymer conductors are able to transport electrical current because their respective atomic and molecular orbitals, as defined by their particular quantum state and wave function, are able to accept electrons from neighboring atoms. This requires the presence of unoccupied atomic or molecular orbitals that are similar to the occupied atomic or molecular orbitals with regard to energy.

As the temperature of a conductor decreases, the relative separation of the orbital energy levels decreases. Theoretically, at a sufficiently low temperature, the orbital energy separation becomes so low that there is effectively no barrier to the transfer of electrons between orbitals, and, at this point, the material is superconducting. This theory is the logical outcome of the band theory of solids, in which the various quantum states of the component atoms of a material essentially combine to define quantum states that span the entire material, rather than just individual atoms.

Quantum mechanics predicts the formation of Bose condensates by the same means, as the quantum states of a number of individual atoms combine to form what has been termed a *superatom*. Phonons are predicted to have the same behavior, since quantum vibrational states that span an entire mass of condensed matter are believed to have a role in superconductivity. In a superconductor, there is complete continuity across the quantum orbitals such that they form a complete eigensystem rather than a large collection of individual atomic eigensystems.

CURRENT RELATIONSHIPS
Electrical current in conventional conductors is described by a simple relationship called Ohm's law, which relates the current in an electrical circuit to the voltage that is applied to the circuit and the resistance that exists within the circuit. The relationship is given by the following formula:

$$E = I \times R,$$

where E is the voltage, I is the current and R is the resistance.

When the resistance is zero, as in a superconductor, this relationship breaks down, since it would suggest that the current could become infinitely high for any applied voltage, which in turn implies that

---

SAMPLE PROBLEM

The normal conductivity of a superconductor can be restored at any specific temperature by an applied magnetic field, according to the relationship

$$B_c(T) = B_c(0)[1 - (T/T_c)^2]$$

where B_c is the critical magnetic field strength, T is the temperature, and T_c is the critical temperature for the material. For the element tantalum T_c = 4.5K and BC(0) = 83mT (milliTesla).

Calculate $B_c(T)$ the magnetic field strength required to convert superconducting tantalum to a normally conducting state, at a temperature of 2K.

Answer:
Using the equation

$$B_c(T) = B_c(0)[1 - (T/T_c)^2]$$

$$= (83\text{mT})[1 - (2/4.5)^2]$$

$$= 66.6\text{mT}$$

---

the voltage source could supply an infinite number of electrons. Logically, this is an impossible condition, because the number of electrons that are available to flow through the circuit is finite.

Fortunately, this can be interpreted in different ways. Certainly electrons will flow continuously through the circuit while the applied voltage exists, and because the circuit returns the electrons to the source as required by the conservation of charge, the supply of electrons could be thought of as infinite in a sense. Another way of interpreting the condition of zero resistance is this: The transfer of electronic charge is instantaneous, or in other words, requires zero time. In a practical sense, the relationship breakdown means that a current set in motion in a superconducting circuit will continue to flow essentially forever, in the absence of any resistance.

TYPES OF SUPERCONDUCTORS
Superconductors are generally designated as Type 1 or Type 2. Type 1 materials have normal conductivity

and resistance, but at a sufficiently low temperature, they exhibit a sharp transition into the superconducting state. Typically the required temperatures are very low. Lead, for example, has the highest superconductivity transition temperature of the Type 1 materials, at just 7.88K (− 445°F). Aluminum becomes superconducting when the temperature is reduced to just 1.175K, while rhodium must be cooled to a mere 0.000375K to become superconducting.

Type 2 superconductors are, with very few exceptions, metallic compounds and alloys. These exotic materials exhibit superconductivity across a broad range of temperatures from near 0K to as high as 158°C (316°F).

APPLICATIONS

An important feature of superconductors is that they exhibit "perfect diamagnetism" and thus produce an equal and opposite magnetic field to an applied magnetic field. This can be demonstrated by attempting to place a permanent magnet atop a superconductor. The magnet will induce an equal and opposite magnetic field in the superconductor and will levitate. This effect is put to use in maglev trains such as Japan's bullet trains that can run at high speed with essentially only the friction of the air to oppose their movement. Another application is lossless electrical energy transmission, which has been tested on a small scale and is expected to become a very important technology as new superconducting materials continue to be developed. Many types of industrial, analytical, and research devices use electromagnets as a vital component of their construction. Superconducting electromagnets often provide more stable and easily controlled alternatives to conventional electromagnets.

—*Richard M. Renneboog M.Sc.*

FURTHER READING

Askerzade, Iman. *Unconventional Superconductors : Anisotropy and Multiband Effects*. Berlin: Springer-Verlag, 2014. Print. Springer Ser. in Materials Science.

Aynajian, Pegor. *Electron-phonon Interaction in Conventional and Unconventional Superconductors*. Berlin: Springer-Verlag, 2013. Print. Springer Theses.

Bhattacharya, Raghu, and M. P. Paranthaman, eds. *High Temperature Superconductors*. Weinheim: Wiley-VCH, 2010. Print.

Lebed, Andrei, ed. *The Physics of Organic Superconductors and Conductors*. Berlin, Heidelberg: Springer-Verlag Berlin Heidelberg, 2008. Print. Springer Ser. in Materials Science.

Moshchalkov, Victor V., and Joachim Fritzsche. *Nanostructured Superconductors*. Singapore: World Scientific, 2011. Print.

SWITCHES

FIELDS OF STUDY

Electronics

SUMMARY

Switches are electrical components that control the flow of electricity to lights and other devices using either mechanical action or electromagnetic force. They range from simple to complex and have applications from turning on household lights to controlling individual functions in electronic devices.

PRINCIPAL TERMS

- **circuit:** a closed path along which electricity travels.
- **conductors:** substances that can transport electricity.
- **continuity:** a clear path for electricity from point A to point B.
- **insulators:** substances that cannot transport electricity.
- **mechanical switches:** devices that use moving parts and direct physical force to bring contacts together to let electricity flow.
- **nonmechanical switches:** devices that use electromagnetism to open and close a circuit.

- **poles:** contact points that complete the circuits in switches.
- **throws:** sets of input/output wires in a switch.

Controlling Electricity

Switches are one of the most basic forms of controlling electricity. At their simplest, switches turn an electrical current on or off. The standard light switch, common in homes around the world, was invented in 1884. Switches occur in simple devices such as lamps, but they can also control individual functions within complex electronic devices like computers, stereos, or DVD players.

A common component of an electrical circuit is a switch. Switches provide the ability to easily complete or disconnect a circuit. Some examples of switch types include toggle switches, rocker switches, push-button switches, and slide switches. By Mcc-Ri (Own work) [CC BY-SA 3.0 (http://creativecommons.org/licenses/by-sa/3.0)], via Wikimedia Commons.

How a Switch Works

Electricity travels along any substance (such as copper wire) that can act as a conductor. Broken or damaged conductors can prevent electrical continuity, or flow. If the conductor forms a closed circuit, or loop, between a power source and a connected device, the device receives power. A switch acts like a gate, opening and closing the circuit. An insulator wraps around a conductor to prevent electricity from leaving its intended path.

Types of Switches

Many switches are mechanical, opening or closing a circuit with direct physical force. This is what happens when a light switch on the wall is flipped up or down. A mechanical switch may be operated manually, be triggered by other components, or respond to a sensor. Switches can also be nonmechanical. Electromagnetism opens and closes their circuits.

The most basic attributes of a switch are its poles (the number of circuits it controls) and throws (the number of electrical inputs and outputs it has). A light that can be switched on and off from two locations is single pole, double throw (SPDT). A switch that can control both a ceiling light and fan is double pole, single throw (DPST).

Simple Idea, Many Applications

Switches are an integral part of everyday electronics and equipment in homes and businesses. They have applications in a broad range of technologies from simple to complex. One area of innovation is "smart homes," with appliances that can be controlled by mobile phones or that respond to human behavior via advanced switches. Also, the development of microscopic switches means that electronic devices continue to shrink in size.

—*J. D. Ho, MFA*

Bibliography

Cipriani, Jason. "Let There Be Light." *Wired*. Condé Nast, 24 Sept. 2013. Web. 25 Mar. 2015.

Ju, Ann. "Popular Origami Pattern Makes the Mechanical Switch." Phys.org. Phys.org, 10 Mar. 2015. Web. 25 Mar. 2015.

Lowe, Doug, and Dickon Ross. *Electronics All-in-One for Dummies*. Hoboken: Wiley, 2012. Print.

Novet, Jordan. "Patent Details Google's Ideas for Smart Home Doorknobs, Doorbells, Wall Switches, and More." *VentureBeat*. VentureBeat, 6 Mar. 2015. Web. 25 Mar. 2015.

Parker, Steve. *Electricity*. Rev. ed. New York: DK, 2013. Print.

Ultimate Guide to Wiring. 7th ed. Upper Saddle River: Creative Homeowner, 2010. Print.

Westcott, Sean, and Jean R. Westcott. *The Complete Idiot's Guide to Electronics 101.* New York: Alpha, 2011. Print.

SYMMETRY

FIELDS OF STUDY

Particle Physics; Nanotechnology; Geophysics

SUMMARY

The arrangement of atoms in condensed matter (solids) is ultimately determined by the electronic configuration and relative sizes of the atoms. Crystals, regular solids and even single molecules have three-dimensional shapes that are constructed about different planes and axes of symmetry such that the shape on one side of a plane reflects the shape on the opposite side. The notion of symmetry is related to the notion of invariants. An invariant is a quantity that does not change when a symmetry operation is performed. The symmetry operations that leave a structure invariant form a mathematical group. The study of invariants is an important part of atomic, nuclear and particle physics.

PRINCIPAL TERMS

- **center of mass:** the point in an object or system around which the mass of said object or system is evenly distributed.
- **degrees of freedom:** the number of physical parameters required to specify the position and configuration of a particle or other body.
- **frame of reference:** a set of coordinate axes that serve to describe position or movement of an object with reference to that coordinate system.
- **inversion:** the reversal of the way an image looks after an interaction with a mirror or lens.
- **matter:** anything that can be characterized as having mass and can be measured by some criterion of measurement.

CONDENSED MATTER

Condensed, or solid, matter has a three-dimensional shape, regardless of the scale. Single molecules are composed of individual atoms constrained in a specific three-dimensional arrangement by the chemical bonds between them. Those bonds are, in their turn, constrained to specific geometric arrangements about their respective atomic nuclei according to the tenets of quantum mechanics. On a macroscopic scale, solids such as crystals exhibit very well-defined structures with plane faces and edges that form specific angles. Such structures typically have a number of planes and axes of symmetry within the frame of reference of the center of mass of the structure.

SYMMETRY OPERATORS

There are only two basic symmetry operations: reflection through a plane, and rotation about an axis. The combination of reflection and rotation is inversion. A three-dimensional structure can have several separate symmetry operations to describe the number of degrees of freedom of the structure. The number of symmetry operators that exist within a structure comprise a "point group" that effectively describes the overall shape of the structure. A cube, for example,

---

SAMPLE PROBLEM

Graphene consists of a planar array of fused benzene rings. The benzene ring consists of six carbon atoms in a planar regular hexagonal arrangement. Determine the number of symmetry operators in the benzene molecule.

Answer:

As a regular six-sided polygon, there is a plane of symmetry in the plane of the benzene ring itself. A plane of symmetry bisects the ring through each of three opposing pairs of carbon atoms, and through the mid-points of each opposing pair of carbon-carbon bonds. Each plane also contains an axis of rotation. There is another axis of rotation perpendicular to the plane of the ring. There are therefore a total of 7 planes of symmetry and 7 axes of rotation.

has three axes of rotation orthogonal to each other through the center of mass of the cube, as well as three planes of reflection in the same orientation. Rotation about any of the axes produces an identical structure. Similarly, the shape of the structure on one side of a plane of reflection is precisely mirrored by the structure on the opposite side of the plane, again producing an identical structure. An object with spherical symmetry has an infinite number of planes and axes of symmetry through its center of mass.

Describing the Physical World

Symmetry is a useful way of classifying physical structures according to their common shapes. The properties of nanoparticles and their interactions with electromagnetic waves are determined to a large extent by their particular symmetry. Many different scientific fields also use symmetry descriptors in classifying various objects of study and analysis such as the shapes of protein molecules in biochemical analysis or mineralogical traces in forensic examinations.

—*Richard M. Renneboog M.Sc.*

Bibliography

Costa, Giovanni and Fogli, Gianluigi. *Symmetries and Group Theory in Particle Physics. An Introduction to Space-Time and Internal Symmetries* New York, NY: Springer, 2012. Print.

El-Batanouny, Michael and Frederick Wooten. *Symmetry and Condensed Matter Physics: A Computational Approach* New York, NY: Cambridge University Press, 2008. Print.

Robinson, Matthew. *Symmetry and the Standard Model: Mathematics and Particle Physics.* New York, NY: Springer, 2011. Print.

Sundermeyer, Kurt. *Symmetries in Fundamental Physics.* 2nd ed., New York, NY: Springer, 2014. Print.

SYSTÈME INTERNATIONAL (SI) UNITS

FIELDS OF STUDY

Classic Mechanics; Thermodynamics

SUMMARY

The *Système International* is a system of units of measurement based on universally constant physical properties rather than on arbitrary definitions. The units have identical values everywhere. Historically, the system was developed in 1795 by the French Academy of Science to facilitate communication of scientific and technical information, and has become the standard of science, technology and commerce all around the world.

PRINCIPAL TERMS

- **coulomb:** the basic unit of charge in the International System of Units (SI).
- **hertz:** the SI unit of frequency; one hertz (Hz) is equal to one cycle (complete orbit) per second.
- **joule:** abbreviated J, the International System of Units unit of work and energy. One joule is equal to the work done by a force of one newton acting across a distance of one meter.
- **kilogram:** the base unit of mass in the International System of Units (SI), equal to 1,000 grams (2.2 pounds).
- **meter:** the SI base unit of distance (or length) measurement.
- **newton:** abbreviated N, the International System of Units unit of force. One newton is equal to the force required to accelerate a one-kilogram mass at one meter per second per second.
- **radians:** (rad) the SI unit of measure for angles, based on relationship between the radius and circumference of a circle. A full circle is composed of 2π radians.
- **watt:** the SI unit for measuring power (work over time), defined as one joule per second.

The Système International (SI)

The *Système International*, commonly called the "metric system" in English-speaking countries, was the first system of measurement designed to relate to unchanging physical properties. Prior to its development, an essentially random assortment of measurements was used, that required tradespeople and merchants to understand and convert between measurements that were often arbitrarily defined. The increasing importance of exchanging scientific

> **SAMPLE PROBLEM**
>
> Given a silicon chip with a surface area of 10 cm by 10 cm, how many transistor structures can be etched onto the surface if each transistor structure is 5 nm by 5 nm?
>
> **Answer:**
>
> Each transistor structure occupies (5 nm X 5 nm =) 25 nm².
>
> The surface are of the chip is (10 cm X 10 cm =) 100 cm².
>
> Convert these to standard measurement in meters, as follows:
>
> $5 nm = 5 \times 10^{-9} m$, so $25 nm^2 = 25 \times 10^{-18} m^2$,
>
> and
>
> $5 nm = 5 \times 10^{-9} m$, so $25 nm^2 = 25 \times 10^{-18} m^2$,
>
> and
>
> $10 cm = 10^{-2} m$, so $100 cm^2 = 100 \times 10^{-4} m^2 = 10^{-2} m^2$
>
> To determine the number of transistor structures that can be etched into the available area, divide the total surface area by the area required for each transistor structure, as follows:
>
> $(10^{-2} m^2)/(25 \times 10^{-18} m^2) = 4 \times 10^{14}$
>
> Thus 4×10^{14} transistor structures are possible in the available area.

and technical information encouraged the standardization of measurements, and led directly to the development of the *Système International* by the French Academy of Science in 1795.

By adopting a standardized system of measurement, based on unchanging physical properties, users could be assured that a quantity or property measured and stated in one part of the world would be exactly the same in any other part of the world. This standardization was invaluable in the advancement of physics and chemistry in particular. The variability innate in non-standard measurement is readily demonstrated by considering the term horsepower, which was replaced by the joule as the basic unit of measurement. One horsepower was defined as the work done by a horse in moving a specific weight through a specific distance. The questions that inevitably must be asked are "whose horse?", "what weight?" and "what distance?" since all of these factors are highly variable.

THE BASIC SI UNITS

Smaller and larger quantities are conveniently stated in units that are factors or multiples of ten of the basic unit in the *Système International*. These are indicated by appending the appropriate standard prefix to the basic unit name. The prefixes are taken from Greek, and each indicates the number of factors or multiples of the basic unit. The prefixes deci-, centi-, milli-, micro-, nano-, pico- and femto- are used to indicate that the basic unit has been divided into ten, one hundred, one thousand, one million, one billion, one trillion or one quadrillion equal parts. In the other direction, the prefixes deca-, hecta-, kilo-, mega-, giga- and tera- are used to indicate that the basic unit has been multiplied by ten, a hundred, a thousand, a million and a trillion times.

The numbers are made more accessible by the use of scientific notation, which states smaller and larger numbers as a simple number multiplied by 10 raised to the appropriate power index. Distances are given using the meter as the basic unit, with smaller distances being stated in millimeters, centimeters, etc., and longer distances in kilometers. Similarly, the basic SI unit for mass and weight is the gram, although the systematic companion unit to the meter is the kilogram, equal to 1000 grams.

All properties that are cyclic in nature, such as electromagnetic and sound waves, or that have a rotational component, are stated using the hertz as the basic unit of frequency, with one hertz corresponding to exactly one complete cycle of the property. Rotational and cyclic motions are measured in terms of radians, with one complete cycle corresponding to 2π radians. Electrical charge is stated using the coulomb as the basic unit. Work, measured in joules, consumes power, measured in watts. Work, regardless of the measurements being used, is defined as a force acting through a distance. The SI unit of force is the newton.

UNIVERSALITY OF SI UNITS

Since the units of the *Système International* are based on constant physical properties rather than variable definitions, they are applicable everywhere in the universe. Failure to observe the constancy of the units, however, by using non-standard units at the same time can be an expensive mistake. A case in point: a multi-million dollar probe sent to explore Mars was destroyed when it crashed because one part of the control program was written using non-standard units and sent incorrect information to the part of the program that was written using standard SI units.

—*Richard M. Renneboog M.Sc.*

BIBLIOGRAPHY

Cardarelli, François. *Encyclopaedia of Scientific Units, Weights, and Measures: Their SI Equivalences and Origins*. London; New York: Springer, 2003. Print.

Gupta, S. V. *Units of Measurement Past, Present and Future. International System of Units*. Berlin: Springer Berlin, 2013. Print.

Kijewski, Wacek. *SI Units, Conversion and Measurement Skills*. London [u.a.]: Minerva, 1999. Print.

Taylor, B. N., Ambler Thompson, and International Bureau of Weights and Measures. *The International System of Units (SI)*. Gaithersburg, MD; Washington: U.S. Dept. of Commerce, National Institute of Standards and Technology; For Sale by the Supt. of Docs., U.S. G.P.O., 2008. Print.

TEMPERATURE AND INTERNAL ENERGY

FIELDS OF STUDY

Thermodynamics; Classical Mechanics

SUMMARY

The temperature of an object or substance is actually the average thermal energy of all its particles. The standard unit for temperature is the kelvin (K). Thermal energy, like all forms of energy, is measured in joules (J). It is in fact a subtype of kinetic energy, based on the motion of the particles that make up matter.

PRINCIPAL TERMS

- **enthalpy:** a measure of the total internal energy (thermal energy) of a system, the product of its volume and pressure.
- **heat:** the active process of energy transfer due to changes in an object's thermal energy.
- **ideal gas law:** a law stating that the pressure (P) and volume (V) of an ideal gas are directly related to its number of particles (n), its temperature (T), and the ideal gas constant (R).
- **kinetic energy:** the energy an object possesses due to its motion.
- **potential energy:** the energy stored within an object due to its position (e.g., gravitational pull on a stationary object above the ground) or its configuration (e.g., an electrical charge or chemical makeup).
- **static energy:** electrical energy resulting from an imbalance in electrical charges.
- **temperature gradient:** a measurement of the rate of temperature change over distance.
- **thermal energy:** energy generated by the movement of particles within an object or substance.

A Measure of Internal Energy

Temperature does not simply indicate how hot or cold a given object or substance is. An object's temperature, measured in kelvins (K), degrees Celsius (°C), or degrees Fahrenheit (°F), is actually a measure of the average thermal energy contained in that object. The hotter the object, the higher its temperature and the more energy contained within. Thermodynamics is the branch of physics concerned with the study of thermal energy.

The language of thermodynamics can be confusing, as much of its terminology has general-use definitions as well. Specifically, temperature measures the average thermal energy contained in the particles of an object. Like all forms of energy, it is measured in joules (J). Recall that all matter, even solid materials, consists of molecules and atoms with lots of space between them. Even the carbon atoms in a diamond, locked into a crystalline structure of incredible hardness, constantly vibrate imperceptibly. Because thermal energy is generated by the movement of individual particles, it is closely related to kinetic energy. Indeed, thermal energy is sometimes called "thermal kinetic energy." However, "thermal energy" refers specifically to the average kinetic energy of the particles in an object, while "kinetic energy" refers to any energy associated with motion, whether of an entire object or of its individual molecules.

When the temperature of an object changes, the particles in it speed up or slow down accordingly. A temperature gradient describes the direction and rate of temperature change in terms of temperature per unit of distance. In International System of Units (SI) units, this is measured in kelvins per meter (K/m). Heat, in thermodynamics, is a process of energy transfer or loss, not a property of an object.

Common Temperature Scales

The three most common temperature scales are the Kelvin scale, the Celsius scale, and the Fahrenheit scale. The kelvin is the standard unit for scientific work, although Celsius is often used as well. The Kelvin scale is distinguished by the fact that 0 kelvin corresponds to absolute zero, the point at which all motion is thought to cease. Water freezes at 273 kelvins and boils at 373 kelvins. Celsius is based on the freezing and boiling points of water, set at 0 and 100 degrees Celsius, respectively. The Fahrenheit scale, used for everyday temperature readings in the United States, was developed based on the lowest temperature to which brine could be cooled and the average human body temperature. On the modern scale, pure water freezes at 32 degrees Fahrenheit and boils at 212 degrees Fahrenheit. Equations for converting between the temperature scales are below:

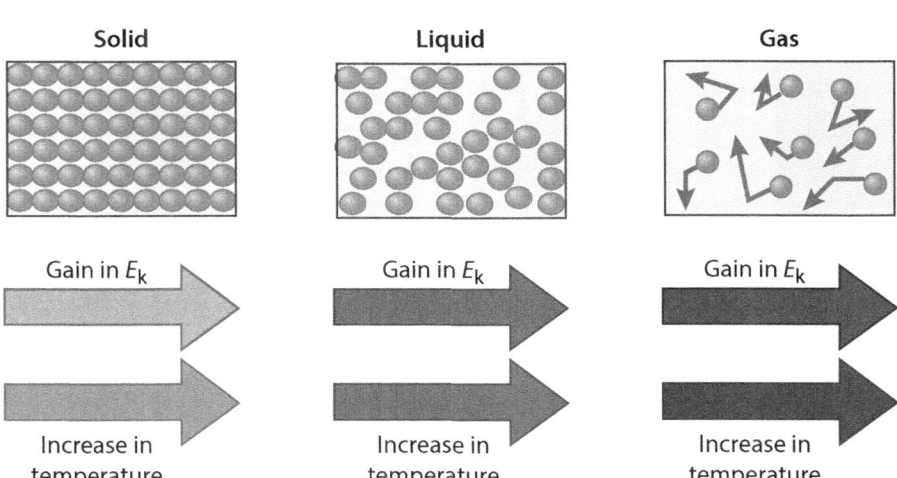

The energy of molecules in matter, also known as internal energy (E_k), changes with state changes. Solids have molecules that vibrate in place, liquids have molecules that move about more freely, and gases have molecules that move about at high energy. Temperature is a measure of this internal energy; higher temperatures indicate higher internal energy of the molecules. Thus, when matter increases in temperature, its internal energy increases, and it changes state from solid to liquid to gas.

$$K = °C + 273.15 = \frac{°F - 32}{1.8} + 273.15$$

$$°C = K - 273.15 = \frac{°F - 32}{1.8}$$

$$°F = (°C)(1.8) + 32 = (K - 273.15)(1.8) + 32$$

Conservation of Energy

The law of conservation of energy states that in an isolated system—one from which neither matter nor energy can escape—energy is conserved. The universe is, in theory, the ultimate isolated system. Therefore, according to this law, energy in the universe can be neither created nor destroyed, only transformed or transferred.

When a system loses thermal energy as heat, that energy is transformed into another form of energy. Often, heat is generated as a by-product when other forms of energy, such as static energy, are transferred. Resistance causes some static energy being transferred along wires to be transformed into thermal energy and lost to the surrounding environment as heat. Lightbulbs transform electrical energy into heat and light.

Energy can exist in a variety of forms, some of which cannot be directly observed. Potential energy is stored energy an object possesses due to its position or configuration. The chemical structure of food holds potential energy that is released by digestion. The body eventually turns a portion of this potential energy into the kinetic energy of moving limbs and beating hearts.

When the transfer of energy results in displacement, work has been done. Work, like energy, is measured in joules. Transferring one unit of energy is equivalent to performing one unit of work. Many devices use heat to perform work. The engine of a car uses heat from a spark plug to release the fuel's chemical potential energy through combustion. This released energy increases the kinetic energy of the gas particles in the piston chamber. These particles bounce off the walls of the chamber with increasing force, pushing the piston. The pistons transmit their kinetic energy to the driveshaft, which transmits it to the wheels. Ultimately the wheels impart kinetic energy to the entire vehicle, producing forward motion. For the sake of study, an engine can be considered an isolated system, retaining all matter and energy within it. However, real-world factors such as refueling and exhaust emissions (input and output of matter) actually make it an open system.

> **SAMPLE PROBLEM**
>
> (A) An air temperature of 90 degrees Fahrenheit is considered quite hot. What is the equivalent temperature in Celsius, the scale used in European countries? What is it in kelvins, the standard unit of temperature?
>
> (B) Absolute zero is 0 kelvin. What is the corresponding temperature in Celsius? What is it in Fahrenheit?
>
> **Answer:**
> To convert between temperature scales, use the equations above. Note that the math for converting to or from Fahrenheit is slightly more complicated due to the fact that the individual degrees are not the same size as degrees Celsius and kelvins.
>
> (A) To convert from Fahrenheit to Celsius and to kelvins, use the following equations:
>
> $$°C = \frac{°F - 32}{1.8}$$
>
> $$K = \frac{°F - 32}{1.8} + 273.15$$
>
> Simply plug in the Fahrenheit value given and solve:
>
> $$°C = \frac{90\ °F - 32}{1.8}$$
>
> $$°C = \frac{58}{1.8}$$
>
> $$°C \approx 32.22$$
>
> Ninety degrees Fahrenheit is approximately equal to 32.22 degrees Celsius.
>
> $$K = \frac{90\ °F - 32}{1.8} + 273.15$$
>
> $$K = \frac{58}{1.8} + 273.15$$
>
> $$K \approx 305.37$$
>
> It is also approximately equal to 305.37 kelvins.
>
> (B) To convert from kelvins to Celsius and to Fahrenheit, use the following equations:
>
> $$°C = K - 273.15$$
>
> $$°F = (K - 273.15)(1.8) + 32$$
>
> Simply plug in the kelvin value given and solve:
>
> $$°C = 0\ K - 273.15$$
>
> $$°C = -273.15$$
>
> *(Sample problem continued on next page)*

> *(SAMPLE PROBLEM CONTINUED)*
>
> Zero kelvin is equal to −273.15 degrees Celsius.
>
> $$°F = (0\ K - 273.15)(1.8) + 32$$
>
> $$°F = -491.67 + 32$$
>
> $$°F = -459.67$$
>
> It is also equal to −459.67 degrees Fahrenheit.

PARTICLES IN MOTION: THE IDEAL GAS LAW

Gases are particularly illustrative when studying the relationship between temperature, thermal energy, kinetic energy, and particle motion in a substance. In a gas, the particles are free to bounce around. Therefore, a gas will expand to fill any container of any shape. In addition, gas particles are in constant motion, bouncing off the walls of the container as well as each other. This causes the gas to exert outward pressure on the container.

Heating a gas raises not only its temperature but also the pressure exerted on its container. This relationship is one of several laid out in the ideal gas law:

$$PV = nRT.$$

This law describes the relationships between pressure (P), typically measured in kilopascals (kPa); volume (V), measured in cubic meters (m³); the number of particles (n), measured in moles (mol); and the temperature (T), measured in kelvins. It also includes the ideal gas constant (R), equal to approximately 8.314 J/mol·K.

The relationships to temperature can be better seen when the equation is rewritten as follows:

$$T = PV/nR.$$

From this equation, it is clear that increasing either the pressure or the volume of a gas will raise its temperature. Conversely, increasing the number of particles will cause the temperature to drop. This makes intuitive sense, given the definition of temperature. Adding more particles will lower the average kinetic energy of each particle.

Although the particles of solids and liquids are more rigidly bound than those of gases, the same relationships hold for the other phases of matter. Increasing the pressure on ice, for instance, causes it to melt more quickly. A larger block of ice has more particles, so it takes longer to melt than a smaller block would under the same conditions.

The enthalpy (H) of a system is equal to its total internal (thermal) energy (U) plus the product of its volume and pressure. Only when the system's overall energy changes can the enthalpy be measured. This is written as

$$\Delta H = \Delta U + \Delta PV$$

or

$$\Delta H = \Delta U + P\Delta V.$$

Temperature is directly related to enthalpy. When the temperature of a system rises, so too does its thermal energy, and thus so does its enthalpy.

TEMPERATURE IN THE EVERYDAY

Knowing the temperature of various objects and of the environment is immensely useful. Extremes of temperature in either direction can be dangerous, causing injuries or maladies such as burns, heatstroke, or frostbite. Understanding the various relationships that underpin thermodynamics is also helpful, albeit in more subtle ways. The ideal gas law explains why covering a pot causes it to boil faster (increased pressure raises temperature). Knowing that heat is often a by-product of other energy-transfer processes can help diagnose wiring problems in electronics when they seem to be running hotter than usual. It can even lead to a deeper understanding of human health. One major reason mammals eat (i.e., consume potential chemical energy) so often is to maintain a high internal body temperature. This benefits mammals' disease resistance, metabolism, and homeostasis in both hot and cold environments.

—*Kenrick Vezina, MS*

BIBLIOGRAPHY

Allain, Rhett. "What's the Difference between Work and Potential Energy?" *Wired.* Condé Nast, 1 July 2014. Web. 16 Sept. 2015.

"Energy, Kinetic Energy, Work, Dot Product, and Power." MIT OpenCourseWare. Massachusetts Inst. of Technology, 13 Oct. 2004. Web. 16 Sept. 2015.

Henderson, Tom. "Thermal Physics." *Physics Classroom.* Physics Classroom, 1996–2015. Web. 21 Sept. 2015.

Hurley, Katherine, and Jennifer Shamieh. "Enthalpy." ChemWiki. University of California, Davis, n.d. Web. 22 Sept. 2015.

Nave, Carl R. "Temperature." *HyperPhysics.* Georgia State U, 2012. Web. 16 Sept. 2015.

Shankar, Ramamurti. "Lecture 5: Work-Energy Theorem and Law of Conservation of Energy." *Open Yale Courses.* Yale U, 2006. Web. 21 Sept. 2015.

"SI Units: Temperature." NIST Physical Measurement Laboratory. Natl. Inst. of Standards and Technology, 14 May 2015. Web. 21 Sept. 2015.

TERM SYMBOL

FIELDS OF STUDY

Quantum Physics

SUMMARY

The term symbol is a shorthand notation that identifies the electron distribution in an atom and its corresponding energy state. The notation specifies the angular momentum quantum numbers corresponding to the quantum state of the particular atom.

PRINCIPAL TERMS

- **angular momentum:** the rotational momentum of an object around an axis, defined as the product of its moment of inertia and its angular velocity.
- **electron:** a negatively charged subatomic particle that is often bound to the positive charge of the nucleus but can also exist in a free state in an atom.
- **neutrons:** subatomic particles that, with protons, make up the mass of an atom's nucleus; they have functionally the same weight as protons but no electric charge.
- **protons:** subatomic particles that, with neutrons, make up the mass of an atom's nucleus; they have functionally the same weight as neutrons but hold a positive electric charge.
- **quantum mechanics:** the branch of physics that deals with matter interactions on a subatomic scale, based on the concepts that energy is quantized, not continuous, and that elementary particles exhibit wavelike behavior.
- **quantum state:** the condition of a physical system as defined by its associated quantum attributes.

ATOMS

Atoms are described in simple terms as being composed of protons, neutrons and electrons. Each electron in an atom possesses a specific energy in accord with quantum mechanics and the atom itself is described as being in a particular quantum state. The quantum state depends on the specific distribution of electrons in the atom and their particular energies. Each electron is assigned a specific value of angular momentum in accord with the Pauli exclusion principle, which states that two electrons in the same quantum level (generally referred to as an orbital) must have opposite spin states and therefore opposite values of angular momentum.

The Notation
The general form of the notation is as follows:

$$2^{S+1}L_J,$$

where S is the total spin quantum number, L is the total orbital angular momentum, and J is the state of the orbital defined by L. The value of L is given in spectroscopic notation, which identifies the orbitals by their letter designations: s, p, d, f, g, h, j and so on. The quantum disposition of the electrons in the system can be influenced by external factors such as an interaction with light, such that the "spin" of the electrons is altered. The paired electrons can thus be in either a singlet or a triplet state, depending on the value of S.

For example, the carbon atom has six electrons distributed among the $1s$, $2s$ and $2p$ orbitals. This is normally written simply as

$$1s^2 2s^2 2p^2.$$

The addition of the term symbol to this statement identifies the particular quantum state of the carbon atom. The triplet state of the carbon atom would then be written as

$$1s^2 2s^2 2p^2 3P^2$$

and the ground state of the carbon atom would be written as

$$1s^2 2s^2 2p^2 3P_0.$$

The notation is used to identify the energy levels and transitions between them in spectroscopic data.

Final Analysis
The term symbol notation is not encountered in the everyday practice of chemistry, except in the fine detail analysis of spectroscopic data. It is, however, essential in the communication of precise physical information in astronomy and in particle physics.

—*Richard M. Renneboog M.Sc.*

Bibliography
Atkins, P. W., Julio De Paula, and Ronald Friedman. *Physical Chemistry: Quanta, Matter, and Change*. 2nd ed. New York: W.H. Freeman, 2014. Print.

International Union of Pure and Applied Chemistry. Commission on Spectrochemical and Other Optical Procedures and Analysis. *Nomenclature, Symbols, Units and Their Usage in Spectrochemical Analysis: Rules Approved 1975*. Oxford; New York: Pergamon, 1976. Print.

Kuhn, Hans, Horst-Dieter Försterling, and David Hennessey Waldeck. *Principles of Physical Chemistry*. 2nd ed. Hoboken, N.J: John Wiley, 2009. Print.

Mortimer, Robert G. *Physical Chemistry*. 3rd ed. Burlington, MA: Elsevier Academic, 2008. Print.

TORQUE

FIELD OF STUDY
Classical Mechanics

SUMMARY
Torque is the rotational force applied to an object. The magnitude of torque depends on the rotational mass of the object and its distance from its axis, or point of rotation. Torque is relevant to many industrial and household machines.

PRINCIPAL TERMS
- **angular momentum:** the rotational momentum of an object around an axis, defined as the product of its moment and its angular velocity.
- **axis:** the center around which an object rotates.
- **cross product:** an operation, broadly analogous to multiplication, performed on two vectors in a three-dimensional space that results in a third vector that is perpendicular to both; if both vectors have the same direction or if one of them has a value of zero, the cross product will be zero.
- **fulcrum:** the supporting point around which a lever pivots.

- **mechanical advantage:** a measurement of the increase in force achieved by applying a mechanical tool or device to an existing system.
- **moment:** a combination of a physical quantity and a distance with respect to a fixed axis; the physical quantities of an object as measured at some distance from that axis.
- **vector:** a quantity that has direction as well as magnitude.

LINEAR AND ANGULAR MOMENTUM

All moving objects have momentum. An object's momentum is simply a measure of how much motion it has. The greater the momentum in a given direction, the more the object tends to continue moving in that direction. If an object is moving in a straight line, its linear momentum is equal to the product of its mass and its velocity.

Momentum and velocity are both vector quantities. Therefore, an object moving linearly has either a positive or negative velocity, depending on which way it is moving relative to its starting point. For example, if a car drives north at a velocity of 20 meters per second, north can be considered the positive direction. Then, if the car turns around and drives back to its starting point at the same speed, it will have a velocity of −20 meters per second.

However, most objects do not move only in straight lines. Many rotate around some kind of a fixed point, or axis. An object rotating around an axis has angular momentum. Angular momentum, like linear momentum, is the product of an object's mass and its velocity, but it is measured in different units. An object's angular mass, or moment of inertia, is how much it resists changing its angular velocity around its axis. Its angular velocity is the rate at which its angle around the axis changes. Thus, the angular momentum of an object is equal to the product of its moment of inertia and its angular velocity.

According to Isaac Newton's second law of motion, momentum stays constant if no outside force acts on a given system, as the rate of change in momentum is zero. However, once an outside force interacts with a system, momentum can increase or decrease, depending on the direction and magnitude of the applied force. When the system involves rotation, that outside force is called torque.

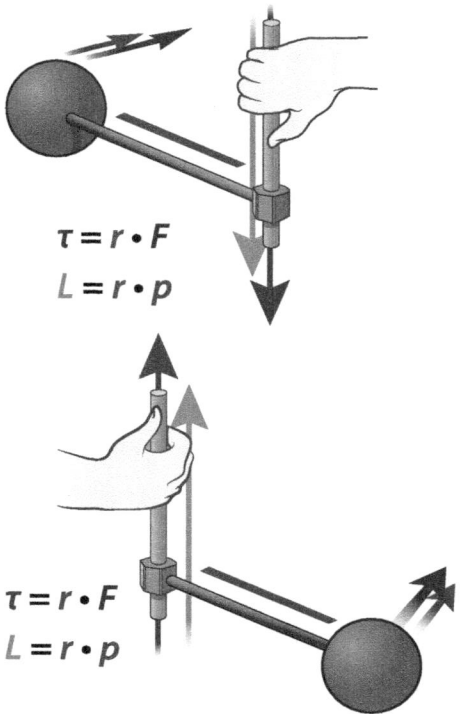

A lever rotated about an axis will experience torque, or a moment of force, which is dependent on the force (F) applied to the lever and the length (r) of the lever. According to the right-hand rule, the clockwise rotation of the lever results in an angular momentum (L) and torque (τ) in the upward direction.

TORQUE, FORCE, AND DISPLACEMENT VECTORS

Torque is the degree to which a force causes an object to rotate around an axis. It can also be thought of as an object's change in angular momentum, or a "twist" applied to a moving object. For example, a force applied to a door will cause the door to rotate around its hinges rather than moving in a straight line. The hinges serve as the door's axis. The rotational part of the applied force is the torque.

Mathematically, torque is calculated as

$$T = F \times r$$

where T is torque, F is the rotational force, and $\mathbf{r}$ is the displacement vector, which measures the distance between the axis and the point of force application. (Vector quantities are represented by bolded variables to distinguish them from scalar quantities, which have magnitude but no direction.) The unit of

PRINCIPLES OF PHYSICS   Torque

SAMPLE PROBLEM

A bolt is tightened with a torque of 17 newton-meters (N·m). Two different wrenches are available to turn the bolt. One is 20 centimeters (cm) long, and the other 30 cm long. If force is applied to both wrenches at an angle of 30 degrees, what is the difference in the amount of force needed to turn the bolt with each wrench?

Answer:

To find the difference in force between the two wrenches, the force needed for each wrench to turn the bolt must first be calculated. Because the turning force is applied at an angle of 30 degrees, the perpendicular component of the force is equal to the sine of 30 degrees:

$$\sin 30° = \sin(\pi/6) = 1/2$$

Substitute all the known quantities for the 20-centimeter wrench into the equation for magnitude of torque:

$$T = Fr\sin\theta$$

$$17 \text{ N·m} = F(20 \text{ cm})(1/2)$$

Convert the centimeters to meters and then solve for F:

$$17 \text{ N·m} = F(0.2 \text{ m})(1/2)$$

$$17 \text{ N·m} = F(0.1 \text{ m})$$

$$(17 \text{ N·m}) / (0.1 \text{ m}) = F$$

$$170 \text{ N} = F$$

The 20-centimeter wrench requires an application of 170 N of force. Repeat the calculations for the 30-centimeter wrench:

$$17 \text{ N·m} = F(0.3 \text{ m})(1/2)$$

$$17 \text{ N·m} = F(0.15 \text{ m})$$

$$(17 \text{ N·m}) / (0.15 \text{ m}) = F$$

$$113.33 \text{ N} = F$$

Calculate the difference:

$$170 \text{ N} - 113.33 \text{ N} = 56.67 \text{ N}$$

The 30-centimeter wrench requires 56.67 N less force to turn the bolt.

torque is the newton-meter (N·m), a compound unit of force and distance. By convention, torque is considered negative if the direction of rotation is clockwise and positive if it is counterclockwise.

Torque is important to many household tools and machines. Most of these are based on the six classical simple machines that use mechanical advantage to multiply force. Levers offer the best example of torque. A lever consists of a flat surface attached to a fulcrum. When no net torque is applied, the surface balances on the fulcrum. However, if more force is applied to one side than the other, it will produce a torque that causes the surface to rotate. The position of the fulcrum is key. As the distance from the fulcrum to the point of force application increases, the amount of force required to rotate the plane around the fulcrum decreases. This is because the lever applies a rotational torque that multiplies the force. For example, pushing on a door (a type of lever) very close to its hinges takes much more force than pushing farther away from the hinges. The longer the radius, the greater the torque applied. Similarly, another simple machine, the screw, can translate the rotational force of torque into an amplified linear force.

Direction and Magnitude: Cross Products and Angles

When multiplying vector quantities, direction and magnitude must be calculated separately. Because torque is the cross product of two vectors (F and r) in a three-dimensional space, its direction is perpendicular to both. For example, if the string of a yo-yo is pinned to a tabletop and a force is pushing the yo-yo to rotate around the pin, the yo-yo's displacement vector points along the string and its force points along its spinning edge. The torque vector is thus perpendicular to both of them, pointing straight up in the air.

When calculating the magnitude of torque, the angle at which it is applied must be taken into account. By definition, the vector F in the torque equation represents only the part of the force perpendicular to the displacement vector r. The full equation for the magnitude of torque, not taking direction into account, is

$$T = Fr\sin\theta$$

where θ is the angle between the vectors F and r. For example, if seventeen newtons of force are applied to a door at a point 1.5 meters from its hinges, and the angle of force application is sixty degrees from the flat of the door, then the magnitude of the torque would be calculated as follows:

$$T = Fr\sin\theta$$
$$T = (17\text{ N})(1.5\text{ m})(\sin 60°)$$
$$T = (25.5\text{ N}\cdot\text{m})\left(\frac{\sqrt{3}}{2}\right)$$
$$T = 22\text{ N}\cdot\text{m}$$

If the force is applied perpendicular to door, however, so that θ equals ninety degrees, then $\sin\theta$ is equal to 1 and can be disregarded.

Calculating Torque

Before calculating the torque of an object, the axis of rotation must be determined. For example, if the object is a lever, such as a see-saw, then the axis of rotation is the fulcrum. Next, the directions and magnitudes of the forces acting on the object must be identified. If two children are sitting on opposite sides of the see-saw, each child exerts a downward force on the lever that is a combination of the child's mass and the force of gravity. If the system in which the object exists is in equilibrium—for example, if the see-saw is perfectly balanced—then the net torque is be zero. If the system is not in equilibrium and an external force is acting on it to produce rotation, the torque can be calculated by using the equations above.

Torque in Daily Life

Torque is critical to many aspects of engineering and mechanical design, from static structures to complex machines. A car's engine, for example, produces torque from the combustion of gas. That torque turns the crankshafts, which then turn the wheels. The amount of torque ascribed to an engine reflects how quickly it can accelerate the car. That torque must overcome friction and air drag, among other forces, in order to move the car forward.

Torque can also be used to calculate an engine's power (P), or the amount of work it can perform over time, if the angular velocity (ω) of its output shaft is also known:

$$P = T\omega$$

This is especially useful for finding the power output of an electric motor.

For static structures, connecting elements (e.g., metal bolts that hold sheets of steel together) must be able to withstand the torque of potential outside forces, such as the weight of cars on the bridge or the strength of a storm's winds against the building.

—Lindsay K. Brownell, MS

BIBLIOGRAPHY

Cooley, Brian. "CNET on Cars: Car Tech 101—Horsepower vs. Torque." *CNET*. CBS Interactive, 26 Feb. 2013. Web. 7 Aug. 2015.

Khan, Salman. "Moments, Torque, and Angular Momentum." *Khan Academy*. Khan Acad., 23 May 2008. Web. 7 Aug. 2015.

Nave, Carl R. "Torque." *HyperPhysics*. Georgia State U, 2012. Web. 7 Aug. 2015.

Richmond, Michael. "Torque." *University Physics I*. Rochester Inst. of Technology, 15 Feb. 2002. Web. 7 Aug. 2015.

Roldán Cuenya, Beatriz. "Chapter 11: Torque and Angular Momentum." *Physics for Scientists and Engineers I*. University of Central Florida, 2013. Web. 7 Aug. 2015.

"Torque and Levers." *Siyavula Textbooks: Grade 11 Physical Science*. Rice U, 29 July 2011. Web. 7 Aug. 2015.

TURBINES

FIELDS OF STUDY

Classical Mechanics

SUMMARY

Turbines are an important component of many machines, large and small. They use rotational motion to convert energy from wind, water, or steam into electricity. They can also use fuel to power a diverse array of engines, such as those in cars, motorcycles, jets, rockets, and ships.

PRINCIPAL TERMS

- **force:** the ability to produce motion, often calculated by multiplying the mass of an object by its acceleration.
- **impulse turbine:** a turbine set in motion by the velocity of a fluid hitting each blade.
- **kinetic energy:** the work capacity of an object in motion.
- **potential energy:** the stored work capacity of an object.
- **power:** the rate at which work is done.
- **reaction turbine:** a turbine set in motion by the pressure and flow of a fluid.
- **torque:** a twisting force that produces rotational motion.
- **work:** the energy transferred to an object by a force, calculated by multiplying the force applied to an object by the distance it has moved the object or the amount of resistance it has overcome.

THE POWER OF ROTATIONAL MOTION

The word "turbine" comes from the Latin word *turbo*, which refers to something that spins. First used in the 1800s, the word described machines that harnessed energy from steam. Similar devices, such as water wheels and windmills, have existed since ancient times, but the shift to the term "turbine" corresponded to major advances in technology.

In order for an object to move, something must transfer energy to it. Anything that uses energy to change an object's state of motion is called a force. If a force moves an object, it has done work. The rate at which work occurs is called power, calculated by multiplying force by speed. Greater force allows work to be done more quickly. Lifting a person straight up off the ground requires a large amount of force. A lever allows force to be applied over a greater distance, which means that less force is required at any given moment. A see-saw illustrates this principle. Two people on different ends of a see-saw can easily lift one another into the air, even if they cannot lift the other person's weight in their arms. Turbines use torque, a force that produces rotational motion around a central point. Using torque to accomplish a task is similar to using a lever.

Turbines

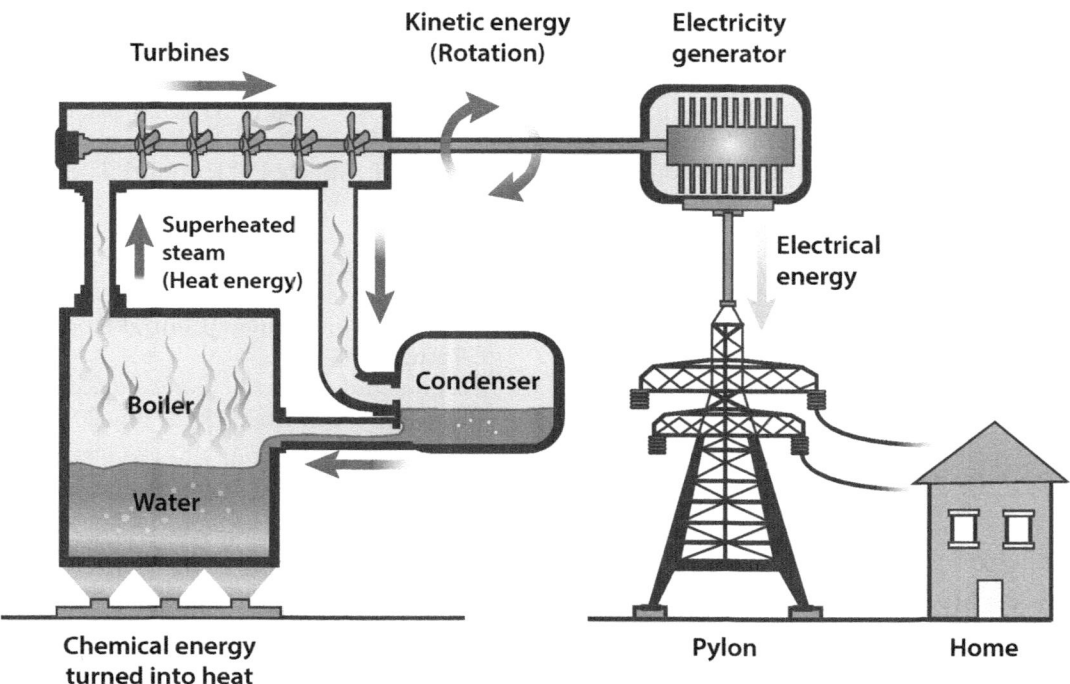

Electricity can be generated by a turbine condensing and/or converting energy into a rotational force that turns a generator. The energy source for a turbine can be wind, water, or thermal. The thermal energy can come from a number of energy sources, including chemical reactions such as nuclear decay, coal or oil burning, or geothermal energy.

Turbines vary greatly depending on their type and purpose, but all turbines have several basic components. Blades (also called vanes) catch the wind or water or respond to the pressure of water or steam. A child's toy, the pinwheel, demonstrates the way a wind turbine works. A rotor is the central part of a turbine to which the blades are attached. The rotor is connected to a shaft inside the turbine. There may be one or two shafts that spin to power a generator, which in turn creates electricity.

Making Energy More Usable

A turbine is a way to transform energy from an unusable form (such as water, wind, or steam) to a usable form (electricity). Moving water and air have kinetic energy because they are in motion. If water is stored behind a dam, it has potential energy because it is being contained, but it has the potential to move. If the dam is removed, the water's potential energy becomes kinetic energy as the water flows again. When harnessed by a turbine, both water and air possess the ability to perform work. When water flows through turbines, it powers them with its kinetic energy. All types of turbines are essentially engines that use kinetic energy to cause rotational motion around a central axis. This movement can then power larger systems, such as geothermal or nuclear power plants, as well as water- or wind-powered systems.

The workings of turbines powered by wind and water are visible at wind farms or dams, but it is more difficult to see the workings of turbines powered by steam. As gases heat up, they expand, putting pressure on whatever is containing them. An empty milk jug placed outside on a hot day may pop its top when the air inside it expands too much. This same principle propels the blades of a turbine. Steam entering the turbine is under pressure because it is in an enclosed space, and that pressure forces the blades of the turbine to move.

Steam turbines are used in many types of power plants. In a geothermal power plant, hot water is drawn from underground reservoirs, such as hot springs or geysers, to power turbines that then create electricity. It is also possible to power turbines by concentrating solar energy to heat water and make steam. In a nuclear facility, steam created from nuclear reactions powers turbines to generate electricity.

Turbines may also increase the power of combustion engines that run on gasoline, as in a jet engine or a turbocharged automobile. A gas turbine, like a steam turbine, relies on the ability of gases to expand when heated. Gas turbines are used in jet engines and ships. In an airplane, the action of the turbine creates exhaust that propels the plane forward. In a ship, the turbine drives a propeller, which in turn moves the ship. In these cases, the turbine is not converting moving wind or water to electricity but rather using energy to increase power.

IMPULSE AND REACTION TURBINES
Turbines are designed with different environments and needs in mind. A turbine powered by wind is much different than one powered by water. Even among turbines designed for use with water, there is a great deal of variation because water itself varies in volume, speed, and force. When designing hydroturbines (water-powered turbines), engineers consider the water's flow rate (volume per second) and its head (level or depth). High head is deep water, and low head is shallower water.

The two types of hydroturbine are impulse turbines and reaction turbines. Some turbines, called impulse-reaction turbines, are a mixture of these two types, which means they can adapt to a broad range of conditions.

An impulse is a force that acts for a short time to produce a particular change in momentum. Because of the way an impulse turbine is designed, force hits each blade in sequence, causing it to move. As the force hits each blade, the rotor assembly spins. An old-fashioned water wheel demonstrates the physics behind impulse turbines: the water hits one blade at a time, turning the wheel. In a reaction turbine, the blades move in response to the steady application of force—that is, they react. Instead of water hitting each blade in sequence, it flows over the entire assembly, ideally with constant pressure.

An impulse turbine works best when hit at high velocity. Water hits each turbine blade with a lot of force and propels it. A reaction turbine does not require as much force to be set in motion. An impulse turbine would be suited to harnessing the energy from a small, fast-moving river, while a larger, deeper, slow-moving river would be better served by a reaction turbine. In real life, rivers can change seasonally or in response to storms, so many hydroturbines use combination impulse-reaction turbines in order to work with variable flow.

A wind turbine may be thought of as a reaction turbine because wind moves over the entire turbine, setting it in motion. The two types of wind turbine are vertical axis and horizontal axis. Horizontal-axis turbines resemble pinwheels and operate in much the same way. A vertical-axis turbine operates like a kitchen mixer; both the axis and the blades are oriented vertically, with the blades curving to catch the wind.

TURBINES AND THE FUTURE
Turbines facilitated the advent of the Industrial Revolution, and they continue to serve an important role in industry and technology. They are a part of almost every engine, from jets to cars. Turbines are also an important part of both alternative and fossil-fuel-based energy production. As a result, many innovators are working to streamline turbine operation and expand how turbines are used. For instance, in the wind industry, engineers are developing lightweight turbines that can hover high in the air to catch better air currents. They are also experimenting with adding a second rotor to traditional wind turbines in order to increase efficiency. In the water-power industry, turbines already generate power from rivers and tides and will be a part of generating power from tidal lagoons. In addition to industrial applications, engineers are developing hydroturbines for small-scale individual use, such as generating power for personal electronic devices.

—*J. D. Ho, MFA*

BIBLIOGRAPHY
Breeze, Paul. *Power Generation Technologies*. Philadelphia: Elsevier, 2014. Print.
Chiras, Daniel D, Mick Sagrillo, and Ian Woofenden. *Power from the Wind*. Gabriola Island: New Soc., 2009. Print.
"Engineers Study the Benefits of Adding a Second, Smaller Rotor to Wind Turbines." Phys.org. Phys.org, 10 Mar. 2015. Web. 14 Apr. 2015.
"How Do Wind Turbines Work?" *Office of Energy Efficiency & Renewable Energy*. US Dept. of Energy, n.d. Web. 14 Apr. 2015.

Madrigal, Alexis C. "How to Make a Wind Turbine That Flies." *Atlantic.* Atlantic Monthly, 16 July 2012. Web. 14 Apr. 2015.

Owano, Nancy. "Blue Freedom Uses Power of Flowing Water to Charge." Tech Xplore. Science X, 26 Mar. 2015. Web. 14 Apr. 2015.

Soares, Claire. *Gas Turbines: A Handbook of Air, Land and Sea Applications.* Waltham: Butterworth, 2014. Print.

"Types of Hydropower Turbines." *Office of Energy Efficiency & Renewable Energy.* US Dept. of Energy, n.d. Web. 14 Apr. 2015.

Woodford, Chris, and Jon Woodcock. *Cool Stuff 2.0 and How It Works.* London: DK, 2007. Print.

ULTRASOUND

FIELDS OF STUDY
Acoustics; Harmonics

SUMMARY

Ultrasound is sound at a frequency higher than humans can hear. Hearing range varies across ages and individuals, but most young adults cannot hear sounds above twenty thousand hertz. Thus, any sound wave with a frequency above this is considered ultrasound. Ultrasound is widely used in medical imaging, sonar, industrial maintenance, and chemistry. It is also important for animals that use echolocation, such as bats and dolphins.

PRINCIPAL TERMS

- **frequency:** the number of cycles completed in a given unit of time.
- **infrasound:** a sound wave with a frequency below twenty hertz, the lower limit of human hearing.
- **sonar:** a method of using sound waves to "see," typically in an underwater environment, by sending and receiving pulses of sound; originally an acronym for "sound navigation ranging."
- **sonography:** the use of ultrasound in medicine to produce images of internal structures, such as tendons, muscles, and organs.
- **ultrasonic:** describes a sound, or a device that makes use of sound, with a frequency above twenty thousand hertz, the upper limit of human hearing.
- **ultrasound identification:** a real-time locating system that uses ultrasound to track the location of objects or individuals on small spatial scales.
- **ultrasonic impact treatment:** the use of ultrasonic vibrations to strengthen metals.
- **ultrasonic testing:** the use of ultrasound to test materials for internal flaws.

ABOVE HUMAN HEARING

The sounds that humans can hear are limited to the audible range of frequencies, between twenty and twenty thousand hertz. Within this range, higher frequencies are heard as higher pitched, and lower frequencies are heard as lower pitched. Anything above the audible range is considered ultrasonic, and any sound that falls below it is called infrasound. Some animals use ultrasound to navigate their environment, a technique known as "echolocation." The principle behind echolocation is one that scientists have co-opted to great effect. Ultrasonic acoustics follows the same principles as all other acoustics. The same physics is used to describe all waves, whether they be sound, light, or mechanical waves in the ocean. What people hear as sounds are vibrations caused by waves traveling through air, water, or some other medium.

NATURAL ULTRASOUND

The study of acoustics dates back at least to ancient Greece, when Pythagoras (ca. 580–500 BCE) wrote about the mathematics underpinning the music made by stringed instruments. Scientists would not come to appreciate the full spectrum of sound until much later. Yet millions of years before human civilization, some animals had evolved to use ultrasound as an important tool for survival. Many can hear sounds well outside the human range of hearing. They can communicate with one another at frequencies inaudible to the human ear.

Some species even use ultrasound to navigate. In 1794, Italian biologist Lazzaro Spallanzani (1729–99) demonstrated that bats can navigate in complete darkness using ultrasonic echolocation. Bats produce pulses of ultrasound, which bounce off the surrounding objects and return to their source, creating an echo. The time it takes for this echo to reach the bat's ears tells it not only the distance but also the direction of any nearby obstacles—or prey. Many whales and dolphins use a similar technique underwater.

LOW RANGE, HIGH RESOLUTION

When using sound for detection, in theory, any frequency can be used, whether infrasonic, audible, or ultrasonic. However, there is a trade-off. Infrasound tends to have very good range, but it has poor resolution, meaning that it fails to detect small objects. Ultrasound, in contrast, has limited range but excellent resolution. It is this ability to detect fine detail that makes ultrasound so useful in a variety of fields, including warfare, medicine, and industry.

The first human-made ultrasonic device was a dog whistle. It was invented in 1876 by famed English polymath Francis Galton (1822–1911), who was testing the hearing abilities of various animals. The first notable application of ultrasound was conceived in 1916 by French physicist Paul Langevin (1872–1946), who used waves of ultrasound to detect submarines underwater—an early precursor to sonar.

In addition to submarines, modern militaries also need to detect much smaller objects, such as divers, mines, and other such obstacles. Thus, military sonar uses a variety of both infrasonic and ultrasonic devices for detection. Some anglers, archaeologists, and ecologists also use underwater ultrasonic radar to search for fish, artifacts, or marine mammals.

On land, ultrasound identification (USID) uses constant pulses of ultrasound to track individuals or objects in small-scale environments. It is ideal for situations in which other locating systems, such as radio-frequency identification (RFID), would meet with too much interference. USID is commonly used in hospitals to track patients and make sure they do not wander.

Another common use of ultrasound in hospitals is medical sonography. Ultrasound is used to examine patients' tissues and organs without causing harm or risking invasive surgery. Most famously, ultrasound is used by obstetricians to examine fetuses during pregnancy. It can also be used to detect tumors or other abnormalities.

INDUSTRIAL ULTRASOUND

Ultrasonic testing is used in industry and materials science to search for problems that are otherwise invisible or difficult to detect. For example, by sending pulses of ultrasound through a pipe and measuring the echoes, engineers can determine how thick the pipe is and whether it is corroding.

The intense, high-frequency vibrations produced by focused ultrasound have other practical uses. They can be used to clean surfaces or mix substances. Ultrasonic impact treatment is a technique used in metallurgy to strengthen metals. For example, it can be used to restore the integrity of welds during bridge maintenance. In chemistry, ultrasound can be used to initiate or speed up chemical reactions by counteracting the forces that hold liquid molecules together. The frequency range of ultrasound is much larger than audible sound—there is no strict upper limit for ultrasound, after all—and its wide range of applications reflects this. Playing a role in everything

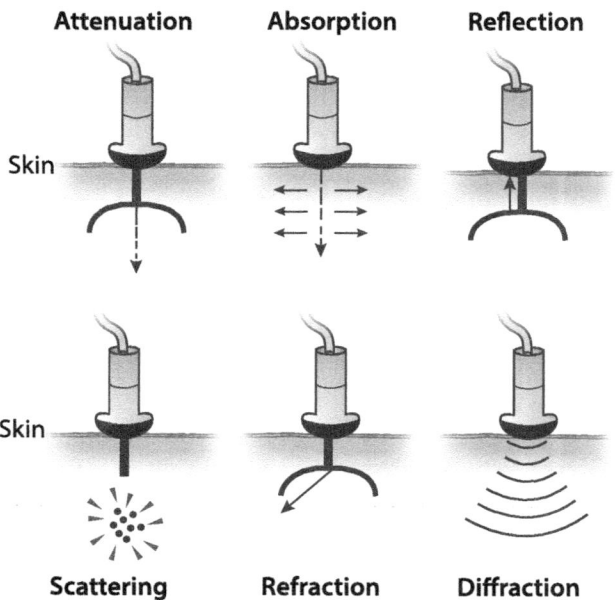

Boundary	% Reflected
Fat/Muscle	1.08
Fat/Kidney	0.6
Soft tissue/Water	0.2
Bone/Fat	49
Soft tissue/Air	99

Ultrasound is sound with a frequency greater than twenty thousand hertz, above the range of human hearing. It is often used in the medical field. When ultrasound waves contact body tissues, they can attenuate (weaken), reflect, refract, diffract, scatter, or be absorbed. As the waves cross a boundary between two different body tissues, the percent reflected indicates the nature of the two tissues.

from pregnancy to plumbing, ultrasonic technology is a silent but essential part of modern life.

—Kenrick Vezina, MS

BIBLIOGRAPHY

Abu-Zidan, Fikri M., Ashraf F. Hefny, and Peter Corr. "Clinical Ultrasound Physics." *Journal of Emergencies, Trauma, and Shock*, 4.4 (2011): 501–3. Print.

Berg, Richard E. "Sound." *Encyclopaedia Britannica*. Encyclopaedia Britannica, 6 Mar. 2014. Web. 23 Mar. 2015.

Corcoran, Aaron J., Jesse R. Barber, and William E. Conner. "Tiger Moth Jams Bat Sonar." *Science* 325.5938 (2009): 325–27. Print.

Demi, Marcello. "The Basics of Ultrasound." *X-Ray and Ultrasound Imaging*. Ed. Daniele Panetta and Demi. Waltham: Elsevier, 2014. 297–322. Print. Vol. 2 of Comprehensive Biomedical Physics. Anders Brahme, gen. ed. 10 vols. 2014.

"Physics of Ultrasound." *BATS: Better Anaesthesia through Sonography*. BATS Research Group and Medware, 2006. Web. 23 Mar. 2015.

Stiles, Timothy A. "Ultrasound Imaging as an Undergraduate Physics Laboratory Exercise." *American Journal of Physics* 82.5 (2014): 490–501. Print.

ULTRAVIOLET RADIATION

FIELDS OF STUDY

Electromagnetism; Atomic Physics; Nuclear Physics

SUMMARY

Ultraviolet (UV) radiation is a subtype of electromagnetic radiation with a wavelength between 100 and 400 nanometers. It falls between x-rays (shorter wavelength, more energy) and visible light (longer wavelength, less energy) on the electromagnetic spectrum. As it applies to humans, UV radiation is responsible for tans, sunburns, and, in some, cases skin cancer.

PRINCIPAL TERMS

- **carcinogenic:** capable of causing cancer.
- **electromagnetic spectrum:** the full range of electromagnetic radiation, sorted into segments with similar properties by wavelength; ultraviolet radiation occupies one of these segments.
- **nonionizing radiation:** radiation that lacks the energy necessary to knock electrons free when it hits an atom; all electromagnetic radiation with a longer wavelength than ultraviolet is nonionizing.
- **UV degradation:** the damage caused by ultraviolet radiation when it strikes certain materials.
- **vacuum ultraviolet radiation:** a subtype of ultraviolet radiation with wavelength under 200 nanometers.
- **wavelength:** the distance between crests (or troughs) of a wave; the length of one complete cycle. Longer wavelengths correspond to lower frequencies and less energy, and vice versa.

THE ELECTROMAGNETIC SPECTRUM

Every type of electromagnetic radiation (EMR) is categorized on the electromagnetic (EM) spectrum by wavelength. A wavelength is the distance between two crests or two troughs in a wave. Wavelengths are inversely proportional to the frequency of and energy transmitted by the wave. The EM spectrum is typically arranged starting with short wavelengths (high frequency and high energy) and extending to very long wavelengths (low frequency and low energy). Ultraviolet (UV) light has a wavelength between 100 and 400 nanometers. It falls between x-rays and visible light on the EM spectrum.

On Earth, the sun is the primary source of ultraviolet radiation. The sun emits ultraviolet radiation in large amounts alongside visible light and heat (infrared radiation). The ozone layer is a portion of the atmosphere especially effective at absorbing UV radiation. It protects life on the planet's surface from potentially harmful overexposure. Therefore, the ozone layer is of special concern for public health.

IONIZING VERSUS NONIONIZING RADIATION

UV radiation sits at the boundary between two types of radiation: ionizing radiation and nonionizing radiation. UV radiation and all EMR with a shorter

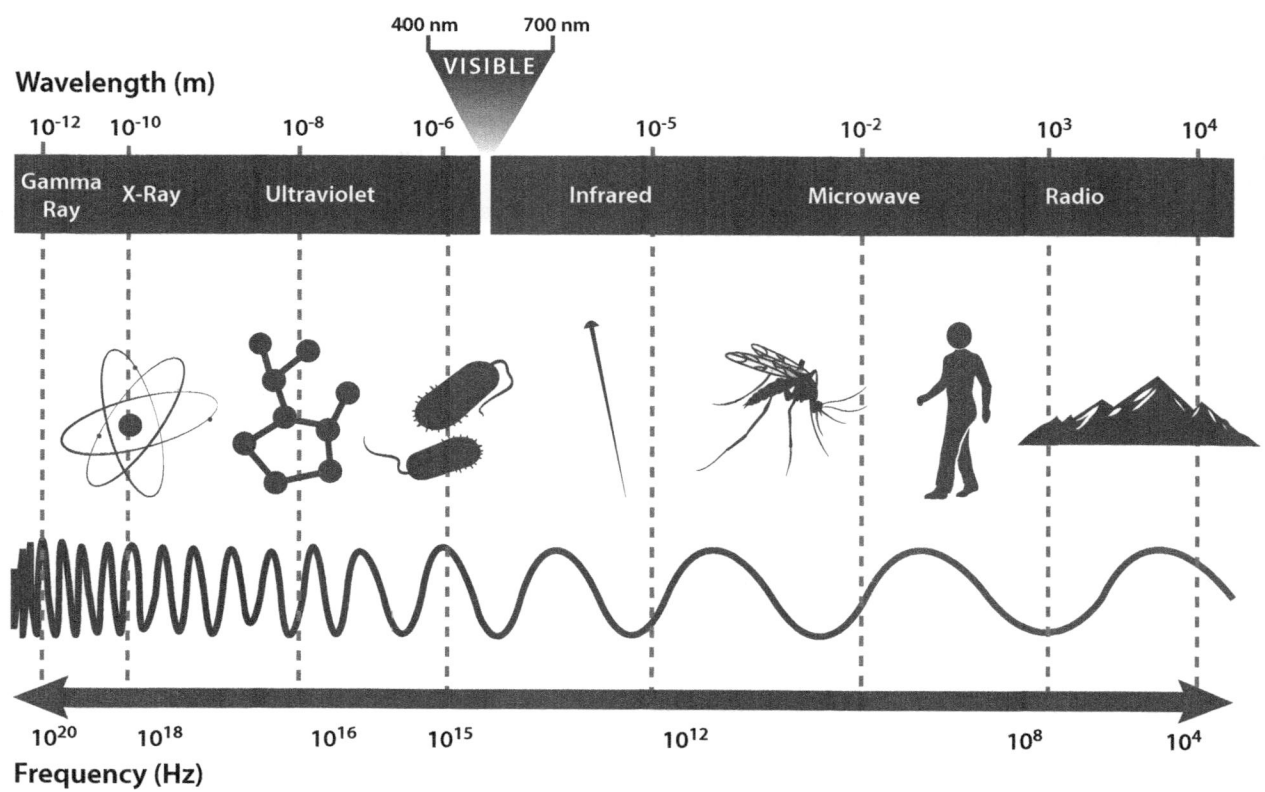

The electromagnetic spectrum is the range of all possible electromagnetic radiation wavelengths. Ultraviolet radiation falls just outside the visible light range. Wavelengths for UV light are between 4.0×10^{-7} m and 1.0×10^{-8} m, around the size of molecules. Radio waves can be as large as mountains, and gamma radiation has a short wavelength, smaller than atoms.

wavelength is ionizing. This means it has enough energy to knock electrons loose from an atom. In doing so, UV radiation alters the way it interacts with other nearby atoms. If ionizing radiation penetrates DNA, it can cause errors in cell replication, which can lead to cancer. Therefore, all ionizing radiation may be carcinogenic with enough exposure. However, the risk varies greatly by type of radiation and type of exposure. Nonionizing radiation has wavelengths greater than UV. This is visible light and above on the EM spectrum. Nonionizing radiation can still be dangerous (infrared radiation can cause burns, for example), but it does not alter the structure of DNA.

Biologists break UV radiation into three different subcategories by wavelength: near (315–400 nanometers), actinic (200–315 nanometers), and vacuum (100–200 nanometers). Near UV is too weak to be anything but poorly absorbed by living organisms. Actinic UV is strong enough to penetrate and be absorbed by organisms. Despite its high energy, vacuum UV has such a small wavelength that it is often absorbed by the atmosphere and other environmental obstructions. Thus, it has a minimal impact on living organisms. It is not only organic material that UV radiation can damage, however. Prolonged exposure to sunlight leads to UV degradation. In this case, UV radiation weakens or breaks down materials over time. This accounts for the "bleaching" effect of the sun on clothes. The pigments that give the clothing its color are broken down by long-term exposure to UV radiation.

The Importance of UV Radiation

Despite the danger of overexposure, UV radiation is vital to the human body's ability to create vitamin D and can be therapeutic for individuals with skin diseases such as psoriasis. UV exposure is also therapeutic for those with seasonal affective disorder, a form of depression linked to the decrease in daylight during winter months at middle and high latitudes.

Although "visible" light is considered a separate category on the EM spectrum, some organisms—particularly bees and other insects—can see into the UV range. Flowers viewed through UV-sensitive cameras display unique patterns invisible to the human eye. Thus, UV radiation, and the ability to detect it, is as important as visible light for some pollinators. In practical terms, high-energy UV rays can effectively kill bacteria for sanitation purposes without as large a risk from exposure as more powerful radiation such as x-rays. Furthermore, exposure to UV light naturally causes certain materials to fluoresce, or glow. It is this principle that allows crime-scene investigators to use UV wands to expose certain bodily fluids. Thus, despite its potential for harm, UV radiation continues to serve an important role in many applications.

—*Kenrick Vezina, MS*

BIBLIOGRAPHY

Allen, Jeannie. "Ultraviolet Radiation: How It Affects Life on Earth." *Earth Observatory*. NASA, 6 Sept. 2001. Web. 7 May 2015.

"Anatomy of an Electromagnetic Wave." *Mission: Science*. NASA, 13 Aug. 2014. Web. 7 May 2015.

"Electromagnetic Radiation." *Encyclopaedia Britannica*. Encyclopaedia Britannica, 26 Nov. 2014. Web. 7 May 2015.

"Ultraviolet Waves." *Mission: Science*. NASA, 2010. Web. 27 May 2015.

"UV Radiation." *SunWise Program*. US Environmental Protection Agency, 12 Aug. 2013. Web. 27 May 2015.

"What Is Ultraviolet (UV) Radiation?" *Skin Cancer Prevention and Early Detection*. Amer. Cancer Soc., 20 Mar. 2015. Web. 27 May 2015.

V

VELOCITY OF SOUND

FIELDS OF STUDY

Acoustics; Classic Mechanics; Fluid Mechanics

SUMMARY

The velocity of sound is a characteristic property of a particular conductive medium. Sound waves travel longitudinally as successions of compression and rarefaction whose displacement from equilibrium varies in a sinusoidal manner. The density of the medium determines the velocity of sound within that medium. The detection of sound waves has many applications in medicine, engineering and other fields.

PRINCIPAL TERMS

- **acoustics:** the study of sound; also, the qualities of a space that affect how sound is heard within that space.
- **density:** a measure of the mass of a quantity of matter relative to the volume of space that it occupies.
- **longitudinal wave:** a type of wave wherein the medium is displaced in a direction parallel to the movement of energy, as in the case of sound waves.
- **reflection:** the rebounding of a wave from a surface or boundary between two media, causing it to travel back through the original medium.
- **refraction:** the alteration of a wave's path, speed, and wavelength when it passes from one medium to another.
- **sinusoidal:** having a shape or pattern of behavior that can be described by a sine wave function.
- **sound barrier:** the effects created when an object travels faster than the local speed of sound.
- **wave propagation:** the manner in which a wave travels.

The Nature of Sound

Philosophical questions aside, sound is produced by any disturbance that initiates compression waves in the surrounding medium. Whether or not the waves are interpreted as sound by humans or other living creatures does not change the nature of the sound waves.

A compression wave begins when some perturbation results in compression of the molecules of the surrounding medium. The compressed molecules exert that force against the molecules that are adjacent to them, and wave propagation continues in this manner throughout the medium. As the medium, material is compressed at the wavefront, it is rarefied at the same time in the space behind the wavefront. The medium recovers as the compressed and rarefied areas regain their normal distribution of atoms or molecules, and the movement of the sound wave in any particular direction is as a longitudinal wave in which the direction of the maximum and minimum amplitudes is collinear with the direction of movement of the wave itself.

In contrast, a transverse wave is characterized by the amplitudes of the wave being at right angles to the direction of motion of the wave. The amplitudes of both types of waves, however, can be described as having sinusoidal behavior. The science of acoustics studies the behavior and applications of sound waves in different media. It is a fundamental principle that the velocity of sound waves is directly proportional to the density of the medium through which they travel.

One of the most noticeable effects of sound is the Doppler effect, in which the pitch or tone of the sound is perceived at higher apparent frequencies when the sound source is moving toward the listener, and at lower apparent frequencies when the sound source is moving away from the listener. While the velocity of the sound waves is the same in either case, the effect can be thought of as due to the compression waves being pushed closer together by the motion of the sound source in the first case, and pulled farther apart in the second case.

Density, Reflection and Refraction

Sound waves are subject to reflection and refraction. An echo is a familiar example of a reflected sound wave, while the effects of refraction are noticeable as the different nature of sounds made within one medium but heard in another. The effect can be easily demonstrated by striking two hard objects such as rocks together in water while listening to the sound that comes through the air. If this is carried out in the reverse order, by striking the objects together in air and listening for the sound they make while you are submerged, the effect of the different densities of the two mediums is immediately apparent.

Transition of sound waves from a denser medium to a less dense medium is more effective than transition of sound waves from a thin medium to a denser medium. The denser the medium, the more effectively the sound is transmitted from one point to another. This effect is also easily demonstrated by tapping two objects together while holding them away, and then by tapping them together while holding one of them against an ear. The sound will be much clearer and sharper, as well as louder, in the latter case.

Sound is carried in solid, or condensed, matter, mediums by phonons. A phonon is a quantized vibrational state of matter. Phonons of high frequency are recognized as heat, while phonons of longer wavelength are responsible for the transmission of sound in solids. The vibrational modes of phonons begin and propagate in the same manner as compression waves in other mediums. Their quantization is related directly to the integral distances between atoms in the unit cell structure of the solid material. Crystals like sodium chloride and other minerals typically have distinct regular unit cell structures due to the stacking of atoms within the crystalline material. Similarly, metals are characterized by very distinctive and regular arrangements of atoms, and are correspondingly some of the best mediums for the transmission of sound.

The Sound Barrier

One of the most persistent limitations of aircraft in the first half of the twentieth century, and one that has serious ramifications still in the present day, is the sound barrier. The particular effect of aircraft traveling at the speed of sound and above is the formation of a shock wave in the medium (air) that can cause damaging turbulence and prevent the proper induction of air into the engines for combustion. The formation of a shock wave in any medium by an object passing through it is a common feature of all fluid media, and the term *sound barrier* strictly applies to gas-phase media, most particularly air. In denser media such as water exceeding the sound barrier results in cavitation, or the formation of voids, in the medium. Like the turbulence that would form in air, cavitation can result in significant damage and at the very least is a practical limitation characteristic of the medium.

Practical Applications of Sound

The velocity of sound in media of different densities has significant use in medical diagnostic imaging as the sonograph or ultrasound image. By manipulating the slight differences in frequency due to the Doppler effect, and utilizing the differences in velocity of sound reflected and refracted by tissues of different densities, sonography can generate real-time images of organs such as the heart while they are functioning. Ultrasound is also used to produce images of internal soft tissue structures such as blood clots in veins and babies in the uterus of their mothers. Echo location and range finding are important functions in carrying out searches below the surface of a body

---

SAMPLE PROBLEM

Thunder is caused by the explosive expansion of air that has been superheated by a bolt of lightning. Assuming the light from the bolt is instantaneously visible due to the much higher velocity of light than of sound, calculate the length of time that the sound of thunder will be heard at a distance of 25 kilometers from the point of a lightning strike. Use 330 m.sec^{-1} ($1,000 \text{ ft.sec}^{-1}$) as the velocity of sound in air.

Answer:

The distance that the sound must travel is 25 km, or 25,000 m.

At a speed of 330 m.sec^{-1} it will take ($25,000 \text{ m} / 330 \text{ m.sec}^{-1}$) = 75.76 seconds for the sound to reach an observer.

of water. Echo location is also a prominent feature of communication for different animal species, and has great potential for development as an assistive method for humans as well.

—*Richard M. Renneboog M.Sc.*

BIBLIOGRAPHY

Ingard, K. Uno. *Notes on Acoustics*. Hingham, Mass.: Infinity Science, 2008. Print.

Kuttruff, Heinrich. *Acoustics: An Introduction*. Boca Raton: CRC, 2014. Print.

Berg, Richard E., and David G. Stork. *The Physics of Sound*. 3rd ed. San Francisco; Toronto: Pearson Addison Wesley, 2005. Print.

Kruth, Patricia, and Henry Stobart, eds. *Sound*. Cambridge; New York: Cambridge University Press, 2007. Print.

Rossing, Thomas D., F. Richard Moore, and Paul A. Wheeler. *The Science of Sound*. 3rd ed. N.p.: Pearson, 2014. Print.

VELOCITY VS. SPEED

FIELDS OF STUDY

Classical Mechanics

SUMMARY

Speed and velocity are two closely related quantities, both of which measure an object's movement relative to time. Speed is fully described by a simple numerical value; velocity is speed with added information about the direction of movement. Using equations developed by English physicist Isaac Newton, simple calculations of speed and velocity are useful in day-to-day activities like planning car trips or tracking the distance covered while jogging.

PRINCIPAL TERMS

- **average:** in physics, the overall value for a given quantity, obtained by comparing initial and final values of a measurement against another unit or quantity; for instance, average speed compares the total distance traveled to the total time taken to move that distance.
- **displacement:** the absolute distance between where a moving object begins and where it ends, regardless of the object's path. Displacement may be equal to or less than distance traveled.
- **instantaneous:** denotes a measurement taken at a specific point in time.
- **rate:** the ratio of a unit, such as distance or weight, relative to a period of time.
- **scalar:** a quantity that is fully described by a numerical value alone, such as speed, length, or mass.
- **slope:** a line on a graph that indicates the rate of change over time; for instance, on a plot of the distance traveled by an object over time, the slope equals the speed of travel and its shape can convey information about acceleration.
- **vector:** quantities that require a direction, along with a numerical value, to be fully described, such as velocity or acceleration.

GOING WHERE AND HOW FAST?

Knowing how quickly an object is moving, and in what direction, is very useful. It is so useful that many people have an intuitive sense of these things. Drivers constantly estimate speed and velocity, watching the approach of a nearby car at an intersection to determine whether it is safe to cross, for instance. The formalized versions of these calculations underpin much of transportation and navigation. A car's onboard GPS uses information about a car's velocity, the position of its destination, and the total distance of various routes to give arrival estimates; a hiker with a detailed map and accurate compass may do much the same.

Speed and velocity fall under a subfield of classical mechanics known as "kinematics." Kinematics focuses on the motion of objects and can trace its roots back to classical Greek philosophers and mathematicians like Aristotle. The kinematics most used today was defined by English physicist Isaac Newton (1642–1727) in eighteenth century. In the twenty-first century, the most accurate systems of kinematics rely on relativistic quantum mechanics, based on the general relativity theory proposed by German-born American physicist Albert Einstein (1879–1955). Quantum mechanics are vital in explaining the universe at very small and very large scales, but under typical conditions, Newtonian kinematics are sufficient.

SPEED VERSUS VELOCITY

In everyday conversation, the terms "velocity" and "speed" are often used interchangeably to mean the rate at which an object is moving. In physics, though, these terms have distinct meanings. Speed is the simpler of the two, a ratio of distance traveled to the time spent traveling. It is a scalar quantity, fully described by a simple number. Velocity, however, requires a direction, making it a vector quantity.

The average speed and velocity of an object are easy to calculate using the beginning and end points of its movements and the time of its travel. In the case of an object moving at a uniform speed and uniform direction, the average speed and velocity will accurately describe the object at any instant of its movement. Very few objects move with uniform speed and direction, however.

A football pass, for example, is thrown in an arc. At first it moves forward and upward, but eventually the force of gravity begins to pull it back down again. It slows at the top of its arc and speeds up again as gravity begins working with the down-forward movement of the ball instead of against it. At different moments along the ball's trajectory, its instantaneous speed might be faster or slower than the average. Its instantaneous velocity might not only be faster or slower, but it may also be in different directions.

Plotting the distance an object has traveled against the time elapsed while it travels is a useful way to visualize changes in an object's speed over time. (Note that average acceleration is the change in velocity over a period of time, or $\bar{a} = \Delta v / \Delta t$.) The slope of

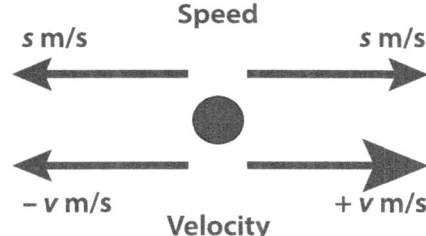

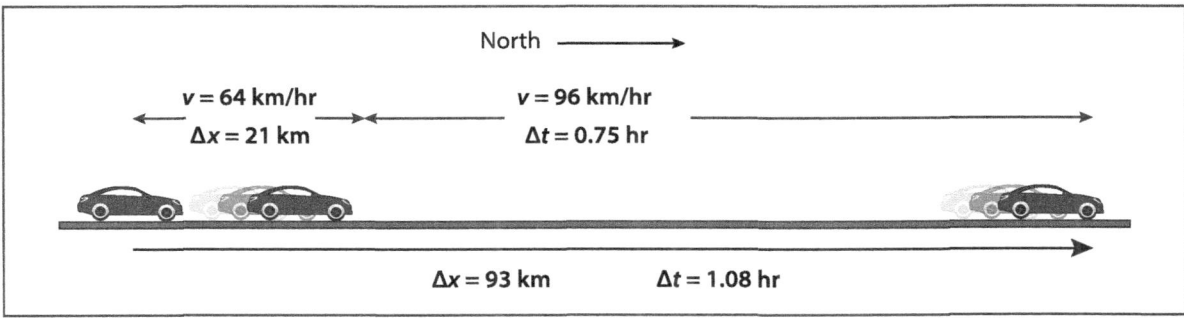

$$s = \frac{|\Delta x|}{\Delta t} = \frac{|\text{Distance}|}{|\text{Time}|} = \frac{93 \text{ km}}{1.08 \text{ hr}} = 86 \text{ km/hr}$$

$$v = \frac{|\Delta x|}{\Delta t} = \frac{\text{Displacement}}{\text{Time}} = \frac{93 \text{ km north}}{1.08 \text{ hr}} = 86 \text{ km/hr north}$$

Speed (*s*) and velocity (*v*) are similar calculations, however velocity includes direction as well as magnitude. Average speed is calculated by dividing the absolute value of the distance traveled (|Δ*x*|) by the time it took to travel (Δ*t*). Velocity is calculated by dividing an object's displacement (Δ*x*) by the time it took to be displaced (Δ*t*). When measuring displacement, the direction must be noted either by generic directions (positive or negative values) or by geographical directions (north).

the line formed by this plot indicates the speed at any given instant.

The key element in distinguishing speed from velocity is displacement. Whereas speed is equal to distance divided by time, velocity is equal to displacement divided by time. Displacement is the absolute distance between the start and end points of an object's motion. An object moving around a circle, such as the moon in orbit around the earth, is one example. Consider the formula for average velocity, where $(\bar{v})$ is average velocity, r is displacement, and t is time taken to complete one orbit.

$$\bar{v} = \frac{\Delta r}{t}$$

One orbit takes 27.3 days, or 2,358,720 seconds. Displacement (Δr) is equal to the ending position minus the starting position, which here can be inferred as zero. In a circle, the start and end points of movement are the same. As demonstrated below, this means an object in perfectly circular motion has zero velocity.

$$\bar{v} = \frac{\Delta r}{t} = \frac{0}{2,358,720 \text{ s}} = 0$$

Obviously, however, the moon is not sitting still. Its average speed $((\bar{s}))$ can still be calculated, using the circumference of the circle it travels as a measure of distance (Δd). In this case, plug in the distance traveled by the moon during one full orbit:

$$\bar{s} = \frac{\Delta d}{t}$$

The moon's orbit is about 384,400 kilometers, or 384,400,000 meters. Plugging in these values gives the following:

$$\bar{s} = \frac{\Delta d}{t} = \frac{384,400,000 \text{ m}}{2,358,720 \text{ s}} = 162.97 \text{ m/s}$$

These rough calculations show that it is possible for an object to be moving quite fast while still having an average velocity of zero.

Similarly, it is easy to imagine a displacement that is smaller than the distance traveled without being zero. A hike to a mountaintop is rarely a straight line—the displacement from the trailhead to the summit may only be a few thousand meters directly north, but hikers can easily travel two or three times that distance on trails that meander back and forth on their way to the top.

How Kinematics Are Used

More complicated forms of kinematics account for forces like friction and the resistance of the

---

SAMPLE PROBLEM

A woman is driving to a new restaurant 6 miles northwest of her home in the city center. The reservations are for 7 p.m. She leaves home at 6 p.m., but along the way, she needs to take several turns and detours. When she arrives, she checks the odometer: she has traveled exactly 30 miles. Looking at the clock reveals the time to be 6:50 p.m. What was the average speed of the vehicle during this trip, in meters per second? What is the average velocity, in meters per second with direction?

Answer:

First, make note of what is known: the trip began at 6 p.m. and ended at 6:50 p.m. Subtracting the end time from the start time, one finds that the trip took 50 minutes, or 3,000 seconds (50 min × 60 s/min).

The odometer measures the distance (Δd) traveled as 30 miles. Convert this to meters.

30 miles × 1,609.347 meters/mile = 48,280.41 meters

For average speed, use the following formula:

$$\bar{s} = \frac{\Delta d}{t}$$

Plug in the known values for Δd and t, and solve:

$$\bar{s} = \frac{\Delta d}{t} = \frac{48,280.41 \text{ m}}{3,000 \text{ s}} = 16.09 \text{ m/s}$$

To calculate average velocity (<insert equation POP_VelocityVersusSpeed_equation9>), distance is insufficient. Displacement (Δr) is needed. Displacement measures the absolute distance (in a straight line) between two points and their direction relative to one another. The odometer does not measure displacement, it measures the total distance the car traveled over the road, including all turns and detours. Looking at a map, however, shows that the restaurant is only 6 miles (9,656.06 meters) northwest of the woman's home. Plugging in this value for displacement yields:

$$\bar{v} = \frac{\Delta r}{t} = \frac{9,656.06 \text{ m NW}}{3,000 \text{ s}} = 3.22 \text{ m/s NW}$$

atmosphere; engineers use these sophisticated equations to determine highly precise quantities such as the exact velocity needed for a spacecraft to successfully enter orbit without pulling free of gravity or falling to the surface below. For the average person, being able to comfortably estimate how long it will take to make it across the state in a given time is useful enough.

—Kenrick Vezina, MS

BIBLIOGRAPHY

Allain, Rhett. "What's the Difference between Speed and Velocity?" *Wired*. Condé Nast, 16 June 2014. Web. 12 May 2015.

Henderson, Tom. "1-D Kinematics." *Physics Classroom*. Physics Classroom, 2015. Web. 12 May 2015.

Ohanian, Hans C., and John T. Markert. *Physics for Engineers and Scientists*. 3rd extended ed. New York: Norton, 2007. Print.

Simanek, Donald E. "Kinematics." *A Brief Course in Classical Mechanics*. Lock Haven U, Feb. 2005. Web. 12 May 2015.

"Velocity (Mechanics)." *Encyclopaedia Britannica*. Encyclopaedia Britannica, 10 Dec. 2014. Web. 12 May 2015.

Williams, David R. "Moon Fact Sheet." *Lunar and Planetary Science*. NASA Goddard Space Flight Center, 25 Apr. 2014. Web. 12 May 2015.

VOLUME AND CAPACITY

FIELDS OF STUDY

Classical Mechanics; Fluid Mechanics

SUMMARY

Volume is the three-dimensional space occupied by an object. The internal volume of a container determines its capacity. The volume of various three-dimensional shapes can generally be calculated using simple formulas based on the area of one side multiplied by the height of the shape. Volume is a fundamental parameter for understanding density and other physical properties.

PRINCIPAL TERMS

- **buoyancy:** upward force exerted by a fluid on a submerged or floating object.
- **density:** the amount of mass per unit volume.
- **ideal gas law:** a law that describes the relationship between the temperature (T), pressure (p), volume (V), and number of particles (n) of an ideal gas, written as $pV = nRT$.
- **pressure:** the force exerted per unit area.
- **surface area:** the total area of the outward-facing surface of a three-dimensional object.
- **temperature:** the average kinetic energy of particles that make up a substance.

A THREE-DIMENSIONAL MEASUREMENT

The volume of an object is the amount of three-dimensional space it occupies. Volume is measured in units of cubic distance, such as cubic meters or cubic feet. It is a necessary parameter for determining spatial relationships. When volume is used to determine storage capability, it is often called capacity.

Volume is also necessary for determining the density of an object. Density quantifies how much matter is contained in a certain area. High-density items, such as solid lead, have lots of atoms or molecules in a very small space. Loosely arranged particles, such as oxygen molecules in the atmosphere, have a low density.

The dynamics of fluids (gases and liquids) are distinct from those of solids. Fluid dynamics are highly contingent on volume and density, making these two measurements important when engineering ways to contain, transfer, and use fluids.

FLUIDS, BUOYANCY, AND THE IDEAL GAS LAW

All fluids have buoyancy, which is a force that counters gravity and pushes upward on submerged objects. An object's ability to float depends on the densities of both the object and the fluid. If the object is less dense than the fluid in which it is submerged, it will float.

All fluids exert pressure on their surroundings, whether those surroundings are the container that holds the fluid or an object submerged in it. Pressure

Volume and capacity

is distinct from buoyancy; rather than pushing upward, for example, it pushes inward on a submerged object from all sides. The strength of this pressure is influenced by the density of the fluid. Denser fluids tend to exert more pressure.

The two types of fluids have slightly different relationships to volume and pressure. A liquid is able to vary its shape but not its volume, while a gas is able to vary both its shape and its volume. This means that a gas will expand to fill whatever container it is in. The unique ability of a gas to vary its density is described by the ideal gas law. This law relates temperature (T) in kelvins (K), pressure (p) in pascals (Pa), volume (V) in cubic meters (m^3), and number of gas particles (n) in moles (mol). The law is written as

$$pV = nRT$$

where R is the gas constant, equal to 8.3144621 joules per mole-kelvin (J/mol·K).

The ideal gas law assumes that all the molecules of a gas bounce around and off of each other and the container with perfect elasticity, completely uninfluenced by the attractive forces that hold molecules of solids and liquids together. While there is no such thing as a truly ideal gas, the relationships the law describes are still useful.

Pressure is directly proportional to the number of gas particles and the temperature of the gas. Tightly packed particles press outward with more force than do loosely packed particles. Similarly, the temperature of a gas is a measure of the kinetic energy of the particles. Heating up a gas causes the particles to move more quickly. When they bounce off each other and their container, they exert more force.

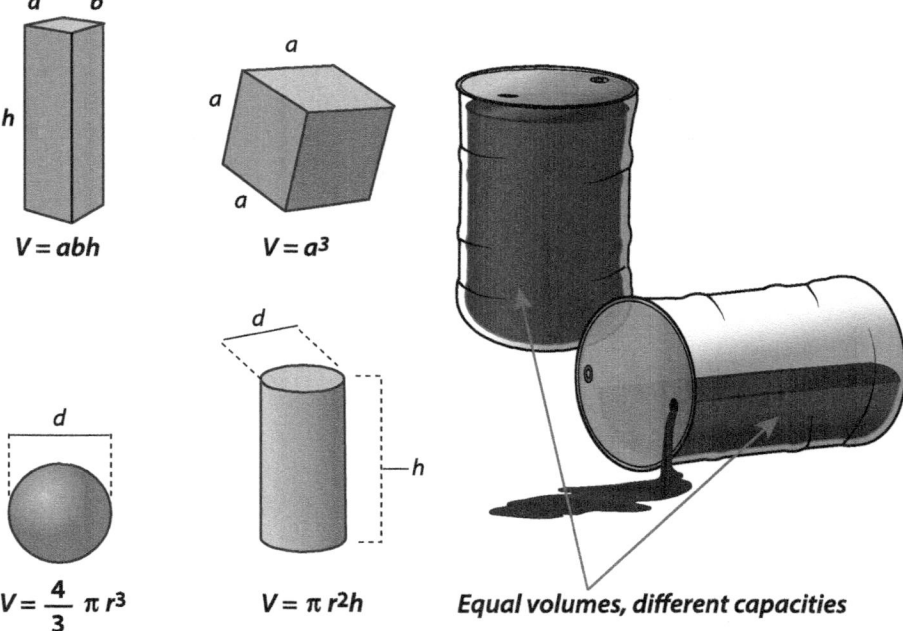

Every shape has a different formula for calculating its volume. Volume of a cylinder is measured by multi-plying the cylinder's radius squared by its height and by π. Capacity is not always the same as the volume of a container. Capacity is the amount of liquid a container can hold. In the example of two oil drums, one upright and one tipped over, the tipped drum has a smaller capacity because it can only be filled to the level of the opening without spilling.

Pressure is inversely proportional to volume. If a gas squeezes into a smaller space, its particles will bounce off each other and the container more frequently, thus increasing the pressure.

CALCULATING CAPACITY

The volumes of common three-dimensional shapes can be determined using simple measurements and the following formulas:

The volume of a cylinder is equal to pi (π) times the radius of one circular surface squared (r^2) times the height of the cylinder (h):

$$V_c = \pi r^2 h.$$

The volume of a rectangular prism, such as a box, is equal to the width of one side (w) times the length of that side (l) times the height of the prism (h):

$$V_r = wlh$$

The volume of a sphere is equal to four-thirds times pi (π) times the radius of the sphere cubed (r^3):

$$V_s = \frac{4}{3}\pi r^3$$

Using these equations, it is possible to calculate the capacity of many everyday objects.

SURFACE AREA VS. VOLUME
The surface area of a three-dimensional shape bears a unique relationship to its volume. Surface area increases as a squared value (m^2), whereas volume increases as a cubed value (m^3). Thus, increasing the overall size of a shape causes its volume to increase much more quickly than its surface area. This is significant because surface area represents the area of interaction between an object and its environment. A very large object has a proportionately very small surface area, meaning the environment's influence on its interior may be slow. For example, an elephant is very large but has relatively little surface area, so its skin takes a long time to heat up in the sun and a long time to cool down. Because of this, elephants have evolved large ears to provide a surface that they can use to regulate their temperature. Conversely, very small objects have a very high surface-area-to-volume ratio. This is why babies are more susceptible than adults to chills and overheating.

VOLUME IN EVERYDAY LIFE
Volume quantifies space, an important and limited resource in everyday life. It is also a basic measurement for determining other characteristics of objects. Volume and density are necessary measurements for determining the behavior of a gas and the force it exerts on its environment. Furthermore, the mathematical ratio of surface area to volume plays a major role in biology and chemistry by influencing the rate of interaction between an object and its environment.

—*Kenrick Vezina, MS*

BIBLIOGRAPHY
Faber, Thomas E. "Fluid Mechanics." *Encyclopaedia Britannica*. Encyclopaedia Britannica, 12 Aug. 2011. Web. 12 June 2015.
Hodanbosi, Carol. "Fluids Pressure and Depth." Ed. Jonathan G. Fairman. Glenn Research Center. NASA, 12 June 2014. Web. 12 June 2015.
Hutchinson, John S. "The Ideal Gas Law." *Connexions*. Rice U, 16 Jan. 2005. Web. 12 June 2015.
Stein, James D. *Cosmic Numbers: The Numbers That Define Our Universe*. New York: Basic, 2011. Print.
Weisstein, Eric W. "Volume." Wolfram MathWorld. Wolfram Research, n.d. Web. 12 June 2015.
Yevick, David, and Hannah Yevick. *Fundamental Math and Physics for Scientists and Engineers*. Hoboken: Wiley, 2015. Print.

SAMPLE PROBLEM

A cylindrical oil-based home heating tank is 3 meters tall. The radius of each circular end of the cylinder is 0.5 meter. What is the heater's capacity in cubic meters?

Answer:
Plug the values above into the formula for the volume of a cylinder, and solve:

$$V_c = \pi r^2 h$$

$$V_c = \pi \times (0.5 \text{ m})^2 \times 3 \text{ m}$$

$$V_c = 2.36 \text{ m}^3$$

The volume of the tank is approximately 2.36 cubic meters. Note that for many three-dimensional shapes, calculating the volume is as simple as calculating the area of one end and multiplying this value by the height of the shape. In the above example, the quantity πr^2 is simply the area of one of the circles that serves as a base for the cylinder.

W

WATT

FIELDS OF STUDY

Classical mechanics

SUMMARY

The watt is the standard unit for measuring power, which is defined as work or energy transfer over time. One watt is equivalent to one joule of energy transferred or consumed per second. Everyday appliances such as toasters use wattage to indicate the power needed to operate the device, the power it outputs, or the maximum power it can utilize without damage.

PRINCIPAL TERMS

- **derived unit:** a unit created by combining two or more other units.
- **displacement field:** in electrodynamics, the electric field produced solely by free charges (e.g., free electrons).
- **International System of Units (SI):** a standardized set of units and measures used by scientists worldwide, based on and largely synonymous with the metric system.
- **joule:** the SI derived unit of energy, work, or heat.
- **Ohm's law:** an empirical law stating that the current, or flow of electrical charge, between two points is directly proportional to the voltage, or difference in electric potential, between those points.
- **power:** the work done or energy transferred over time.
- **power rating:** the maximum electrical power a device can use without being damaged.

QUANTIFYING ENERGY TRANSFER OVER TIME

Watts (W) are the International System of Units (SI) unit for measuring power—that is, work done or energy transferred over time. The power rating of a device indicates the maximum wattage it can use without damage. One watt is equivalent to one joule (J) of energy transferred per second. Because it is based on a combination of two or more SI units, the watt is considered a derived unit. It is named after Scottish engineer James Watt (1736–1819), who famously developed radical improvements to the steam engine. Wattage is most commonly used to indicate the energy consumption of electronics.

HOW ELECTRICAL POWER IS GENERATED

Electrical power is generated by the transfer of electrical energy, usually in the form of electrons, between atoms. Electrons flow down the atoms of a copper wire because there is a large difference between the electric potential of the source (e.g., a battery) and that of the destination (e.g., a light bulb). The energy

FROM	TO				
	BTUs/hr	Joules/s	Horsepower	Kilowatts	Watts
BTUs/hr	1	0.293	3.928×10^{-4}	2.93×10^{-4}	0.293
Joules/s	3.413	1	1.340×10^{-3}	1×10^{-3}	1
Horespower	2.546×10^{3}	746	1	0.746	746
Kilowatts	3.413×10^{3}	1000	1.34	1	1000
Watts	3.413	1	1.340×10^{-3}	1×10^{-3}	1

Conversion table for units of power shows the conversions among BTUs/hr, joules/s, kilowatts, watts, and horsepower.

carried by the electrons is either transferred or transformed in order to do work. An incandescent bulb works by transforming the energy carried by the electrons into light, which is a form of electromagnetic energy. The electric field generated by free electrons and other free charges moving through conductive materials is called a displacement field.

Calculating Wattage

Electric current, or the rate at which electric charge is transferred across a material, obeys Ohm's law, named after German physicist Georg Ohm (1789–1854). The SI unit of electrical resistance, the ohm (Ω), is also named after him. Ohm's law states that current (I) is directly proportional to the voltage (V), or difference in electrical potential across the material, divided by the resistance (R) of that material:

$$I = \frac{V}{R}$$

The term "wattage" simply means power (P) as measured in watts. Power can be expressed a variety of ways:

$$P = IV$$
$$P = I^2 R$$
$$P = \frac{V^2}{R}$$

Given a nine-volt battery connected to itself by a loop of copper wire with a resistance of eighteen ohms, one can determine the amount of electrical power produced by the battery sending a current through the wire:

$$P = \frac{V^2}{R}$$
$$P = \frac{9^2}{18} = \frac{81}{18}$$
$$P = 4.5 \text{ W}$$

Wattage Today

Wattage is a vital parameter in the design and use of high-performance electronics, which often have very sensitive components. Understanding wattage and its relationship to voltage and resistance will only become more useful in an increasingly electric world.

—*Kenrick Vezina, MS*

Bibliography

"DC Circuit Theory." *Basic Electronics Tutorials*. Wayne Storr, n.d. Web. 8 June 2015.

"Electric Current." *Encyclopaedia Britannica*. Encyclopaedia Britannica, 10 Sept. 2014. Web. 8 June 2015.

Gibilisco, Stan. *Electricity Demystified*. 2nd ed. New York: McGraw, 2012. Print.

Henderson, Tom. "Ohm's Law." *The Physics Classroom*. Physics Classroom, n.d. Web. 1 June 2015.

Hughes, John M. *Practical Electronics: Components and Techniques*. Sebastopol: O'Reilly, 2015. Print.

Kelly, P. F. *Electricity and Magnetism*. Boca Raton: CRC, 2015. Print.

WAVELENGTH

FIELDS OF STUDY

Electromagnetism; Electronics

SUMMARY

This article discusses aspects of physical and electromagnetic waves, with regard to their wavelength. Waves are described as either longitudinal or transverse. The time required for one cycle of any wave is the period of the wavelength. Wavelengths are related to both the frequency and the speed at which waves travel through a medium.

PRINCIPAL TERMS

- **aperture:** an adjustable opening in a barrier through which light or other electromagnetic emission can pass.
- **crest:** the highest point of a wave from its neutral value.
- **electromagnetic spectrum:** the continuous range, or continuum, of the frequencies of electromagnetic waves.
- **frequency:** the number of cycles of a property that occur in a certain amount of time.

- **resolution:** the ability of a detector to differentiate or separate different wavelengths.
- **phase:** a stage in a wave property; typically used to describe the relationship of two or more waves.
- **trough:** the lowest point of a wave from its neutral value.
- **velocity:** the speed and direction of motion.

WAVE CHARACTERISTICS

All waves can be thought of as cyclic disturbances in some property. The number of times that the cycle occurs in a certain amount of time is the frequency of the wave. When there is no disturbance, a property (such as a wave) has a neutral value. For waves, the value of the property alternates continuously between equal positive and negative variations. The crest (high point) and trough (low point) seen in water waves are classic examples. The neutral value of a water wave is a perfectly smooth surface. As a water wave progresses, the level rises to a maximum value at the crest, then falls to an equal distance below the neutral point in the trough. Such a "sinusoidal" wave (one that is shaped like a sine curve) is at its neutral value three times in each cycle: at the beginning, the middle, and the end. Wavelength is the distance between adjacent crests or troughs.

Sound waves behave in a similar manner. Sound waves and water waves are examples of "longitudinal" waves. Their particles displace in either the same or the opposite direction (depending on the type) of the wave. A medium's effect on a wave can be seen in the compression-rarefaction sequence. Compression displaces matter from its neutral value. Rarefaction displaces the matter in the opposite sense. Graphs of displacement in both sound waves and water waves exhibit sinusoidal shapes.

The velocity of longitudinal waves depends on the density of the medium through which they travel. The denser the medium, the faster longitudinal waves travel through. The speed of sound in air, for example, is 331 meters per second (1,087 feet per second) at 0 degrees Celsius (32 degrees Fahrenheit) and 342 meters per second (1,123 feet per second) at 18 degrees Celsius (65 degrees Fahrenheit). In water, the speed of sound is about 1,140 meters per second (4,724 feet per second). The density of liquid water is about 770 times that of air.

The waves of the electromagnetic spectrum are described as "transverse" waves because the direction of their displacement is at right angles to the direction of their spread. Electromagnetic (EM) waves are not as simply described as longitudinal or physical waves because they are independent of matter and do not spread in the same way. An EM wave is perhaps best thought of as the vector combination of an electric value and a magnetic value with a single velocity. All EM waves travel at the speed of light. The magnitudes of the electric and magnetic components of an EM wave exhibit are sinusoidal, as is typical of other kinds of waves and wavelike cyclic behaviors.

THE ELECTROMAGNETIC SPECTRUM

The EM spectrum is a continuum of wavelengths, or frequencies, ranging from zero to infinity. The complete absence of any EM wave is the zero point, having neither frequency nor wavelength. At the opposite extreme, frequencies and wavelengths are theorized to become so compacted as to be indistinguishable from solid matter. Visible light, detectable by the human eye, makes up just one small part of the EM spectrum, with wavelengths ranging from 770 nanometers to 400 nanometers. (A nanometer is one-billionth of a meter.) The corresponding frequencies range from about 1013 to 1016 hertz (Hz). (One hertz is one cycle per second.) By comparison, x-rays have frequencies of about 1018 Hz, and gamma rays have frequencies of 1020 Hz and beyond.

The frequency (f) and the wavelength (λ) of EM radiation are inversely related. The greater the frequency is, the shorter the wavelength. The common

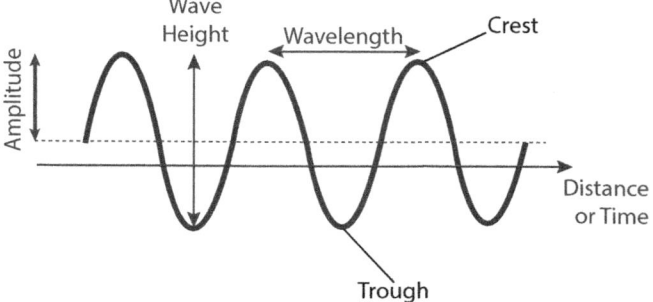

A wave's wavelength is measured from one crest to a consecutive crest or from one trough to a consecutive trough. The electromagnetic spectrum is a wide range of waves, each with different wavelengths. Some electromagnetic wavelengths are visible to the human eye, just as some sound waves have wavelengths audible to the human ear.

factor for all EM radiation is its velocity (v), the speed of light. Unlike longitudinal waves, the velocity of EM waves is constant rather than dependent on the medium. As a result, considering the speed of EM waves is generally not useful in practical applications. Instead, an EM wave is referred to by its frequency in hertz or its wavelength in meters. In spectroscopy, the designation of "wave number" is commonly used. The wave number of an EM frequency is the reciprocal of the wavelength, normally stated in 1/cm.

APPLICATIONS

EM radiation has a number of applications determined by the wavelength range of the appropriate frequencies. Analytical applications rely on the interaction between the particular EM radiation and the electron cloud in atoms and molecules. Lower frequencies have longer wavelengths and can be used in communications. Adjusting amplitude and phase relations of EM signals enables the speed-of-light transmission of information by radio waves and microwaves. The wavelengths of this region of the EM spectrum are mostly too large to interact effectively with atoms and molecules. Such interaction cannot happen until the energy and wavelengths of the EM waves become similar in dimension to those of the atoms and molecules. Thus, microwaves effectively transmit energy into materials placed in a microwave oven by stimulating the vibrational energy of water molecules and certain kinds of chemical bonds in the material.

Wavelength continues to decrease into and across the infrared and visible region of the EM spectrum. All wavelengths of EM radiation from this point on have application in analytical methods because they can interact with matter in specific ways. Infrared, visible, and ultraviolet wavelengths are readily absorbed and emitted by atoms and molecules. Electrons' patterns of absorbing or emitting EM wavelengths are routinely analyzed and used for identification and monitoring methods. Above the ultraviolet range of wavelengths are the x-ray wavelengths. These wavelengths are too short to interact with electrons effectively. However, they are closer in size to atomic nuclei and can interact with the nuclear structure of atoms. Thus, they are able to pass through solid matter fairly readily and can be diffracted by the nuclei of atoms. X-ray diffraction is used to analyze crystal structures and solid surfaces. Above the x-ray range of wavelengths are the gamma-ray wavelengths. Because of their extremely high energy and short wavelength, gamma rays destroy matter, and their use is very limited outside of astronomy.

WAVELENGTH AND RESOLUTION

The wavelengths of EM radiation are typically detected by electronic devices that convert the analog frequency of the radiation into a digital equivalent. Such detectors use an aperture to restrict the EM radiation that enters the device, where it is then resolved into component wavelengths. The resolution of the devices depends strictly on the ability of gratings and other components to differentiate the wavelengths of the light or other EM radiation that they receive. The finer the resolution, the more precise the incoming information is. High-definition cameras, microscopes, and telescopes, for example, provide clearer images than devices with lower resolution.

—*Richard M. Renneboog, MSc*

BIBLIOGRAPHY

"An Introduction to Waves." *GCSE Bitesize.* BBC News, 2014. Web. 25 Feb. 2015.

Anderson, Rosaleen J., David J. Bendell, and Paul W. *Groundwater. Organic Spectroscopic Analysis.* Cambridge: Royal Soc. of Chemistry, 2004. Print.

Kirkland, Kyle. *Light and Optics.* New York: Facts On File, 2007. Print.

Kumar, B. N. *Basic Physics for All.* Lanham: University Press of America, 2009. Print.

Nave, C. R. "Traveling Wave Relationship." *HyperPhysics.* Dept. of Physics and Astronomy, Georgia State U, 2014. Web. 25 Feb. 2015.

Rogers, Alan. *Essentials of Photonics.* 2nd ed. Boca Raton: CRC, 2008. Print.

Shipman, James T., Jerry D. Wilson, and Charles A. Higgins Jr. *An Introduction to Physical Science.* 14th ed. Boston: Brooks/Cole, 2015. Print.

Smith, Brian C. *Quantitative Spectroscopy: Theory and Practice.* San Diego: Elsevier, 2002. Print.

Tilley, Richard. *Colour and the Optical Properties of Materials.* 2nd ed. Chichester: Wiley, 2011. Print.

WAVE PROPERTIES

FIELDS OF STUDY

Classical Mechanics; Electromagnetism; Acoustics

SUMMARY

Waves, whether mechanical or electromagnetic, share certain common properties. All waves have a wavelength, a frequency, and an amplitude, which together determine the ability of a wave to transmit energy and to displace the medium through which it travels.

PRINCIPAL TERMS

- **diffraction:** a change in the direction of a wave as it passes around an obstruction or through an opening.
- **interference:** how waves interact when they meet in the same medium. Waves whose crests align will reinforce one another (constructive interference); waves where one's crests align with the other's troughs will dampen one another (destructive interference).
- **longitudinal wave:** a type of wave wherein the medium is displaced in a direction parallel to the movement of energy, as in the case of sound waves.
- **reflection:** the bouncing back of a wave after it hits a barrier, as when light reflects off a mirror.
- **refraction:** the alteration of a wave's path, speed, and wavelength when it passes from one medium to another.
- **transverse wave:** a type of wave wherein the medium is displaced in a direction perpendicular to the movement of energy, as in the case of waves on the surface of water.

TYPES OF WAVES

There are two major ways to categorize waves. One way is based on whether the wave can transmit energy through a vacuum. Waves that can are classified as electromagnetic waves, such as light, while waves that depend on a physical medium such as air or water are mechanical waves, such as sound waves or ocean waves. The other major categorization is based on the movement of the wave relative to the direction in which energy is transferred. Transverse waves move perpendicular to the direction of energy transfer. An example of this is when water moves up and down in ripples on a pond, while the waves travel horizontally across the surface. Longitudinal waves, also called "compression waves," move parallel to the direction of energy transfer. An example of this is when a Slinky is stretched out horizontally and one end is quickly pushed and then pulled, sending waves along the toy's length. In a transverse wave, the highest and lowest points of the waves are called crests and troughs, respectively. In a longitudinal wave, the points of maximum compression are called compressions, and points of minimal compression (i.e., maximum spread) are called rarefactions.

WAVELENGTH, FREQUENCY, AND AMPLITUDE

All waves are described by three properties: wavelength, frequency, and amplitude. In a transverse wave, wavelength is the distance between two peaks or two troughs; in a longitudinal wave, it is the distance between two compressions or two rarefactions. Frequency measures how often a complete wave cycle occurs in a given period of time, typically one second. Frequency and wavelength are inversely related to each other. Waves with high frequencies have short wavelengths, and waves with low frequencies have long wavelengths. The energy-transmission ability of a wave is also related to both of these properties: high-energy waves have short wavelengths and high frequencies; low-energy waves have long wavelengths and low frequencies.

Amplitude is a measure of the intensity of a wave in terms of how much it deforms the medium through which it is traveling. In a transverse wave, it is the distance between the resting equilibrium of the medium (the baseline) and the top of a crest. (An alternate way to measure amplitude is the vertical distance from the crest to the trough, called peak-to-peak amplitude.) In a longitudinal wave, amplitude is measured as the displacement of the particles of the medium at rest relative to a compression.

REFLECTION, DIFFRACTION, AND REFRACTION

When waves encounter barriers or pass from one medium to another, several phenomena can occur.

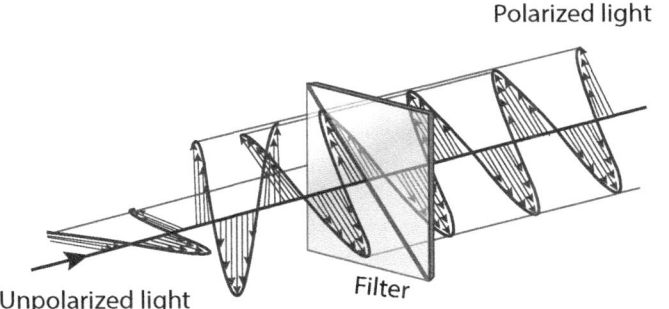

The behavior of waves can change when the wave comes in contact with a different medium or contacts another wave. Waves of light oscillate in many directions, on many planes, as unpolarized light. When light contacts a polarizing filter only the light waves that oscillate in the same plane as the filter will pass through. The light emitting through this filter is called polarized light.

Reflection occurs when a wave hits a barrier and is bounced back. Examples of this include a sound wave echoing off the side of a cliff and light bouncing off a mirror. Diffraction occurs when a wave passes by an obstacle or through an opening and bends its path as a result, as when ripples in a pond bend around rocks obstructing their path. Refraction occurs when a wave passes from one medium to another, causing its direction, speed, and wavelength to change. An example of this is light passing from air into water. This is why a straw sticking out of a glass of water looks as though it is bent at the surface of the water. Dispersion is a special type of refraction that splits a wave into several wavelengths, as when a prism refracts white light and produces a rainbow.

WAVE INTERACTIONS

Waves may also interact with one another when traveling in the same medium. This is interference, which may be constructive or destructive. If the waves align so that their peaks and troughs are paired, they merge to become a single wave with greater amplitude. This is called constructive interference. If the waves are aligned so that the peak of one wave aligns to the trough of another, they merge and dampen each other, reducing or even canceling their amplitudes. This is called destructive interference.

—*Kenrick Vezina, MS*

BIBLIOGRAPHY

"Anatomy of an Electromagnetic Wave." *Mission:Science.* NASA, 13 Aug. 2014. Web. 29 June 2015.

"Frequency and Period." *SparkNotes: SAT Physics.* SparkNotes, 2011. Web. 29 June 2015.

Fritzsche, Hellmut. "Electromagnetic Radiation." *Encyclopædia Britannica.* Encyclopædia Britannica, 26 Nov. 2014. Web. 29 June 2015.

Russell, Daniel A. "Longitudinal and Transverse Wave Motion." *Acoustics and Vibration Animations.* Pennsylvania State U, 18 Feb. 2015. Web. 29 June 2015.

"Waves." *The Physics Classroom.* Physics Classroom, n.d. Web. 29 June 2015.

Zielinski, Ellen, Courtney Faber, and Marissa H. Forbes. "Lesson: Waves and Wave Properties." *TeachEngineering.* Regents of the University of Colorado, 18 July 2014. Web. 29 June 2015.

WEAK FORCE

FIELDS OF STUDY

Quantum Electrodynamics; Quantum Field Theory; Quantum Mechanics; Relativity; String theory; Superstring theory

SUMMARY

The weak force is one of the four fundamental forces that affect all matter in the universe. The other three forces work to hold things together; the weak force plays a role in decay or breaking things apart. It works with neutrino interactions and causes radioactive decay. Its range is very short and, compared to the other forces, it is very weak, though it is still stronger than gravity.

PRINCIPAL TERMS

- **alpha decay:** a form of radioactive decay in which a radioactive atom's nucleus splits and discharges

an alpha particle, made up of two protons and two neutrons
- **beta particle:** an electron that is emitted from an unstable nucleus via radioactive decay.
- **beta radiation:** beta rays emitted during beta decay, a subtype of radioactive decay.
- **boson:** an elementary particle that carries a specific type of force, or a composite particle containing an event number of fermions, having an integer spin.
- **radioactive decay:** the loss of energy and matter from an unstable nucleus in the form of ionizing radiation.
- **weak interaction:** interaction between subatomic particles at a short distance that is influenced by the weak nuclear force, one of the four fundamental forces in nature.

The Weak Force and Its Properties

The weak force plays an important role in all types of radioactive decay, including alpha decay, beta decay, gamma decay, and other decay processes that fundamental particles experience. Alpha decay takes place when a helium nucleus forms and is emitted from an atom, beta decay happens when an electron or positron is emitted from an atom, and gamma decay takes place when a high-energy photon is emitted from an atom.

The weak force has a small range of influence, smaller than 1 fm, and is much weaker than the strong force, about 10^{-7} weaker. Many believe that the four fundamental forces are all interrelated and are all part of a single force that has yet to be discovered. Therefore, sometimes the weak force is consolidated with electromagnetism and called the electroweak interaction.

Each of the four fundamental forces has its own carriers. For the weak force, the carriers are bosons called W and Z. W can be positively or negatively charged while Z has no charge and is neutral. These carriers are gigantic when compared to the other types of carriers for the other forces, and this is one reason why this force is weaker than the strong force, which has much lighter carriers. Steven Weinberg, Sheldon Salam, and Abdus Glashow predicted that W and Z bosons existed in the 1960s, but they were not discovered until 1983 at CERN.

Enrico Fermi, the Italian physicist, theorized in 1933 that an as yet undiscovered force, the weak interaction, was responsible for beta decay. This happens when a neutron in the nucleus of an atom changes into a proton by expelling an electron, or a beta particle. This force could change a neutron into a proton, an electron, and a neutrino. Through this process, one element can morph into another element when one of its neutrons changes into a proton or vice versa.

When a charged W boson is emitted, the weak force can change the color or flavor of a quark, which then causes a proton to change into a neutron or vice versa, depending on the charge of the W boson. This decay reaction is what causes nuclear fusion, the reaction that makes stars burn. That, in turn, creates heavier elements that cause supernova explosions and create the building blocks for everything—planets, people, plants, animals, everything on Earth. The effect of the neutral Z boson is more difficult to detect.

The weak force causes the reaction in hydrogen bombs. When two protons are smashed together with enough energy to overcome their mutual repulsion, the strong force can bind them together. However, this creates an unstable atom with two protons. The weak force then comes into play—one of the protons undergoes beta decay, causing the reaction.

Without the weak force, elements would be more stable. For example, elements like uranium could be safely handled. However, without the weak force and its role in beta radiation, the Sun would lose its power, and Earth, and everything on it, would die.

Why is the Weak Force Important?

Understanding the weak force leads to a better understanding of how fundamental particles behave. It is the force that, due to its role in decay, powers stars (including the Sun) and forms elements, including everything on Earth. It is the basis of much of the natural radiation that is present in the universe. Because of its role in decay, it is also the force that allows radiocarbon dating.

—*Marianne Moss Madsen, MS*

Further Reading

Georgi, Howard. *Weak Interactions and Modern Particle Theory*. Mineola, N.Y.: Dover Publications, 2009. Print.

Greene, Brian. *The Elegant Universe Superstrings, Hidden Dimensions, and the Quest for the Ultimate Theory*. New York: Vintage, 2010. Print.

Lang, Thomas G. *Our Fluid Universe: A Unified Theory of Physics that Physically Describes Photons, Matter, Gravity, Electromagnetic Forces, the Strong and Weak Forces, Quantum Theory, and Much More*. CreateSpace Independent Publishing Platform. 2014. Electronic.

Murray, Raymond L., and Keith E. Holbert. *Nuclear Energy an Introduction to the Concepts, Systems, and Applications of Nuclear Processes*. 7th ed. Amsterdam: Elsevier, 2015. Print.

Quigg, Chris. *Gauge Theories of the Strong, Weak, and Electromagnetic Interactions*. Princeton: Princeton University Press, 2013. Print.

Susskind, Leonard, and George Hrabovsky. *The Theoretical Minimum: What You Need to Know to Start Doing Physics*. New York: Basic, 2013. Print.

Tyson, Neil DeGrasse,, and Donald Goldsmith. *Origins: Fourteen Billion Years of Cosmic Evolution*. New York: W.W. Norton, 2014. Print.

Vok, Tim. *Physics and the Weak Nuclear Force*. Amazon Digital Services, LLC. 2015. Electronic.

WORK AND FORCE

FIELDS OF STUDY

Classical Mechanics

SUMMARY

Work is the transformation of energy from one form into another. For instance, a water wheel turns the potential energy of water into the rotational energy of the spinning wheel and ultimately into the kinetic energy of a millstone grinding grain. Force is the way this energy is transferred between objects. The relationship between force, mass, and acceleration is the basis of Isaac Newton's second law of motion and mechanical engineering from antiquity to the present.

PRINCIPAL TERMS

- **centripetal force:** for an object moving in a circular path, the force pulling the object toward the center of the circle.
- **displacement:** the absolute distance and direction between the starting and end points of an object's motion, which ignores any twists or turn the object's path may take; it is always equal to or less than the total distance traveled.
- **joule:** the International System of Units standard unit for energy; one joule (J) is equal to the energy transferred (i.e., work done) when applying a force of one newton across a distance of one meter.
- **mechanical advantage:** the amplification of force provided by a device or machine such as a lever or pulley.
- **newton:** the International System of Units standard unit of force; one newton (N) is the force required to accelerate a 1 kilogram object at a rate of 1 meter per second squared (1 kg·m/s^2).
- **newton-meter:** the International System of Units standard unit for torque; one newton-meter (N·m) is equal to the torque resulting at the axis from the force of 1 newton applied perpendicularly to an attached 1-meter-long moment arm (i.e., a lever).
- **power:** the rate at which work is done, or at which energy is consumed; the International System of Units standard unit for power is the watt (W), which is equivalent to one joule per second (J/s).
- **torque:** the measure of how much a force acting on an object causes it to rotate.

Energy, Force, and Work

Energy is a fundamental property of matter. In functional terms, it is a measure of how much work a system (such as a car engine or a human body) can perform. According to the first law of thermodynamics, the total energy of the universe is constant: it can be neither created nor destroyed. Instead, energy is transformed or transferred. When energy is transformed or transferred, work is being done. In physics, work does not need to be useful or desirable to be considered work. When the energy from a heavy wind blows over a tree, the tree is doing work.

Force is the vector by which energy is transferred. It is a push or a pull acting upon an object. When a cue ball hits another billiard ball, for example, it imparts a force to the ball it hits. This transfers some of the energy of the cue ball to the target, and the force causes the target to move. When the force (transferred energy) is sufficient to move an object some distance—to displace it—it is said to be doing work. Specifically, when dealing with forces, displacement, not distance, is the relevant value. Displacement measures the absolute distance relative to an object's starting point and includes information about the direction it has traveled. Both displacement and force always have a direction attached. Furthermore, if there is no displacement after a force is applied to the object, the force fails to do work. An example of this would be a building, which experiences a constant downward force exerted on it by the earth's gravity. The earth's crust is exerting an equal but opposite force upward, so there is no net movement, and gravity is performing no work on the building.

Kinematics

Kinematics is the name for the subfield of classical mechanics concerned with energy, motion, force, and work. It has its origins in the thinking of classical Greek philosophers and mathematicians like Aristotle (ca. 384–322 BCE). Early modern astronomers such as Tycho Brahe (1546–1601), Galileo Galilei (1564–1642), and Johannes Kepler (1571–1630) contributed to the understanding of kinematics by studying the motion of the heavenly bodies. In the seventeenth century, Isaac Newton (1642–1727) wrote the *Principia*, which established the formal groundwork for what would be dubbed "classical mechanics" during the twentieth century. Classical mechanics are often referred to as "Newtonian mechanics." Over the course of the twentieth century, classical mechanics was superseded by the theory of general relativity and the new field of quantum mechanics as the most accurate methods of understanding kinematics, but classical mechanics still sees wide use in everyday situations.

Newton's Laws

Newtown's three laws of motion are as follows. First, an object in motion tends to stay in motion, and an object at rest tends to stay at rest unless acted upon by an external force. Second, the force (F) necessary to move an object is the product of the mass of the object (m) and the acceleration needed (a):

$$F = ma$$

Third, for every action there is an equal and opposite reaction.

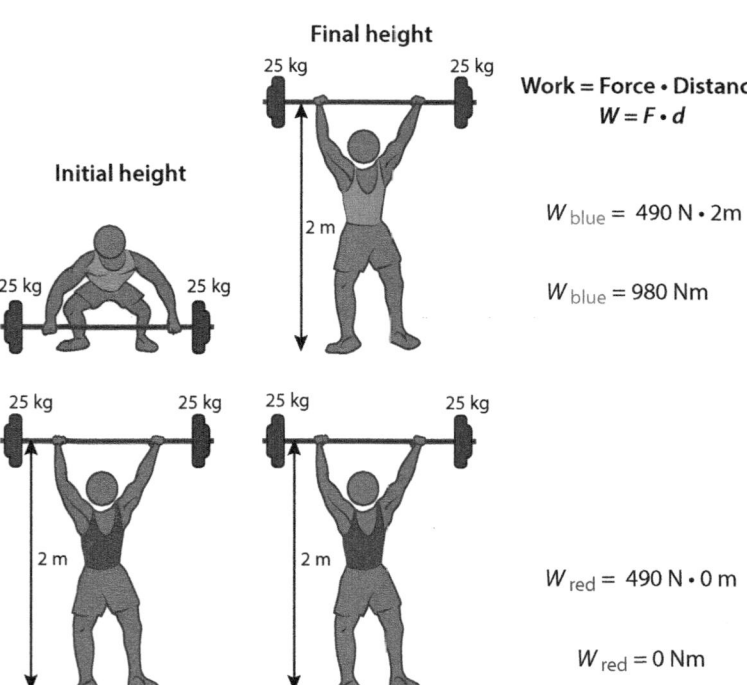

Two weight lifters are using a barbell of the same mass (m). One lifts the barbell from the ground to a height of 2 meters. The other is holding the barbell at a height (d) of 2 meters. Both weight lifters are exerting the same force (F) of 490 Newtons on the barbell. Force is calculated by multiplying the mass by the acceleration (in this case the gravitational acceleration, 9.8 m/s²). But the work done by each weight lifter is different. Work is calculated by multiplying force by the distance the object is moved. The weight lifter that raises the barbell does 980 Nm of work, the weight lifter that holds the barbell at the same height does no work.

SAMPLE PROBLEM

In the Japanese sport of Sumo, two wrestlers compete to push each other out of a circular ring. Whoever pushes the other out first, wins. A standard sumo ring is 4.55 meters (14.9 feet) in diameter. The announcer declares the weight of each contestant: the bigger man weighs 160 kilograms (about 353 pounds), and the smaller man weighs 140 kilograms (about 309 pounds). Both men begin in the center of the ring.

The match begins. After thirty seconds of struggle, the bigger man steps out of the ring. A visual estimation puts his speed as he stumbles back across the line at a bit less than walking speed, about 1 meter per second (2.24 miles per hour). How much force, in newtons, did the smaller man use to push the bigger man from the ring?

Solve using Newton's second law. Recall that acceleration is the change in velocity (Δv) over time (t):

$$a = \Delta v / t = (v_{end} - v_{start}) / t$$

Answer:

The mass of the bigger man is provided, but his acceleration must be calculated. Logically, his starting velocity is zero. His ending velocity is an estimated 1 meter per second away from the center of the ring. Plug these values into the formula for acceleration and solve:

$$a = \Delta v / t$$

$$= (v_{end} - v_{start}) / t$$

$$= (1 \text{ m/s} - 0 \text{ m/s}) / 30 \text{ s}$$

$$= (1 \text{ m/s}) / 30 \text{ s}$$

$$\approx 0.033 \text{ m/s}^2$$

Next, substitute the acceleration back into Newton's second law:

$$F = ma$$

$$= 160 \text{ kg} \times 0.033 \text{ m/s}^2$$

$$\approx 5.33 \text{ kg·m/s}^2 = 5.33 \text{ N}$$

Note that in this instance, the mass of the smaller man and the distance traveled are irrelevant; what matters are the speed and direction in which the mass was moved, not what did the pushing or how far the mass was pushed.

At any given moment, a building on earth is experiencing the effects of all three laws: a building appears to be stationary, but as the earth spins on its axis, it imparts the energy of its motion to the building through the foundation that is sunk into the soil. Why does a building not simply fly off the surface after receiving the force from the earth's spin, the same way a tennis ball on a string goes flying if spun and released? Gravity is a centripetal force pulling the building's center of mass toward the center of the earth, bending its path of movement so that it does not move in a straight line but bends in a circle conforming to the rotation of the planet. This is the first law in action.

Technically, everything on the surface of the planet is constantly falling in an arc toward the center— accelerated by the force of gravity. According to the second law, this acceleration coupled with the mass of a very large object such as a skyscraper raises the question: What keeps the object from falling through the earth's crust toward its center?

Here the third law comes into play. For all the force generated by the falling building, the earth's crust counterbalances the mass with an equal and opposite force.

Newton's laws are useful not only in the summary. They are excellent for making predictions about the world under everyday circumstances.

MEASURING ENERGY CONSUMPTION
Work and the energy used to perform it are valuable resources. Much of the work that takes place around us is fueled by electrical energy provided by utilities. Power is a measurement of work done or energy used over time, and power is what is most often measured when utilities keep track of energy consumption for billing purposes. Most homes, for instance, have an external device that tracks the home's use of power in kilowatt-hours. A single kilowatt of power is equal to one thousand joules of energy consumed per second, and a kilowatt-hour measures that consumption in hour-long intervals. Power is also useful as a measure of the rate at which a system can perform work, so car engines and household machines often provide horsepower ratings. A horsepower is a nonstandard measurement originally used to compare the amount of power produced by an engine compared to the power a horse used in pulling.

ENERGY, WORK, AND FORCE IN EVERYDAY LIFE
All powered devices operate on the principles of force and work. Automobiles make use of the explosive force of gasoline combustion to move pistons in cylinders, each of which is fired in a cycle timed to turn a crankshaft that eventually is translated into the rotational work of the driveshaft. This rotational work is known as torque. Torque is important in almost every modern engine and is measured in newton-meters (N·m), which measure the amount distance of rotational movement relative to the force applied to a lever-like machine to produce rotation. The way the firing of the pistons becomes an amplified rotational force is functionally same as the way the steering wheel moves the steering rod, turning a relatively small force into a larger one through mechanical advantage. This advantage is the same advantage one gets by using a lever. Imagine a cylinder with a long rod sticking out at a right angle. Applying a force to the outer part of the attached rod would rotate the central cylinder much more than applying the same force to the rod but closer to the cylinder. Cars use the mechanical advantage offered by levers and wheels to amplify the force generated by exploding gasoline, creating a lot of torque with fairly little energy.

THE FUTURE OF KINEMATICS
Twenty-first-century kinematics based on general relativity has transcended the classical mechanics used here in terms of precision and accuracy. Yet even astrophysicists and aeronautics engineers, who must fully map out and understand all of the forces acting on an object, use classical kinematics to estimate and validate their work. The simplicity of Newton's equations and their accuracy in day-to-day situations means they will see continued use for years to come.

—*Kenrick Vezina, MS*

BIBLIOGRAPHY
Coolman, Robert. "What Is Classical Mechanics?" *Live Science*. Purch, 12 Sept. 2014. Web. 30 Apr. 2015.
Henderson, Tom. "Work, Energy, and Power." *Physics Classroom*. Physics Classroom, 2015. Web. 3 June 2015.
"Kinematics." *Encyclopaedia Britannica*. Encyclopaedia Britannica, 5 June 2013. Web. 3 June 2015.
Oxford Dictionary of Physics. 7th ed. Oxford: Oxford University Press, 2015. Print.
Reuleaux, Franz. *Kinematics of Machinery: Outlines of a Theory of Machines*. North Chelmsford: Courier, 2012. Print.
Simanek, Donald. *A Brief Course in Classical Mechanics*. Lock-Haven U, Feb. 2005. Web. 28 Apr. 2015.

WORK-ENERGY THEOREM

FIELDS OF STUDY

Classical Mechanics; Thermodynamics

SUMMARY

The work-energy theorem describes the relationship between work performed on an object and the kinetic energy of that object. It states that when some net amount of work is performed on an object, that object's kinetic energy will change. This theorem can be written mathematically to relate an object's mass and the change in its velocity to the amount of work performed—a useful way of connecting the motion and mass of an object to its capacity for work.

PRINCIPAL TERMS

- **conservation of energy:** the principle that energy in the universe can be neither created nor destroyed, only transformed and transferred.
- **displacement:** the absolute distance an object moves from its starting point, regardless of the path it travels.
- **kinematics:** a subfield of classical mechanics that studies the motion of objects without reference to the forces that cause this motion.
- **kinetic energy:** the energy contained in an object due to its motion.
- **net force:** the sum of all forces acting on an object.
- **potential energy:** the energy stored within an object or system due to its position or configuration relative to the forces acting on it.
- **total mechanical energy:** the sum of the kinetic energy and the potential energy an object possesses as a result of work done on it.
- **work:** the successful displacement of an object caused by the application of a force.

Work and Kinetic Energy

The work-energy theorem describes the relationship between kinetic energy (the energy of an object in motion) and work (the displacement of an object by a force). It states that when work is performed on an object, the kinetic energy of that object will change. When the kinetic energy of an object changes, it moves. So, in simple terms, performing work on an object causes it to move.

Because the work-energy theorem is concerned only with masses and velocities, not with forces, it is considered part of the field of kinematics. Kinematics is a subfield within classical mechanics that studies the motion of objects without regard for the forces causing the motion. Classical mechanics, in turn, is the branch of physics concerned with the physical laws that govern both the motion of objects and the forces that move them. Isaac Newton (1642–1727) laid the foundations for modern classical mechanics with his three laws of motion, published in the late seventeenth century.

Work, Energy, and Force

Mathematically, the work-energy theorem is represented by the following equation, where W is the total work performed on an object, ΔK is the change in the object's kinetic energy, m is the mass of the object, and v_i and v_f are its initial and final velocities, respectively:

$$W = \Delta K = 1/2\, m v_f^2 - 1/2\, m v_i^2.$$

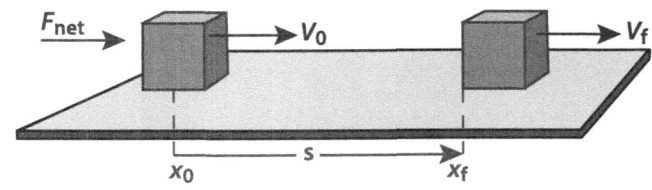

$W_{net} = F_{net}\,(s)$
$W_{net} = F_{net}\,(x_f - x_0)$
$W_{net} = 1/2\, m v_f^2 - 1/2\, m v_0^2$
$W_{net} = \Delta K = K_f - K_0$

The principle of the theorem relates the energy associated with a body in motion to the work done by an applied force. It states that the net work done on a body is equal to the change in kinetic energy. A series of equation transformations show how net work, equal to net force multiplied by displacement of the object, is also equal to the change in kinetic energy, or the initial kinetic energy subtracted from the final kinetic energy.

Thus, the total work done is equal to the total change in kinetic energy. In this sense, work can be thought of as the transfer of energy, if that transfer of energy results in displacement. Indeed, in the International System of Units (SI), the unit for both work and energy is the joule (J).

Consider a game of billiards. When the cue ball is in motion, it has kinetic energy. When it collides with another ball, it transfers some of its kinetic energy to the second ball. The force of the collision performs work on the second ball, causing it to move. This interaction underlines the relationship between energy, work, and force. In SI units, one joule represents the amount of work done or energy transferred when one newton (N) of force acts over a distance of one meter (m). In other words, if the cue ball exerts one newton of force on the second ball, causing it to be displaced by one meter, then one joule of energy has been transferred from the cue ball to the second ball, and one joule of work has been performed.

Forces in physics are interactions. According to Newton's second law, the net force (F; sum of the forces) acting on an object is equal to the object's mass (m) times its resulting acceleration (a):

$$F = ma.$$

In turn, the work (W) done by that force is equal to the net force (F) applied times the resulting displacement (s) of the object:

$$W = Fs.$$

Displacement is the absolute distance and direction an object has moved from its starting position, ignoring the path taken. Therefore, a car that drove in a perfect circle and stopped exactly where it started would have a displacement of zero, no matter how large the circle it traveled. Similarly, a car that drove ten miles east, made a U-turn, and drove back five miles west would only have a displacement of five miles east, even though it traveled fifteen miles total.

The equation for work reveals that in order for a force to have performed work on an object, the displacement of that object must have a nonzero value. In other words, the object has to have moved. If an applied force does not result in displacement, no work has been done.

SAMPLE PROBLEM

A seventy-kilogram sprinter completes a hundred-meter dash in ten seconds. As she crosses the finish line, her coach uses an infrared speedometer to measures her speed at twelve meters per second. How much work did the sprinter perform during the race?

Answer:

Start by making note of the given information: the sprinter's mass (70 kg), the distance traveled (100 m), the time taken to travel this distance (10 s), and the sprinter's velocity when she crossed the finish line (12 m/s). (Although velocity consists of both speed and direction of travel, and the speedometer only measured the sprinter's speed, it can be assumed that her direction of travel remained constant.) She would have started the race from a dead stop, so her initial velocity was 0 m/s. Recall that the work-energy theorem relates work not only to an object's kinetic energy but also to its mass and change in velocity:

$$W = \Delta K = \frac{1}{2} m v_f^2 - \frac{1}{2} m v_i^2$$

Because no information is provided about the sprinter's kinetic energy, the change in kinetic energy (ΔK) can be disregarded. Plug in the given information:

$$W = \frac{1}{2}(70 \text{ kg})(12 \text{ m/s})^2 - \frac{1}{2}(70 \text{ kg})(0 \text{ m/s})^2$$

The second element of the equation ($1/2 m v_i^2$) can also be ignored, because the initial velocity is zero, and any value multiplied by zero is zero. Simplify the equation and solve:

$$W = \frac{1}{2}(70 \text{ kg})(12 \text{ m/s})^2$$

$$W = \frac{1}{2}(70 \text{ kg})(144 \text{ m}^2/\text{s}^2)$$

$$W = \frac{1}{2}(10{,}080 \text{ kg} \cdot \text{m}^2/\text{s}^2)$$

$$W = 5{,}040 \text{ kg} \cdot \text{m}^2/\text{s}^2$$

The sprinter performed 5,040 kilogram–square meters per second squared (kg·m²/s²) of work. One kg·m²/s² is simply one joule, expressed in SI base units. Therefore, it took 5,040 J of work for the sprinter to accelerate her body from a standstill at the starting line to the 12 m/s she was traveling when she crossed the finish line.

Conservation of Energy

The law of conservation of energy states that in an isolated system, energy is conserved. An isolated system is one from which neither matter nor energy can escape. The universe is, in theory, the ultimate isolated system. Thus, according to this law, energy in the universe is never created or destroyed; it can only be transformed or transferred. The work-energy theorem is an extension of the law of conservation of energy, rewritten in a usable form.

Not all energy is kinetic energy. Energy can exist in a variety of forms. One such form is potential energy, which is energy that is stored in an object or system until it can be converted to another form of energy to do work. Potential energy itself comes in different forms, such as gravitational potential energy and chemical potential energy. The human body makes use of the chemical potential energy that exists in food due to its molecular configuration. When food is digested, it undergoes chemical reactions that break down its molecules and convert some of this chemical potential energy into the thermal energy of body heat and the kinetic energy of moving limbs and beating hearts. A combustion engine similarly converts the chemical potential energy of the fuel into the kinetic energy of the moving pistons that drive the engine.

The principle of conservation of energy is useful when examining any isolated system. Consider the billiards example again. The billiards table can be treated as an isolated system, because the balls stay on the table and any energy from the environment (heat from overhead lights, the kinetic energy in a gust of air) has such a small effect that it can be ignored. Therefore, when two balls collide, the total amount of energy in the system must remain the same before and after the impact. Kinetic energy is simply transferred from one ball to the other. A miniscule amount might be converted to thermal energy due to friction with the table.

Often, when considering some kinematic interaction, it is useful to know the total mechanical energy of the objects at play. The total mechanical energy of an object or system is simply the sum of its potential and kinetic energies. In the real world, total mechanical energy is not typically conserved, because friction must be taken into account. Consider a driver in a speeding car who suddenly slams on the brakes. As the car's tires stop rotating and start sliding across the surface of the pavement, they generate friction. Friction converts kinetic energy into thermal energy, which is not a form of mechanical energy. This energy then dissipates away from the tire tracks into the surrounding environment.

The Work-Energy Theorem in Everyday Life

The work-energy theorem is useful whenever the effect of work on the motion of an object is of interest. For example, understanding how the chemical potential energy in a fuel source performs work when it is released and converted into kinetic energy is an essential part of engineering efficient combustion engines. The fuel has to contain enough energy to move the pistons without breaking them.

Countless other devices in modern life also convert potential energy into kinetic energy in order to perform work. Everyday examples include vacuum cleaners, clocks, and fans. By expressing the relationship between energy transfer (work) and the motion (kinetic energy) of these objects in easy-to-measure terms (mass and velocity), the work-energy theorem makes engineering these devices possible.

—*Kenrick Vezina, MS*

Bibliography

Allain, Rhett. "What's the Difference between Work and Potential Energy?" *Wired*. Condé Nast, 1 July 2014. Web. 22 Sept. 2015.

Boleman, Michael. "Experiment # 6: Work-Energy Theorem." *Mr. Boleman's Course Information*. University of South Alabama, n.d. Web. 22 Sept. 2015.

"Energy, Kinetic Energy, Work, Dot Product, and Power." *MIT OpenCourseWare*. Mass. Inst. of Technology, 13 Oct. 2004. Web. 22 Sept. 2015.

Henderson, Tom. "Kinematics." *Physics Classroom*: Physics Classroom, 2013. Digital file.

Nave, Carl R. "Work, Energy and Power." *HyperPhysics*. Georgia State U, 2012. Web. 22 Sept. 2015.

Shankar, Ramamurti. "Lecture 5: Work-Energy Theorem and Law of Conservation of Energy." *Open Yale Courses*. Yale U, 2006. Web. 22 Sept. 2015.

Simanek, Donald E. "Kinematics." *Brief Course in Classical Mechanics*. Lock Haven U, Feb. 2005. Web. 22 Sept. 2015.

X-RAY RADIATION

FIELDS OF STUDY

Electromagnetism; Atomic Physics; Nuclear Physics

SUMMARY

X-ray radiation is a type of electromagnetic radiation with a wavelength between 0.1 and 10 nanometers. It falls between gamma rays (shorter wavelength, more energy) and ultraviolet rays (longer wavelength, less energy) on the electromagnetic spectrum. X-rays occur naturally both on Earth and in space. Artificially created x-rays are useful in physics and medicine, among other fields.

PRINCIPAL TERMS

- **bremsstrahlung:** the electromagnetic radiation released by charged subatomic particles when they hit another charged particle; the momentum lost in the collision is converted into energy transmitted as radiation.
- **Compton scattering:** the phenomenon that when electromagnetic radiation of very short wavelength (such as x-rays or gamma rays) hits an electron, it behaves like a particle hitting another particle and is deflected at an angle, which changes the wavelength of the radiation after the interaction.
- **electromagnetic spectrum:** the full range of electromagnetic radiation, sorted into segments with similar properties by wavelength; x-rays occupy one of these segments.
- **non-ionizing radiation:** electromagnetic radiation that lacks the energy necessary to knock electrons free when it hits an atom; all types of electromagnetic radiation with a longer wavelength (less energy) than ultraviolet is non-ionizing.
- **photoabsorption:** the absorption of the energy of electromagnetic radiation into matter; different substances absorb radiation at different rates.
- **Rayleigh scattering:** the scattering of electromagnetic radiation when it encounters particles much smaller than the radiation's wavelength, as, for example, the scattering of visible light in the atmosphere.
- **wavelength:** the distance between crests (or troughs) of a wave; the length of one complete cycle. Longer wavelengths correspond to lower frequencies and less energy, and vice versa.

X-Rays and the Electromagnetic Spectrum

Every type of electromagnetic radiation (EMR) is categorized on the electromagnetic (EM) spectrum by wavelength. Wavelength—the distance between two peaks or two troughs of a wave—is inversely proportional to the frequency of and energy transmitted by the wave. The EM spectrum is arranged from short wavelengths (high frequency or energy) to very long wavelengths (low frequency or energy). X-ray light falls between gamma radiation and ultraviolet light on the EM spectrum.

Gamma radiation has a wavelength of less than 0.01 nanometers. (One nanometer is 1×10^{-9} meters.) It comes from gamma decay, a type of radioactive decay. In gamma decay, a high-energy particle called a photon is ejected from the nucleus of an atom.

X-rays have wavelengths between 0.1 and 10 nanometers. High-energy, short-wavelength x-rays are called "hard x-rays." These belong to the same part of the EM spectrum as gamma rays but come from electrons, not atomic nuclei. X-rays are naturally produced by some radioactive elements, radon gas, and superhot gases in space. They are also created by devices like x-ray vacuum tubes.

Ultraviolet (UV) radiation has wavelengths of 10 to 400 nanometers. The most familiar source of UV rays is the sun. UV rays can cause skin to tan, burn, or even develop cancer.

The longer-wavelength forms of EMR include visible light (comprising the colors seen by the

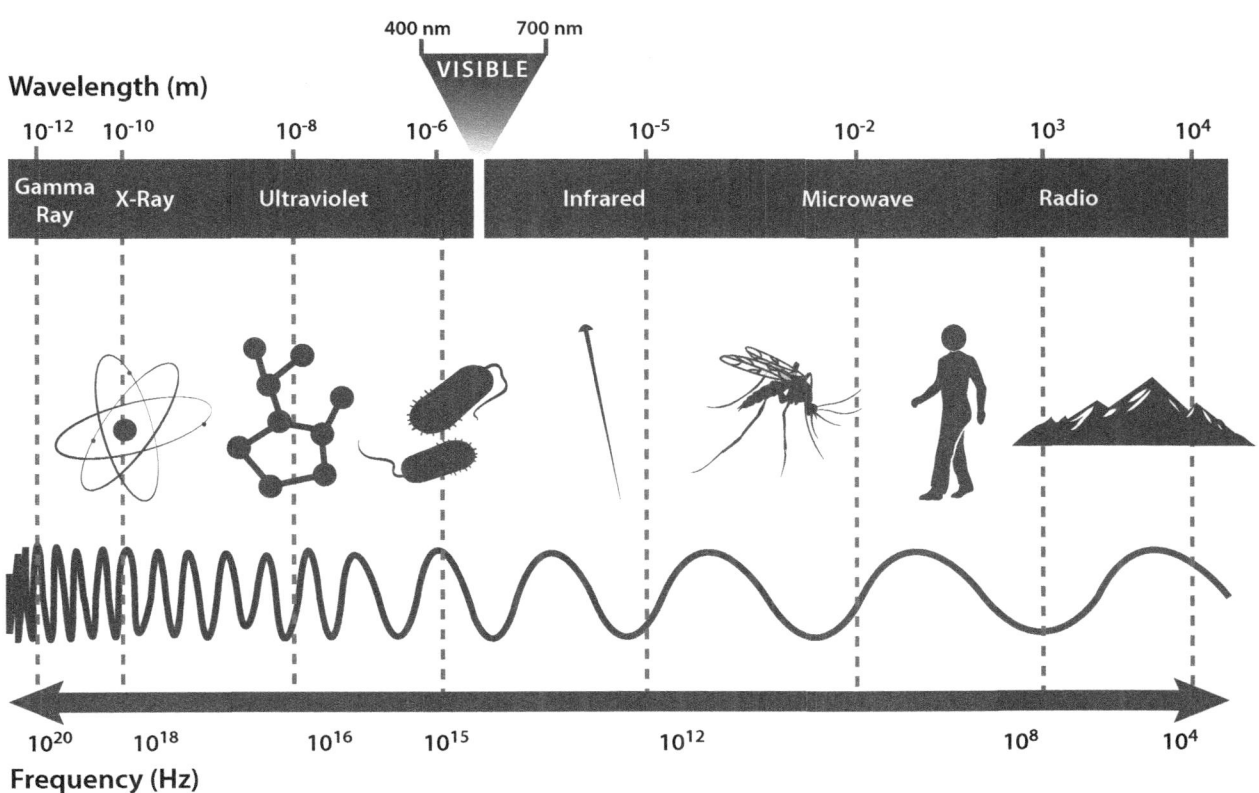

The electromagnetic spectrum is the range of all possible electromagnetic radiation wavelengths. X-rays fall between gamma rays and UV light. Wavelengths for x-rays are between 0.1 and 10 nanometers, around the size of atoms. Radio waves can be as large as mountains, and gamma radiation has a short wavelength, smaller than atoms.

human eye), infrared, radio and microwaves, and longwave radiation.

Ionizing versus Non-Ionizing Radiation

X-rays and gamma rays are often what come to mind when people talk about the dangers of radiation exposure. Both are types of ionizing radiation. Indeed, UV light and all higher-energy EMR is ionizing. This means it has enough energy that when it strikes an atom, it can knock electrons loose, altering the way the atom interacts with other nearby atoms. If ionizing radiation penetrates DNA, it can cause errors in cell replication that lead to cancer. Therefore, all ionizing radiation can cause cancer with enough exposure, though the risk varies greatly by type of radiation and type of exposure. Non-ionizing radiation—EMR with wavelengths longer than ultraviolet—can still be dangerous (for instance, infrared can cause burns), but it will not alter DNA structure.

Medical X-Rays

X-rays have short enough wavelengths to pass through many types of matter without much distortion yet can interact with certain materials, making them useful tools for probing the unseen. Photoabsorption refers to the transfer of energy from EMR photons to the electrons in a material upon impact. Bone has a higher photoabsorption rate than skin and soft tissues do. Thus, x-rays pass through the human body mostly unobstructed except when they hit bone or other calcium-heavy areas. This is how medical x-rays work. To protect sensitive tissues from DNA damage, patients wear lead-lined aprons during medical x-ray scans because lead is impervious to x-rays.

X-rays of the kind used in x-ray machines are often produced by an x-ray tube, which uses the phenomenon of bremsstrahlung to generate the radiation. Bremsstrahlung, "braking radiation" in German, is the radiation released when charged subatomic particles like electrons slow down quickly after contact with atomic nuclei. X-ray tubes use electric fields to fire electrons into a metal at high speed, causing bremsstrahlung x-rays to be emitted.

X-RAYS AND THE SUBATOMIC WORLD

Two important types of scattering may occur when x-rays encounter subatomic particles like those in the atmosphere. Rayleigh scattering occurs when EMR encounters particles much smaller than its wavelength, causing it to change course without losing energy or being absorbed. It is not a significant obstacle to x-rays due to x-rays' short wavelengths. Compton scattering occurs when x-rays hit an electron. The electron is knocked free of its atom, and the x-ray loses a photon of energy and its wavelength increases. The discovery and description of Compton scattering using x-rays by American physicist Arthur H. Compton (1892–1962) in the 1920s proved that EMR can behave like a particle as well as a wave. This wave-particle duality of EMR is key to modern quantum mechanics, a branch of physics that studies subatomic particles.

USES OF X-RAYS

X-rays are present in the cosmic rays that hit Earth from space, generated by super-hot gases like those in the sun. Sources of x-rays are relatively rare on Earth, though they can be produced by radioactive elements and radon gas. Most often, the x-rays that humans encounter come from humanmade devices like x-ray tubes used in medical detection or targeted cancer treatment or scanners used for airport security checkpoints. Scientists also use x-rays to study the chemical makeup of materials from crystals to stars and other celestial bodies.

—*Kenrick Vezina, MS*

BIBLIOGRAPHY

Lucas, Jim. "What Are X-Rays?" *Live Science*. Purch, 12 Mar. 2015. Web. 8 June 2015.

"Medical X-Ray Imaging." FDA. US Food and Drug Administration, 19 May 2015. Web. 8 June 2015.

Nave, Carl R. "Compton Scattering." *HyperPhysics*. Dept. of Astronomy and Physics, Georgia State U, 2012. Web. 8 June 2015.

Phillips, Melba. "Electromagnetic Radiation." *Encyclopaedia Britannica*. Encyclopaedia Britannica, 26 Nov. 2014. Web. 8 June 2015.

"X-Rays." Medline Plus. National Institutes of Health, US Natl. Lib. of Medicine, 24 Sept. 2013. Web. 8 June 2015.

"X-Rays." *Mission: Science*. NASA, 2010. Web. 8 June 2015.

X-Rays, Gamma Rays, and Cancer Risk. Amer. Cancer Soc., 24 Feb. 2015. Web. 8 June 2015.

Appendices

THE STANDARD MODEL

The *Periodic Table of the Elements* was first proposed by Dmitri Mendeleev in 1869. It provided a systematic basis for understanding the chemical elements and their compounds. With the discovery of the electron and the development of quantum mechanics half a century later, the electron theory of matter provided a coherent explanation of how the electrical force was responsible for the properties of atoms and their combinations—molecules. The electron theory of matter seemed to explain everything outside the nucleus of the atom. In the words of Nobel Laureate Richard Feynman, "Outside the nucleus we seem to know everything." (As it turned out Feynman's observation was a bit premature; neutrinos are not massless as was thought in the 1960's. When we view them as they travel through space, we can see the effects of their rest mass and a new type of field is needed to explain the origins of that mass. Gravitational waves have only just been detected, but the gravitational force is the weakest by far, so new developments may yet be waiting.)

There were still more questions. What was going on inside the nucleus, that made protons and neutrons stick to each other? What about those strange antiparticles that matched normal matter in almost every respect but charge? Were the proton and neutron all that elementary or did they rattle when colliding with another particle, suggesting another layer of structure? What of the unusual particles created in those collisions? Within a few years of experimentation the number of new short-lived particles outnumbered the known chemical elements. Was there a set of underlying rules?

The answers to those questions constitute the standard model of particle physics. The standard model has reached a level of maturity culminating in the discovery of the Higgs boson on July 4, 2012, but still leaves some unanswered questions: Are gravity and electromagnetism related in some way? Einstein thought so. Why is there so much normal matter in the universe and so little antimatter? Exactly how does the nuclear weak force relate to the nuclear strong force?

The basic ingredients of the standard model are: (1) a common way of looking at the interactions among elementary particles, and (2) a system for classifying the particles based on symmetry and the quantum concept of spin.

Diagramming the Interaction of Particles

To begin a discussion of the standard model, it is helpful to create a mental picture or model of interacting particles. Consider two children on skateboards tossing a ball back and forth. The ball is one of the particles that carry the force that one nuclear particle exerts on another and the children are the nuclear particles that interact via the fundamental forces.

Historically the picture became clearest first for electrons and positrons interacting by the electromagnetic force. The positron, or anti-electron, was predicted by Paul Dirac in his search for a version of the Schrödinger equation that would be invariant under the Lorentz transformations of special relativity. Dirac found that the simplest equation that met the requirements of special relativity required that the wave function be a four component vector, two of the components describing the two allowed directions of spin for the electron and the other two describing the two allowed directions of spin for the antiparticle. The anti-electron or positron was found by 1932 in cosmic ray showers and soon thereafter in radioactive decays of certain isotopes.

A comprehensive theory of quantum electrodynamics (or QED) was worked out in the immediate postwar years by American physicists Richard P. Feynman and Julian Schwinger and the Japanese Sin-Itiro Tomonaga who shared the 1965 Nobel Prize in Physics for their work. While the theory can be cast in many mathematically equivalent forms Feynman pointed out a particularly appealing form that permitted an easy interpretation, as the sum of readily visualized scattering event, with each event representable by a diagram. One could in fact dispense with much of the mathematics and focus on the Feynman diagrams.

Experiments with the hadron family were next to yield to diagrammatic analysis. The lowest mass members of the hadron family were the proton and neutron. The remaining members have very short lifetimes and, generally, could only be made in the very powerful particle accelerators built in the 1950's and later. As the experimental data accumulated, it became clear that all hadrons were composed of three smaller particles, which would be called quarks.

To explain why the groupings of three could not be separated, the quarks were assigned "colors", the anti-quarks complimentary "colors." Only combinations of quarks that were effectively white could be detected. Quarks exerted forces on each other by exchanging "mesons," which were combinations of a quark and an antiquark of the complementary color. In a nucleus the attractive force between protons and neutrons resulted from the exchange of mesons. Within each hadron the quarks were bound together by the exchange of "gluons," so that the attractive force between nucleons was in fact a consequence of exchange of force-carrying particles as well.

INTERACTIONS: FOUR FORCES

Presentations of modern physics has reduce the number of fundamental forces in the observable universe to the following four:
- gravitational force
- electromagnetic force
- weak nuclear force
- strong nuclear force

The gravitational force is one of two long-range forces. It is by far the weaker of the two, being some forty orders of magnitude weaker than the electrical force between elementary particles bearing charges. It is however the only force of significance over astronomical distances. It will be discussed at length in the third book in this series, *Principles of Astronomy*.

The electromagnetic force and the weak nuclear force are the only forces (other than gravity) felt by the lowest mass family of particles.

PARTICLE FAMILIES: LEPTONS

The positron, electron-sized antiparticle to the electron, was predicted by Paul Dirac in 1930 and discovered by 1932. The muon, predicted by Hideki Yukawa in 1935 to explain the strong force between nucleons, was discovered in 1936 and turned out not to be involved in the strong force at all, but was rather a more massive version of the electron. The tau particle was discovered in the mid 1970's.

The electron, muon and tau particle each have charge (particles have a –e charge and antiparticles have +e) and they each have been assigned corresponding neutrinos and antineutrinos: $\overline{\nu_e}, \overline{\nu_\mu},$ and $\overline{\nu_\tau}$.

Generation	Lepton	Q	L_e	L_m	L_t
1	E	-1	1	0	0
	n_e	0	1	0	0
2	m	-1	0	1	0
	n_m	0	0	1	0
3	t	-1	0	0	1
	n_t	0	0	0	1

All leptons have spin (h/4ϖ).

Here Q is the charge in units of the elementary charge e, and the L's are quantum numbers. There are also 6 antileptons with opposite charge and negative quantum numbers. It appears to be a law of physics that the quantum numbers must add up on either side of an equation. Thus when a neutron decays by the weak interaction,

$$n \rightarrow p + e + \overline{\nu_e} \ \overline{\nu_e}.$$

An electron antineutrino must appear on the right hand side to conserve the electron lepton number.

PARTICLE FAMILIES: QUARKS

As time went on, physicists were increasingly convinced that the number of types of quarks should parallel the number of members of the lepton family. By analogy, but also summarizing a wealth of experimentation, we believe that the more massive nuclear particles are composed of three quarks each. We assign each quark a flavor (arbitrarily named) and we have the following:

Generation	Flavor	Charge
1	d (down)	-1/3
	u (up)	2/3
2	s (strange)	-2/3
	c (charm)	3/3
3	t (top)	-1/3
	b (bottom)	2/3

In addition, there are six antiquarks: anti-up, anti-down, and so on. At any instant the quarks and antiquarks are assigned "colors," using the primary colors—red, green and blue—for the quarks and their complementary colors—cyan, magenta and yellow—for the antiquarks. Observable combinations of quarks must be the equivalent of "white," and so we have a combination of three primary colors for the quarks that make up colors for the quarks that make up antimatter, and a combination of a normal quark and a complementary-colored quark for the mesons that carry the nuclear force.

Protons are two up quarks and a down quark, held together by even smaller particles called gluons. Neutrons are two down quarks and an up quark held together by gluons. The strong force that protons and neutrons exert on each other is carried by the three ϖ mesons, ϖ^+, ϖ^0, and ϖ^-, where the ϖ^+ meson combines an up quark with an anti-down so the proton becomes a neutron and vice versa.

Spin ½ combinations of three quarks constitute an octet of elementary particles of which only the proton and neutron have appreciable lifetimes. There are ten spin 3/2 combinations, all of which have very short lifetimes.

Mesons are bosons that combine quarks and antiquark. There are numerous combinations with spin 0 and spin 1. (Ambitious students who know a little mathematics are referred to *Introduction to Elementary Particles* by Griffiths for further detail.)

Before discussing the weak force, a little more needs to be said about spin and symmetry.

SPIN AND STATISTICS: A CLOSER LOOK
Perhaps the key discovery that makes subatomic physics so different from the ordinary physics of bricks and cables is that of the intrinsic angular momentum of elementary particles and the fact that this angular momentum is quantized. Niels Bohr seems to be the first individual to invoke quantization of angular momentum in his model of the hydrogen atom. In the Bohr model the angular momentum was set equal to $nh/2\varpi$ for the nth energy level. While the Bohr model was incorrect in many respects, it agreed with the observed energy levels for the hydrogen atom exactly. In the famous Stern-Gerlach experiment it was shown that for a beam of sodium atoms passing through an inhomogeneous magnetic field, the z-component of angular momentum was +- h/ϖ, a value that could not be explained in classical terms. The Stern-Gerlach experiment convinced physicists that particles could carry their own angular momentum and that this intrinsic, or "spin," angular momentum combined with the orbital angular momentum according to the quantum rules.

The Pauli exclusion principle states that no two electrons in the same atom can have the same four quantum numbers. A deeper form of the Pauli principle states that the wave function, Ψ, of any quantum mechanical system must change into itself or minus itself whenever the coordinates of identical particles are interchanged. This is because the probability of finding the particle at a given location depends on $|\Psi|^2$ at that location and in quantum mechanics, identical particles are truly indistinguishable. Particles whose spin angular momentum is an even multiple of $h/4\varpi$ are called bosons and obey Bose-Einstein statistics. Particles whose spin angular momentum is an odd multiple of $h/4\varpi$ are called fermions and obey Fermi-Dirac statistics. The proton, neutron and electron are each fermions, the photons (quanta of light energy) are bosons.

Fermions intrinsically avoid each other. The probability of finding two electrons of the same type in the same location and with spins in the same direction is rigorously zero. Sometimes this tendency of fermions to avoid each other is called exchange repulsion. The reason ordinary objects cannot interpenetrate each other is their exchange repulsion. On the other hand, there is no limitation to the number of bosons that can occupy the same quantum state. As the temperature is lowered, a system of bosons will all occupy the lowest quantum state, This phenomenon is known as a Bose condensation. It is responsible for laser beams, super conduction, and for pairs of electrons behaving as bosons, the properties of superfluid ^{4}H, and so on.

GAUGE PARTICLES: CARRIERS OF THE ELECTROMAGNETIC AND WEAK NUCLEAR FORCES
The modern theory of electromagnetism and light considers each charged particle to be surrounded by a cloud of virtual photons. While it is tempting to view two electrons repelling each other by a repulsive force acting at a distance, it is equally as correct to view the particles as surrounded by clouds of virtual photons which carry the force from one charged particle to another. These photons become real when one of the electrons is accelerated in a collision.

The nature of electromagnetism is such that the electric and magnetic fields are not uniquely determined by the electric charges and currents that are responsible for them, but in fact there is a certain flexibility: one can add terms to the electric potential, providing one adds a corresponding correction to the potential that describes the magnetic field. This sort of invariance of the physics under a mathematical transformation is called a gauge invariance. It can be shown that if a gauge invariance is found, a corresponding conservation law for a physical quantity will also be found. The gauge invariance of the electromagnetic field is the conservation of electrical charge.

THE ELECTROWEAK FORCE

For a number of reasons it is convenient to view the weak force, and indeed the electromagnetic force, as involving the exchange of four bosons: the photon, and a triple W^+, W^- and Z^0—the intermediate vector bosons. The intermediate vector bosons have masses much larger than the proton mass and exist for very short times, thus the short range of the weak force.

Among the unanswered questions in the standard model is the origin of mass for the W^+, Z^0 and W^- particles that carry the weak interaction. One possible solution was proposed by Higgs, who suggested that there was yet another field with unique properties, now universally known as the Higgs field. This field, unlike the electromagnetic field, was always positive. The Higgs boson, discovered on July 4, 2012, seemed to fill the bill. The next few years may well determine whether or not this discovery could represent the completion of the standard model.

NOBEL NOTES: SUB-ATOMIC AND SUB-NUCLEAR PHYSICS

The committees charged with selecting the Nobel winners in physics and chemistry have given particular attention to atomic processes and to the radiation which we now associate with nuclear transitions. The very first Nobel Prize in Physics in 1901 went to Conrad Röntgen for the discovery of x-rays. X-rays result from electrons in the inner shells of atoms. In 1903 Antoine Henri Becquerel, Marie Curie, and Pierre Curie were cited for the discovery of natural radioactivity in uranium salts. J. J. Thompson was honored in 1905 for the discovery of the electron. In 1908, Ernest Rutherford, who would shortly discover proof of the existence of the atomic nucleus, was awarded the Nobel Prize for Chemistry. In 1911, it was Marie Curie's work in the chemistry of radio-isotopes that received the Nobel Prize for Chemistry.

The Thirty Years that Shook Physics

In 1921 Albert Einstein won the Nobel Prize for his explanation of the photoelectric effect; his work on the theory of relativity was still considered too controversial for an award. (Interestingly, although this award was not based on his remarkable theory of relativity, it was a foregone conclusion that Albert Einstein would win the Nobel Prize, and thus, the money from the award was included in his divorce settlement with Mileva Marić in 1919.) Einstein had by that time worked out his general theory of relativity, confirmed by a British expedition to South Africa in 1919 where they measured gravitational shifts in the apparent positions of the stars during a solar eclipse. Confirmation of the theory also came from its ability to explain the quite small precession of the perihelion of the planet Mercury. The most dramatic prediction of Einstein's general relativity, the existence of gravitational radiation, would have to wait until 2015 to achieve observational confirmation.

In the 1920s and 1930s, the emphasis for the awards was placed an improved theoretical understanding of the behavior of matter on the microscale and the growing number of elementary particles. The 1918 physics prize went to Max Planck, who had started the quantum revolution by assuming that the energy of electromagnetic radiation energy was quantized. The 1922 prize went to Niels Bohr, who found that he could predict the energy levels of the hydrogen atom (and 1-electron ions) but not of atoms with more than one electron. The 1923 prize went to an American, Robert Mulliken, for determining the size of the electron's charge. By 1929, a workable solution for many electron atoms was found, beginning with the work of Prince Louis-Victor Pierre Raymond de Broglie, which established that the wavelength for material particles was inversely proportional to their momentum. That de Broglie's work was sound was established eight years later by the Nobel Prize-winning work of Davisson and Thomson at Bell Labs (1937) when they diffracted electrons in a manner parallel to that of Sir William Henry Bragg and William Lawrence Bragg (Nobel Prize, 1915). This work fully established the wave–particle duality that had been predicted by Einstein nearly a decade earlier. Quantitative predictions of all atomic energy levels were made possible by the work of Heisenberg, for which he received the Nobel Prize in 1932. An alternative formulation was discovered by Schrödinger at about the same time. The existence of two such different mathematical approaches was puzzling for a time, but then Paul Adrian Maurice Dirac was able to show that the two approaches were in fact equivalent, despite apparent differences. Dirac and Schrödinger were both honored for their efforts by the Nobel Prize for Physics in 1933.

Physicist George Gamow has labeled the period from 1900-1930 as "the thirty years that shook physics." Physics research tended to concentrate at a few centers in Europe: at Cambridge University in England, at Gottingen in Germany and at Neil's Bohr's Institute in Denmark. Anyone serious about learning physics, and with the financial means to do so, went to one of these three locations. The international language of physics was referred to as "broken" German. By the end of the Second World War, however, things had changed dramatically. As Europe was slowly rebuilding, the United States came to play a significant role in physics research, and the international language of science had become a type of "broken" English. Dutch scientist C. C. G. Gorter went as far as to suggest formalizing the grammar of "broken" as an aid to scientific communication.

THE DIFFERENCE BETWEEN PHYSICISTS AND "MERE MORTALS"

The leading physicists of the time might be thought of as rare creatures of the mind, incapable of existing in the same realm as mere mortals. To some extent, their behaviors bore out this suspicion.

Paul Dirac, for example, was noted for his extreme taciturnity. For some unknown reason, his father had insisted that conversations within the home be conducted in French. Exactly how this affected the young Paul, one can only speculate. Dirac eventually retired to the United States to become a professor at the University of Florida where he was said to enjoy the climate as well as attempting to stimulate his students' interest in physics. His students, however, all of whom had all purchased his classic *Principles of Quantum Mechanics*, wanted to know why his classes consisted of Dirac essentially writing the book out for them on the blackboard. His response was simply that it had taken him many years to figure out how to present the material properly, and there was no reason to change now.

J. Robert Oppenheimer came from a rather prosperous New York family and would eventually be named leader of the Manhattan project. Although he did not win the Nobel Prize himself, he clearly assumed a leadership position within the physics community in the United States. He got his education at University of Göttingen in Germany. All went well until it was time to graduate, when it was discovered that he had never officially enrolled at the university. Only the intervention of his advisor, Professor Max Born, who would himself be awarded the Nobel Prize in 1954, prevented his having to return to the United States without a diploma. Oppenheimer was a unique research director, with awesome responsibilities but no formal training in management. In the recent biography of Apple, Inc.'s co-founder and chief executive, Steve Jobs, written by Walter Isaacson, Oppenheimer was identified as one of Job's role models.

Wolfgang Pauli was yet another physicist about whom his colleagues had plenty of stories to tell. According to the "folk wisdom" surrounding physicist, the general pattern is that a physicist tends to specialize in either theory or experiment; the corollary to that belief is that letting a theorist into one's laboratory is an invitation to disaster. The greater the "power" of theoretical physicist, the greater the risk. Wolfgang Pauli, one of the greatest theorists in modern physics, got blamed for more than one situation in which an experimental apparatus broke down suddenly, unexpectedly, and without explanation. Experimental physicists took to calling this the "Pauli effect." George Gamow recounts in particular the story of two physicists, in two different cities in Europe, discussing by phone the fact that a particular apparatus had suddenly broken down in a third city. They later learned that, on the day of the breakdown, Pauli had been in the same city on that particular day, albeit only briefly while changing trains.

Richard Feynman, is probably the best known American-born physicist. No doubt at least some of his fame came about because of exploits at the Los Alamos Laboratory during the Second World War like this one: Many of the scientific staff were been given safes to protect their critical data. To the consternation of the security personnel, Feynman developed a reputation for breaking into his colleagues' safes and leaving drawings of a smiley face with the caption "Guess who?" Feynman never fit the university professor stereotype and was proud that he did not. He declined several honorary doctorates, including one from Princeton University, his alma mater. He also declined election to the National Academy of Sciences. When word reached him that he had won the Nobel Prize, his first impulse was to decline it, but then realized he would gain even more notoriety as "the man who turned down the Nobel," so he accepted. As an author later in life, his memoirs found their way onto the best-seller lists. Following his career as author, he developed an interest in Tannu Touva, a tiny Asian country, which had been absorbed into the then Soviet Union. By sheer persistence he and his friends made contact with some of the few Touvans who understood written English, and thus discovered the rich culture of the Asian steppes. Eventually he and a few of his friends got approval to organize a program and exhibit at the Los Angeles County Museum on the *Nomads of the Steppes*. Alas, Feynman succumbed to cancer before he could visit Touva himself.

ELEMENTARY PARTICLE PHYSICS

A number of Nobel Prizes commemorate developments in elementary particle physics. C. T. R. Wilson received part of the 1927 Prize for the invention of the cloud chamber that was used by Carl Anderson in the discovery of the positron, leading to his Nobel Prize award in 1936. In 1960, the Nobel Prize in Physics went to Donald Glaser of the University of California, Berkeley, for the invention of the bubble chamber, which allows particle physicists to photograph the bubbles trails left in supercooled liquid helium after the passage of a charged particle. Glaser claimed that he got the idea watching bubbles rise in the glass of beer that he was drinking. Indeed, the first bubble chamber was constructed using beer as a working fluid.

Particle Accelerators

The building of particle accelerators also got some attention from the Nobel committee. The first particle accelerator was the Van de Graff Generator in which a static charge could be generated by friction or other means and transferred to a nearly closed conductor. The charge would distribute itself on the outside of a sphere so as to minimize the electric field at the interior and allow charge accumulation until the atmosphere began to break down.

The next advance in particle acceleration occurred with the development of the cyclotron by E. O. Lawrence, who was honored with the Nobel Prize in 1939. The cyclotron took advantage of the fact that the orbital period of a charged particle, moving at right angles to a static magnetic field, did not depend on its velocity. Thus the particle could be accelerated by the same potential difference a great many times. One can achieve even higher energies in a synchrotron in which the magnetic field changes over time to compensate for the effects of relativity. In 1959 the Nobel Prize for Physics went to Emilio Segre and Owen Chamberlain for their creation of antiprotons.

In physics, experimental results often lead to modifications of theory and modifications of theory often lead to experimental results, such as the discovery of the positron, mu meson and antiproton. In 1969, Murray Gell-Man received the Nobel award for the quark model, a theory which allowed for six quarks and could account for all the hadrons and muons known at the time. To allow for an equal number of lepton flavors, theorists Sheldon Glashow, Abdus Salam and Stephen Weinberg proposed in 1979 that the electromagnetic and weak forces could be manifestations of a single electroweak force with four boson carriers: the photon, the neutral Z particle and the W+ and W- bosons. Then, in 1983 Carlo Rubia and Simon van der Meer discovered these particles experimentally. In so doing they solved a long-standing astrophysical problem, illustrated in the third volume in this series: *Principles of Astronomy*.

NOBEL LAUREATES

The Nobel Prize in Physics has been awarded 109 times to 201 Nobel Laureates between 1901 and 2015. John Bardeen is the only Nobel Laureate who has been awarded twice, in 1956 and 1972. This means that a total of 200 individuals have received the Nobel Prize in Physics.

2015	Takaaki Kajita and Arthur B. McDonald	"for the discovery of neutrino oscillations, which shows that neutrinos have mass"
2014	Isamu Akasaki, Hiroshi Amano and Shuji Nakamura	"for the invention of efficient blue light-emitting diodes which has enabled bright and energy-saving white light sources"
2013	François Englert and Peter W. Higgs	"for the theoretical discovery of a mechanism that contributes to our understanding of the origin of mass of subatomic particles, and which recently was confirmed through the discovery of the predicted fundamental particle, by the ATLAS and CMS experiments at CERN's Large Hadron Collider"
2012	Serge Haroche and David J. Wineland	"for ground-breaking experimental methods that enable measuring and manipulation of individual quantum systems"
2011	Saul Perlmutter, Brian P. Schmidt and Adam G. Riess	"for the discovery of the accelerating expansion of the Universe through observations of distant supernovae"
2010	Andre Geim and Konstantin Novoselov	"for groundbreaking experiments regarding the two-dimensional material graphene"
2009	Charles Kuen Kao	"for groundbreaking achievements concerning the transmission of light in fibers for optical communication"
	Willard S. Boyle and George E. Smith	"for the invention of an imaging semiconductor circuit - the CCD sensor"
2008	Yoichiro Nambu	"for the discovery of the mechanism of spontaneous broken symmetry in subatomic physics"
	Makoto Kobayashi and Toshihide Maskawa	"for the discovery of the origin of the broken symmetry which predicts the existence of at least three families of quarks in nature"

354

2007	Albert Fert and Peter Grünberg	"for the discovery of Giant Magnetoresistance"
2006	John C. Mather and George F. Smoot	"for their discovery of the blackbody form and anisotropy of the cosmic microwave background radiation"
2005	Roy J. Glauber	"for his contribution to the quantum theory of optical coherence"
	John L. Hall and Theodor W. Hänsch	"for their contributions to the development of laser-based precision spectroscopy, including the optical frequency comb technique"
2004	David J. Gross, H. David Politzer and Frank Wilczek	"for the discovery of asymptotic freedom in the theory of the strong interaction"
2003	Alexei A. Abrikosov, Vitaly L. Ginzburg and Anthony J. Leggett	"for pioneering contributions to the theory of superconductors and superfluids"
2002	Raymond Davis Jr. and Masatoshi Koshiba	"for pioneering contributions to astrophysics, in particular for the detection of cosmic neutrinos"
	Riccardo Giacconi	"for pioneering contributions to astrophysics, which have led to the discovery of cosmic X-ray sources"
2001	Eric A. Cornell, Wolfgang Ketterle and Carl E. Wieman	"for the achievement of Bose-Einstein condensation in dilute gases of alkali atoms, and for early fundamental studies of the properties of the condensates"
2000		"for basic work on information and communication technology"
	Zhores I. Alferov and Herbert Kroemer	"for developing semiconductor heterostructures used in high-speed- and opto-electronics"
	Jack S. Kilby	"for his part in the invention of the integrated circuit"
1999	Gerardus 't Hooft and Martinus J.G. Veltman	"for elucidating the quantum structure of electroweak interactions in physics"
1998	Robert B. Laughlin, Horst L. Störmer and Daniel C. Tsui	"for their discovery of a new form of quantum fluid with fractionally charged excitations"

1997	Steven Chu, Claude Cohen-Tannoudji and William D. Phillips	"for development of methods to cool and trap atoms with laser light
1996	David M. Lee, Douglas D. Osheroff and Robert C. Richardson	"for their discovery of superfluidity in helium-3
1995		"for pioneering experimental contributions to lepton physics
	Martin L. Perl	"for the discovery of the tau lepton
	Frederick Reines	"for the detection of the neutrino
1994		"for pioneering contributions to the development of neutron scattering techniques for studies of condensed matter
	Bertram N. Brockhouse	"for the development of neutron spectroscopy
	Clifford G. Shull	"for the development of the neutron diffraction technique
1993	Russell A. Hulse and Joseph H. Taylor Jr.	"for the discovery of a new type of pulsar, a discovery that has opened up new possibilities for the study of gravitation
1992	Georges Charpak	"for his invention and development of particle detectors, in particular the multiwire proportional chamber
1991	Pierre-Gilles de Gennes	"for discovering that methods developed for studying order phenomena in simple systems can be generalized to more complex forms of matter, in particular to liquid crystals and polymers
1990	Jerome I. Friedman, Henry W. Kendall and Richard E. Taylor	"for their pioneering investigations concerning deep inelastic scattering of electrons on protons and bound neutrons, which have been of essential importance for the development of the quark model in particle physics
1989	Norman F. Ramsey	"for the invention of the separated oscillatory fields method and its use in the hydrogen maser and other atomic clocks
	Hans G. Dehmelt and Wolfgang Paul	"for the development of the ion trap technique
1988	Leon M. Lederman, Melvin Schwartz and Jack Steinberger	"for the neutrino beam method and the demonstration of the doublet structure of the leptons through the discovery of the muon neutrino
1987	J. Georg Bednorz and K. Alexander Müller	"for their important break-through in the discovery of superconductivity in ceramic materials

1986	Ernst Ruska	"for his fundamental work in electron optics, and for the design of the first electron microscope
	Gerd Binnig and Heinrich Rohrer	"for their design of the scanning tunneling microscope
1985	Klaus von Klitzing	"for the discovery of the quantized Hall effect
1984	Carlo Rubbia and Simon van der Meer	"for their decisive contributions to the large project, which led to the discovery of the field particles W and Z, communicators of weak interaction
1983	Subramanyan Chandrasekhar	"for his theoretical studies of the physical processes of importance to the structure and evolution of the stars
	William Alfred Fowler	"for his theoretical and experimental studies of the nuclear reactions of importance in the formation of the chemical elements in the universe
1982	Kenneth G. Wilson	"for his theory for critical phenomena in connection with phase transitions
1981	Nicolaas Bloembergen and Arthur Leonard Schawlow	"for their contribution to the development of laser spectroscopy
	Kai M. Siegbahn	"for his contribution to the development of high-resolution electron spectroscopy
1980	James Watson Cronin and Val Logsdon Fitch	"for the discovery of violations of fundamental symmetry principles in the decay of neutral K-mesons
1979	Sheldon Lee Glashow, Abdus Salam and Steven Weinberg	"for their contributions to the theory of the unified weak and electromagnetic interaction between elementary particles, including, inter alia, the prediction of the weak neutral current
1978	Pyotr Leonidovich Kapitsa	"for his basic inventions and discoveries in the area of low-temperature physics
	Arno Allan Penzias and Robert Woodrow Wilson	"for their discovery of cosmic microwave background radiation
1977	Philip Warren Anderson, Sir Nevill Francis Mott and John Hasbrouck van Vleck	"for their fundamental theoretical investigations of the electronic structure of magnetic and disordered systems
1976	Burton Richter and Samuel Chao Chung Ting	"for their pioneering work in the discovery of a heavy elementary particle of a new kind

Year	Laureate(s)	Citation
1975	Aage Niels Bohr, Ben Roy Mottelson and Leo James Rainwater	"for the discovery of the connection between collective motion and particle motion in atomic nuclei and the development of the theory of the structure of the atomic nucleus based on this connection"
1974	Sir Martin Ryle and Antony Hewish	"for their pioneering research in radio astrophysics: Ryle for his observations and inventions, in particular of the aperture synthesis technique, and Hewish for his decisive role in the discovery of pulsars"
1973	Leo Esaki and Ivar Giaever	"for their experimental discoveries regarding tunneling phenomena in semiconductors and superconductors, respectively"
	Brian David Josephson	"for his theoretical predictions of the properties of a supercurrent through a tunnel barrier, in particular those phenomena which are generally known as the Josephson effects"
1972	John Bardeen, Leon Neil Cooper and John Robert Schrieffer	"for their jointly developed theory of superconductivity, usually called the BCS-theory"
1971	Dennis Gabor	"for his invention and development of the holographic method"
1970	Hannes Olof Gösta Alfvén	"for fundamental work and discoveries in magnetohydro-dynamics with fruitful applications in different parts of plasma physics"
	Louis Eugène Félix Néel	"for fundamental work and discoveries concerning antiferromagnetism and ferrimagnetism which have led to important applications in solid state physics"
1969	Murray Gell-Mann	"for his contributions and discoveries concerning the classification of elementary particles and their interactions"
1968	Luis Walter Alvarez	"for his decisive contributions to elementary particle physics, in particular the discovery of a large number of resonance states, made possible through his development of the technique of using hydrogen bubble chamber and data analysis"
1967	Hans Albrecht Bethe	"for his contributions to the theory of nuclear reactions, especially his discoveries concerning the energy production in stars"
1966	Alfred Kastler	"for the discovery and development of optical methods for studying Hertzian resonances in atoms"
1965	Sin-Itiro Tomonaga, Julian Schwinger and Richard P. Feynman	"for their fundamental work in quantum electrodynamics, with deep-ploughing consequences for the physics of elementary particles"

1964	Charles Hard Townes, Nicolay Gennadiyevich Basov and Aleksandr Mikhailovich Prokhorov	"for fundamental work in the field of quantum electronics, which has led to the construction of oscillators and amplifiers based on the maser-laser principle
1963	Eugene Paul Wigner	"for his contributions to the theory of the atomic nucleus and the elementary particles, particularly through the discovery and application of fundamental symmetry principles
	Maria Goeppert Mayer and J. Hans D. Jensen	"for their discoveries concerning nuclear shell structure
1962	Lev Davidovich Landau	"for his pioneering theories for condensed matter, especially liquid helium
1961	Robert Hofstadter	"for his pioneering studies of electron scattering in atomic nuclei and for his thereby achieved discoveries concerning the structure of the nucleons
	Rudolf Ludwig Mössbauer	"for his researches concerning the resonance absorption of gamma radiation and his discovery in this connection of the effect which bears his name
1960	Donald Arthur Glaser	"for the invention of the bubble chamber
1959	Emilio Gino Segrè and Owen Chamberlain	"for their discovery of the antiproton
1958	Pavel Alekseyevich Cherenkov, Il'ja Mikhailovich Frank and Igor Yevgenyevich Tamm	"for the discovery and the interpretation of the Cherenkov effect
1957	Chen Ning Yang and Tsung-Dao (T.D.) Lee	"for their penetrating investigation of the so-called parity laws which has led to important discoveries regarding the elementary particles
1956	William Bradford Shockley, John Bardeen and Walter Houser Brattain	"for their researches on semiconductors and their discovery of the transistor effect
1955	Willis Eugene Lamb	"for his discoveries concerning the fine structure of the hydrogen spectrum
	Polykarp Kusch	"for his precision determination of the magnetic moment of the electron

1954	Max Born	"for his fundamental research in quantum mechanics, especially for his statistical interpretation of the wavefunction"
	Walther Bothe	"for the coincidence method and his discoveries made therewith"
1953	Frits Zernike	"for his demonstration of the phase contrast method, especially for his invention of the phase contrast microscope"
1952	Felix Bloch and Edward Mills Purcell	"for their development of new methods for nuclear magnetic precision measurements and discoveries in connection therewith"
1951	Sir John Douglas Cockcroft and Ernest Thomas Sinton Walton	"for their pioneer work on the transmutation of atomic nuclei by artificially accelerated atomic particles"
1950	Cecil Frank Powell	"for his development of the photographic method of studying nuclear processes and his discoveries regarding mesons made with this method"
1949	Hideki Yukawa	"for his prediction of the existence of mesons on the basis of theoretical work on nuclear forces"
1948	Patrick Maynard Stuart Blackett	"for his development of the Wilson cloud chamber method, and his discoveries therewith in the fields of nuclear physics and cosmic radiation"
1947	Sir Edward Victor Appleton	"for his investigations of the physics of the upper atmosphere especially for the discovery of the so-called Appleton layer"
1946	Percy Williams Bridgman	"for the invention of an apparatus to produce extremely high pressures, and for the discoveries he made therewith in the field of high pressure physics"
1945	Wolfgang Pauli	"for the discovery of the Exclusion Principle, also called the Pauli Principle"
1944	Isidor Isaac Rabi	"for his resonance method for recording the magnetic properties of atomic nuclei"
1943	Otto Stern	"for his contribution to the development of the molecular ray method and his discovery of the magnetic moment of the proton"
1939	Ernest Orlando Lawrence	"for the invention and development of the cyclotron and for results obtained with it, especially with regard to artificial radioactive elements"
1938	Enrico Fermi	"for his demonstrations of the existence of new radioactive elements produced by neutron irradiation, and for his related discovery of nuclear reactions brought about by slow neutrons"

Year	Laureate(s)	Citation
1937	Clinton Joseph Davisson and George Paget Thomson	"for their experimental discovery of the diffraction of electrons by crystals
1936	Victor Franz Hess	"for his discovery of cosmic radiation
	Carl David Anderson	"for his discovery of the positron
1935	James Chadwick	"for the discovery of the neutron
1933	Erwin Schrödinger and Paul Adrien Maurice Dirac	"for the discovery of new productive forms of atomic theory
1932	Werner Karl Heisenberg	"for the creation of quantum mechanics, the application of which has, inter alia, led to the discovery of the allotropic forms of hydrogen
1930	Sir Chandrasekhara Venkata Raman	"for his work on the scattering of light and for the discovery of the effect named after him
1929	Prince Louis-Victor Pierre Raymond de Broglie	"for his discovery of the wave nature of electrons
1928	Owen Willans Richardson	"for his work on the thermionic phenomenon and especially for the discovery of the law named after him
1927	Arthur Holly Compton	"for his discovery of the effect named after him
	Charles Thomson Rees Wilson	"for his method of making the paths of electrically charged particles visible by condensation of vapour
1926	Jean Baptiste Perrin	"for his work on the discontinuous structure of matter, and especially for his discovery of sedimentation equilibrium
1925	James Franck and Gustav Ludwig Hertz	"for their discovery of the laws governing the impact of an electron upon an atom
1924	Karl Manne Georg Siegbahn	"for his discoveries and research in the field of X-ray spectroscopy
1923	Robert Andrews Millikan	"for his work on the elementary charge of electricity and on the photoelectric effect
1922	Niels Henrik David Bohr	"for his services in the investigation of the structure of atoms and of the radiation emanating from them
1921	Albert Einstein	"for his services to Theoretical Physics, and especially for his discovery of the law of the photoelectric effect

1920	Charles Edouard Guillaume	"in recognition of the service he has rendered to precision measurements in Physics by his discovery of anomalies in nickel steel alloys
1919	Johannes Stark	"for his discovery of the Doppler effect in canal rays and the splitting of spectral lines in electric fields
1918	Max Karl Ernst Ludwig Planck	"in recognition of the services he rendered to the advancement of Physics by his discovery of energy quanta
1917	Charles Glover Barkla	"for his discovery of the characteristic Röntgen radiation of the elements
1915	Sir William Henry Bragg and William Lawrence Bragg	"for their services in the analysis of crystal structure by means of X-rays
1914	Max von Laue	"for his discovery of the diffraction of X-rays by crystals
1913	Heike Kamerlingh Onnes	"for his investigations on the properties of matter at low temperatures which led, inter alia, to the production of liquid helium
1912	Nils Gustaf Dalén	"for his invention of automatic regulators for use in conjunction with gas accumulators for illuminating lighthouses and buoys
1911	Wilhelm Wien	"for his discoveries regarding the laws governing the radiation of heat
1910	Johannes Diderik van der Waals	"for his work on the equation of state for gases and liquids
1909	Guglielmo Marconi and Karl Ferdinand Braun	"in recognition of their contributions to the development of wireless telegraphy
1908	Gabriel Lippmann	"for his method of reproducing colours photographically based on the phenomenon of interference
1907	Albert Abraham Michelson	"for his optical precision instruments and the spectroscopic and metrological investigations carried out with their aid
1906	Joseph John Thomson	"in recognition of the great merits of his theoretical and experimental investigations on the conduction of electricity by gases
1905	Philipp Eduard Anton von Lenard	"for his work on cathode rays

1904	Lord Rayleigh (John William Strutt)	"for his investigations of the densities of the most important gases and for his discovery of argon in connection with these studies"
1903	Antoine Henri Becquerel	"in recognition of the extraordinary services he has rendered by his discovery of spontaneous radioactivity"
	Pierre Curie and Marie Curie, née Sklodowska	"in recognition of the extraordinary services they have rendered by their joint researches on the radiation phenomena discovered by Professor Henri Becquerel"
1902	Hendrik Antoon Lorentz and Pieter Zeeman	"in recognition of the extraordinary service they rendered by their researches into the influence of magnetism upon radiation phenomena"
1901	Wilhelm Conrad Röntgen	"in recognition of the extraordinary services he has rendered by the discovery of the remarkable rays subsequently named after him"

PRE-NOBEL NOTABLES

	Pythagoras (ca. 580– ca. 500 BCE)	interest in mathematics leads to idea of "proof"
	Aristotle (ca. 384-ca. 322 BCE)	Earth was the center of the universe and that every celestial object moved around Earth in perfect circles
	Archimedes of Syracuse (ca. 287–212 BCE)	principle of buoyancy
1861-1862	James Clerk Maxwell (1831–1879)	Maxwell's equations
1573	Danish astronomer Tycho Brahe (1546–1601)	*De nova stella* (*On the New Star*) refutes Aristotelian concept of unchanging celestial realm
1609	Johannes Kepler (1571–1630)	*Astronomia nova* is published with laws of planetary motion
1627	René Descartes (1596–1650)	Cartesian coordinate system
1638	Galileo Galilei (1564–1642)	*Dialog Concerning Two New Sciences*
1660	Robert Hooke (1635–1703) English physicist	Hooke's law
1662	Robert Boyle (1627–91)	Boyle's law
1687	Isaac Newton (1642–1727)	*Philosophiae Naturalis Principia Mathematica* (1687), also known as the *Principia*
1714	Roger Cotes (1682–1716) British mathematician	published a paper titled "Logometria"
1733	Leonhard Euler (1707–83)	Euler's laws of motion
1735	Henri Pitot (1695– 1771)	Pitot tube, presented at the Academy of Science
1738	Daniel Bernoulli (1700–1782)	*Hydrodynamica* (measuring the pressure of fluids)
1776	James Watt (1736–1819)	Steam engine
1785	Charles-Augustin de Coulomb (1736–1806) French physicist	Coulomb's constant
1788	Joseph-Louis Lagrange (1736–1813)	Lagrangian mechanics reformulated classical mechanics
1794	Lazzaro Spallanzani (1729–99)	demonstrated that bats can navigate in complete darkness using ultrasonic echolocation

1807	Thomas Young (1773–1829)	derives a formula for energy; first to use term energy in modern sense
1809	Joseph Louis Gay-Lussac (1778–1850)	Guy-Lussac law
1810	Amedeo Avogadro (1776–1856)	Avogadro's Law
1823	André-Marie Ampere (1775–1836)	Ampere's law (electromagnetic circuits)
1824	Nicola Léonard Sadi Carnot (1796–1832).	theoretical model of an engine that is still used today
1825	Karl Friedrich Gauss (1777–1855) German theorist	Gauss's laws formulated
1827	Georg Simon Ohm (1787–1854)	Ohm's law
1831	Michael Faraday (1791–1867)	showed that a moving magnet induces an electric current in a conductor
1835	Joseph Henry (1797–1878)	invented electromechanical relay
1840	Germain Hess (1802–50)	Hess's law
1840	James Prescott Joule FRS (1818 to 1889)	kinetic theory of heat
1842	Christian Doppler (1803–53)	Doppler effect, or Doppler shift 1842
1848	William Thomson, 1st Baron Kelvin (1824–1907)	proposed absolute temperature scale (Kelvin), uses the term "kinetic"
1853	William John Macquorn Rankine (1820–72)	proposed the term "potential energy" in 1853
1859	Robert Bunsen (1811–99)	created spectroscope
1876	Francis Galton (1822–1911)	first human-made ultrasonic device was a dog whistle.
1877	Ludwig Boltzmann (1844–1906)	proposed quantum.
1878	Jacques Charles (1746–1832)	Charles law

1879	Josef Stefan (1835–93) Ludwig Boltzmann (1844–1906)	Stefan-Boltzmann law
1879	Heinrich Hertz (1857–94)	Proves existence of electromagnetic waves
1893	Wilhelm Wien (1864–1928)	Wien's displacement law to calculate the emission of a blackbody
1895	Wallace Clement Sabine (1868–1919)	Improves acoustics in Fogg lecture hall, Harvard, and founded field of architectural acoustics
1900	Max Planck (1858–1947)	Planck's law Planck's law of blackbody radiation
1900	Paul Ulrich Villard (1860–1934)	Discovered gamma rays

PHYSICS CONSTANTS

Symbol	Name	Value
c	speed of light in a vacuum	299,792,458 m/s
G	gravitational constant	6.67384×10^{-11} N m^2/kg^2
h	Planck's constant	$6.62606957 \times 10^{-34}$ J s
		$4.135667516 \times 10^{-15}$ eVs
hc		$1.98644521 \times 10^{-25}$ J m 1239.84193 eV nm
$\hbar$	$h/2\pi$ (reduced Planck's constant; Dirac's constant)	$1.054571726 \times 10^{-34}$ J s
		$6.58211928 \times 10^{-16}$ eV s
ε_0	electric constant permitivitty of free space; vacuum permitivitty	$8.854187817 \times 10^{-12}$ C^2/N m^2
μ_0	magnetic constant permeability of free space; vacuum permeability	$4\pi \times 10^{-7}$ T m/A
e	elementary charge	$1.602176565 \times 10^{-19}$ C
m_e	electron mass	$9.10938291 \times 10^{-31}$ kg 0.510998928 MeV 0.00054857990946 u
m_p	proton mass	$1.672621777 \times 10^{-27}$ kg 938.272046 MeV 1.00727646812 u
m_n	neutron mass	$1.674927351 \times 10^{-27}$ kg 939.565379 MeV 1.00866491600 u
m_u	atomic mass constant	$1.660538921 \times 10^{-27}$ kg 931.494061 MeV 1 u
N_A	Avogadro's constant	$6.02214129 \times 10^{23}$ 1/mol
k	Boltzmann's constant	$1.3806488 \times 10^{-23}$ J/K
R	gas constant	8.3144621 J/mol K
σ	Stefan-Boltzmann constant	5.670373×10^{-8} W/m^{2}K^4
b	Wien displacement constant	2.8977721 mm/K 58.789254 GHz/K

Symbol	Name	Value
g	standard gravity	9.80665 m/s^2
atm	standard atmosphere	$101,325 \text{ Pa}$
H_0	Hubble constant*	$69.3 \; 2.25 \times 10^{-18}$ km/s/Mpc s^{-1}

Source: NIST, exception *NASA

PHYSICS LAWS

Ampère's Law
The line integral of the magnetic flux around a closed curve is proportional to the algebraic sum of electric currents flowing through that closed curve; or, in differential form curl $B = J$. This was later modified to add a second term when it was incorporated into Maxwell's equations.

Archimedes' Principle
A body that is submerged in a fluid is buoyed up by a force equal in magnitude to the weight of the fluid that is displaced, and directed upward along a line through the center of gravity of the displaced fluid.

Avogadro's Hypothesis (1811)
Equal volumes of all gases at the same temperature and pressure contain equal numbers of molecules. It is, in fact, only true for ideal gases.

Bernoulli's Equation
In an irrotational fluid, the sum of the static pressure, the weight of the fluid per unit mass times the height, and half the density times the velocity squared is constant throughout the fluid.

Boyle's Law (1662); Mariotte's law (1676)
The product of the pressure and the volume of an ideal gas at constant temperature is a constant.

Bragg's Law (1912)
When a beam of X-rays strikes a crystal surface in which the layers of atoms or ions are regularly separated, the maximum intensity of the reflected ray occurs when the complement of the angle of incidence, *theta*, the wavelength of the X-rays, *lambda*, and the distance between layers of atoms or ions, *d*, are related by the equation $2\,d \sin theta = n\,lambda$.

Causality Principle
The principle that cause must always preceed effect. More formally, if an event A ("the cause") somehow influences an event B ("the effect") which occurs later in time, then event B cannot in turn have an influence on event A. That is, event B must occur at a later time t than event A, and further, all frames must agree upon this ordering.

Centrifugal Pseudoforce
A pseudoforce on an object when it is moving in uniform circular motion. The "force" is directed outward from the center of motion.

Charles' Law (1787)
The volume of an ideal gas at constant pressure is proportional to the thermodynamic temperature of that gas.

Complementarity Principle
The principle that a given system cannot exhibit both wave-like behavior and particle-like behavior at the same time. That is, certain experiments will reveal the wave-like nature of a system, and certain experiments will reveal the particle-like nature of a system, but no experiment will reveal both simultaneously.

Compton Effect (1923)
An effect that demonstrates that photons (the quantum of electromagnetic radiation) have momentum. A photon fired at a stationary particle, such as an electron, will impart momentum to the electron and, since its energy has been decreased, will experience a corresponding decrease in frequency.

Conservation Laws
Conservation of mass-energy
The total mass-energy of a closed system remains constant.
Conservation of electric charge
The total electric charge of a closed system remains constant.
Conservation of linear momentum
The total linear momentum of a closed system remains constant.
Conservation of angular momentum
The total angular momentum of a closed system remains constant.

Constancy Principle
One of the postulates of A. Einstein's special theory of relativity, which puts forth that the speed of light in vacuum is measured as the same speed to all observers, regardless of their relative motion.

Continuity Equation
An equation which states that a fluid flowing through a pipe flows at a rate which is inversely proportional to the cross-sectional area of the pipe. It is in essence a restatement of the conservation of mass during constant flow.

Copernican Principle (1624)
The idea, suggested by Copernicus, that the Sun, not the Earth, is at the center of the Universe. We now know that neither idea is correct.

Coriolis Pseudoforce (1835)
A pseudoforce which arises because of motion relative to a frame of reference which is itself rotating relative to a second, inertial frame. The magnitude of the Coriolis "force" is dependent on the speed of the object relative to the noninertial frame, and the direction of the "force" is orthogonal to the object's velocity.

Correspondence Principle
The principle that when a new, more general theory is put forth, it must reduce to the more specialized (and usually simpler) theory under normal circumstances. There are correspondence principles for general relativity to special relativity and special relativity to Newtonian mechanics, but the most widely known correspondence principle is that of quantum mechanics to classical mechanics.

Coulomb's Law
The primary law for electrostatics, analogous to Newton's law of universal gravitation. It states that the force between two point charges is proportional to the algebraic product of their respective charges as well as proportional to the inverse square of the distance between them.

Curie's Law
The susceptibility of an isotropic paramagnetic substance is related to its thermodynamic temperature T by the equation $KHI = C / T$.

Doppler Effect
Waves emitted by a moving object as received by an observer will be blueshifted (compressed) if approaching, redshifted (elongated) if receding. It occurs both in sound as well as electromagnetic phenomena.

Einstein's Mass-Energy Equation
The energy E of a particle is equal to its mass m times the square of the speed of light c, giving rise to the best known physics equation in the Universe: $E = mc^2$.

Faraday's Law
The line integral of the electric field around a closed curve is proportional to the instantaneous time rate of change of the magnetic flux through a surface bounded by that closed curve; in differential form curl $E = -dB/dt$, where here d/dt represents partial differentiation.

Faraday's Laws of electrolysis
Faraday's first law of electrolysis
The amount of chemical change during electrolysis is proportional to the charge passed.
Faraday's second law of electrolysis
The charge Q required to deposit or liberate a mass m is proportional to the charge z of the ion, the mass, and inversely proportional to the relative ionic mass M; mathematically $Q = F m z / M$.

Faraday's first law of electromagnetic induction
An electromotive force is induced in a conductor when the magnetic field surrounding it changes.
Faraday's second law of electromagnetic induction
The magnitude of the electromotive force is proportional to the rate of change of the field.
Faraday's third law of electromagnetic induction
The sense of the induced electromotive force depends on the direction of the rate of the change of the field.

Fermat's Principle
The principle states that the path taken by a ray of light between any two points in a system is always the path that takes the least time.

Gauss' Law
The electric flux through a closed surface is proportional to the algebraic sum of electric charges contained within that closed surface; in differential form div E = rho, where *rho* is the charge density.

Gauss' Law for Magnetic Fields
The magnetic flux through a closed surface is zero; no magnetic charges exist; in differential form div B = 0.

Hall Effect
When charged particles flow through a tube which has both an electric field and a magnetic field (perpendicular to the electric field) present in it, only certain velocities of the charged particles are preferred, and will make it un-deviated through the tube; the rest will be deflected into the sides.

Hooke's Law
The stress applied to any solid is proportional to the strain it produces within the elastic limit for that solid. The constant of that proportionality is the Young modulus of elasticity for that substance.

Huygens' Principle
The mechanical propagation of a wave (specifically, of light) is equivalent to assuming that every point on the wavefront acts as point source of wave emission

Ideal Gas Law
An equation which sums up the ideal gas laws in one simple equation $PV = nRT$.

Joule-Thomson Effect; Joule-Kelvin Effect
The change in temperature that occurs when a gas expands into a region of lower pressure.

Lambert's Laws
Lambert's first law
The illuminance on a surface illuminated by light falling on it perpendicularly from a point source is proportional to the inverse square of the distance between the surface and the source.
Lambert's second law
If the rays meet the surface at an angle, then the illuminance is proportional to the cosine of the angle with the normal.
Lambert's third law
The luminous intensity of light decreases exponentially with distance as it travels through an absorbing medium.

Lenz's Law (1835)
An induced electric current always flows in such a direction that it opposes the change producing it.

Maxwell's Equations (1864)
Gauss' law
The electric flux through a closed surface is proportional to the algebraic sum of electric charges contained within that closed surface; in differential form div E = rho, where *rho* is the charge density.
Gauss' law for magnetic fields
The magnetic flux through a closed surface is zero; no magnetic charges exist. In differential form div B = 0.
Faraday's law
The line integral of the electric field around a closed curve is proportional to the instantaneous time rate of change of the magnetic flux through a surface bounded by that closed curve; in differential form curl $E = -dB/dt$,..
Ampère's law, modified form
The line integral of the magnetic field around a closed curve is proportional to the sum of two terms: first, the algebraic sum of electric currents flowing through that closed curve; and second, the instantaneous time rate of change of the electric flux through a surface bounded by that closed curve; in differential form curl $H = J + dD/dt$,.
In addition to describing electromagnetism, his equations also predict that waves can propagate through the electromagnetic field, and would always propagate at the the speed of light in vacuum.

Newton's Law of universal gravitation
Two bodies attract each other with equal and opposite forces; the magnitude of this force is proportional to the product of the two masses and is also proportional to the inverse square of the distance between the centers of mass of the two bodies; $F = (GmM/r^2)e$, where m and M are the masses of the two bodies, r is the distance between. the two, and e is a unit vector directed from the test mass to the second.

Newton's Laws of motion
Newton's first law of motion
A body continues in its state of constant velocity (which may be zero) unless it is acted upon by an external force.
Newton's second law of motion
For an unbalanced force acting on a body, the acceleration produced is proportional to the force impressed; the constant of proportionality is the inertial mass of the body.
Newton's third law of motion
In a system where no external forces are present, every action force is always opposed by an equal and opposite reaction force.

Ohm's Law (1827)
The ratio of the potential difference between the ends of a conductor to the current flowing through it is constant; the constant of proportionality is called the resistance, and is different for different materials.

Pascal's Principle
Pressure applied to an enclosed incompressible static fluid is transmitted undiminished to all parts of the fluid.

Planck Equation
The quantum mechanical equation relating the energy of a photon E to its frequency ν: $E = h\nu$.

Reflection Law, Snell's Law
For a wavefront intersecting a reflecting surface, the angle of incidence is equal to the angle of reflection, in the same plane defined by the ray of incidence and the normal.

Refraction Law
For a wavefront traveling through a boundary between two media, the first with a refractive index of n_1, and the other with one of n_2, the angle of incidence θ is related to the angle of refraction ϕ by $n_1 \sin\theta = n_2 \sin\phi$.

Relativity Principle
The principle, employed by Einstein's relativity theories, that the laws of physics are the same, at least qualitatively, in all frames. That is, there is no frame that is better (or qualitatively any different) from any other. This principle, along with the constancy principle, constitute the founding principles of special relativity.

Stefan-Boltzmann Law
The radiated power P (rate of emission of electromagnetic energy) of a hot body is proportional to the radiating surface area, A, and the fourth power of the thermodynamic temperature, T. The constant of proportionality is the Stefan-Boltzmann constant. Mathematically $P = e\,\sigma\,A\,T^4$, where the efficiency rating e is called the emissivity of the object.

Thermodynamic Laws
First law of thermodynamics
The change in internal energy of a system is the sum of the heat transferred to or from the system and the work done on or by the system.
Second law of thermodynamics
The entropy—a measure of the unavailability of a system's energy to do useful work—of a closed system tends to increase with time.
Third law of thermodynamics
For changes involving only perfect crystalline solids at absolute zero, the change of the total entropy is zero.

Uncertainty Principle (1927)
A principle, central to quantum mechanics, which states that two complementary parameters (such as position and momentum, energy and time, or angular momentum and angular displacement) cannot both be known to infinite accuracy; the more you know about one, the less you know about the other.

van der Waals force
Forces responsible for the non-ideal behavior of gases, and for the lattice energy of molecular crystals. There are three causes: dipole-dipole interaction; dipole-induced dipole moments; and dispersion forces arising because of small instantaneous dipoles in atoms.

Wave-Particle Duality
The principle of quantum mechanics which implies that light (and, indeed, all other subatomic particles) sometimes act like a wave, and sometime act like a particle, depending on the experiment you are performing. For instance, low frequency electromagnetic radiation tends to act more like a wave than a

particle; high frequency electromagnetic radiation tends to act more like a particle than a wave.

Wiedemann-Franz Law
The ratio of the thermal conductivity of any pure metal to its electrical conductivity is approximately constant for any given temperature. This law holds fairly well except at low temperatures.

GLOSSARY

absorption coefficient: a value characteristic of a particular medium that represents the amount of light or sound it absorbs from a wave passing through it.

acceleration: the rate at which the velocity of an object increases over time.

accuracy: the extent to which measurements of a property differ from its actual value.

acoustics: the study of sound; also, the qualities of a space that affect how sound is heard within that space.

activation energy: the minimum energy required for a chemical reaction to take place.

active galactic nucleus: the region in the center of certain types of galaxies that emits massive amounts of energy across most, if not all, of the electromagnetic spectrum; believed to result from a supermassive black hole.

actual mechanical advantage: the ratio comparing the output force of a machine to the input force, taking into account friction and other factors that limit the efficiency of real-world machines. A mechanical advantage of more than one indicates an amplification of force.

albedo: the portion of electromagnetic energy that is reflected when its waves encounter a surface or boundary; often used to describe solar radiation reflecting off Earth or another body in space.

alpha decay: a form of radioactive decay in which a radioactive atom's nucleus splits and discharges an alpha particle, made up of two protons and two neutrons.

alpha particle: a particle consisting of two protons and two neutrons, identical to a helium nucleus, which is emitted from an unstable nucleus via radioactive decay.

alpha radiation: alpha (α) particles typically emitted during alpha decay, a subtype of radioactive decay.

Ampère's law: the rule stating that the strength of a magnetic field about a current-carrying conductor is directly proportional to the magnitude of the current.

amplifier distortion: an imperfect reproduction of the original signal when transmitting sound electronically.

amplitude: a quantifying wave property measured from a point of rest to a point of maximum displacement; related to sound power and intensity of sound waves.

angle: the separation along a curved path between two rays originating at the same point.

angular magnification: the angle subtended at the eye by a magnified image of an object divided by the angle of the object being viewed by the naked eye without magnification.

angular momentum: the rotational momentum of an object around an axis, defined as the product of its moment of inertia and its angular velocity.

angular velocity: the speed and direction of movement of a rotating object.

antineutrino: a subatomic particle with a neutral charge, the antiparticle to the neutrino, which is emitted during beta decay.

antinode: a point of maximum amplitude in a standing wave.

antiphase: waves with a phase difference of 180 degrees relative to one another. The crests of one wave align with the troughs of the other and vice versa.

aperture: an adjustable opening in a barrier through which light or other electromagnetic emission can pass.

arclength: the length of an arc defined by an angle and two radii.

attenuation: the loss of energy from a wave passing through a medium due to absorption or scattering.

average: in physics, the overall value for a given quantity, obtained by comparing initial and final values of a measurement against another unit or quantity; for instance, average speed compares the total distance traveled to the total time taken to move that distance.

axis: the center around which an object rotates.

background radiation: the total amount of ionizing radiation to which Earth is constantly exposed from both natural and artificial sources.

ballistic trajectory: the motion described by a projectile traveling by inertia in a gravitational field.

barrel distortion: the optical distortion caused by a wide-angle lens that makes an image seem inflated at the center, as though stretched across a barrel.

barrel of oil equivalent: the energy output of burning one barrel (42 gallons, or 159 liters) of crude oil; equal to 5.8×10^6 BTU or 6.1×10^9 joules.

beat frequency: the apparent frequency at which waves from two or more sources create constructive interference.

Beer-Lambert law: a formula that relates the attenuation of an electromagnetic wave in a given medium to the thickness of that medium and the concentration of attenuating materials within it.

Bernoulli's equation: used to relate and determine velocity, acceleration, and density in fluid mechanics.

beta particle: an electron that is emitted from an unstable nucleus via radioactive decay.

beta radiation: beta (β) rays emitted during beta decay, a subtype of radioactive decay.

bias: the extent to which a measurement differs from the real value of the property being measured, or an intrinsic factor in a method that consistently causes such deviation.

blackbody radiation: radiation emitted by a body solely as a result of its temperature, regardless of its composition or any previously absorbed radiation.

blueshift: the apparent shortening of the wavelength of electromagnetic radiation due to the movement of the radiation source toward the observer.

Boltzmann factor: the ratio of the probability of finding a particle or system at a certain energy level to the probability of finding it at another energy level; proportional to the probability of the system being in a particular quantum state.

bond energy: the energy required to break the chemical bonds in one mole of molecules and separate them into their component atoms.

boson: an elementary particle that carries a specific type of force, or a composite particle containing an even number of fermions, having an integer spin.

bosonic string: in string theory, the structure of bosons as one-dimensional strings rather than dimensionless points.

bra-ket notation: a system of mathematical notation developed by physicist Paul Dirac to manipulate very large vector equations more easily.

bremsstrahlung: radiation generated when a particle is slowed via a collision with another particle; from the German word for "braking radiation."

British thermal unit (BTU): a nonstandard unit of energy measurement supposedly equivalent to the heat needed to raise the temperature of one pound of water by one degree Fahrenheit.

buoyancy: upward force exerted by a fluid on a submerged or floating object.

calorie: the amount of energy needed to raise the temperature of one gram of water by one degree Celsius (1.8 degrees Fahrenheit) at one atmosphere of pressure, equivalent to approximately 4.2 joules.

capacitor: an electrical part consisting of two conductors separated from each other by a nonconductor, allowing it to store electrical charge temporarily.

carcinogenic: capable of causing cancer.

Cartesian coordinate system: a system that uses a pair or trio of numbers to indicate the location of a point in two-dimensional or three-dimensional space relative to a set point of origin.

Cartesian coordinates: a pair or triplet of numbers that indicate the location of a point in a plane or in a three-dimensional space and are the signed distances from the origin.

center of mass: the point in an object or system around which the mass of said object or system is evenly distributed.

center of momentum: the inertial frame of a system where the vector sum of the moments of all of the particles in that system is zero.

centrifugal force: a fictitious force that seems to push a body in circular motion away from the axis of rotation; an artifact of a non-inertial frame of reference, in any inertial reference frame, objects in circular motion are subject to centripetal force.

centripetal force: a force "toward the center" that, in combination with inertia, generates the curved path of an object in circular motion.

Cherenkov radiation: electromagnetic radiation emitted when a charged particle, such as an electron, travels through a given medium faster than light would in that same medium.

chromatic aberration: a distortion created by the separation of white light into its component wavelengths when passing from one medium into another, such as through a lens.

circle of confusion: an imperfect image produced by a lens due to the rays of light passing through it not converging at a perfect point.

circuit: a closed path along which electricity travels.

circumference: the distance around a circle defined by a radius.

closed system: a system that may exchange energy, but not matter, with its surroundings.

collimated rays: parallel rays of light that propagate with minimal spreading.

collision: an interaction in which two or more bodies come into contact and briefly exert force on each other.

color charge: a property of quarks that distinguishes quarks and gluons from each other.

color force: the force of the strong interaction that operates on the quark level.

coma: the extended geometric image formed along the optical axis of a lens by light entering the lens obliquely.

compression: the pushing forces applied to an object in order to diminish its size or volume.

compressive strength: the ability of a material to resist deformation or structural failure when experiencing a force of compression (squeezing).

Compton scattering: the collision of a high-energy photon with a lower-energy electron, causing energy to be transferred from the photon to the electron and changing the angle of the photon's trajectory.

concave: having surfaces that curve inward, like a bowl.

conductor: a material that has a low resistance to electric charges, allowing them to move through it easily.

conformal field theory: a quantum field theory that is independent of the scale of the system and supports only massless excitations.

conservation of energy: a fundamental law of physics that states that the amount of energy in a system remains constant over time. Although the energy can be transformed or transferred, it cannot be created or destroyed.

conservation of momentum: in physics, the principle that the total momentum in a closed system is always constant.

constructive interference: when two or more waves of the same phase combine to form a larger amplitude.

consumed energy: the amount of energy used or transferred over a period of time, often measured in kilowatt-hours.

continuity: a clear path for electricity from point A to point B.

continuous energy levels: the idea that there are limitless levels of energy between each point on a continuum.

converging: the joining of light rays after an interaction with a mirror or a lens.

convex: having surfaces that curve outward, like a ball.

cosine: a trigonometric function describing the relationship between sides of a right triangle; the cosine of an angle is equal to the length of the side adjacent to the angle divided by the length of the hypotenuse.

cosmic rays: extremely high-energy subatomic particles, mainly protons and atomic nuclei, that originate outside of Earth's atmosphere from largely unknown sources.

coulomb: the basic unit of charge in the International System of Units (SI).

Coulomb's law: a scientific law stating that the electric flux at any defined surface is the vector sum of the electric field strength at every point on that surface.

crest: the highest point of a wave from its neutral value; the distance between the crest or trough of a wave and the wave's neutral value is called the amplitude.

cross product: an operation, broadly analogous to multiplication, performed on two vectors in a three-dimensional space that results in a third vector that is perpendicular to both; if both vectors have the same direction or if one of them has a value of zero, the cross product will be zero.

current: the rate at which an electric charge, usually in the form of electrons, moves through a wire or other conductive material.

damped oscillator: an oscillator that is subject to friction or other braking forces.

deceleration: the rate at which the velocity of an object decreases over time.

decibel: a logarithmic unit that describes the power of a given sound in relation to the threshold of human hearing; abbreviated dB.

definite pitch: a sound in which the pitch is easily detected.

degrees of freedom: the number of physical parameters required to specify the position and configuration of a particle or other body.

demand: the load on an electrical supply system over time.

density: a measure of the mass of a quantity of matter relative to the volume of space that it occupies.

depth of field: the range of distance over which objects appear sharp or in focus.

derived unit: unit of measure that is described in terms of two or more base units; for example, meters are a base unit of distance, whereas square meters are derived units of area calculated from meters of height multiplied by meters of depth.

destructive interference: when two or more waves of different phases combine to form a smaller amplitude.

diaphragm: a circular structure with an aperture that controls the amount of light entering an instrument.

diffraction: a change in the direction of a wave as it passes around an obstruction or through an opening.

dimension: a direction in which an object can move. In spatial physics, the three dimensions are traditionally represented by the symbols x, y, and z.

diodes: devices in which an anode and a cathode, or the transistor equivalent, control the direction of current flow.

directly related: a relationship between two parameters that exists if increasing one increases the other according to a set multiplier.

discrepancy: the difference between the measured value of a property and its real value, or between nonidentical measurements of the same property.

displacement: in fluid mechanics, the process by which a body immersed in a fluid pushes the fluid out of the way and occupies the space in its stead. The volume of the displaced fluid is equal to the volume of the displacing body. The absolute distance and direction between the starting and end points of an object's motion, which ignores any twists or turn the object's path may take; it is always equal to or less than the total distance traveled.

displacement field: in electrodynamics, the electric field produced solely by free charges (e.g., free electrons).

dissipation: the irreversible loss of energy from a system.

distinguishability: the ability of particles with the same energy to be differentiated from one another.

distribution function: a mathematical function that describes the probability that a certain variable, such as the energy state of a particle, will have a given value or range of values.

diverging: the separation of light rays after an interaction with a mirror or a lens.

dot product: the product of the lengths of two vectors and the cosine of the angle between them; also called the inner product or the scalar product.

dummy load: a device applied to an electrical circuit or other system in order to provide a corresponding load without performing an output function.

dynamic equilibrium: the state in which reversible reactions occur at equal rates in opposite directions, balancing each other and resulting in no net change.

dynamic systems: systems that are subject to change.

dynamometer: a device that measures mechanical power or force, often used for the output of engines and other similar devices.

dyne: a unit of force in the centimeter-gram-second (CGS) unit system, equal to $10-5$ newtons, or 10 micronewtons.

efficiency: the measure of how effective a machine is at transforming or transferring energy, quantified as the ratio of the actual performance of the machine to an idealized, theoretical version of it. A perfect machine would have an efficiency value of one.

efficiency: the measure of how effective a machine is at transforming or transferring energy, quantified as the ratio of the actual performance of the machine to an idealized, theoretical version of the same machine. A perfect machine would have an efficiency value of one.

eigenfunction: any mathematical function for which the operation being considered yields the original function multiplied by a constant (an eigenvalue). Eigenfuctions are defined only with respect to the operator for which the eigenfunction is considered

eigenstate: the state for which the value of a measurable, observable operator is defined without uncertainty

eigensystem: the set of all eigenvectors of a matrix paired with their respective eigenvalues.

eigenvalue: the mathematical constant that, when multiplied by an eigenfunction, yields the result of that operation performed on the function.

electric flux: the measure of the strength of an electric field across an area with different electrical potentials.

electric potential energy: energy that is present in particles due to their charge and closeness to other charges.

electrical efficiency: the ratio of the power applied to an electrical circuit to the power delivered by a particular device in the circuit.

electromagnet: a device that becomes magnetic due to the presence of an electric current.

electromagnetic field: a physical field consisting of a combined electric field (generated by stationary electric charges) and magnetic field (generated by moving electric charges) that affects the behavior of charged objects in its vicinity.

electromagnetic spectrum: the full range of electromagnetic radiation, sorted into segments with similar properties by wavelength; x-rays occupy one of these segments.

electromotive force: the energy per unit of charge supplied by an electric source such as a battery, measured in volts.

electron: a negatively charged subatomic particle that is often bound to the positive charge of the nucleus but can also exist in a free state in an atom.

electron volt: the amount of energy carried by a single electron moved across an electric potential difference of one volt; equal to 1.6×10^{-19} joules.

electroweak force: the force responsible for nuclear transformations.

elementary particle: one of the fundamental constituents of matter.

elements: the parts of an electrical circuit, each of which has a specific function.

ellipse: an oval surrounding two focal point points where the sum of the distances from any point on the oval to the two focal points is constant.

elongation: the lengthening of an object under stress.

endothermic: describes a chemical reaction or process that requires the input of energy from an external source in order to proceed.

energy: a property of matter and objects that can be transferred and transformed but never created or destroyed, sometimes described as the ability to do work; measured in joules (J).

enthalpy: a measure of the total internal energy (thermal energy) of a system, the product of its volume and pressure.

entropy: a measure of a system's level of disorder, which in physics refers to the potential amount of states the molecules of the system may assume.

equilibrium: the condition of a thermodynamic system in which there is no energy flow.

exothermic: describes a chemical reaction or process that results in an output of energy from the system.

extrinsic value: any characteristic property of matter that is due to the interaction of the matter with an external factor.

Faraday's law: the scientific law stating that when a magnetic field and a conductor move relative to each other, a voltage is induced in the conductor, the magnitude of which depends directly on the relative speed of the field and conductor.

fermion: one of two main classes of particles, characterized by adherence to Fermi-Dirac statistics; includes quarks, leptons, and any particles that contain an odd number of quarks or leptons, such as protons and neutrons.

ferromagnetic: describing material that can be made permanently magnetic by the presence of a magnetic field, such as iron and other ferrous (iron-like) metals.

fixed reference frame: a frame of reference that is fixed to the environment and not to the subject being observed; sometimes specified as "Earth-fixed" or "space-fixed."

flow rate: the amount of fluid that flows in a given period of time; expressed as a numerical quantity.

focal length: the distance from the focal point of a lens or mirror to the center of the lens or mirror.

focal point: the place where light from a source converges after it reflects off a mirror or refracts through a lens.

force: a push, pull, or any other interaction that affects the motion of an object. Force is measured in newtons (N).

force field: the effect of a field force and a function of the relative position of the object from the force field source; the object and the source do not need to be in physical contact for this effect to occur.

frame of reference: a set of coordinate axes that serve to describe position or movement of an object with reference to that coordinate system.

free fall: falling only under the influence of gravity.

frequency: the number of complete wavelengths that occur within one unit of time, typically expressed as hertz (Hz; cycles per second).

friction: the force created by the resistance to relative motion between solid surfaces.

frictional resistance: the force created by the resistance to relative motion between solid surfaces; it is normally proportional to the roughness of the surfaces as well as the force squeezing the surfaces together.

fulcrum: the supporting point around which a lever pivots.

fundamental frequency: the lowest frequency of a resonating medium; also called the first harmonic.

gamma radiation: electromagnetic radiation with a wavelength shorter than 1×10^{-11} meters; gamma (γ) rays typically emitted during gamma decay, a subtype of radioactive decay.

gamma ray: a high-energy photon that is usually emitted by an unstable nucleus via radioactive decay; it may also be produced by various other means, including particle-antiparticle annihilation and the interaction of atmospheric particles with cosmic rays.

gamma-ray burst: a high-energy flash of gamma radiation produced by a violent explosion in a distant galaxy; believed to be the brightest and most energetic electromagnetic event in the universe since the big bang.

Gaussian surface: any closed or finite hypothetical surface.

Geiger-Müller probe: a device that can be used to detect alpha radiation as well as beta radiation, gamma radiation, and x-rays.

general relativity: the theory stating that gravity is a geometric property of space and time.

gram: the elementary particle that carries the strong interaction between quarks. Abbreviated g, a standard unit of mass in the International System of Units; derived from the kilogram (kg, thousands of grams), which is the base unit of mass.

gravitational acceleration: the rate of increase of velocity experienced by matter under the influence of a gravitational field; often referred to as the force of gravity.

gravitational force: the pull that objects exert on each other due to their masses and the separation between their centers of mass.

graviton: the gauge boson that mediates the gravitational force.

gravity: the force that describes the attraction between one body and another.

ground: a direct connection to a larger body by which excess current is carried away, preventing errant electrical potentials from being generated in the circuit.

hadron: a subatomic particle that is made of either three quarks or one quark and one antiquark and held together by the strong force.

half-life: the average time it takes for half of the unstable nuclei in a radioactive element to undergo radioactive decay, transforming into a lighter element and giving off radiation.

harmonic oscillations: the motion resulting from a system that when displaced from its equilibrium point exerts a restoring force that is proportional to the displacement.

harmonics: the study of the interaction of wave phenomena.

harmony: the combination of two or more different musical notes played simultaneously that create a pleasant sound.

heat: the active process of energy transfer due to changes in an object's thermal energy.

heat transfer: transfer of thermal energy from one region or system to another.

heat value: a measurement of the energy released as heat when a specific amount of a specific substance is burned; typically applied to fuels and foods in units of energy or mass.

hertz: the SI unit of frequency; one hertz (Hz) is equal to one cycle (complete orbit) per second.

Hess's law: the principle that the overall enthalpy change of a given reaction remains the same regardless of the number of reaction steps involved and is equal to the sum of the enthalpy changes of any and all intermediate steps; also called Hess's law of constant heat summation.

Higgs field: a theorized field that, through interactions with matter, causes that matter to have mass.

Hooke's law: the law stating that the deformation of an elastic object, such as a spring, is directly proportional to the force acting on the object, as long as the object's elastic limit is not exceeded.

horsepower: an alternative unit of power, in theory based on the amount of power an average horse can produce; commonly used to describe engines.

horsepower hour: a nonstandard unit of power supposedly equivalent to the amount of work a horse does over an hour.

hot conductor: in an electrical circuit, the conductor that brings current into a component element (the conductor having the higher electrical potential).

ideal gas law: a law stating that the pressure (P) and volume (V) of an ideal gas are directly related to its number of particles (n), its temperature (T), and the ideal gas constant (R), written as $PV = nRT$.

ideal mechanical advantage: the ratio comparing the output force of a machine to the input force, ignoring friction and other factors that limit the efficiency of real-world machines. A mechanical advantage of more than one indicates an amplification of force.

impedance: the opposition to electrical current flow produced by a voltage.

imperfect system: any system that functions with less than 100 percent efficiency.

impulse turbine: a turbine set in motion by the velocity of a fluid hitting each blade.

incompressible: unable to be pressed or squeezed.

indefinite pitch: a sound in which the pitch is not easily detected.

inductance: the property by which a current is created in conductors to resist a change in the magnetic field through the conductor.

inductor: a conducting coil in which the current generates a proportional magnetic field that, in turn, impedes the flow.

inertia: the principle that an object at rest tends to stay at rest and an object in motion tends to stay in motion unless acted on by an outside force.

inertial reference frame: a means of describing relative motions through space according to Newtonian mechanics.

infrasound: a sound wave with a frequency below twenty hertz, the lower limit of human hearing.

initial velocity: the velocity of an object at the start of some interval of time.

input: the force (or energy) that is "put in" to a machine; for example, the horizontal force of wind provides the input for a windmill.

instantaneous: denotes a measurement taken at a specific point in time.

instantaneous phase: the time-variant angle of a sinusoidal function.

instantaneous velocity: the velocity (speed and direction of travel) of an object in motion at any one instant of time.

insulator: a material that has a high resistance to electric charges, preventing them from moving through it easily.

insulators: substances that cannot transport electricity.

intensity: in acoustics, sound power per unit area; measured in decibels or watts per square meter.

interference: the distortion of a wanted signal by an unwanted one, such as an unwanted radio station "bleeding into" another because both are broadcasting on the same wavelength.

internal energy: for an ideal gas, the energy of its atoms and molecules that is directly proportional to the absolute temperature.

International System of Units (SI): a standardized system of units and measures based on the metric system, used worldwide to enable clear and precise communication in the sciences and other disciplines.

intrinsic value: any characteristic property of matter that is due solely to the nature of the matter itself.

inverse Compton scattering: a collision between a high-energy electron and a low-energy photon that results in energy being transferred to the photon.

inversely related: a relationship between two parameters that exists if increasing one decreases the other according to a set multiplier.

inverse-square law: radiation emanating from a single point has an intensity inversely proportional to the square of the distance the radiation has traveled from its source.

inversion: the reversal of the way an image looks after an interaction with a mirror or lens.

isotopes: variants of a chemical element with differing numbers of neutrons

joule: abbreviated J, the International System of Units unit of work and energy. One joule is equal to the work done by a force of one newton acting across a distance of one meter.

just-noticeable difference (JND): the amount a parameter, such as sound intensity, must be changed so that the difference is noticeable at least half the time.

kilocalorie: a nonstandard unit commonly used to measure energy content in food.

kilogram: the base unit of mass in the International System of Units (SI), equal to 1,000 grams (2.2 pounds).

kilowatt: a unit measuring work over time, equivalent to 1,000 watts, or joules per second

kilowatt-hour: a unit for measuring electricity consumption, equal to one thousand watts of power consumed over one hour, or 3.6×10^6 joules.

kinematics: a subfield of classical mechanics that studies the motion of objects without reference to the forces that cause this motion.

kinetic energy: energy due to any kind of motion, be it rotation, vibration, or translation.

kinetic theory of gases: the theory that atomic and molecular motion determines the behavior of gases.

Lagrangian points: positions where the gravitational pull of two large masses equals the centripetal force needed for a small object to maintain a stable position in orbit with the two large masses.

laminar flow: the even and stable flow of a fluid; opposite of turbulent flow.

law of conservation of energy: the total amount of energy in a closed system is always the same; therefore, energy can neither be created nor destroyed.

law of the lever: the law that states if the fulcrum of the lever is closer to the output end than the input end, it will generate a mechanical advantage.

Lenz's law: the principle that the current induced in a conductor by a moving magnetic field flows in the direction opposite to the motion of the magnetic field.

lepton: an elementary particle that has no color and thus cannot form nuclei.

lift: the force that directly opposes the weight of an object and the force of gravity and holds the object aloft.

light wave: an oscillation in an electromagnetic field.

line spectrum: the lines of color that represent the characteristic frequencies at which atoms emit electromagnetic radiation.

linear magnification: the ratio of the apparent height of an object's magnified image to the actual height of the object.

linear momentum: an object's mass times its velocity; often called simply "momentum," as the basic concept is defined in terms of an object moving in a straight line.

logarithm: the power to which a fixed numerical base (the default is 10) must be raised to produce a given number.

longitudinal wave: a type of wave wherein the medium is displaced in a direction parallel to the movement of energy, as in the case of sound waves.

Lorentz transformation: a means of describing relative motions through space according to the mathematics of relativity.

loudness: the intensity of sound waves, which depends on the wave's amplitude; measurements of loudness or volume are expressed in decibels.

magnetic flux: the measure of the strength of the magnetic field surrounding an active electrical conductor or a magnet, equal to the product of the external magnetic field (B), the area of the affected surface (A), and the cosine of the angle between them ($\cos\theta$).

magnetic induction: the generation of magnetism within a magnetizable material by proximity to a magnetic field.

magnitude: the size, or numerical measurement, of a vector.

mass: how much matter there is in an object; measured in kilograms Mass determines the effects of gravitation and inertia. Unlike weight, which is dependent on gravitation, an object's mass remains constant throughout the universe.

mass-spring system: a system consisting of an elastic object connected to an object with mass.

matrix: a mathematical notation in which a series of coordinates is written in an array and can then be manipulated according to set rules.

matter: anything that can be characterized as having mass and can be measured by some criterion of measurement.

maximum: the point of greatest wave height, also known as a crest or peak. It is the point of greatest positive displacement from the position of rest in a wave.

Maxwell's equations: the mathematical relationships governing electromagnetism, as formulated by James Clerk Maxwell.

measurement: quantifying an observation (e.g., the length of a person's foot) using discrete units (e.g. meters); alternately, the unit or system used to do so.

mechanical advantage: the amplification of force provided by many machines, measured as the ratio of the output force to the input force.

mechanical efficiency: the ratio of the power applied to a mechanical system relative to the power delivered by the system.

mechanical resonance: the increase in the amplitude of motion by a medium in response to a periodic applied force or a sympathetic vibration close to the fundamental frequency.

mechanical switches: devices that use moving parts and direct physical force to bring contacts together to let electricity flow.

meter: the SI base unit of distance (or length) measurement.

minimum: the point of greatest wave depth, also known as a trough or valley. It is the point of greatest negative displacement from the position of rest in a wave.

mixing: being able to transform a dynamic system with multiple phase spaces in its initial state in more than one way over time to reach a target state that has completely overlapping phase spaces.

mole: a quantity of a pure substance made up of exactly $6.02214129 \times 10^{23}$ atoms or molecules, the mass of which in grams has the same value as the mass of one atom or molecule of the substance in unified atomic mass units.

moment: a combination of a physical quantity and a distance with respect to a fixed axis; the physical quantities of an object as measured at some distance from that axis.

momentum: an intrinsic property of matter, the product of mass and velocity.

M-theory: a physical model of the structure of the universe that encompasses all consistent versions of superstring theory.

mutation: in biology, a change in the structure of a gene. Ionizing radiation such as beta radiation can alter the electric charge of the atoms in a gene.

net force: the sum of all of the forces acting on an object; note that forces with equal magnitude but opposite directions negate each other. An object moves in the direction of the net force acting on it.

neutral conductor: in an electrical circuit, the conductor that accepts the current coming out of the component elements (the conductor having the lower electrical potential).

neutrons: subatomic particles that, with protons, make up the mass of an atom's nucleus; they have functionally the same weight as protons but no electric charge.

newton: abbreviated N, the International System of Units unit of force. One newton is equal to the force required to accelerate a one-kilogram mass at one meter per second per second.

Newton meter: a device used to measure force; also called a "force meter" or a "force gauge."

Newton's laws of motion: three laws devised by physicist and mathematician Isaac Newton to describe the motion of objects in relation to the forces acting on them.

Newton's laws of motion: three laws, defined by Isaac Newton, that describe the motion and forces acting on large objects moving at speeds much smaller than the speed of light.

newton-meter: the International System of Units standard unit for torque; one newton-meter (N·m) is equal to the torque resulting at the axis from the force of 1 newton applied perpendicularly to an attached 1-meter-long moment arm (i.e., a lever).

node: a point of minimum amplitude (typically zero) in a standing wave.

noise: in acoustics, any sound whether wanted or not; may also describe background sound that obscures a desired signal or sound.

nonionizing radiation: radiation that lacks the energy necessary to knock electrons free when it hits an atom; all electromagnetic radiation with a longer wavelength than ultraviolet is nonionizing.

non-ionizing radiation: electromagnetic radiation that lacks the energy necessary to knock electrons free when it hits an atom; all types of electromagnetic radiation with a longer wavelength (less energy) than ultraviolet is non-ionizing.

nonmechanical switches: devices that use electromagnetism to open and close a circuit.

nonviscous fluid: a fluid that flows without friction.

normal force: the force exerted on an object perpendicular to the surface of contact.

nucleon: a type of baryon, either a proton or a neutron, that is found in atom's nucleus.

octave: a progression of eight harmonic tones.

Ohm's law: an empirical law stating that the current, or flow of electrical charge, between two points is directly proportional to the voltage, or difference in electric potential, between those points.

open system: a system in which both matter and energy can be exchanged between the system and its surroundings.

optical system: a system of mirrors, lenses, or prisms that can be used for imaging in an instrument such as a telescope or microscope.

optics: the science of the interaction of light with lenses and mirrors.

oscillation: a variation between maximum and minimum values of displacement from a neutral value.

output: the force (or energy) produced by a machine. The machine transforms the input into the output; for instance, a windmill transforms the force of wind (input) into the circular motion of a millstone for grinding (output).

overtone: a frequency that is a whole-number multiple of the fundamental frequency of a resonating medium.

parabolic: refers to a shape (a parabola) that can be described by an equation of the form $y = ax^2 + b$.

parallel: circuits or segments of circuits in which any two different elements are joined at two common points side by side.

peak amplitude: the value of the amplitude at its maximum displacement from the neutral value of the wave.

peak-to-peak amplitude: the absolute value of the sum of the peak positive and peak negative amplitudes; the distance between the crest and the trough of a wave.

pendulum: a suspended mass that can undergo regular oscillations.

perception of sound: the ability to perceive mechanical waves in one's environment (e.g., air, water) as sounds, limited by the physiology of the ear and the brain's ability to interpret information from sound waves.

period: the length of time for one complete cycle of a wave or other cyclic property to occur.

periodicity: the extent to which a property repeats over time; regular recurrence.

permeability: the ability with which a magnetized material supports a magnetic field.

permittivity: a measure of the ability of an electric field to penetrate a medium.

perpendicular: set at 90 degrees to a line or surface, forming a right angle.

perturbation: a disturbance in the motion or orbit of a massive body due to the gravitational pull of or impact with another object.

perturbation theory: the application of known solutions to simple quantum mechanical systems (such as that of the hydrogen atom) to more complex systems that cannot be resolved easily.

phase: a stage in a wave property; typically used to describe the relationship of two or more waves.

phase offset: also called phase difference; the time interval or phase angle that results when one wave is ahead of or behind another.

phase space: a space containing all possible states of the particles in a given system, wherein each state is represented by a single point.

phon scale: a unit of measure for the perceived loudness of sounds; compares all sounds to a baseline sound of 1,000 hertz (1 kilohertz) frequency.

photoabsorption: the absorption of the energy of electromagnetic radiation into matter; different substances absorb radiation at different rates.

photoelectric effect: a phenomenon that describes the emission of electrons from matter (typically metal) upon exposure to electromagnetic radiation.

photon: a massless elementary particle that is the smallest possible unit, or quantum, of light and other electromagnetic radiation.

pi: the ratio of the circumference of a circle to its diameter, symbolically represented as π. Its numerical value is approximately 3.14159.

pincushion distortion: the optical distortion caused by a telephoto lens that makes an image seem bunched up in the center.

pitch: the perceived relative frequency of audible sound waves.

Planck's law: a mathematical description of the amount of radiation emitted at different frequencies by a blackbody at a given temperature.

polar coordinates: a pair of numbers indicating a point's length (radius) from a fixed center, called the pole or origin, and the angle between the radial and the polar axes.

polarization: the process by which the motion of electromagnetic waves, which is perpendicular to the direction of energy transfer, is brought into alignment in the same plane.

poles: contact points that complete the circuits in switches.

position: in quantum mechanics, an electron's location in space relative to the nucleus of an atom.

positron: a subatomic particle that has the same mass as an electron as well as an equal-but-opposite electric charge, that is, a positive charge.

potential energy: the energy stored within an object due to its position (e.g., gravitational pull on a stationary object above the ground) or its configuration (e.g., an electrical charge or chemical makeup).

pound: a standard unit of mass in the imperial and US customary measurement systems, equivalent to 0.45359237 kilograms; also used as a unit of force.

power: the rate of work (energy transfer) over time; the International System of Units unit of power is the watt (W), which equals one joule per second (J/s).

power rating: the maximum electrical power a device can use without being damaged.

precision: the extent to which different measurements of the same property differ from one another.

pressure: the force exerted per unit area.

probability density function: the math function that describes the probability of an electron being found in a defined region of space about an atomic nucleus.

protons: subatomic particles that, with neutrons, make up the mass of an atom's nucleus; they have functionally the same weight as neutrons but hold a positive electric charge.

pulse: a lone disturbance passing through a medium from one place to another, similar to a wave but not cyclical and repeating.

quantization distortion: a distortion common in digital signal processing, such as encoding music as audio files, that results from mapping an original signal of possibly infinite values to a signal with a smaller, countable set of values.

quantum chromodynamics: a quantum field theory that describes the interactions of quarks and gluons, subatomic particles that are responsible for the strong interaction.

quantum field theory: a theory that explains interactions between subatomic particles as the result of a field extending between them.

quantum mechanics: the branch of physics that deals with matter interactions on a subatomic scale, based on the concepts that energy is quantized, not continuous, and that elementary particles exhibit wavelike behavior.

quantum state: the condition of a physical system as defined by its associated quantum attributes.

quark: an elementary fermion that combines with other quarks to form a baryon, such as a proton or a neutron, or with an antiquark to form a particle called a meson.

radian: abbreviated rad, the International System of Units standard unit of angular measure, the length of the corresponding arc in a unit circle. A full circle is composed of 2π radians.

radians: (rad) the SI unit of measure for angles, based on relationship between the radius and circumference of a circle.

radiant energy: energy consisting of electromagnetic radiation; an important mechanism for the transfer of energy in or out of some systems.

radiation: energy transmitted via electromagnetic waves (e.g., light, heat, x-rays) or subatomic particles (e.g., alpha particles, beta particles).

radio waves: electromagnetic radiation with a wavelength between 1×10^{-3} and 1×10^{5} meters; able to travel long distances without being broken up by atmospheric interference.

radioactive decay: the loss of energy and matter from an unstable nucleus in the form of ionizing radiation.

radioisotope: a chemical element with unstable nuclei that give off radiation due to variations in the number of neutrons they contain.

radius: distance from the center of a circle to any point along the circle.

random error: a measurement error that is due to unpredictable and inconsistent factors that do not affect all measurements equally.

range: the horizontal distance that a projectile can travel before striking the ground.

rate: the ratio of a unit, such as distance or weight, relative to a period of time.

Rayleigh scattering: the scattering of electromagnetic radiation when it encounters particles much smaller than the radiation's wavelength, as, for example, the scattering of visible light in the atmosphere.

reaction turbine: a turbine set in motion by the pressure and flow of a fluid.

reciprocal: the inverse of a value, calculated as 1 divided by the value.

redshift: the apparent lengthening of the wavelength of electromagnetic radiation due to the movement of the radiation source away from the observer.

reference frame: the velocity of an object relative to the objects around it and the point of observation.

reflection: the rebounding of a wave from a surface or boundary between two mediums, causing it to travel back through the original medium.

refraction: the alteration of a wave's path, speed, and wavelength when it passes from one medium to another.

relativistic beaming: the effect in which a luminous beam appears brightest when pointing directly at an observer.

renormalization: a mathematical procedure by which processes occurring on very different scales can be aligned.

resistor: a device or material that resists the flow of electrons through it.

resolution: the ability of a detector to differentiate or separate different wavelengths.

resonance: the response of an elastic body to a force acting on the body at its natural frequency.

revolution: describes circular motion wherein an object circles around an external axis (e.g., the moon orbiting Earth); contrast to rotation, wherein the axis is internal (e.g., the moon spinning about its axis).

right-hand rule: a technique used to find the direction of magnetic forces and induced currents by using the right hand. The direction of the current is expressed by the fingers, and the direction of the magnetic field is expressed by the extended thumb. If a wire is grasped in the right hand with the thumb pointing in the direction of current flow, the fingers will point in the direction of the magnetic field around the wire.

rigid body: an idealization of a solid object that assumes that it cannot be deformed by the forces acting on it.

root-mean-square amplitude: for sinusoidal wave systems, the square root of the sum of squared amplitude values divided by the number of amplitude values.

rotational energy: the kinetic energy of a spinning object.

rotational speed (rpm): the number of times an object rotates about a fixed axis in a set amount of time, typically measured in revolutions per minute (rpm).

scalar: a quantity that is fully described by a numerical value alone, such as speed, length, or mass.

scale: in music, a group of notes arranged by pitch or frequency. A description of the area under observation in broad approximation.

Schrödinger equation: a wave equation that describes the quantum state of a particle or system and can be used to predict its most likely future behavior.

sensitivity: the ability of an ear or a mechanical device to pick up and interpret sound; highly sensitive devices will have a smaller just-noticeable difference values.

series: circuits or segments of circuits in which any two different elements are joined only at one common point end to end.

signal-to-noise-and-distortion (SINAD) ratio: the ratio of total signal power received to noise and distortion received, indicating how well a signal or waveform has been reproduced.

simple machine: the simplest devices capable of generating a mechanical advantage.

simple machine: a simple mechanical device that redirects or amplifies force.

sine: a trigonometric function describing the relationship between sides of a right triangle; the sine of an angle is equal to the length of the side opposite the angle divided by the length of hypotenuse.

sinusoidal: having a shape or pattern of behavior that can be described by a sine wave function.

sinusoidal function: a curve that is like a sine wave but experiences a shift in amplitude or phase.

slope: a line on a graph that indicates the rate of change over time; for instance, on a plot of the distance traveled by an object over time, the slope equals the speed of travel and its shape can convey information about acceleration.

slug: a unit of mass in the foot-pound-second (FPS) system, defined as the amount of mass that will experience an acceleration of one foot per second squared (1 ft/s^2) when acted upon by one pound-force (lbF); equivalent to approximately 14.5939 kilograms or 32.1740 pounds.

Snell's law: a mathematical description of the refraction of light as it goes from one medium to another.

sonar: a method of using sound waves to "see," typically in an underwater environment, by sending and receiving pulses of sound; originally an acronym for "sound navigation ranging."

sone: a unit for quantifying perceived loudness; one sone equals forty phons.

sonography: the use of ultrasound in medicine to produce images of internal structures, such as tendons, muscles, and organs.

sound barrier: the effects created when an object travels faster than the local speed of sound.

special relativity: the theory that states that for all inertial nonaccelerating reference frames, the laws of motion remain the same, and that speed of light in a vacuum is the same for all observers, regardless of the observer's movement relative to the source of light or the movement of the source itself.

specific gravity: the ratio of the density of a substance to that of a standard reference substance; also known as relative density.

speed: the distance traveled per unit of time.

speed of sound: a sound wave's traveled distance per unit of time.

spherical aberration: the blurring or distortion of an image produced by a spherical lens or mirror due to differences in the refraction of light that enters the optical system along the optical axis and light that enters closer to the edge of the lens.

spin: an intrinsic form of angular momentum carried by elementary particles, composite particles (hadrons), and atomic nuclei.

spring constant: a mathematical value that defines the stiffness of a spring. A characteristic factor of a particular spring that determines the expansion or contraction of the spring when displaced by a specific force.

standard enthalpy of formation: the change in enthalpy that results from the formation of one mole of a given substance, with the reactants and the products in their standard states.

standard model: a generally accepted unified framework of particle physics that explains electromagnetism, the weak interaction, and the strong interaction as products of interactions between different types of elementary particles.

standard state: a set of standard thermodynamic reference conditions for a given substance, generally calculated at a temperature of 298.15 kelvins (25 degrees Celsius or 77 degrees Fahrenheit) and a pressure of 105 pascals (14.5 pounds per square inch).

standard temperature and pressure (STP): standard reference conditions when dealing with gases, defined by the International Union of Pure and Applied Chemistry as a temperature of 273.15 kelvins (0 degrees Celsius or 32 degrees Fahrenheit) and pressure of 101.3 kilopascals (1 atmosphere); used in chemistry and physics to establish a standardized set of conditions for experimentation.

standing wave: a wave pattern that maintains a static series of nodes and antinodes within a specific wavelength.

state variables: external factors, such as temperature and pressure, that determine the physical state of matter.

static energy: electrical energy resulting from an imbalance in electrical charges.

static pressure: the pressure of a fluid on an object when that object is at rest relative to the fluid.

Stefan-Boltzmann law: a mathematical description of the total radiant energy emitted by a blackbody, relating it to the temperature of the blackbody raised to the fourth power.

strong force: the force that binds nucleons together in a nucleus.

strong interaction: the fundamental process of particle interaction that binds quarks into hadrons and hadrons into nuclei.

superposition: in quantum mechanics, the concept that a particle exists in all possible states at the same time until either its position or its energy is known.

supersymmetry: the theory that all bosons have partner fermion particles

surface area: the total area of the outward-facing surface of a three-dimensional object.

synchrotron radiation: radiation emitted by the acceleration of a charged particle traveling in a magnetic field at near light speed.

system boundary: a physical or conceptual delineator between a system and the outside environment.

system efficiency: the proportion of a system's input it converts to the intended output.

system elements: the individual components that work together and make up an overall system used to complete a task.

system input: a force, such as voltage or torque, applied to a system.

system output: the work that the system performs.

systematic error: a measurement error that is due to intrinsic, mechanical, or environmental factors that affect all measurements equally.

tangent: in a right triangle, the ratio of the length of the side opposite an acute angle to the length of the side adjacent to the same angle.

temperature: the average kinetic energy of particles that make up a substance.

temperature gradient: a measurement of the rate of temperature change over distance.

tensile strength: the ability of a material to resist structural failure when experiencing a force of tension (pulling).

tension: the force directed along the length of a wire, string, or cable pulled at opposite ends.

terminal velocity: the velocity at which the acceleration due to gravity of an object falling freely through a fluid medium, such as air or water, is exactly balanced by the deceleration of the object due to resistance from that medium.

therm: the energy released by burning 100 cubic feet (2.83 cubic meters) of natural gas; equal to 1.02×10^5 BTU or 1.08×10^8 joules.

thermal conductivity: the ease with which heat propagates through a material; measured in watts per meter kelvin.

thermal efficiency: the ratio of the work performed by a system relative to the heat energy that is supplied to the system.

thermal energy: energy generated by the movement of particles within an object or substance.

thermal radiation: electromagnetic radiation generated by charged particles in matter being moved around by heat; typically associated with infrared radiation and a frequency of 7×10^{-7} meters to 1×10^{-3} meters.

thermodynamic processes: processes of change within a system that are related to its entropy and enthalpy.

thermodynamics: the study of the relationships between heat, energy, and work in a system.

throws: sets of input/output wires in a switch.

timbre: the quality of a tone or sound that is unique from its loudness and pitch and allows one to distinguish between different sound sources.

ton of coal equivalent: the energy released by burning 1 US short ton (0.91 metric ton) of coal; equal to 19.49×10^6 BTU or 21.28×10^{10} joules.

tone: a vocal or musical sound with a specific pitch and loudness.

torque: the tendency of a force to cause an object to rotate, defined mathematically as the rate of change of the object's angular momentum; also called moment of force.

total mechanical energy: the sum of all the kinetic and potential energies of an object in a closed system.

trajectory: the path of a thrown or falling object, such as a baseball.

transverse: type of wave that displaces its medium perpendicular to the direction of energy transfer. Ocean waves, for example, move water vertically but transmit energy horizontally across the surface.

transverse magnification: also known as lateral magnification; synonymous with linear magnification.

transverse wave: a type of wave wherein the medium is displaced in a direction perpendicular to the movement of energy, as in the case of waves on the surface of water.

trough: the lowest point of a wave from its neutral value.

turns: the number of times that a conductor is wrapped around to form a helical coil.

two-body problem: a mathematical description of the motion in space of two rigid, point-like objects interacting with each other.

ultrasonic: describes a sound, or a device that makes use of sound, with a frequency above twenty thousand hertz, the upper limit of human hearing.

ultrasonic impact treatment: the use of ultrasonic vibrations to strengthen metals.

ultrasonic testing: the use of ultrasound to test materials for internal flaws.

ultrasound identification: a real-time locating system that uses ultrasound to track the location of objects or individuals on small spatial scales.

ultraviolet (UV) radiation: electromagnetic radiation with more energy than visible light but less than x-rays, in the wavelength range of 4×10^{-7} to 1×10^{-7} meters.

ultraviolet catastrophe: the erroneous prediction, based on the laws of classical physics, that a blackbody would emit an infinite amount of energy at short wavelengths, starting around the ultraviolet region.

uncertainty principle: the idea, proposed by Werner Heisenberg, that one can determine with high precision either the position of a particle at a given time or its momentum, but not both.

UV degradation: the damage caused by ultraviolet radiation when it strikes certain materials.

vacuum ultraviolet radiation: a subtype of ultraviolet radiation with wavelength under 200 nanometers.

vector: a geometric concept describing both the magnitude of a property and the direction in which it operates, to be fully described, such as velocity or acceleration.

velocity: a vector quantity that includes both the speed and the direction of motion.

velocity: a vector quantity that describes the rate of displacement over time. The rate of change of position in a specified direction of motion

vertex: the uppermost point in a ballistic trajectory.

vignetting: a darkening or shading of an image's edges compared to the center of the image.

virtual image: an image that forms at the point where the paths of rays cross when projected backward from a lens.

visible light: electromagnetic radiation that human eyes can see, with a wavelength between 4×10^{-7} to 7×10^{-7} meters.

voltage: the difference in electric potential between two points, measured in volts; electric current flows naturally from the higher-voltage point to the lower-voltage point.

volume: the space occupied by an object or substance, measured in cubic meters (m^3); together with mass, it determines the density of an object and its specific gravity. Alternative units include liters (L) or cubic feet (ft^3).

warping: the digital manipulation of an image that mimics distortion and can be used to correct for distortion or to introduce various creative effects.

watt: the SI unit for measuring power (work over time), defined as one joule per second.

wave function: a function that describes the quantum state of a system and represents the probability of finding the system in a given state at a given time.

wave propagation: the manner in which a wave travels.

wavelength: in any wave system, the distance from one point in a wave to the equivalent point in the next wave, typically measured between successive peak values.

wave-particle duality: the idea that a particle can behave as either a particle or a wave, depending on how and when it is being observed.

weak interaction: interaction between subatomic particles at a short distance that is influenced by the weak nuclear force, one of the four fundamental forces in nature.

weight: the downward force imparted to an object by gravity acting on its mass; measured according to the International System of Units (SI) in newtons, though it is normally expressed in units of mass for everyday objects on Earth.

work: the use of energy to move an object over a distance by means of the application of force.

work: a force successfully moving an object, or the successful transfer of energy. The International System of Units unit of work is the joule.

work-energy theorem: the principle that the work performed on an object is equal to the change in that object's kinetic energy.

x-ray: electromagnetic radiation with a wavelength between 1×10^{-10} and 1×10^{-8} meters.

BIBLIOGRAPHY

Abu-Zidan, Fikri M., Ashraf F. Hefny, and Peter Corr. "Clinical Ultrasound Physics." *Journal of Emergencies, Trauma, and Shock*, 4.4 (2011): 501–3. Print.

Ackerman, Steven A., and John A. Knox. *Meteorology: Understanding the Atmosphere*. 3rd ed. Sudbury: Jones, 2012. Print.

Aczel, Amir D. *Uranium Wars: The Scientific Rivalry That Created the Nuclear Age*. New York: St. Martin's, 2009. Print.

Albin, Edward F. *Earth Science Made Simple*. New York: Broadway, 2004. Print.

Alexander, Charles, and Matthew Saddiku. *Fundamentals of Electric Circuits*. 5th ed. New York: McGraw, 2012. Print.

Al-Khalili, Jim. *Quantum: A Guide for the Perplexed*. London: Phoenix, 2012. Print.

Allen, Elizabeth. *The Manuel of Photography*. 10th ed. Oxford: Focal, 2011. Print.

Allen, James P. *Biophysical Chemistry*. Hoboken: Wiley, 2008. Print.

Amaldi, Ugo, and Adele La Rana. *Particle Accelerators: From Big Bang Physics to Hadron Therapy*. Cham [Switzerland]: Springer, 2015. Print.

Amer. Industrial Hygiene Assn. *Radio-Frequency and Microwave Radiation*. 3rd ed. Fairfax: Amer. Industrial Hygiene Assn., 2004. Print.

Anders, André. *Cathodic Arcs From Fractal Spots to Energetic Condensation*. New York, NY: Springer US, 2009. Print. Springer Ser. on Atomic, Optical, and Plasma Physics.

Anderson, David F., and Scott Eberhardt. *Understanding Flight*. 2nd ed. New York: McGraw-Hill, 2009. Print.

Anderson, G. M. *Thermodynamics of Natural Systems*. Cambridge: Cambridge University Press, 2005. Print.

Anderson, Rosaleen J., David J. Bendell, and Paul W. Groundwater. *Organic Spectroscopic Analysis*. Cambridge: Royal Soc. of Chemistry, 2004. Print.

Andrews, David G. *An Introduction to Atmospheric Physics*. 2nd ed. New York: Cambridge University Press, 2010. Print.

Ang, Tom. *Digital Photography Masterclass*. 2nd Amer. ed. New York: DK, 2013. Print.

Anton, Howard. *Elementary Linear Algebra*. 11th ed. Hoboken: Wiley, 2013. Print.

Archimedes. *The Works of Archimedes*. Ed. Sir Thomas Heath. Newburyport: Dover Publications, 2013. Print. Dover Books on Mathematics.

Aruldhas, G. *Quantum Mechanics*. 2nd ed. New Delhi: PHI, 2009. Print.

Askerzade, Iman. *Unconventional Superconductors : Anisotropy and Multiband Effects*. Berlin: Springer-Verlag, 2014. Print. Springer Ser. in Materials Science.

Atkins, P. W., and P. W. Atkins. *The Laws of Thermodynamics : A Very Short Introduction*. Oxford; New York: Oxford University Press, 2010. Print.

Atkins, P. W., Julio De Paula, and Ronald Friedman. *Physical Chemistry: Quanta, Matter, and Change*. 2nd ed. New York: W.H. Freeman, 2014. Print.

Atkins, Peter, and Julio de Paula. *Atkins' Physical Chemistry*. 10th ed. Oxford: Oxford University Press, 2014. Print.

Atkins, Peter, and Julio de Paula. *Atkins' Physical Chemistry*. 10th ed. Oxford: Oxford University Press, 2014. Print.

Atkins, Peter. *The Laws of Thermodynamics: A Very Short Introduction*. New York: Oxford University Press, 2010. Print.

Aynajian, Pegor. *Electron-phonon Interaction in Conventional and Unconventional Superconductors*. Berlin: Springer-Verlag, 2013. Print. Springer Theses.

Baggott, J. E. *The Quantum Story: A History in 40 Moments*. New York: Oxford University Press, 2011. Print.

Bakshi, U. A., A. V. Bakshi, and K. A. Bakshi. *Electronic Measurement Systems*. 2nd rev. ed. Pune: Technical, 2009. Print.

Balkan, N., ed. *Hot Electrons in Semiconductors : Physics and Devices*. Oxford; New York: Clarendon ; Oxford University Press, 1998. Print.

Band, Y. B. *Light and Matter: Electromagnetism, Optics, Spectroscopy and Lasers*. Chichester: Wiley, 2006. Print.

Bass, Michael, ed. *Handbook of Optics*. 3rd ed. 5 vols. New York: McGraw, 2010. Print.

Beament, James. *How We Hear Music: The Relationship between Music and the Hearing Mechanism*. Rochester: Boydell, 2001. Print.

Becker, Katrin, Melanie Becker, and John Schwarz. *String Theory and M-Theory. A Modern Introduction.* New York: Cambridge University Press, 2007. Print.

Beddard, G. S. *Applying Maths in the Chemical and Biomolecular Sciences: An Example-based Approach.* Oxford; New York: Oxford University Press, 2009. Print.

Beech, Martin. *The Pendulum Paradigm: Variations on a Theme and the Measure of Heaven and Earth.* Boca Raton: Brown, 2014. Print.

Bejan, Adrian. *Convection Heat Transfer.* Hoboken: Wiley, 2013. Digital file.

Bemelmans, Josef, Giovanni P. Galdi, and Mads Kyed. "Fluid Flows around Floating Bodies, I: The Hydrostatic Case." *Journal of Mathematical Fluid Mechanics* 14.4 (2012): 751–70. Print.

Ben-Naim, Arieh. *A Farewell to Entropy: Statistical Thermodynamics Based on Information: S=logW.* Singapore: World Scientific, 2011. Print.

Bennett, Jeffrey O., et al. *The Essential Cosmic Perspective.* 7th ed. Boston: Pearson, 2015. Print.

Berg, Richard E., and David G. Stork. *The Physics of Sound.* 3rd ed. San Francisco; Toronto: Pearson Addison Wesley, 2005. Print.

Bewoor, Anand K., and Vinay A. Kulkarni. *Metrology & Measurement.* New Delhi: Tata, 2009. Print.

Bhattacharya, Raghu, and M. P. Paranthaman, eds. *High Temperature Superconductors.* Weinheim: Wiley-VCH, 2010. Print.

Biermann, Peter L., et al. "Active Galactic Nuclei: Sources for Ultra High Energy Cosmic Rays!" *Institute for Nuclear Theory.* University of Washington, 20 Feb. 2008. Web. 10 Aug. 2015.

Bird, John. *Electrical and Electronic Principles and Technology.* 5th ed. London [u.a.]: Routledge, 2014. Print.

Blum, Walter, Werner Riegler, and Luigi Rolandi. *Particle Detection with Drift Chambers.* Berlin: Springer, 2010. Print.

Blumenhagen, Ralph, Dieter Lüst, and Stefan Theisen. *Basic Concepts of String Theory.* New York: Springer, 2013. Print.

Bodanis, David. $E = mc^2$: *A Biography of the World's Most Famous Equation.* Toronto: Random, 2001. Print.

Bolton, J. *Classical Physics of Matter.* Philadelphia: Inst. of Physics Pub., 2000. Print.

Borgnakke, Claus, and Richard E. Sonntag. *Fundamentals of Thermodynamics.* 8th ed. Hoboken: Wiley, 2013. Print.

Bowman, Gary E. *Essential Quantum Mechanics.* New York: Oxford University Press, 2008. Print.

Braslavsky, Silvia E. *Glossary of Terms Used in Photochemistry.* 3rd ed. Spec. issue of *Pure and Applied Chemistry* 79.3 (2007): 293–465. PDF file.

Breeze, Paul. *Power Generation Technologies.* Philadelphia: Elsevier, 2014. Print.

Brookes, A. M. P. *Basic Electric Circuits.* 2nd ed. Elmsford: Pergamon, 2014. Print.

Brown, Laurie M. *Feynman's Thesis: A New Approach to Quantum Theory.* Singapore: World Scientific, 2010. Print.

Brumfiel, Geoff. "Common Interpretation of Heisenberg's Uncertainty Principle Is Proved False." *Scientific American.* Scientific Amer., 11 Sept. 2012. Web. 10 June 2015.

Bub, Jeffrey. *Interpreting the Quantum World.* 1997. New York: Cambridge University Press, 1999. Print.

Calle, Carlos I. *Superstrings and Other Things: A Guide to Physics.* 2nd ed. Boca Raton: CRC, 2010. Print.

Calmet, Xavier, ed. *Quantum Aspects of Black Holes.* Cham: Springer, 2015. Print.

Cao, Tian Yu. *From Current Algebra to Quantum Chromodynamics: A Case for Structural Realism.* New York: Cambridge University Press, 2010. Print.

Capecchi, Danilo. *The Problem of the Motion of Bodies: A Historical View of the Development of Classical Mechanics.* Cham: Springer, 2014. Print.

Cappelli, Andrea, et al. *The Birth of String Theory.* New York: Cambridge University Press, 2012. Print.

Cardarelli, François. *Encyclopaedia of Scientific Units, Weights, and Measures: Their SI Equivalences and Origins.* London; New York: Springer, 2003. Print.

Ceraolo, Massimo, and Davide Poli. *Fundamentals of Electric Power Engineering: From Electromagnetics to Power Systems.* Hoboken: Wiley, 2014. Print.

Chabay, Ruth W., and Bruce A. Sherwood. *Matter and Interactions.* 4th ed. Hoboken: Wiley, 2015. Print.

Chaichian, Masud, Hugo Perez Rojas, and Anca Tureanu. *Basic Concepts in Physics: From the Cosmos to Quarks.* New York: Springer, 2014. Print.

Chen, Y. T., and Alan Cook. *Gravitational Experiments in the Laboratory.* New York: Cambridge University Press, 2005. Print.

Cheng, Ta-Pei. *Einstein's Physics: Atoms, Quanta, and Relativity Derived, Explained, and Appraised*. Oxford: Oxford University Press, 2014. Print.

Chiras, Daniel D, Mick Sagrillo, and Ian Woofenden. *Power from the Wind*. Gabriola Island: New Soc., 2009. Print.

Choppin, Gregory R., Jan-Olov Liljenzin, Jan Rydberg, and Christian Ekberg. *Radiochemistry and Nuclear Chemistry*. 4th ed. Amsterdam: Academic, 2013. Print.

Chow, Tai L. *Introduction to Electromagnetic Theory: A Modern Perspective*. Boston: Jones, 2006. Print.

Clark, Ronald William. *Edison: The Man Who Made the Future*. London: Bloomsbury Reader, 2011. Print.

Clery, Daniel. *A Piece of the Sun: The Quest for Fusion Energy*. New York: Overlook, 2014. Print.

Cleveland, Cutler J. "Horsepower." *Encyclopedia of Earth*. Environmental Information Coalition, 8 Oct. 2007. Web. 13 Mar. 2015.

Coddington, Richard C. "Inertial Frame, Euler's First Law." *Department of Agricultural and Biological Engineering*. University of Illinois at Urbana-Champaign, 2015. Web. 27 Aug. 2015.

Concise Dictionary of Physics. Hyderabad: V & S, 2012. Print.

Conrady, A. E. *Applied Optics and Optical Design*. Newburyport: Dover, 2013. Print.

Conte, Mario, and William W. MacKay. *An Introduction to the Physics of Particle Accelerators*. 2nd ed. Hackensack, NJ: World Scientific, 2013. Print.

Cooper, Christopher. *The Basics of Electric Current*. New York: Rosen, 2015. Print.

Cooper, Malcolm, et al. *X-Ray Compton Scattering*. Oxford: Oxford University Press, 2004. Print. Oxford Ser. on Synchrotron Radiation 5.

Corcoran, Aaron J., Jesse R. Barber, and William E. Conner. "Tiger Moth Jams Bat Sonar." *Science* 325.5938 (2009): 325–27. Print.

Costanti, Felice. "The Golden Crown: A Discussion." *The Genius of Archimedes: 23 Centuries of Influence on Mathematics, Science and Engineering; Proceedings of an International Conference Held at Syracuse, Italy, June 8–10, 2010*. Ed. Stephanos A. Paipetis and Marco Ceccarelli. Dordrecht: Springer, 2010. 215–26. Print.

Creath, Katherine, Chandrasekhar Roychoudhuri, and Al F. Kracklauer. *The Nature of Light : What Is a Photon?* Boca Raton: CRC, 2008. Print.

Cross, Rod. *Physics of Baseball and Softball*. New York: Springer, 2011. Print.

Cutnell, John D., Kenneth W. Johnson, Shane Stadler, and David Young. *Physics*. 10th ed. Hoboken, NJ: Wiley, 2015. Print.

Dahl, Jens Peder,. *Introduction to the Quantum World of Atoms and Molecules*. New Jersey: World Scientific, 2009. Print.

Darrigol, Olivier. *A History of Optics from Greek Antiquity to the Nineteenth Century*. New York: Oxford University Press, 2012. Print.

De Pree, Christopher Gordon, and Ira Maximilian Freeman. *Physics Made Simple*. New York: Broadway, 2004. Print.

DeLaney, Thomas F., and Hanne M. Kooy. *Proton and Charged Particle Radiotherapy*. Philadelphia: Wolters Kluwer Health/Lippincott Williams & Wilkins, 2008. Print.

Demi, Marcello. "The Basics of Ultrasound." *X-Ray and Ultrasound Imaging*. Ed. Daniele Panetta and Demi. Waltham: Elsevier, 2014. 297–322. Print. Vol. 2 of *Comprehensive Biomedical Physics*. Anders Brahme, gen. ed. 10 vols. 2014.

Denny, Mark. *Their Arrows Will Darken the Sun: The Evolution and Science of Ballistics*. Baltimore: Johns Hopkins University Press, 2011. Print.

Devaney, Robert L. *An Introduction to Chaotic Dynamical Systems*. 2nd ed. Boulder: Westview, 2005. Print.

Dick, Rainer. *Advanced Quantum Mechanics: Materials and Photons*. New York: Springer-Verlag, 2014. Print.

Dorf, Richard C., and James A. Svoboda. *Introduction to Electric Circuits*. 9th ed. Hoboken: Wiley, 2014. Print.

Dresig, Hans, and Franz Holzweißig. *Dynamics of Machinery: Theory and Applications*. Berlin: Springer, 2010. Print.

Dugdale, J. S. *Entropy and Its Physical Meaning*. London; Bristol, PA: Taylor & Francis, 1996. Print.

Duree, Galen, Jr. *Optics for Dummies*. Hoboken: Wiley, 2011. Print.

Eckert, Michael. *The Dawn of Fluid Dynamics: A Discipline between Science and Technology*. Weinheim: Wiley, 2006. Print.

Einstein, Albert. *Relativity: The Special and the General Theory*. Trans. Robert W. Lawson. London: Folio Soc., 2004. Print.

Ellis, John. "The Discovery of the Gluon." *50 Years of Quarks*. Ed. Harald Fritzsch and Murray Gell-Mann. Hackensack: World Scientific, 2015. 189–98. Print.

Enss, Christian. *Cryogenic Particle Detection*. 2nd ed. Berlin: Springer, 2005. Print.

Ersoy, Okan K. *Diffraction, Fourier Optics and Imaging*. Hoboken: Wiley, 2007. Print.

Evans, Robert L. *Fueling Our Future: An Introduction to Sustainable Energy*. Cambridge: Cambridge University Press, 2008. Print.

Everest, F. Alton, and Ken C. Pohlmann. *Master Handbook of Acoustics*. 6th ed. New York: McGraw, 2015. Print.

Fayer, Michael D. *Absolutely Small: How Quantum Theory Explains Our Everyday World*. New York: AMACOM, 2010. Print.

Fernando, H. J. S. *Handbook of Environmental Fluid Dynamics*. Boca Raton: CRC, 2013. Print.

Feynman, Richard P., and Ralph Leighton. *What Do YOU Care What Other People Think?: Further Adventures of a Curious Character*. New York: Norton, 1988. Print.

Feynman, Richard P., Ralph Leighton, and Edward Hutchings. *"Surely You're Joking, Mr. Feynman!": Adventures of a Curious Character*. New York: W.W. Norton, 1985. Print.

Feynman, Richard P., Robert B. Leighton, and Matthew L. Sands. *The Feynman Lectures on Physics*. Vol. 1-3. Reading, Mass.: Addison-Wesley Pub., 1963. Print.

Finn, J. Michael. *Classical Mechanics*. Sudbury, MA: Jones and Bartlett, 2010. Print.

Finston, David R., and Patrick J. Morandi. *Abstract Algebra: Structure and Application*. Cham: Birkhäuser, 2014. Print.

Fisher, Len. *How to Dunk a Doughnut: The Science of Everyday Life*. New York: Arcade, 2003. Print.

Fitzpatrick, Richard. *Maxwell's Equations and the Principles of Electromagnetism*. Hingham: Infinity Science, 2008. Print.

Fleisch, Daniel, and Laura Kinnaman. *A Student's Guide to Waves*. New York: Cambridge University Press, 2015. Print.

Fleisher, Paul. *Doppler Radar, Satellites, and Computer Models: The Science of Weather Forecasting*. Minneapolis: Lerner, 2011. Print.

Fleishman, Gregory D., and Igor N. Toptygin. *Cosmic Electrodynamics: Electrodynamics and Magnetic Hydrodynamics of Cosmic Plasmas*. New York: Springer, 2013. Print.

Ford, Kenneth W. *101 Quantum Questions: What You Need to Know about the World You Can't See*. Cambridge: Harvard University Press, 2011. Print.

Foulkes, Frank R. *Physical Chemistry for Engineering and Applied Sciences*. Boca Raton: CRC, 2013. Print.

Freegarde, Tim. *Introduction to the Physics of Waves*. Cambridge: Cambridge University Press, 2013. Print.

Gatcum, Chris. *The Beginner's Photography Guide*. New York: DK, 2013. Print.

Gattringer, Christof, and Christian B. Lang. *Quantum Chromodynamics on the Lattice: An Introductory Presentation*. Heidelberg: Springer, 2010. Print.

Ghatak, Ajoy. *Optics*. 5th ed. New Delhi: McGraw, 2012. Print.

Giambattista, Alan, and Betty McCarthy Richardson. *Physics*. 3rd ed. New York: McGraw, 2015. Print.

Gibbons, Patrick C. *Physics*. 2nd ed. Hauppauge, N.Y.: Barron's, 2008. Print. Barrons EZ 101 Study Keys.

Gibilisco, Stan. *Electricity Demystified*. 2nd ed. New York: McGraw, 2012. Print.

Gilbert, P. U. P. A., and W. Haeberli. *Physics in the Arts*. Burlington: Academic, 2008. Print.

Giordano, Nicholas J. *College Physics: Reasoning and Relationships*. Belmont: Brooks, 2010. Print.

Goldsmith, Mike. *Sound: A Very Short Introduction*. New York: Oxford University Press, 2015. Print.

Goldstein, Dennis H. *Polarized Light*. 3rd ed. Boca Raton: CRC, 2011. Print.

Grant, I. S., and W. R. Phillips. *Electromagnetism*. 2nd ed. Chichester: John Wiley, 2011. Print.

Greene, Brian. *The Elegant Universe: Superstrings, Hidden Dimensions, and the Quest for the Ultimate Theory*. New York: Vintage, 2010. Print.

Gregory, R. Douglas. *Classical Mechanics: An Undergraduate Text*. New York: Cambridge University Press, 2006. Print.

Greiner, Walter, ed. *Nuclear Physics: Present and Future*. Cham: Springer, 2015. Print.

Gribbin, John, and Mary Gribbin. *Richard Feynman: A Life in Science*. New York: Dutton, 1997. Print.

Griffiths, David. *Introduction to Elementary Particles*. 2nd rev. ed. Weinheim: Wiley, 2008. Print.

Grissom, Thomas. *The Physicist's World: The Story of Motion and the Limits to Knowledge.* Baltimore: Johns Hopkins University Press, 2011. Print.

Gros, Claudius. *Complex and Adaptive Dynamical System: A Primer.* 4th ed. Cham: Springer, 2015. Print.

Gross, Charles A. *Electric Machines.* Boca Raton: CRC, 2007. Print.

Gross, Dietmar, et al. *Engineering Mechanics 3: Dynamics.* 2nd ed. Berlin: Springer, 2014. Print.

Grupen, Claus, and Irène Buvat. *Handbook of Particle Detection and Imaging.* Berlin: Springer, 2012. Print.

Grupen, Claus, Boris A. Shwartz, and Helmuth Spieler. *Particle Detectors.* Cambridge: Cambridge UP, 2011. Print.

Gubser, Steven S. *The Little Book of String Theory.* Princeton: Princeton University Press, 2010. Print.

Gupta, S. V. *Mass Metrology.* Heidelberg: Springer, 2012. Print.

Gupta, S. V. *Units of Measurement Past, Present and Future: International System of Units.* Berlin: Springer Berlin, 2013. Print.

Halliday, David, Robert Resnick, and Jearl Walker. *Fundamentals of Physics.* 10th ed. Hoboken, NJ: Wiley, 2014. Print.

Halpern, Paul. *Einstein's Dice and Schrödinger's Cat: How Two Great Minds Battled Quantum Randomness to Create a Unified Theory of Physics.* New York: Basic, 2015. Print.

Hamaguchi, Chihiro. *Basic Semiconductor Physics.* 2nd ed. Berlin: Springer Berlin, 2014. Print.

Hamilton, Sue L. *Forensic Ballistics: Styles of Projectiles.* Edina: ABDO, 2008. Print.

Hansen, Klavs. *Statistical Physics of Nanoparticles in the Gas Phase.* Dordrecht: Springer, 2013. Print.

Hartmann, William M. *Principles of Musical Acoustics.* New York: Springer, 2013. Print.

Haynie, Donald T. *Biological Thermodynamics.* Cambridge: Cambridge University Press, 2008. Print.

Heisenberg, Werner. *The Physical Principles of the Quantum Theory.* Trans. Carl Eckart and F. C. Hoyt. 1930. Mineola: Dover, 1949. Print.

Herman, Russell L. *A Course in Mathematical Methods for Physicists.* Boca Raton: CRC, 2014. Print.

Hicks, Tyler G. *Handbook of Mechanical Engineering Calculations.* 2nd ed. New York: McGraw, 2006. Print.

Hiebl, Ewald, and Maurizio Musso, eds. *Christian Doppler: Life and Work, Principle and Applications; Proceedings of the Commemorative Symposia in 2003, Salzburg, Prague, Vienna, Venice.* Pöllauberg: Living Ed., 2007. Print.

Hillesheim, Heather E. *Sound and Light.* New York: Chelsea House, 2012. Print.

Huang, Kerson. *Quantum Field Theory from Operators to Path Integrals.* 2nd ed. Weinheim: Wiley, 2010. Print.

Huber, Martin C. E., et al., eds. *Observing Photons in Space: A Guide to Experimental Space Astronomy.* 2nd ed. New York: Springer, 2013. Print.

Hughes, John M. *Practical Electronics: Components and Techniques.* Sebastopol: O'Reilly, 2015. Print.

Humy, Fernand Emile D'. *The Birth of the Vacuum Tube: "The Edison Effect."* New York: Newcomen Society of England, American Branch, 1949. Print.

Ingard, K. Uno. *Notes on Acoustics.* Hingham, Mass.: Infinity Science, 2008. Print.

Ingledew, John, and Lorentz Gullachsen. *Photography.* 2nd ed. London: King, 2013. Print.

International Union of Pure and Applied Chemistry. Commission on Spectrochemical and Other Optical Procedures and Analysis. *Nomenclature, Symbols, Units and Their Usage in Spectrochemical Analysis: Rules Approved 1975.* Oxford; New York: Pergamon, 1976. Print.

Itzykson, Claude, and Jean-Bernard Zuber. *Quantum Field Theory.* 1980. Mineola: Dover, 2005. Print.

Jauch, Josef M. *Theory of Photons and Electrons.* Berlin [S.l.]: Springer-Verlag, 2012. Print.

Jha, A. K. *A Textbook of Applied Physics.* 2nd ed. New Delhi: I. K. International, 2012. Print.

Johnson, Charles S. *Science for the Curious Photographer: An Introduction to the Science of Photography.* Natick: Peters, 2010. Print.

Johnson, Erin R., and Axel D. Becke. *Density Functionals: Thermochemistry.* Berlin: Springer, 2015. Print.

Johnson, Rebecca L. *Atomic Structure.* Minneapolis: Twenty-First Century, 2008. Print.

Kafatos, Minas C., and Robert Nadeau. *The Conscious Universe: Part and Whole in Modern Physical Theory.* New York: Springer, 1990. Print.

Kakalios, James. *The Amazing Story of Quantum Mechanics: A Math-Free Exploration of the Science That Made Our World.* New York: Penguin, 2011. Print.

Kamal, Anwar. *Nuclear Physics*. Berlin: Springer, 2014. Print.

Kaviany, M. *Heat Transfer Physics*. New York: Cambridge University Press, 2014. Print.

Keller, Ole. *Light: The Physics of the Photon*. Boca Raton: CRC, 2014. Print.

Kelly, P. F. *Electricity and Magnetism*. Boca Raton: CRC, 2015. Print.

Kenkel, John. *Analytical Chemistry for Technicians*. 4th ed. Boca Raton: CRC, 2014. Print.

Kennedy, J. E., G. R. Ter Haar, and D. Cranston. "High Intensity Focused Ultrasound: Surgery of the Future?" *British Jour. of Radiology* 76.909 (2014): 590–99. Print.

Kijewski, Wacek. *SI Units, Conversion and Measurement Skills*. London [u.a.]: Minerva, 1999. Print.

Kinard, Jeff. *Artillery: An Illustrated History of Its Impact*. Santa Barbara: ABC-CLIO, 2007. Print.

King, George C. *Vibrations and Waves*. New York: Wiley, 2013. Print.

Kirkland, Kyle. *Light and Optics*. New York: Facts On File, 2007. Print.

Kirkland, Kyle. *Physical Sciences: Notable Research and Discoveries*. New York: Facts on File, 2010. Print. Frontiers of Science.

Kirshner, Robert P. *The Extravagant Universe: Exploding Stars, Dark Energy, and the Accelerating Cosmos*. Princeton: Princeton University Press, 2004. Print.

Kitamura, Toyoyuki. *Liquid Glass Transition: A Unified Theory from the Two Band Model*. Chennai: Elsevier, 2013. Print.

Klein, Herbert Arthur. *Science of Measurement: A Historical Survey*. 1974. New York: Dover, 2012. Digital file.

Knight, Randall Dewey, Brian Jones, and Stuart Field. *College Physics: A Strategic Approach*. San Francisco: Pearson/Addison Wesley, 2007. Print.

Knight, Randall Dewey. *Physics for Scientists and Engineers: A Strategic Approach*. San Francisco: Pearson/Addison Wesley, 2004. Print.

Koks, Don. *Explorations in Mathematical Physics: The Concepts behind an Elegant Language*. New York: Springer, 2006. Print.

Kouveliotou, Chryssa, Ralph A. M. J. Wijers, and Stanford E. Woosley, eds. *Gamma-Ray Bursts*. New York: Cambridge University Press, 2012. Print. Cambridge Astrophysics Ser. 51.

Krappe, Hans J., and Krzysztof Pomorski. *Theory of Nuclear Fission: A Textbook*. Berlin [u.a.]: Springer, 2012. Print.

Kreuzer, H. J., and Isaac Tamblyn. *Thermodynamics*. Singapore; Hackensack, N.J.: World Scientific, 2010. Print.

Kruth, Patricia, and Henry Stobart, eds. *Sound*. Cambridge; New York: Cambridge University Press, 2007. Print.

Kubat, Milan. *Power Semiconductors*. Reprint of 1984 1st ed. Berlin: Springer-Verlag, 2013. Print.

Kuhn, Hans, Horst-Dieter Försterling, and David Hennessey Waldeck. *Principles of Physical Chemistry*. 2nd ed. Hoboken, N.J: John Wiley, 2009. Print.

Kumar, B. N. *Basic Physics for All*. Lanham: UP of America, 2009. Print.

Kuttruff, Heinrich. *Acoustics: An Introduction*. Boca Raton: CRC, 2014. Print.

Lahiri, Amitabha, and Palash B. Pal. *A First Book of Quantum Field Theory*. 2nd ed. Boca Raton: CRC, 2005. Print.

Laikin, Milton. *Lens Design*. 4th ed. Boca Raton: CRC, 2007. Print.

Lancaster, Tom, and Stephen J. Blundell. *Quantum Field Theory for the Gifted Amateur*. Oxford: Oxford University Press, 2014. Print.

Lang, Thomas G. *Our Fluid Universe: A Unified Theory of Physics that Physically Describes Photons, Matter, Gravity, Electromagnetic Forces, the Strong and Weak Forces, Quantum Theory, and Much More*. CreateSpace Independent Publishing Platform. 2014. Electronic.

Lebed, Andrei, ed. *The Physics of Organic Superconductors and Conductors*. Berlin, Heidelberg: Springer-Verlag Berlin Heidelberg, 2008. Print. Springer Ser. in Materials Science.

Leggett, Anthony J. *Quantum Liquids: Bose Condensation and Cooper Pairing in Condensed-matter Systems*. Oxford [u.a.]: Oxford U, 2015. Print.

Lemons, Don S. *Mere Thermodynamics*. Baltimore: Johns Hopkins University Press, 2009. Print.

Lerner, Lawrence S. *Physics for Scientists and Engineers*. Sudbury: Jones, 1996. Print.

Levi, Mark. *The Mathematical Mechanic: Using Physical Reasoning to Solve Problems*. Princeton: Princeton University Press, 2009. Print.

Levin, Kathryn, Alexander L. Fetter, and Dan M. Stamper-Kurn, eds. *Ultracold Bosonic and Fermionic Gases*. Amsterdam: Elsevier, 2012. Print.

Lewerenz, Hans-Joachim. *Photons in Natural and Life Sciences: An Interdisciplinary Approach.* Berlin: Springer, 2012. Print.

Lide, David R., ed. *CRC Handbook of Chemistry and Physics.* 95th ed. Internet vers. N.p.: Taylor, 2015. Web. 23 Feb. 2015.

Linder, Bruno. *Elementary Physical Chemistry.* Hackensack: World Scientific, 2011. Print.

Longair, Malcolm. *Quantum Concepts in Physics: An Alternative Approach to the Understanding of Quantum Mechanics.* New York: Cambridge University Press, 2013. Print.

Lou, Liang-fu. *Introduction to Phonons and Electrons.* Singapore; River Edge, N.J.: World Scientific, 2003. Print.

Lowe, Doug, and Dickon Ross. *Electronics All-in-One for Dummies.* Hoboken: Wiley, 2012. Print.

Loyd, David H. *Physics Laboratory Manual.* 4th ed. Boston: Brooks, 2014. Print.

Lutz, Josef, Heinrich Schlangenotto, Uwe Scheuermann, and Doncker De Rik. *Semiconductor Power Devices Physics, Characteristics, Reliability.* Berlin: Springer, 2014. Print.

Lynch, David K., and William Livingston. *Color and Light in Nature.* 2nd ed. New York: Cambridge University Press, 2001. Print.

MacDougal, Douglas W. *Newton's Gravity: An Introductory Guide to the Mechanics of the Universe.* New York: Springer, 2012. Print.

Madureira, Nuno Luis. *Key Concepts in Energy.* Cham: Springer, 2014. Print.

Mandl, Franz. *Quantum Mechanics.* 1992. New York: Wiley, 2013. Print.

Mansfield, Michael, and Colm O'Sullivan. *Understanding Physics.* 2nd ed. Hoboken: Wiley, 2012. Print.

Martellucci, Sergio, Arthur N. Chester, Alain Aspect, and Massimo Inguscio. *Bose-Einstein Condensates and Atom Lasers.* Boston, MA: Springer US, 2002. Print.

Matolyak, John, and Ajawad Haija. *Essential Physics.* Boca Raton: CRC, 2013. Print.

Matsushita, Teruo. *Electricity and Magnetism: New Formulation by Introduction of Superconductivity.* Tokyo: Springer, 2014. Print.

Matthews, Michael R., Colin F. Gauld, and Arthur Stinner, eds. *The Pendulum: Scientific, Historical, Philosophical & Educational Perspectives.* Dordrecht: Springer, 2005. Print.

Mehra, Jagdish. *The Beat of a Different Drum: The Life and Science of Richard Feynman.* Oxford [England]; New York: Clarendon ; Oxford University Press, 1994. Print.

Meyers, Robert A. *Mathematics of Complexity and Dynamical Systems.* New York: Springer, 2012. Print.

Mortimer, Robert G. *Physical Chemistry.* 3rd ed. Burlington, MA: Elsevier Academic, 2008. Print.

Moshchalkov, Victor V., and Joachim Fritzsche. *Nanostructured Superconductors.* Singapore: World Scientific, 2011. Print.

Moskowitz, Clara. "Hottest Particle Soup May Reveal Secrets of Primordial Universe." *LiveScience.* Purch, 13 Aug. 2012. Web. 21 Aug. 2015.

Munson, Bruce Roy, et al. *Fundamentals of Fluid Mechanics.* 7th ed. Hoboken: Wiley, 2013. Print.

Murray, Raymond L., and Keith E. Holbert. *Nuclear Energy an Introduction to the Concepts, Systems, and Applications of Nuclear Processes.* 7th ed. Amsterdam: Elsevier, 2015. Print.

Myers, Rusty L. *The Basics of Physics.* Westport: Greenwood, 2006. Print.

National Geographic Society. *The Science of Everything: How Things Work in Our World.* Washington: Natl. Geographic Soc., 2013. Print.

Newman, Jay. *Physics of the Life Sciences.* New York: Springer, 2008. Print.

Newton, Roger G. *Waves and Particles: Two Essays on Fundamental Physics.* Hackensack: World Scientific, 2014. Print.

Nilsson, James William, and Susan A. Riedel. *Electric Circuits.* New York: Prentice, 2008. Print.

Ohanian, Hans C., and John T. Markert. *Physics for Engineers and Scientists.* 3rd extended ed. New York: Norton, 2007. Print.

Oliveira, Mário J. de. *Equilibrium Thermodynamics.* Berlin: Springer, 2013. Print.

Ostdiek, Vern J., and Donald J. Bord. *Inquiry into Physics.* 7th ed. Boston: Brooks, 2013. Print.

Oxford Dictionary of Physics. 7th ed. Oxford: Oxford University Press, 2015. Print.

Pain, H. J., and Patricia Rankin. *Introduction to Vibrations and Waves.* Hoboken: Wiley, 2015. Print.

Parasiliti, Francesco, and Paolo Bertoldi, eds. *Energy Efficiency in Motor Driven Systems.* New York: Springer, 2003. Print.

Parker, Steve. *Electricity.* Rev. ed. New York: DK, 2013. Print.

Pathria, R. K., and Paul D. Beale. *Statistical Mechanics.* 3rd ed. Burlington: Butterworth, 2011. Print.

Pereyra, Pedro. *Fundamentals of Quantum Physics: Textbook for Students of Science and Engineering.* Berlin: Springer, 2012. Print.

Perko, Lawrence. *Differential Equations and Dynamical Systems.* 3rd ed. New York: Springer, 2001. Print.

Petkov, Vesselin. *Relativity and the Nature of Spacetime.* 2nd ed. New York: Springer, 2011. Print.

Pfeiffer, Friedrich. *Mechanical System Dynamics.* Corrected 2nd ed. Berlin: Springer, 2010. Print. Lecture Notes in Applied and Computational Mechanics.

Pickover, Clifford A. *The Physics Book: From the Big Bang to Quantum Resurrection, 250 Milestones in the History of Physics.* New York: Sterling, 2011. Print.

Pitaevskii, Lev P. *Bose-Einstein Condensation and Superfluidity.* New York: Oxford University Press, 2016. Print.

Pitaevskii, Lev, and Sandro Stringari. *Bose-Einstein Condensation.* Oxford: Clarendon, 2013. Print.

Pochapsky, Thomas C., and Susan Sondej Pochapsky. *NMR for Physical and Biological Scientists.* New York: Taylor & Francis, 2007. Print.

Pounder, Elton R. *Physics of Ice.* London: Pergamon, 1965. Print.

Povh, Bogdan, Klaus Rith, Christoph Scholz, and Frank Zetsche. *Particles and Nuclei: An Introduction to the Physical Concepts.* 7th ed. Berlin; Heidelberg [u.a.]: Springer, 2015. Print.

Prater, Edward L. *Basic Machines.* Pensacola Naval Educ. and Training Professional Dev. and Technology Center, 1994. Construction Knowledge. net. Web. 2 Sept. 2015.

Prestini, Elena. *The Evolution of Applied Harmonic Analysis.* Boston: Birkhauser, 2004. Print.

Pulfrey, David L. *Understanding Modern Transistors and Diodes.* New York: Cambridge University Press, 2010. Print.

Purcell, Edward M., and David Morin. *Electricity and Magnetism.* 3rd ed. Cambridge [etc.]: Cambridge University Press, 2013. Print.

Putten, Anton F. P. Van. *Electronic Measurement Systems: Theory and Practice.* Bristol: Institute of Physics Pub., 1996. Print.

Rabinovich, Semyon G. *Evaluating Measurement Accuracy: A Practical Approach.* New York: Springer, 2010. Print.

Radosevich, James A. *UV Radiation: Properties, Effects, and Applications.* New York: Nova, 2014. Print.

Raghavan, Jayakumar. *Particle Accelerators, Colliders, and the Story of High Energy Physics: Charming the Cosmic Snake.* Berlin: Springer Berlin, 2011. Print.

Randall, Lisa. *Higgs Discovery: The Power of Empty Space.* Ecco. 2013.

Rankine, William John Macquorn. "On the General Law of the Transformation of Energy." *Miscellaneous Scientific Papers: By W. J. Macquorn Rankine, CE, LLD, FRS.* Ed. W. J. Millar. 1881. Lexington: Elibron, 2005. 203–8. Print.

Rasmussen, Seth C. *How Glass Changed the World: The History and Chemistry of Glass from Antiquity to the Thirteenth Century.* New York: Springer, 2012. Print.

Reed, Bruce Cameron. *Quantum Mechanics.* Sudbury: Jones, 2008. Print.

Rees, W. G. *Physical Principles of Remote Sensing.* 3rd ed. New York: Cambridge University Press, 2013. Print.

Reger, Daniel L., Scott R. Goode, and David W. Ball. *Chemistry: Principles and Practice.* 3rd ed. Belmont: Brooks, 2009. Print.

Reisberg, Daniel, ed. *Auditory Imagery.* New York: Psychology, 2014. Print.

Reuleaux, Franz. *Kinematics of Machinery: Outlines of a Theory of Machines.* North Chelmsford: Courier, 2012. Print.

Rexford, Kenneth, and Peter R. Giuliani. *Electrical Control for Machines.* 6th ed. New York: Thomson Learning, 2004. Print.

Ridley, B.K. *Quantum Processes in Semiconductors.* 5th Rev ed. Oxford: Oxford University Press, 2013. Print.

Riggs, Shelton. *The Nuclear Force: The Force Which Binds Quark Trios Into Protons and Neutrons.* Amazon Digital Services, LLC, 2015. Electronic.

Robbins, Allan H., and Wilhelm C. Miller. *Circuit Analysis, Theory and Practice.* 5th ed. Clifton Park: Delmar, 2013. Print.

Robinson, Matthew. *Symmetry and the Standard Model: Mathematics and Particle Physics.* New York: Springer, 2011. Print.

Rogers, Alan. *Essentials of Photonics.* 2nd ed. Boca Raton: CRC, 2008. Print.

Rogers, Donald W. *Concise Physical Chemistry.* Hoboken: Wiley, 2011. Print.

Rorres, Chris. "The Turn of the Screw: Optimal Design of an Archimedes Screw." *ASCE Journal of Hydraulic Engineering* 126.1 (2000): 72–80. PDF file.

Rosenblum, Bruce, and Fred Kuttner. *Quantum Enigma: Physics Encounters Consciousness*. 2nd ed. New York: Oxford University Press, 2011. Print.

Rossing, Thomas D., F. Richard Moore, and Paul A. Wheeler. *The Science of Sound*. 3rd ed. N.p.: Pearson, 2014. Print.

Roychoudhuri, Chandrasekhar, A. F. Kracklauer, and Katherine Creath, eds. *The Nature of Light: What Is a Photon?* Boca Raton: CRC, 2008. Print.

Ruban, A. I., and J. S. B. Gajjar. *Fluid Dynamics: Part I, Classical Fluid Dynamics*. London: Oxford University Press, 2014. Print.

Ruina, Andy, and Rudra Pratap. *Introduction to Statics and Dynamics*. N.p.: Oxford University Press, 2010. PDF file.

Sasián, José. *Introduction to Aberrations in Optical Imaging Systems*. New York: Cambridge University Press, 2013. Print.

Schmidt, Werner, and Asim Kurjak. *Color Doppler Sonography in Gynecology and Obstetrics*. Stuttgart: Thiem, 2005. Print.

Schobert, Harold H. *Energy: The Basics*. New York: Routledge, 2014. Print.

Schulz, Alexander L. *Capacitors: Theory, Types, and Applications*. Hauppauge: Nova Science, 2010. Print.

Schwartz, Matthew D. *Quantum Field Theory and the Standard Model*. New York: Cambridge University Press, 2014. Print.

Schwarz, Patricia M., and John H. Schwarz. *Special Relativity: From Einstein to Strings*. New York: Cambridge University Press, 2004. Print.

Seeler, Karl A. *System Dynamics: An Introduction for Mechanical Engineers*. New York: Springer, 2014. Print.

Serway, Raymond A., and Chris Vuille. *College Physics*. 9th ed. Boston: Brooks, 2012. Print.

Serway, Raymond A., and John W. Jewett, Jr. *Physics for Scientists and Engineers*. 9th ed. Boston: Cengage, 2013. Print.

Shabany, Younes. *Heat Transfer: Thermal Management of Electronics*. Boca Raton: CRC, 2010. Print.

Shamos, Morris H., ed. "Lenz's Law." *Great Experiments in Physics: Firsthand Accounts from Galileo to Einstein*. 1959. New York: Dover, 1987. 159–65. Print.

Shankar, R. *Fundamentals of Physics: Mechanics, Relativity, and Thermodynamics (The Open Yale Course Series)*. New Haven: Yale University Press, 2014. Print.

Sharma, K. K. *Optics Principles and Applications*. Burlington: Academic, 2006. Print.

Shipman, James T., Jerry D. Wilson, and Charles A. Higgins Jr. *An Introduction to Physical Science*. 14th ed. Boston: Brooks/Cole, 2015. Print.

Shireman, Myrl. *Strengthening Physical Science Skills for Middle & Upper Grades*. Greensboro: Twain, 2008. Print.

Shultis, J. Kenneth, and Richard E. Faw. *Fundamentals of Nuclear Science and Engineering*. 2nd ed. Boca Raton: CRC, 2008. Print.

Siegel, Ethan. "What Does Torque in a Car Do?" *Starts with a Bang!* ScienceBlogs, 21 Apr. 2009. Web. 23 July 2015.

Simon, Steven H., *The Oxford Solid State Basics:* New York: Oxford University Press, 2013. Print.

Singer, Neal. *Wonders of Nuclear Fusion: Creating an Ultimate Energy Source*. Albquerque: University of New Mexico, 2011. Print.

Singh, Vijay P. *Entropy Theory and Its Application in Environmental and Water Engineering*. Chichester, Hoboken, NJ: Wiley-Blackwell, 2013. Print.

Sirdeshmukh, D B, L Sirdeshmukh, K G. Subhadra, and C S. Sunandana. *Electrical, Electronic and Magnetic Properties of Solids*. New York, NY: Springer, 2014. Print.

Skrabal, Peter M. *Spectroscopy: An Interdisciplinary Integral Description of Spectroscopy From UV to NMR*. Zürich, SW: Hochschulverlag AG, 2012. Print.

Smart, Lesley, and Elaine Moore. *Solid State Chemistry: An Introduction*. Boca Raton: Taylor & Francis, 2012. Print

Smith, Brian C. *Quantitative Spectroscopy: Theory and Practice*. San Diego: Elsevier, 2002. Print.

Smith, Hal L. *Monotone Dynamical Systems: An Introduction to the Theory of Competitive and Cooperative Systems*. 1995. Providence: Amer. Mathematical Soc., 2008. Print.

Smith, Walter Fox. *Waves and Oscillations: A Prelude to Quantum Mechanics*. New York: Oxford University Press, 2010. Print.

Soares, Claire. *Gas Turbines: A Handbook of Air, Land and Sea Applications*. Waltham: Butterworth, 2014. Print.

Solymar, Laszlo, Donald Walsh, and Richard R A , Syms. *Electrical Properties of Materials*. 9th ed. Oxford: Oxford University Press, 2014. Print.

Somervill, Barbara A. *Distance, Area, and Volume*. Chicago: Heinemann, 2011. Digital file.

Spadafora, Ronald R. "Principles of Mechanics." *Firefighter Exams*. New York: McGraw, 2008. Print.

Spilsbury, Louise, and Richard Spilsbury. *Why Can't I Hear That? Pitch and Frequency*. Chicago: Heinemann, 2014. Print.

Srednicki, Mark. *Quantum Field Theory*. New York: Cambridge U, 2007. Print.

Stahler, Wendy, Dustin Clingman, and Kaveh Kahrizi. *Beginning Math and Physics for Game Programmers*. Indianapolis: New Riders, 2004. Print.

Starzak, Michael E. *Energy and Entropy*. New York: Springer, 2014. Print.

Stein, James D. *Cosmic Numbers: The Numbers That Define Our Universe*. New York: Basic, 2011. Print.

Stiles, Timothy A. "Ultrasound Imaging as an Undergraduate Physics Laboratory Exercise." *American Journal of Physics* 82.5 (2014): 490–501. Print.

Struchtrup, Henning. *Thermodynamics and Energy Conversion*. Berlin: Springer, 2014. Print.

Susskind, Leonard, and Art Friedman. *Quantum Mechanics: The Theoretical Minimum*. New York: Basic, 2014. Print.

Susskind, Leonard, and George Hrabovsky. *The Theoretical Minimum: What You Need to Know to Start Doing Physics*. New York: Basic, 2013. Print.

Swanson, D. G. *Quantum Mechanics: Foundations and Applications*. Boca Raton: CRC, 2007. Print.

Taylor, B. N., Ambler Thompson, and International Bureau of Weights and Measures. *The International System of Units (SI)*. Gaithersburg, MD; Washington: U.S. Dept. of Commerce, National Institute of Standards and Technology; For Sale by the Supt. of Docs., U.S. G.P.O., 2008. Print.

Taylor, John R. *Classical Mechanics*. Sausalito: U Science, 2005. Print.

Teller, Paul. *An Interpretive Introduction to Quantum Field Theory*. Princeton: Princeton University Press, 1995. Print.

Thess, André. *The Entropy Principle: Thermodynamics for the Unsatisfied*. Berlin: Springer-Verlag, 2014. Print.

Thirring, Walter. *Classical Mathematical Physics: Dynamical Systems and Field Theories*. 2nd rev. ed. Berlin; New York: Springer, 2010. Print.

Thompson, Daniel M. *Understanding Audio*. Boston: Berklee, 2005. Print.

Thumann, Albert, and Harry Franz. *Efficient Electrical Systems Design Handbook*. Lilburn: Fairmont, 2009. Print.

Tilley, Richard. *Colour and the Optical Properties of Materials*. 2nd ed. Chichester: Wiley, 2011. Print.

Tipler, Paul Allen, and Gene Mosca. *Physics for Scientists and Engineers*. 6th ed. Vol. 1-2. New York, NY: W.H. Freeman, 2008. Print.

Tucker, Paul G. *Unsteady Computational Fluid Dynamics in Aeronautics*. Dordrecht; New York: Springer, 2014. Print.

Tyson, Neil DeGrasse,, and Donald Goldsmith. *Origins: Fourteen Billion Years of Cosmic Evolution*. New York: W.W. Norton, 2014. Print.

Ultimate Guide to Wiring. 7th ed. Upper Saddle River: Creative Homeowner, 2010. Print.

Uzan, Jean-Philippe, and Bénédicte Leclercq. *The Natural Laws of the Universe: Understanding Fundamental Constants*. Berlin; Heidelberg; New York: Springer: Praxis, 2008. Print.

Vayenas, Constantinos G., and Stamatios N. A. Souentie. *Gravity, Special Relativity, and the Strong Force: A Bohr-Einstein-de Broglie Model for the Formation of Hadrons*. New York: Springer, 2012. Print.

Vlaardingerbroek, Marinus T., and Jacques A. Den Boer. *Magnetic Resonance Imaging: Theory and Practice*. 3rd ed. Berlin [u.a.]: Springer, 2004. Print.

Voldman, Steven H. *ESD: Circuits and Devices*. 2nd ed. Chichester, United Kingdom; Hoboken, NJ: John Wiley & Sons, 2015. Print.

Vukosavic, Slobodan N. *Electrical Machines*. New York: Springer, 2013. Print.

Wagner, Mark. *The Geometries of Visual Space*. Mahwah: Erlbaum, 2006. Print.

Wainwright, J., and Ellis, G. F. R. *Dynamical Systems in Cosmology*. Cambridge: Cambridge University Press, 2005. Print.

Walker, James S. *Physics*. 5th ed. San Francisco: Pearson/Addison Wesley, 2016. Print

Walker, James S., and Gary W. Don. *Mathematics and Music: Composition, Perception, and Performance*. Boca Raton: Taylor & Francis Group, 2013. Print.

Walmsley, Ian. *Light: A Very Short Introduction*. New York: Oxford University Press, 2015. Print.

Weil, John A., and James R. Bolton. *Electron Paramagnetic Resonance: Elementary Theory and Practical Applications*. 2nd ed. Hoboken, N.J.: Wiley-Interscience, 2007. Print.

Westcott, Sean, and Jean R. Westcott. *The Complete Idiot's Guide to Electronics 101*. New York: Alpha, 2011. Print.

White, Colin. *Projectile Dynamics in Sport: Principles and Applications.* New York: Routledge, 2011. Print.

Wiedemann, Helmut. *Particle Accelerator Physics.* 4th ed. Cham: Springer International, 2015. Print.

Wilson, A. G. *Entropy in Urban and Regional Modelling.* New York: Routledge, 2013. Print.

Wilson, E.J.N. *An Introduction to Particle Accelerators.* Oxford: Oxford UP, 2006. Print.

Wilson, Jerry D., and Cecilia A. Hernández-Hall. *Physics Laboratory Experiments.* 7th ed. Boston: Brooks, 2010. Print.

Wolfe, James P. *Imaging Phonons: Acoustic Wave Propagation in Solids.* Cambridge: Cambridge University Press, 2005. Print.

Wolfe, William L. *Optics Made Clear: The Nature of Light and How We Use It.* Bellingham: SPIE, 2007. Print.

Won, Chang-Hee, Cheryl B. Schrader, and Anthony N. Michel, eds. *Advances in Statistical Control, Algebraic Systems Theory, and Dynamic Systems Characteristics.* Boston: Birkhäuser, 2008. Print.

Woodford, Chris, and Jon Woodcock. *Cool Stuff 2.0 and How It Works.* London: DK, 2007. Print.

Woodhouse, N. M. J. *Special Relativity.* London: Springer Verlag, 2003. Print.

Wysession, Michael, David Frank, and Sophia Yancopoulos. *Physical Science: Concepts in Action.* Needham: Prentice, 2004. Print.

Yacobi, B. G. *Semiconductor Materials: An Introduction to Basic Principles.* New York: Springer-Verlag, 2013. Print.

Yacoubian, Araz. *Optics Essentials: An Interdisciplinary Guide.* CRC, 2015. Print.

Yevick, David, and Hannah Yevick. *Fundamental Math and Physics for Scientists and Engineers.* Hoboken: Wiley, 2015. Print.

Yip, Sidney. *Nuclear Radiation Interactions.* Hackensack: World Scientific, 2014. Print.

Yorke, R. *Electric Circuit Theory.* Elmsford: Pergamon, 2013. Print.

Young, Hugh D., and Roger A. Freedman. *University Physics.* 13th ed. Harlow: Pearson Education Limited, 2014. Print.

Young, Hugh D., Roger A. Freedman, A. Lewis Ford, and Francis Weston Sears. *Sears & Zemansky's College Physics.* 14th ed. Boston, Mass: Pearson, 2016. Print.

Zangwill, Andrew. *Physics at Surfaces.* Cambridge [Cambridgeshire]; New York: Cambridge UP, 1988. Print.

Zee, A. *Quantum Field Theory in a Nutshell.* 2nd ed. Princeton: Princeton University Press, 2010. Print.

Zumdahl, Steven S., and Susan A. Zumdahl. *Chemistry.* 9th ed. Belmont: Brooks, 2014. Print.

INDEX

A

aberration 1, 2, 191, 217, 376, 389
absolute zero 43, 109, 110, 160, 303, 372
absorption 4, 5, 6, 22, 36, 40, 41, 200, 201, 202, 225, 226, 227, 230, 258, 267, 268, 269, 282, 342, 359, 374, 375, 386
absorption coefficient 4, 6, 225, 226, 374
accuracy 7, 8, 9, 10, 75, 76, 114, 116, 150, 151, 338, 372, 374
acoustics vii, 4, 6, 185, 224, 245, 246, 283, 285, 315, 320, 366, 374, 382, 385
activation energy 104, 105, 107, 108, 109, 374
active galactic nucleus 67, 374
actual mechanical advantage 49, 50, 274, 278, 374
addition x, 4, 11, 35, 75, 178, 205, 221, 237, 286, 305, 307, 313, 316, 348, 371
albedo 4, 374
alpha decay 10, 11, 133, 211, 264, 333, 334, 374
alpha particle 10, 11, 67, 133, 211, 216, 217, 334, 374
alpha radiation 10, 11, 211, 264, 374, 380
alpha scattering 150
Ampère, André-Marie 117, 119, 279, 280, 281, 365, 369, 371, 374
Ampère's law 119, 279, 280, 281, 365, 371, 374
amplifier 77, 78, 79, 374
amplifier distortion 77, 79, 374
angular forces v, 15
angular magnification 189, 190, 374
angular velocity 19, 20, 21, 42, 51, 53, 110, 111, 112, 123, 124, 170, 220, 267, 270, 306, 307, 308, 310, 374
antenna 22, 23
antineutrino 34, 35, 58, 348, 374
antinode 146, 147, 374
antiparticle 34, 67, 133, 200, 347, 348, 374, 380
antiphase 69, 70, 222, 223, 374
aperture 26, 27, 125, 195, 196, 217, 329, 331, 358, 374, 378
Arago dot v, 24, 25
Archimedes v, 27, 28, 29, 275, 278, 364, 369, 394, 396, 402
Archimedes' principle 369
arc length 14
Aristotle ix, 115, 116, 178, 289, 322, 336, 364
astronomy ix, 21, 23, 80, 133, 180, 262, 264, 285, 307, 331
astrophysics 57, 355, 358

atomic bomb xii, 291
attenuation 4, 5, 6, 22, 23, 375
audible 80, 315, 316, 330, 386

B

background radiation 23, 67, 68, 211, 222, 266, 355, 357, 375
ballistic trajectory 19, 241, 242, 244, 375, 392
band gap 30
bands 1, 30, 170, 177, 191
band theory of solids 30
barrel distortion 77, 78, 375
barrel of oil equivalent 46, 47, 375
baryons 138, 249
BCS-theory 358
beat frequency 80, 81, 130, 375
Beer-Lambert law 4, 6, 375
Bernoulli, Daniel 32, 364
Bernoulli, Johann 32
Bernoulli's equation 32, 33, 375
Bernoulli's principle v, 32, 34
beta decay 34, 35, 36, 133, 264, 334, 374, 375
beta particle 11, 35, 57, 58, 67, 133, 211, 334, 375
beta radiation 10, 11, 34, 35, 197, 211, 264, 265, 334, 375, 380, 384
bias 7, 73, 74, 75, 375
biophysics 245
blackbody radiation vii, 36, 37, 229, 253, 254, 366, 375
black holes 67, 134, 374
blueshift 3, 80, 81, 375
Bohr atom v, 39, 41, 42
Bohr, Niels 42, 151, 253, 349, 351, 358
Boltzmann factor 256, 257, 375
Boltzmann, Ludwig 38, 253, 257, 365, 366
bond energy 104, 105, 375
Bose condensation v, 42, 43, 44, 349, 399
Bose-Einstein (condensates, function) 44, 249, 257, 258, 355, 400, 401
Bose, Satyendra Nath 258
boson v, vii, 43, 120, 137, 138, 152, 153, 154, 249, 256, 260, 289, 334, 347, 350, 353, 375, 381
bosonic string 289, 375
Brahe, Tycho 164, 336, 364
bra-ket notation 45, 46, 88, 89, 221, 256, 258, 375
bremsstrahlung 132, 133, 216, 217, 342, 343, 344, 375

British thermal unit (BTU) v, 46, 47, 162, 163, 168, 375, 391
Bunsen, Robert 37, 365
buoyancy 27, 28, 29, 114, 325, 326, 364, 375

C

calculating system efficiency 48
calorie 46, 47, 163, 167, 375
capacitor 73, 74, 94, 95, 96, 98, 99, 100, 120, 134, 136, 137, 233, 234, 235, 376
capacity vi, xi, 25, 70, 182, 184, 206, 224, 311, 325, 326, 327, 339
carcinogenic 317, 318, 376
Carnot, Sadi 66, 365
Cartesian coordinates 164, 166, 376
cathode rays 150, 195, 216, 220, 362
center of gravity 369
center of mass 110, 111, 113, 139, 141, 164, 168, 169, 170, 232, 298, 299, 338, 376
center of momentum 168, 169, 171, 376
centrifugal 15, 16, 51, 52, 53, 376
centrifugal force 15, 16, 51, 52, 53, 376
centripetal 15, 16, 17, 18, 51, 52, 53, 114, 116, 164, 165, 166, 335, 338, 376, 383
centripetal force 15, 16, 17, 18, 51, 52, 53, 114, 116, 164, 165, 166, 335, 338, 376, 383
chaos 108
Cherenkov radiation 67, 68, 376
chromatic aberration 1, 191, 376
circle of confusion 125, 126, 127, 376
circular motion 15, 16, 17, 18, 51, 52, 53, 111, 112, 113, 146, 165, 270, 272, 274, 275, 324, 369, 376, 385, 388
circumference 15, 16, 17, 24, 53, 76, 77, 123, 129, 165, 222, 262, 263, 270, 277, 299, 324, 376, 386, 387
closed system 19, 54, 60, 62, 86, 90, 173, 174, 214, 217, 220, 231, 369, 372, 376, 377, 383, 391
closed system (isolated system) 54
collimated rays 24, 26, 376
collision 16, 17, 59, 67, 68, 90, 91, 92, 93, 94, 132, 133, 154, 179, 195, 216, 217, 218, 221, 252, 258, 291, 340, 342, 349, 375, 376, 382
color charge 137, 138, 139, 248, 249, 250, 259, 260, 376
color force 248, 249, 250, 259, 260, 292, 293, 376
coma 1, 376
complex system 274

compression 5, 13, 14, 61, 170, 185, 197, 224, 231, 283, 286, 287, 288, 320, 321, 330, 332, 376
compressive strength 197, 376
Compton, Arthur Holly 361
Compton scattering 67, 68, 132, 134, 342, 344, 376, 382
concave 56, 57, 125, 126, 189, 190, 376
conformal field theory 251, 252, 376
conservation of charge 57
conservation of energy 19, 32, 39, 40, 54, 57, 58, 60, 61, 93, 107, 151, 173, 174, 195, 214, 220, 221, 231, 235, 277, 286, 303, 339, 341, 377, 383
conservation of momentum 19, 39, 40, 90, 93, 94, 200, 220, 221, 377
constructive interference 24, 69, 70, 80, 130, 222, 332, 333, 375, 377
consumed energy 167, 168, 377
continuity 74, 294, 295, 296, 297, 377
continuity equation 370
continuous energy levels 36, 38, 377
convection and conduction 64
converging 55, 56, 57, 125, 126, 376, 377
convex v, 1, 56, 57, 78, 125, 126, 189, 190, 377
cooper pair 44, 399
Coriolis force 51
cosine 14, 15, 16, 17, 44, 84, 117, 129, 136, 143, 146, 195, 198, 262, 263, 267, 277, 371, 377, 378, 383
cosmic radiation 67
cosmic rays 67, 68, 69, 132, 133, 134, 200, 344, 377, 380
Cotes, Roger 262, 263, 364
coulomb 57, 58, 97, 98, 136, 137, 226, 234, 235, 299, 300, 377
Coulomb, Charles-Augustin de 234, 364
Coulomb's law 58, 59, 134, 136, 234, 377
crest 4, 12, 13, 14, 69, 70, 80, 128, 147, 223, 265, 283, 329, 330, 332, 377, 384, 385
cross product 18, 307, 310, 377
cycle 14, 22, 64, 70, 103, 128, 129, 143, 144, 147, 184, 198, 223, 226, 240, 265, 270, 283, 285, 299, 300, 317, 329, 330, 332, 338, 342, 381, 386

D

damped oscillator 143, 146, 377
de Broglie, Louis 151, 254
deceleration 115, 133, 180, 241, 242, 377, 391
decibel 184, 185, 186, 187, 246, 282, 284, 377
definite pitch 227, 228, 377
degrees of freedom 107, 108, 251, 252, 298, 377

demand 197, 198, 377
depth of field 26, 127, 377
derived unit 5, 16, 75, 76, 97, 98, 140, 167, 231, 234, 264, 328, 377
Descartes, René 88, 364
destructive interference 24, 69, 70, 81, 130, 177, 222, 332, 333, 378
detector 56, 134, 177, 196, 220, 221, 330, 388
diaphragm 26, 378
diffraction 4, 24, 176, 191, 254, 331, 332, 356, 361, 362, 378
dimension 23, 125, 178, 189, 191, 290, 291, 331, 378
diode xiii, 73, 74, 75, 217
Dirac notation 44, 45
Dirac, Paul 45, 89, 256, 258, 348, 352, 375
directly related 99, 117, 135, 140, 157, 184, 185, 196, 278, 302, 305, 378, 381
discrepancy 7, 10, 38, 154, 378
dispersion 240, 372
displacement field 328, 329, 378
dissipation 90, 378
distinguishability 195, 196, 256, 257, 378
distortion 1, 2, 22, 23, 77, 78, 79, 343, 374, 375, 376, 382, 386, 387, 389, 392
distribution function 152, 256, 257, 258, 378
diverging 55, 56, 57, 126, 378
Doppler, Christian 80, 82, 365, 398
Doppler effect v, 3, 80, 81, 82, 130, 178, 320, 321, 362, 365
Doppler shift 80, 178, 365
dot product 18, 44, 45, 89, 378
dummy load 197, 198, 378
dynamic equilibrium 214, 215, 378
dynamic systems 8, 83, 84, 378
dynamic systems theory 83, 84
dynamometer 155, 156, 378
dyne 206, 378

E
Edison effect v, 86, 87
efficiency v, 36, 48, 49, 50, 66, 109, 168, 214, 215, 237, 238, 272, 273, 274, 278, 313, 372, 374, 378, 379, 381, 384, 390, 391
eigenfunction 44, 45, 88, 152, 378, 379
eigenstate 150, 152, 379
eigensystem 88, 89, 294, 295, 379
eigenvalue 44, 88, 89, 378, 379
eigenvectors 89

Einstein, Albert xi, 3, 39, 60, 114, 116, 151, 171, 177, 254, 258, 289, 322, 351, 361
Einstein's predictions 116
elastic collisions 64, 91, 94
electrical efficiency 48, 49, 379
electrical load v, 197
electric circuit components 94
electric circuit diagrams 94
electric circuits 94, 97, 134, 136, 404
electric flux 134, 135, 136, 371, 377, 379
electric potential 74, 97, 98, 99, 102, 117, 136, 162, 167, 169, 195, 200, 216, 231, 233, 234, 235, 236, 238, 239, 251, 328, 350, 379, 385, 392
electric potential energy 97, 98, 99, 102, 169, 231, 233, 234, 235, 238, 379
electromagnet 195, 196, 279, 280, 281, 379
electromagnetic field xi, 4, 67, 68, 117, 119, 134, 173, 195, 217, 249, 251, 267, 268, 282, 294, 350, 371, 379, 383
electromagnetic spectrum xi, 4, 22, 67, 132, 150, 176, 226, 240, 245, 264, 265, 317, 318, 329, 330, 342, 343, 374, 379
electromagnetic theory xi, 282, 396
electromagnetism xi, 19, 37, 42, 58, 74, 117, 119, 129, 133, 135, 136, 137, 138, 153, 174, 240, 248, 249, 251, 253, 285, 293, 296, 334, 347, 349, 371, 384, 385, 390
electromotive force (emf) 48, 94, 117, 118, 119, 120, 198, 370, 379
electronics xiii, 69, 173, 221, 252, 255, 273, 297, 306, 328, 329, 355, 359
electron spin 201, 267, 268
electron volt 163, 167, 195, 200, 201, 216, 379
electroweak force 152, 154, 350, 353, 379
elementary particle 40, 120, 132, 152, 176, 229, 248, 249, 256, 259, 334, 352, 357, 358, 375, 379, 380, 383, 386
ellipse 164, 165, 379
elongation 286, 287, 379
e=mc² 3, 59, 212
endothermic 104, 105, 379
energy and power 100
energy levels 20, 30, 36, 38, 40, 41, 134, 224, 225, 226, 254, 258, 282, 295, 307, 349, 351, 377
enthalpy 104, 105, 106, 107, 108, 109, 110, 157, 158, 302, 305, 379, 381, 389, 391
entropy 54, 83, 106, 107, 108, 109, 157, 158, 214, 215, 372, 379, 391
Euler, Leonhard 32, 110, 111, 364

Index

Euler's constant (Euler's number) 5
Euler's laws v, 110, 113, 364
exothermic 104, 105, 106, 379
extrinsic value 191, 379

F

falling bodies ix
Faraday, Michael 118, 173, 365
Faraday's law v, 117, 118, 119, 134, 135, 136, 173, 174, 280, 371, 379
feet xi, 6, 13, 14, 23, 46, 47, 50, 71, 76, 90, 115, 116, 117, 129, 130, 221, 272, 286, 325, 330, 337, 391, 392
Fermi-Dirac energy distribution function 258
Fermi, Enrico xiii, 36, 258, 334, 360
fermion 58, 120, 121, 216, 220, 221, 248, 249, 256, 258, 259, 289, 292, 380, 387, 390
ferromagnetic 173, 279, 280, 380
Feynman diagrams 120
fixed reference frame 110, 113, 380
flow rate 32, 33, 313, 380
fluid dynamics 33, 325
flywheel 122
focal length 1, 26, 125, 126, 189, 190, 191, 218, 219, 380
focal point 19, 23, 26, 55, 56, 125, 126, 127, 164, 189, 190, 379, 380
force field 122, 231, 233, 260, 380
frame of reference 110, 115, 168, 169, 258, 298, 370, 376, 380
free body diagram 127
free fall 114, 204, 243, 380
frequency 128
frictional resistance 207, 380
fulcrum 272, 274, 275, 278, 307, 310, 380, 383

fundamental forces 35, 36, 113, 120, 137, 152, 153, 154, 194, 248, 251, 292, 333, 334, 347, 348, 392
fundamental frequency 146, 147, 148, 149, 380, 384, 385
fusion vi, xii, 109, 158, 176, 211, 212, 334

G

Galilei, Galileo x, 51, 55, 115, 178, 285, 336, 364
Galton, Francis 316, 365
gamma radiation 10, 11, 132, 133, 134, 200, 201, 212, 264, 266, 318, 342, 343, 359, 380
gamma ray 67, 121, 200, 201, 202, 380
gamma-ray burst 132, 380

Gaussian surface 135, 380
Gauss, Karl Friedrich 135, 365
Gauss's law v, 134, 135, 136
gear 273, 275, 276, 277, 278, 279
Geiger, Hans 11, 150
Geiger-Müller probe (G-M probe) 10, 12, 380
Gell-Mann, Murray xiv, 137, 139, 259, 292, 358, 397
general relativity 80, 81, 82, 114, 116, 117, 153, 154, 322, 336, 338, 351, 370, 380
geophysics 191, 298
gluon v, 120, 137, 138, 139, 248, 249, 250, 397
gram 46, 47, 71, 167, 194, 206, 300, 375, 378, 380
gravitational acceleration 114, 140, 191, 192, 193, 194, 244, 288, 336, 380
gravitational force 16, 36, 114, 116, 140, 141, 165, 170, 194, 204, 205, 208, 210, 231, 232, 234, 235, 251, 276, 277, 289, 347, 348, 381
gravitational potential energy 61, 62, 63, 101, 102, 139, 140, 141, 170, 232, 233, 234, 341
graviton 120, 137, 289, 381
ground xi, 33, 42, 61, 62, 63, 69, 89, 91, 94, 96, 97, 101, 116, 135, 140, 159, 170, 171, 182, 241, 242, 243, 244, 278, 302, 307, 311, 336, 354, 381, 387, 388

H

hadron xiv, 137, 138, 248, 249, 250, 259, 260, 292, 293, 347, 348, 381
half-life 10, 11, 34, 35, 211, 212, 381
harmonic oscillations 13, 143, 286, 288, 381
harmonic oscillator 13, 143, 144, 146, 288
harmonics v, 12, 80, 83, 128, 143, 146, 147, 149, 224, 262, 285, 315, 381
harmony 227, 228, 381
heat transfer 64, 65, 66, 381
heat value 46, 381
Heisenberg uncertainty principle v, xi, 150, 151
Heisenberg, Werner 42, 46, 150, 253, 255, 257, 291, 392
Henry, Joseph 281, 365
Hertz, Heinrich 129, 177, 265, 366
Hess, Germain Henri 105
Hess's law 104, 105, 106, 365, 381
Higgs boson v, vii, 138, 152, 153, 154, 347, 350
Higgs field 152, 153, 154, 350, 381
Higgs, Peter Ware 152
high energy physics vii, 212, 220, 222
Hooke, Robert 55, 233, 364

Hooke's law 143, 144, 170, 231, 233, 286, 287, 288, 364, 381
horsepower 103, 155, 156, 162, 163, 236, 300, 328, 338, 381
horsepower hour 162, 163, 381
hot conductor 94, 96, 381
Hubble, Edwin 81
Hubble's law 81
Hydrodynamica 32, 364

I
ideal gas law vii, 64, 109, 157, 160, 302, 305, 325, 326, 381
ideal mechanical advantage 49, 50, 274, 278, 381
impedance 197, 198, 199, 381
imperfect system 48, 381
impulse turbine 311, 313, 381
inclined plane 207, 208, 272, 273, 274, 275, 276, 277
incompressible 32, 33, 372, 382
indefinite pitch 227, 228, 382
index of refraction 1, 56, 189, 190
inductance 59, 173, 279, 281, 382
inductor 94, 95, 96, 120, 281, 382
inelastic collisions 90
inertia x, 15, 16, 20, 21, 42, 51, 53, 110, 111, 113, 122, 123, 124, 170, 179, 181, 220, 241, 242, 267, 306, 308, 374, 375, 376, 382, 384
inertial reference frame 1, 2, 382
infrasound 315, 382
initial velocity 17, 18, 180, 241, 242, 243, 340, 382
instantaneous 114, 115, 116, 180, 222, 223, 283, 295, 322, 323, 370, 371, 372, 382
instantaneous phase 222, 223, 382
instantaneous velocity 114, 115, 116, 180, 323, 382
insulators 96, 294, 296, 382
intensity v, 5, 6, 12, 24, 25, 26, 37, 38, 40, 69, 144, 177, 184, 185, 186, 187, 223, 224, 227, 245, 246, 264, 265, 282, 283, 284, 332, 369, 371, 374, 382, 383
interference 4, 22, 23, 24, 69, 70, 77, 78, 79, 80, 81, 128, 130, 147, 176, 177, 222, 223, 264, 266, 316, 332, 333, 362, 375, 377, 378, 382, 387
internal energy vi, 104, 107, 157, 302, 303, 372, 379, 382
intrinsic value 191, 192, 382
inverse Compton scattering 132, 134, 382
inversely related 80, 132, 177, 265, 277, 283, 330, 332, 382
inverse-square law 116, 264, 265, 382
inversion 55, 57, 298, 382

isolated system 39, 54, 169, 214, 303, 341
isotopes 10, 11, 35, 36, 43, 67, 69, 195, 196, 202, 211, 212, 266, 347, 382

J
Jeans, James 38
Joule, James Prescott 162, 231, 365
just-noticeable difference (JND) 184, 245, 382, 388

K
Kelvin, William Thomson, Lord 231
Kepler, Johannes 164, 336, 364
Kepler's laws 164
kilocalorie 162, 163, 383
kilogram 16, 53, 67, 71, 97, 100, 101, 102, 112, 122, 123, 140, 156, 182, 183, 184, 191, 194, 206, 231, 272, 299, 300, 335, 340, 380, 383, 385
kilowatt 100, 103, 167, 168, 338, 377, 383
kilowatt-hour 100, 103, 167, 168, 338, 383
kinematics 18, 19, 60, 113, 122, 178, 180, 184, 239, 271, 279, 322, 323, 324, 325, 336, 338, 339, 341, 383, 401
kinetic theory of gases 107, 108, 109, 157, 158, 160, 383
Kirchhoff, Gustav 37

L
Lagrange, Joseph-Louis 231, 364
Lagrangian points 164, 166, 383
Lambert's laws 371
laminar flow 32, 33, 216, 217, 383
Langevin, Paul 316
laser 42, 43, 241, 281, 349, 355, 356, 357, 359
law of conservation of energy 32, 40, 54, 57, 58, 60, 107, 174, 195, 221, 231, 235, 277, 303, 341, 383
law of cosines 18
law of the lever 278, 383
Lenz's law v, 117, 118, 119, 173, 174, 175, 280, 383
lepton 152, 154, 256, 258, 348, 353, 356, 383
lever 123, 236, 272, 274, 275, 276, 278, 307, 308, 310, 311, 335, 338, 380, 383, 385
lift 32, 33, 50, 101, 102, 162, 182, 183, 184, 197, 275, 276, 278, 279, 311, 383
light physics 245
light wave 2, 4, 5, 383
linear magnification 189, 190, 383, 391
linear momentum 16, 51, 110, 111, 308, 369, 383
linear motion 15, 18, 19, 51, 53, 178, 275
linear system 272, 273

409

line spectrum 229, 253, 383
load v, 49, 96, 101, 181, 182, 183, 184, 197, 198, 199, 275, 277, 278, 377, 378
logarithm 5, 184, 187, 246, 383
longitudinal wave 320, 332, 383
Lorentz transformation 1, 3, 383
loudness 12, 14, 144, 184, 185, 186, 187, 227, 228, 245, 246, 282, 283, 284, 383, 386, 389, 391

M

magnetic flux xi, 117, 118, 119, 136, 174, 195, 196, 267, 268, 281, 369, 370, 371, 383
magnetic induction 117, 118, 119, 120, 134, 280, 281, 383
magnification 1, 126, 189, 190, 240, 374, 383, 391
Marsden, Ernest 150
mass defect 211, 212
mass spectrometry 196, 217
mass-spring system 143, 144, 224, 384
materials physics 30
mathematical physics 3, 43, 85, 399, 403
matrix 21, 45, 46, 88, 89, 152, 201, 258, 294, 379, 384
Maxwell, James Clerk xi, 37, 117, 119, 135, 136, 253, 257, 265, 364, 384
Maxwell's equations xi, 117, 120, 135, 136, 364, 369, 384
mechanical advantage vi, vii, 49, 50, 183, 197, 236, 237, 272, 274, 275, 277, 278, 279, 308, 310, 335, 338, 374, 381, 383, 384, 389
mechanical efficiency 48, 50, 384
mechanical load 181, 182
mechanical resonance 146, 147, 384
mechanical switches 74, 296, 384
mesons 121, 138, 139, 249, 250, 289, 348, 349, 357, 360
methodology 202, 268, 282
mole 104, 105, 106, 157, 158, 159, 160, 193, 194, 206, 326, 375, 384, 389
moment 15, 20, 21, 42, 51, 52, 60, 110, 111, 113, 115, 123, 124, 170, 206, 220, 267, 268, 306, 307, 308, 311, 335, 337, 359, 360, 374, 384, 385, 391
Mössbauer effect 200
M-theory 289, 291, 384
muons 68, 353
mutation 11, 35, 384

N

nanotechnology 224, 225, 298
net force 18, 52, 110, 111, 127, 143, 145, 182, 203, 204, 205, 207, 209, 210, 272, 274, 276, 277, 339, 340, 384
neutral conductor 94, 96, 384
neutrino oscillations vii, 354
Newton, Isaac x, 2, 37, 51, 55, 60, 90, 110, 114, 115, 122, 141, 145, 164, 166, 178, 240, 253, 308, 322, 335, 336, 339, 364, 385
Newton meter 206, 207, 385
Newton's laws of motion 1, 2, 51, 60, 110, 143, 171, 178, 179, 203, 205, 285, 385
node 130, 146, 147, 385, 390
noise 22, 23, 78, 149, 187, 219, 223, 283, 284, 385, 389
non-ionizing radiation 342, 385
nonmechanical switches 74, 296, 385
nonviscous fluid 32, 385
normal force 203, 204, 205, 207, 208, 209, 210, 287, 385
nuclear fission xii, 10, 211, 212
nuclear magnetic resonance (NMR) 201, 268, 282
nuclear physics 211
nuclear power plant 66, 293
nucleon 137, 385

O

octave 147, 149, 228, 385
Ohm, Georg Simon 199, 365
Ohm's law 99, 174, 199, 295, 328, 329, 365, 385
open systems 214
optical system 1, 26, 125, 385, 389
optics xi, 21, 25, 26, 125, 176, 219, 240, 245, 357, 385
oscillation 4, 12, 13, 22, 61, 143, 144, 145, 146, 288, 383, 385
overtone 147, 385

P

pair production 221
parabolic 22, 23, 241, 242, 244, 385
parallel circuits 385
parity 139, 359
particle accelerator 134, 216, 217, 220, 353
particle detector 220, 221
Pauli exclusion principle 20, 42, 43, 248, 257, 258, 268, 306, 349
Pauli, Wolfgang 291, 352, 360
peak amplitude 12, 13, 14, 283, 332, 385

peak-to-peak amplitude 12, 13, 14, 332, 385
pendulum 13, 61, 62, 89, 140, 143, 145, 146, 169, 172, 386
perception of sound 245, 246, 283, 284, 386
periodicity 83, 84, 205, 386
permeability 279, 280, 281, 367, 386
permittivity 59, 135, 136, 386
perpendicular 15, 16, 17, 18, 51, 56, 70, 78, 89, 126, 135, 176, 203, 204, 205, 207, 208, 209, 240, 276, 298, 307, 309, 310, 332, 371, 377, 385, 386, 391
perturbation 120, 164, 289, 291, 320, 386
perturbation theory 289, 291, 386
phase offset 222, 223, 386
phase space 83, 256, 386

Philosophiae Naturalis Principia Mathematica x, 110, 179, 180, 181, 336, 364
phonon 224
phon scale 184, 386
photoabsorption 342, 343, 386
photoelectric effect 40, 41, 176, 177, 225, 226, 227, 229, 230, 254, 351, 361, 386
pi 76, 91, 92, 93, 139, 222, 223, 262, 263, 326, 386
pincushion distortion 77, 78, 386
pitch 3, 14, 80, 148, 227, 228, 245, 320, 377, 382, 386, 388, 391
Pitot, Henri 33, 34, 364
Pitot tube 34, 364
Planck, Max 38, 39, 150, 226, 253, 351, 366
Planck's constant 229
Planck's constant vi, 39, 40, 43, 151, 201, 226, 229, 230, 254, 367
Planck's law 36, 38, 229, 254, 366, 386
plasma physics 358
Plato ix, x, 289
polar coordinates 19, 164, 166, 386
polarization 138, 176, 177, 240, 241, 386
poles 53, 66, 251, 297, 386
positron xi, 35, 58, 121, 133, 220, 221, 334, 347, 348, 353, 361, 386
pound 46, 71, 162, 183, 191, 192, 193, 194, 375, 387, 389
power rating 74, 75, 103, 328, 387
precision v, 7, 8, 10, 41, 42, 75, 76, 150, 151, 217, 221, 253, 281, 338, 355, 359, 360, 362, 387, 392
Principles of Quantum Mechanics, The 45
prisms 1, 176, 240, 241, 326, 333
probability xiii, 20, 21, 40, 42, 44, 150, 152, 255, 256, 257, 258, 294, 349, 375, 378, 387, 392

probability density function (PDF) 150, 387
projectiles 241
psychophysics 245
pulley 50, 127, 128, 181, 182, 183, 236, 272, 274, 275, 276, 278, 279, 335
pulse 11, 283, 387
Pythagoras 315, 364

Q

quantization distortion 78, 79, 387
quantum chromodynamics 138, 216, 217, 249, 251, 259, 260, 292, 293, 387
quantum electrodynamics vii, 121, 249, 251, 347, 358
quantum field theory 30, 152, 216, 217, 220, 221, 248, 249, 250, 251, 252, 253, 259, 292, 293, 333, 376, 387, 398, 399, 402, 403, 404
quantum gravity 117, 290, 291
quantum numbers 20, 45, 221, 267, 268, 291, 306, 348, 349
quantum physics 46, 253, 255, 292
quantum state 19, 20, 30, 40, 42, 44, 45, 88, 150, 152, 200, 201, 202, 221, 224, 226, 253, 256, 257, 267, 268, 294, 295, 306, 307, 349, 375, 387, 388, 392
quantum statistics 256
quantum theory 40, 46, 150, 151, 221, 229, 252, 291, 355
quark xiv, 58, 59, 137, 138, 139, 153, 216, 220, 221, 248, 249, 250, 256, 259, 260, 292, 293, 334, 348, 349, 353, 356, 376, 381, 387

R

radian 14, 15, 17, 51, 123, 124, 129, 222, 262, 263, 264, 270, 387
radiant energy 4, 37, 101, 214, 229, 387, 390
radioactive decay 10, 11, 34, 35, 36, 57, 58, 67, 132, 133, 200, 211, 212, 216, 264, 265, 266, 333, 334, 342, 374, 375, 380, 381, 387
radioisotope 11, 35, 36, 132, 133, 387
radio waves 22, 79, 132, 240, 264, 265, 266, 286, 331, 387
radius 1, 14, 15, 17, 19, 20, 24, 25, 39, 52, 53, 76, 111, 112, 113, 123, 126, 129, 140, 164, 165, 189, 196, 218, 262, 263, 270, 277, 299, 310, 326, 327, 376, 386, 387
random error 7, 8, 387
Rankine, William John Macquorn 231, 365
rarefaction 5, 14, 283, 320, 330
Rayleigh-Jeans law 38
Rayleigh scattering 342, 344, 388

Index

reaction turbine 311, 313, 388
reciprocal 31, 96, 128, 129, 130, 201, 331, 388
redshift 3, 80, 81, 130, 388
reference frame 1, 2, 3, 110, 113, 114, 115, 169, 376, 380, 382, 388
reflection 4, 55, 57, 81, 176, 191, 240, 298, 299, 320, 321, 332, 372, 388
refraction 1, 4, 24, 55, 56, 125, 176, 189, 190, 191, 240, 241, 320, 321, 332, 333, 372, 388, 389
relativistic beaming 1, 3, 388
renormalization 251, 252, 388
resistor xiii, 74, 75, 94, 95, 98, 100, 120, 388
resolution 189, 191, 268, 269, 316, 330, 331, 357, 388
resonance 143, 144, 145, 146, 147, 149, 200, 201, 267, 268, 282, 358, 359, 360, 384, 388
revolution 15, 17, 18, 19, 20, 52, 53, 123, 129, 178, 270, 351, 388
revolutions per minute (rpm) 20, 52, 129, 155, 270, 388
right-hand rule 173, 174, 175, 203, 205, 279, 280, 308, 388
rigid body 110, 111, 252, 388
root-mean-square amplitude (RMS) 12, 388
rotational energy 122, 123, 124, 335, 388
rotational speed (rpm) 103, 122, 123, 155, 156, 270, 388
Rutherford, Ernest xi, 10, 11, 34, 35, 40, 67, 150, 216, 265, 290, 351

S

Sabine, Wallace Clement 6, 366
scalar 44, 45, 91, 154, 203, 249, 308, 322, 323, 378, 388
Schrödinger equation 45, 88, 253, 254, 347, 388
Schrödinger, Erwin 45, 152, 254, 361
Schrödinger's wave mechanics 46
screw 272, 274, 275, 276, 277, 278, 310
semiconductor 30, 31, 59, 74, 75, 96, 217, 354, 355
sensitivity 23, 184, 218, 221, 245, 246, 388
series circuit 389
short ton (in terms of coal) 46, 47, 391
signal-to-noise and distortion ration (SINAD) 22, 23, 78, 389
simple machine 49, 236, 272, 274, 275, 276, 277, 278, 310, 389
simple systems 291, 356
sine 13, 14, 15, 16, 17, 21, 84, 128, 129, 143, 144, 146, 222, 262, 263, 309, 320, 330, 389

sinusoidal 12, 13, 14, 22, 128, 129, 143, 146, 198, 222, 223, 291, 320, 330, 382, 388, 389
sinusoidal function 222, 223, 382, 389
SI units 98, 123, 167, 206, 301, 328, 340
slope 89, 232, 322, 323, 389
slug 71, 72, 192, 194, 389
Snell's law 55, 56, 389
solenoid 21, 221, 279, 280, 281, 282
solid-state physics xiii
sonar 220, 315, 316, 389
sone 184, 187, 389
sonography 315, 316, 321, 389
sound amplitude 282
sound barrier 285, 286, 320, 321, 389
sound intensity v, 6, 184, 187, 245, 282, 382
sound perception 187, 228
Spallanzani, Lazzaro 315, 364
special relativity 1, 3, 168, 347, 370, 372, 389
specific gravity v, 27, 28, 71, 72, 73, 389, 392
spectroscopy 37, 178, 200, 201, 202, 269, 331, 355, 356, 357, 361
speed of sound 82, 129, 130, 224, 285, 286, 320, 321, 330, 389
spherical aberration 1, 389
Spin 18, 267, 349
spring constant 62, 143, 144, 170, 233, 234, 235, 287, 288, 389
springs vi, 62, 144, 170, 224, 231, 235, 286, 287, 288, 312
standard enthalpy of formation 104, 106, 389
standard state 104, 106, 390
standard temperature and pressure (STP) 71, 72, 107, 108, 157, 158, 294, 390
standing wave 146, 147, 149, 374, 385, 390
state variables 107, 108, 157, 158, 160, 390
static energy 302, 303, 390
static pressure 32, 34, 369, 390
statistical mechanics 257
Stefan-Boltzmann law 37, 38, 229, 366, 390
Stefan, Josef 38, 366
string theory 289
strong force 292
strong interaction 19, 42, 58, 216, 248, 249, 250, 251, 259, 260, 292, 355, 376, 380, 387, 390
strong nuclear force 36, 153, 194, 248, 290, 348
superconductor 294
superposition 4, 251, 252, 255, 390
superstring theory 259, 289, 292, 333, 384
supersymmetry 289, 390

surface area 6, 23, 38, 75, 76, 77, 238, 239, 300, 325, 327, 372, 390
switches 74, 95, 146, 296, 297, 384, 385, 386
symmetry 153, 154, 298, 299, 347, 349, 354, 357, 359
synchrotron radiation 132, 133, 216, 217, 390
systematic error 7, 390
system boundary 54, 55, 214, 390
system efficiency v, 48, 214, 215, 390
system elements 272, 390
system input 48, 49, 50, 390
system output 48, 49, 50, 390

T
tangent 14, 262, 263, 390
tau particle 348
temperature gradient 302, 390
tensile strength 197, 390
tension 15, 16, 127, 128, 149, 197, 203, 204, 287, 390
terminal velocity 114, 242, 244, 391
term symbol 306
theory of relativity 60, 171, 172, 217, 351, 369
therm 46, 47
thermal conductivity 64, 65, 66, 86, 373, 391
thermal efficiency 48, 49, 391
thermal energy 64, 87, 90, 99, 107, 158, 302, 303, 305, 312, 341, 379, 381, 391
thermal radiation 36, 37, 38, 264, 266, 391
thermionic emission 86, 87
thermodynamic processes 157, 391
Thompson, J. J. 290, 351
throws 243, 297, 391
timbre 228, 391
tone 149, 227, 228, 245, 246, 270, 320, 391
ton of coal equivalent (TOE) 46, 47, 391
total mechanical energy 60, 61, 62, 63, 169, 171, 339, 341, 391
trajectory 16, 18, 19, 67, 111, 114, 115, 116, 117, 195, 196, 241, 242, 244, 257, 323, 375, 376, 391, 392
transverse 69, 70, 78, 79, 189, 190, 198, 221, 224, 240, 320, 329, 330, 332, 391
transverse magnification 189, 190, 391
transverse wave 69, 70, 78, 240, 320, 332, 391
trough v, 4, 12, 13, 14, 69, 70, 147, 223, 265, 330, 332, 333, 377, 384, 385, 391
turbines 311
two-body problem 164, 391

U
ultrasonic 315, 316, 317, 364, 365, 391
ultrasonic impact treatment 315, 391
ultrasonic testing 315, 391
ultrasound 284, 315, 316, 321, 389, 391
ultrasound identification 315, 316, 391
ultraviolet catastrophe 37, 38, 229, 392
ultraviolet (UV) degradation 317, 318, 392
ultraviolet (UV) radiation 68, 227, 240, 264, 266, 317, 318, 319, 392
uncertainty principle v, xi, 42, 150, 151, 152, 253, 255, 257, 291, 392
unique spring constant (k) 62
universal gas constant (R) 64

V
vacuum ultraviolet radiation 317, 392
velocity of sound 3, 225, 320, 321
velocity vs. speed 322
vertex 129, 189, 242, 244, 392
vignetting 26, 392
Villard, Paul Ulrich 366
virtual image 126, 189, 190, 392
visible light 4, 11, 22, 35, 37, 81, 103, 125, 129, 132, 200, 226, 229, 240, 241, 254, 264, 265, 266, 317, 318, 319, 342, 388, 392
volume and capacity 325
volume of a cylinder 326, 327
volume of a rectangular prism 326
volume of a sphere 326

W
warping 78, 79, 116, 392
watt 5, 23, 65, 66, 101, 102, 103, 155, 162, 163, 167, 168, 236, 238, 274, 277, 299, 328, 335, 387, 392
Watt, James x, 101, 155, 156, 328, 364
wave function 13, 14, 20, 21, 40, 42, 44, 45, 46, 88, 89, 120, 128, 254, 255, 257, 294, 295, 320, 347, 349, 389, 392
wave-particle duality 39, 45, 132, 151, 177, 220, 226, 251, 253, 254, 255, 290, 344, 392
wave propagation 224, 285, 320, 392
wave properties 13, 70, 188, 252, 283, 285, 286, 333
weak interaction vii, 19, 35, 36, 42, 58, 248, 251, 334, 348, 350, 357, 390, 392
weak nuclear force 35, 36, 153, 194, 290, 334, 335, 348, 392
wedge 272, 274, 275, 276, 277, 278
wheel and axle 272, 274, 275, 276, 277, 279

Wien, Wilhelm 38, 362, 366
work and force 335
work-energy theorem 339

X
x-ray 6, 211, 264, 331, 342, 343, 344, 393
x-ray radiation 342

Y
Young, Thomas 177, 231, 290, 365

Z
Zhuangzi 28
Zweig, George 137, 259, 292